Lecture Notes in Computer Science 8576

Commenced Publication in 1973
Founding and Former Series Editors:
Gerhard Goos, Juris Hartmanis, and Jan van Leeuwen

Magnús M. Halldórsson (Ed.)

Structural Information and Communication Complexity

21st International Colloquium, SIROCCO 2014
Takayama, Japan, July 23-25, 2014
Proceedings

 Springer

Volume Editor

Magnús M. Halldórsson
Reykjavik University
School of Computer Science
Menntavegur 1
101 Reykjavik, Iceland
E-mail: mmh@ru.is

ISSN 0302-9743   e-ISSN 1611-3349
ISBN 978-3-319-09619-3   e-ISBN 978-3-319-09620-9
DOI 10.1007/978-3-319-09620-9
Springer Cham Heidelberg New York Dordrecht London

Library of Congress Control Number: 2014944560

LNCS Sublibrary: SL 1 – Theoretical Computer Science and General Issues

Typesetting: Camera-ready by author, data conversion by Scientific Publishing Services, Chennai, India

Printed on acid-free paper

Springer is part of Springer Science+Business Media (www.springer.com)

Preface

The 21st International Colloquium on Structural Information and Communication Complexity (SIROCCO 2014) took place in Takayama, Japan, during July 23–25, 2014.

SIROCCO is devoted to the study of communication and knowledge in distributed systems from both the qualitative and quantitative viewpoints. Special emphasis is given to innovative approaches and fundamental understanding, in addition to efforts to optimize current designs. The typical areas include distributed computing, communication networks, game theory, parallel computing, social networks, mobile computing (including autonomous robots), peer-to-peer systems, communication complexity, fault-tolerant graph theories, and randomized/probabilistic issues in networks.

This year, 51 papers were submitted in response to the call for papers, and each paper was evaluated by three to four reviewers. Over 160 reviews were received. The Program Committee selected 24 papers for presentation at the colloquium and publication in this volume after in-depth discussions.

The conference featured five invited talks by Yuval Emek, Technion; Friedhelm Meyer auf der Heide, Paderborn; Rotem Oshman, Princeton; Gopal Pandurangan, NTU Singapore; and Michel Raynal, IRISA.

The SIROCCO Prize for Innovation in Distributed Computing was awarded this year to Pierre Fraigniaud from Université Paris Diderot and CNRS. A commendation summarizing his many and important innovative contributions to distributed computing appears in these proceedings.

The collaboration of Program Committee members and the external reviewers enabled completing the process of reviewing the papers and discussing them in less than four weeks. We thank them all for their devoted service to the SIROCCO community. We thank the authors of all the submitted papers, without them we could not have prepared a program of such quality.

The preparation of this event was guided by the SIROCCO Steering Committee, headed by Shay Kutten.

The conference was supported by the following organizations:

- KDDI Foundation
- Takayama City
- Nagoya Institute of Technology
- Research Foundation for the Electrotechnology of Chubu
  It was also supported by the research projects:
- MEXT Grant-in-Aid for Scientific Research on Innovative Areas "A Multifaceted Approach toward Understanding the Limitations of Computation",
- MEXT Grant-in-Aid for Scientific Research(B) "Deepening and Expanding Coding Theory based on Probabilistic Methods"

The EasyChair system was used to handle the submission of papers, manage the review process, and generate these proceedings.

July 2014   Magnús Halldórsson

Organization

Program Committee

Hagit Attiya	Technion, Israel
Amotz Bar-Noy	CUNY, USA
Guy Even	Tel Aviv University, Israel
Sándor Fekete	TU Braunschweig, Germany
Paola Flocchini	University of Ottawa, Canada
George Giakkoupis	Inria Rennes, France
Magnús M. Halldórsson	Reykjavik University (Chair), Iceland
Taizuke Izumi	Nagoya Institute of Technology, Japan
Valerie King	University of Victoria, Canada
Friedhelm Meyer auf der Heide	University of Paderborn, Germany
Alessia Milani	University of Bordeaux, France
Calvin Newport	Georgetown University, USA
Hirotaka Ono	Kyushu University, Japan
Peter Robinson	NTU Singapore, Singapore
Jukka Suomela	Aalto University, Finland
Corantin Travers	ENSEIRB-MATMECA, France
Roger Wattenhofer	ETH, Switzerland
Peter Widmayer	ETH, Switzerland

Steering Committee

Guy Even	Tel Aviv University, Israel
Ralf Klasing	CNRS and University of Bordeaux, France
Shay Kutten	Technion (Chair), Israel
Thomas Moscibroda	Microsoft Research, China
Boaz Patt-Shamir	Tel Aviv University, Israel
Masafumi Yamashita	Kyushu University, Japan

Organizing Committee

Taisuke Izumi	Nagoya Institute of Technology (Chair), Japan
Tomoko Izumi	Ritsumeikan University, Japan
Sayaka Kamei	Hiroshima University, Japan
Yoshiaki Katayama	Nagoya Institute of Technology, Japan
Fukuhito Oosita	Osaka University, Japan
Yukiko Yamauchi	Kyushu University, Japan

Additional Reviewers

Abshoff, Sebastian
Bampas, Evangelos
Biely, Martin
Bienkowski, Marcin
Castañeda, Armando
Cord-Landwehr,
  Andreas
D'Angelo, Gianlorenzo
Dobrev, Stefan
Dolev, Shlomi
Duchon, Philippe
Efrat, Alon
Englert, Matthias
Even, Guy
Ghaffari, Mohsen
Gustedt, Jens

Göös, Mika
Jung, Daniel
Kapron, Bruce
Kiyomi, Masashi
Kobayashi, Yusuke
Kortsarz, Guy
Kranakis, Evangelos
Královič, Rastislav
Lampis, Michael
Mallmann-Trenn,
  Frederik
Marathe, Madhav
Markarian, Christine
Markou, Euripides
Medina, Moti
Mihalák, Matúš

Nanongkai, Danupon
Navarra, Alfredo
Nonaka, Yoshiaki
Pacheco, Eduardo
Parter, Merav
Pelc, Andrzej
Rabanca, George
Raynal, Michel
Rybicki, Joel
Schmidt, Christiane
Sundaram, Ravi
Trehan, Chhaya
Wada, Koichi
Wadayama, Tadashi
Widder, Josef

Laudatio

It is a pleasure to award the 2014 SIROCCO Prize for Innovation in Distributed Computing to Pierre Fraigniaud. The prize is awarded for his contribution to the understanding of routing in small world models and social networks, and to the understanding of the trade - offs between information and efficiency in routing in general.

Pierre's contribution to research and his innovation in distributed computing are by now well-known and well documented. His work and publications in our community range from routing and labeling problems, mobility, exploration and information spreading, to wireless networks, and from algorithms to lower bounds. Moreover, this is just a partial list. In particular, Pierre has had a large impact on the research into problems of using information for routing. For example, he dealt with compact routing tables (and more generally, with compact distributed data structures) starting with an early paper in SIROCCO'98 [7], and later in other conferences. Another related example is his work on distributed search.

Out of Pierre's rich contributions to routing, we would like to concentrate on his contribution to routing in social networks. The phenomenon of small worlds was investigated originally in a sociological context (see the famous work of Milgram), rather than in the realm of mathematics or distributed computing. Kleinberg has shown that specific graphs could be enhanced to show this phenomenon. That is, not only did the enhanced graph enjoy a small diameter, but also it was easily navigable. Others have later shown a similar phenomenon for other graphs. One could have suspected that this was a general phenomenon of graphs, unrelated to the sociological context.

In a series of papers [1]-[6] Pierre, with his various coauthors, succeeded in defining the conditions necessary for a graph to become navigable. In particular, the small world phenomenon is *not* a universal property of any graph. On the other hand, he brought evidence supporting the claim that this is an inherent property of social networks. As Pierre showed in his paper "A Doubling Dimension Threshold $\Theta(\log \log n)$ for Augmented Graph Navigability" [1], *not* every graph could be augmented in such a way that it can become navigable. Then, around 2010, Pierre's paper, "On the searchability of small-world networks with arbitrary underlying structure" [5], presented an optimal algorithm for navigation in any graph and analyzed the exact performance of the algorithm. The result of the paper answered the question of navigability in general graphs. As this paper showed, there is a gap between general graphs and social networks. In his SIROCCO paper [3], Pierre also suggested how to test the model of social augmentation. This paper took the ideas of navigability and augmentation of social networks from a mathematical model to a practical big data application by asking how to verify the theory of navigability by augmentation. This paper bridged the boundary between theory and measurement in the real world.

Pierre's results have had a profound impact on the field of social networks and on connecting this field to the field of distributed computing.

References

1. Fraigniaud, P., Lebhar, E., Lotker, Z.: A Doubling Dimension Threshold $\Theta(\log\log n)$ for Augmented Graph Navigability. In: Azar, Y., Erlebach, T. (eds.) ESA 2006. LNCS, vol. 4168, pp. 376–386. Springer, Heidelberg (2006)
2. Fraigniaud, P., Gavoille, C., Kosowski, A., Lebhar, E., Lotker, Z.: Universal augmentation schemes for network navigability: overcoming the sqrt(n)-barrier. In: SPAA 2007, pp. 1–7 (2007)
3. Fraigniaud, P., Lebhar, E., Lotker, Z.: Recovering the Long-Range Links in Augmented Graphs. In: Shvartsman, A.A., Felber, P. (eds.) SIROCCO 2008. LNCS, vol. 5058, pp. 104–118. Springer, Heidelberg (2008)
4. Fraigniaud, P., Giakkoupis, G.: The effect of power-law degrees on the navigability of small worlds: [extended abstract]. In: PODC 2009, pp. 240–249 (2009)
5. Fraigniaud, P., Giakkoupis, G.: On the searchability of small-world networks with arbitrary underlying structure. In: STOC 2010, pp. 389–398 (2010)
6. Fraigniaud, P.: A New Perspective on the Small-World Phenomenon: Greedy Routing in Tree-Decomposed Graphs. In: Brodal, G.S., Leonardi, S. (eds.) ESA 2005. LNCS, vol. 3669, pp. 791–802. Springer, Heidelberg (2005)
7. Fraigniaud, P., Gavoille, C.: A Theoretical Model for Routing Complexity. In: SIROCCO 1998, pp. 98–113 (1998)

Invited Presentations

Invited Presentation

From Turing to the Clouds
(On the Computability Power of
Distributed Systems)

Michel Raynal

Institut Universitaire de France
IRISA, Université de Rennes, France
Department of Computing, Polytechnic University, Hong Kong
`raynal@irisa.fr`

Abstract. One of the main issues addressed in Turing's work was about computability, namely, one of his Holy Grail's quests was to answer the question "What can be mechanically computed?". Since the definition of the Turing's machine and the statement of the Turing-Church's thesis, a lot of great technological advances have modified our view of what is a "computing device". Among them, distributed systems –sometimes cosmetically called "cloud computing"– are among the most pervasive, and force us to think again to the computability power of such systems. This is becoming more and more important because, not only the world is distributed, but more and more applications are distributed.

Hence, a fundamental question is the following one: *What can be computed in a distributed system*? The answer to this question depends on the environment in which evolves the considered distributed system, i.e., the assumptions the system relies on. This environment is very often left implicit and nearly always not formulated in terms of precise underlying requirements. In the extreme case where the environment is such that there is no synchrony assumption and the computing entities may commit failures, many problems become impossible to solve (in these cases, a network of Turing machines where some machines may crash, is less powerful than a single reliable Turing machine). Given a distributed computing problem, it is consequently important to know the weakest assumptions (lower bounds) that give the limits beyond which the considered distributed problem cannot be solved.

This talk is a short introduction to this kind of issues. It first presents concepts and results related to distributed computability, and then briefly addresses distributed complexity issues, which are the two lenses that allows us to understand and master computing. The following table summarizes the main issues encountered in distributed computing seen from these two lenses.

	Synchronous	Asynchronous
Failure-free	complexity	complexity
Failure-prone	complexity	computability

The full content of this talk can be found in [19]. The interested reader will find more elements on computability and complexity issues in distributed computing in the (non-exhaustive) list of papers and books that appears below.

References

1. Attiya, H., Welch, J.L.: Distributed computing: fundamentals, simulations and advanced topics, 2nd edn., p. 414. Wiley-Interscience (2004) ISBN 0-471-45324-2
2. Borowsky, E., Gafni, E., Generalized, F.L.P.: Impossibility Results for t-Resilient Asynchronous Computations. In: Proc. 25th ACM Symposium on Theory of Computing (STOC 1993), pp. 91–100. ACM Press (1993)
3. Fischer, M.J., Lynch, N.A., Paterson, M.S.: Impossibility of distributed consensus with one faulty process. Journal of the ACM 32(2), 374–382 (1985)
4. Fraigniaud, P., Korman, A., Peleg, D.: Towards a complexity theory for local distributed computing. Journal of the ACM 60(5), Article 35, 16 (2013)
5. Gafni, E.: Round-by-round Fault Detectors: Unifying Synchrony and Asynchrony. In: Proc. 17th ACM Symposium on Principles of Distributed Computing (PODC 1998), pp. 143–152. ACM Press (1998)
6. Herlihy, M.P.: Wait-free synchronization. ACM Transactions on Programming Languages and Systems 13(1), 124–149 (1991)
7. Herlihy, M.P., Kozlov, D., Rajsbaum, S.: Distributed computing through combinatorial topology, p. 336. Morgan Kaufmann/Elsevier (2014) ISBN 9780124045781
8. Herlihy, M.P., Rajsbaum, S., Raynal, M.: Power and limits of distributed computing shared memory models. Theoretical Computer Science 509, 3–24 (2013)
9. Herlihy, M., Shavit, N.: The topological structure of asynchronous computability. Journal of the ACM 46(6), 858–923 (1999)
10. Herlihy, M., Shavit, N.: The art of multiprocessor programming, p. 508. Morgan Kaufmann (2008) ISBN 978-0-12-370591-4
11. Lamport, L.: Time, clocks, and the ordering of events in a distributed system. Communications of the ACM 21(7), 558–565 (1978)
12. Linial, N.: Locality in distributed graph algorithms. SIAM Journal on Computing 21(1), 193–201 (1992)
13. Lynch, N.A.: Distributed algorithms, p. 872. Morgan Kaufmann (1996)
14. Peleg, D.: Distributed computing, a locally sensitive approach. SIAM Monographs on Discrete Mathematics and Applications, p. 343 (2000) ISBN 0-89871-464-8
15. Raynal, M.: Communication and agreement abstractions for fault-tolerant asynchronous distributed systems, p. 251. Morgan & Claypool Pub. (2010) ISBN 978-1-60845-293-4
16. Raynal, M.: Fault-tolerant agreement in synchronous message-passing systems, p. 165. Morgan & Claypool Publishers (2010) ISBN 978-1-60845-525-6
17. Raynal, M.: Concurrent programming: algorithms, principles, and foundations, p. 530. Springer (2013) ISBN 978-3-642-32026-2
18. Raynal, M.: Distributed algorithms for message-passing systems, p. 515. Springer ISBN: 978-3-642-38122-5
19. Raynal, M.: What can be computed in a distributed system? In: Bensalem, S., Lakhneck, Y., Legay, A. (eds.) From Programs to Systems. LNCS, vol. 8415, pp. 209–224. Springer, Heidelberg (2014)
20. Santoro, N., Widmayer, P.: Time is not a healer. In: Cori, R., Monien, B. (eds.) STACS 1989. LNCS, vol. 349, pp. 304–316. Springer, Heidelberg (1989)
21. Taubenfeld, G.: Synchronization algorithms and concurrent programming, p. 423. Pearson Education/Prentice Hall (2006) ISBN 0-131-97259-6

Biological Distributed Computing

Yuval Emek

Technion, Israel
yemek@ie.technion.ac.il

Biological systems are built by nature, shaped and optimized through billions of years of evolution, and their exact behavior is typically a mystery that requires an enormous effort to be revealed. Digital systems, on the other hand, are built by engineers from fully specified programs in time frames that rarely last for more than a few years. Therefore, it seems unlikely that tools and methodologies that were originally developed in the context of digital systems may be employed for a successful investigation of biological systems. Surprisingly, the distributed computing toolbox does exactly that!

Ranging from the level of individual cells to that of whole organisms, many biological systems are distributed. Although these systems lack a centralized control, they may be involved in computational processes whose level of complexity is a good match for that of their digital counterparts. Recent advances in technologies for querying gene expression as well as for monitoring interactions at the molecular level facilitate a much deeper understanding of these computational processes. Based on that understanding, we can build more accurate models, which in turn provide us with the opportunity to study biological distributed systems through the lens of theoretical distributed computing.

By applying the toolbox of theoretical distributed computing to various biological problems, new insights can be provided. The last few years have been a flourishing period for scientific work that follows this approach and by now, one can safely argue that a new research area that lies at the interface of biology and theoretical distributed computing has emerged.

Arguably the most influential paper in this research area is that of Afek et al. [2] who discovered that a biological process that occurs during the development of the nervous system of the Drosophila melanogaster is in fact equivalent to solving the maximal independent set (MIS) problem. This has led to an extensive work on MIS under the beeping model [3, 11] that provides a good approximation for the communication model in cellular networks, resulting in improved algorithms and lower bounds [1, 2, 13]. It also led to the development of fundamentally new models for distributed computing in networks of devices whose computation and communication capabilities are much weaker than those assumed by the traditional message passing model [4, 7, 8].

Common to the papers mentioned in the previous paragraph is that they focus on the (distributed) computational process that occurs at the cellular level, i.e., in networks whose basic components are individual cells. However, fascinating distributed computing phenomena prevail in biological systems of a much larger scale as well. Indeed, insect colonies can be regarded as a distributed system whose survival relies on the communication and coordination among its entities. The research lane that inspects basic insect colony tasks such as foraging and navigation from the perspective of theoretical distributed computing has been very fruitful in the last couple of years [5, 6, 9, 10, 12].

In the talk, we will survey the emerging research area of applying the distributed computing toolbox to biological systems with a particular focus on the aforementioned biological domains of cellular networks and insect colonies. The main results of this research area will be discussed and some open questions will be raised. The talk will be self-contained.

References

[1] Afek, Y., Alon, N., Bar-Joseph, Z., Cornejo, A., Haeupler, B., Kuhn, F.: Beeping a maximal independent set. In: Peleg, D. (ed.) Distributed Computing. LNCS, vol. 6950, pp. 32–50. Springer, Heidelberg (2011)

[2] Afek, Y., Alon, N., Barad, O., Hornstein, E., Barkai, N., Bar-Joseph, Z.: A Biological Solution to a Fundamental Distributed Computing Problem. Science 331(6014), 183–185 (2011)

[3] Cornejo, A., Kuhn, F.: Deploying wireless networks with beeps. In: Lynch, N.A., Shvartsman, A.A. (eds.) DISC 2010. LNCS, vol. 6343, pp. 148–162. Springer, Heidelberg (2010)

[4] Dolev, S., Gmyr, R., Richa, A., Scheideler, C.: Ameba-inspired self-organizing particle systems. A manuscript

[5] Emek, Y., Langner, T., Stolz, D., Uitto, J., Wattenhofer, R.: How Many Ants Does It Take To Find the Food? In: 21th International Colloquium on Structural Information and Communication Complexity (SIROCCO), Hida Takayama, Japan (July 2014)

[6] Emek, Y., Langner, T., Uitto, J., Wattenhofer, R.: Solving the ANTS Problem with Asynchronous Finite State Machines. In: Esparza, J., Fraigniaud, P., Husfeldt, T., Koutsoupias, E. (eds.) ICALP 2014, Part II. LNCS, vol. 8573, pp. 471–482. Springer, Heidelberg (2014)

[7] Emek, Y., Uitto, J., Wattenhofer, R.: Failures in the stone age. A manuscript

[8] Emek, Y., Wattenhofer, R.: Stone age distributed computing. In: PODC, pp. 137–146 (2013)

[9] Feinerman, O., Korman, A.: Memory lower bounds for randomized collaborative search and implications for biology. In: Aguilera, M.K. (ed.) DISC 2012. LNCS, vol. 7611, pp. 61–75. Springer, Heidelberg (2012)

[10] Feinerman, O., Korman, A., Lotker, Z., Sereni, J.-S.: Collaborative search on the plane without communication. In: PODC, pp. 77–86 (2012)

[11] Flury, R., Wattenhofer, R.: Slotted Programming for Sensor Networks. In: IPSN (April 2010)

[12] Lenzen, C., Lynch, N., Newport, C., Radeva, T.: Trade-offs between selection complexity and performance when searching the plane without communication. In: PODC (2014)

[13] Scott, A., Jeavons, P., Xu, L.: Feedback from nature: an optimal distributed algorithm for maximal independent set selection. In: PODC, pp. 147–156 (2013)

Table of Contents

Rendezvous

Mobile Agents

Algorithmic Aspects of Resource Management in the Cloud*

Sebastian Kniesburges, Christine Markarian**, Friedhelm Meyer auf der Heide, and Christian Scheideler

Heinz Nixforf Institute & Department of Computer Science,
University of Paderborn, Germany
{seppel,fmadh}@upb.de, {chrissm,scheideler}@mail.upb.de

Abstract. In this survey article, we discuss two algorithmic research areas that emerge from problems that arise when resources are offered in the cloud. The first area, *online leasing*, captures problems arising from the fact that resources in the cloud are not *bought*, but *leased* by cloud vendors. The second area, *Distributed Storage Systems*, deals with problems arising from so-called *cloud federations*, i.e., when several cloud providers are needed to fulfill a given task.

1 Introduction

Cloud Computing offers the opportunity to access IT resources and services such as computing power, storage, network capacities, and software, over the internet. The first commercial cloud, Amazon Web Services [1], was launched in 2006. Since then, many other cloud providers have shown up, among them Salesforce [2], Google App Engine [3], and Microsoft Azur [4]. According to a recent report by Forrester Research [9], the market of cloud computing in 2020 is expected to increase by a factor of six compared to that in 2010. In this survey article, we discuss two algorithmic research areas that emerge from problems that arise when resources are offered in the cloud.

Rather than buying resources, as it is typical for traditional resource providers, the cloud computing paradigm offers the possibility to rent resources for time periods, capacities and qualities tailored to its current needs. As observed in [7], this may lead to significant cost reductions. Algorithmic challenges arising from this property of cloud computing will be our first topic *online leasing*.

A significant amount of research has been done to improve the management of the resources in the cloud (see [19] for a survey). Algorithmic problems motivated by so-called *cloud federations*, i.e., when several cloud providers are needed to fulfill a given task, are addressed in our second topic *Distributed Storage Systems*.

---

* This work was partially supported by the German Research Foundation (DFG) within the Collaborative Research Centre "On-The-Fly Computing" (SFB 901).
** Fellow of the International Graduate School "Dynamic Intelligent Systems".

M. Halldórsson (Ed.): SIROCCO 2014, LNCS 8576, pp. 1–13, 2014.

Target One: Online Leasing. The trend of renting servers for shorter time durations is continuing, as some providers have already started renting machines for less than an hour [9]. This is beneficial both for the *providers* and the *clients*. On the one hand, providers want to minimize idle times of resources, and therefore allow that users may rent them for individual time periods. On the other hand, clients seek to reduce their costs, i.e., clients seek to pay for resources only when they need them. This trend is expected to go on just like in phone companies which nowadays provide more flexible services compared to before, i.e., they moved from charging land-lines per several minutes to charging cell-phones per minute, and nowadays even per second [10]. There have been many attempts to capture these two needs [11–18], one of the most efficient is by allowing competition among providers by reducing providers' costs, which was addressed by Malik et al. in [8]. They propose the *Virtual Cloud Model* which is based on the concept of *Rent Out the Rented Resources*. The idea of this model is to virtualize an already virtualized infrastructure, i.e., a *cloud vendor* rents resources from a third party enterprise, performs virtualization, and then rents them out to clients. In this way, fewer resources are left idle, cloud providers reduce their upfront and administrative costs, and clients rent resources for cheaper prices. (See Fig 1)

Problem Definition: We describe our view of this model as follows. Requests are given by clients, which specify the kind of service demanded and its duration (Ex. 1 minute, 2 minutes,..., 1 hour). Each cloud provider offers a number of

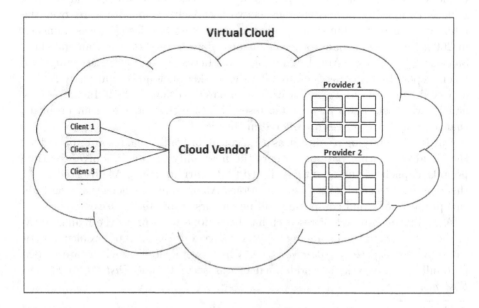

Fig. 1. Virtual Cloud Model

services for different prices. Based on the clients' requests, a cloud vendor *leases* services from providers, i.e., selects the provider to rent from(among those which offer the demanded service) and rents the service from it for some duration (Ex. 1 minute, 2 minutes,..., 1 hour). A service may be used by all clients as long as it is rented. Each period of lease has a cost and the longer the duration of the lease, the cheaper the price per minute. The goal of the cloud vendor is to serve each client while minimizing its leasing costs. Since the cloud vendor does not know in advance which services will be requested and for how long, it might happen that it buys long/expensive leases for some service, just to realize later on that no more requests for this service are issued in subsequent times. Or it buys short leases, just to notice later on that having bought a longer lease would have cost less. Such a scenario can be formalized as an *Online Leasing Problem.*

The first online leasing problem was introduced by Meyerson [21] with the *Parking Permit Problem.* Here, each day, depending on the weather, we have to either use the car (if its rainy) or walk (if its sunny). If we take the car, we must have a valid parking permit. There are K different types of parking permits, each with its own duration and cost. The goal is to buy a set of permits in order to cover all rainy days and minimize the total cost of purchases (and without using weather forecasts). In Section 2, we give an overview of the state of the art of online leasing problems and propose future work.

Target Two: Distributed Storage Systems. The benefits of cloud computing led to a wide acceptance of cloud services as a cost reducing alternative compared to in-house IT solutions. Nevertheless, cloud computing also comes with some downsides and obstacles that have prevented potential cloud customers from moving their applications and data into the cloud. The major obstacles are [32, 33]:

- Availability concerns: Can availability be ensured if the application no longer runs on in-house IT services?
- Data lock-in concerns: Can applications and data easily be moved from one cloud provider to another?
- Security and auditability objections: Can the security of essential data be ensured, and can it be controlled where the data is stored?
- Data transfer bottlenecks: Can the placement of data and applications in the cloud be controlled to reduce the data transfer costs?

One proposed solution for the availability concerns, the data transfer bottleneck and the data lock-in concerns is to use multiple cloud providers instead of one, just as a large service provider in the Internet may use multiple network providers to ensure connectivity [32]. This has led to the concept of cloud federation [34], which is the practice of interconnecting the cloud computing environments of two or more service providers for the purpose of load balancing traffic and accommodating spikes in demand. A cloud federation offers two substantial benefits to cloud providers. On the one hand, it allows providers to earn revenue from computing resources that would otherwise be idle or underutilized, and on the other hand, a cloud federation enables cloud providers to expand their geographic footprints and accommodate sudden spikes in demand without having to build new points-of-presence. Cloud federations are also interesting in the context of security because a possible solution to address the security concerns is to combine a public cloud with a private cloud, where the sensitive data is kept in the private cloud. The ability to form cloud federations is also very useful for cloud vendors since it would allow them to set up their own virtualized infrastructure by renting resources at various cloud providers and implementing their service platform on top of these resources so that the desired service can be offered to the clients.

In recent years several solutions for cloud federations have been proposed [35, 36, 7, 11, 14–17, 37–39]. One of the key research challenges in this field is the question how to connect the different resources available in a cloud federation and how to distribute the data among the resources. One proposed approach is based on the use of *distributed hash tables (DHTs)* [11, 37, 39]. The basic idea behind this approach is to create a new virtual address space and to hash the data as well as the resources to that space in order to determine which resource is supposed to store which data item. The virtual addresses of the resources may also be used to establish connections between these resources so that a fully decentralized DHT can be established.

Problem Definition: We consider the following problem. A cloud vendor that leases a set of n possibly heterogeneous resources by different cloud providers needs to connect these resources to build its own virtual cloud to satisfy the demands of its clients. In our model each resource can be identified by its specific address. Since the cloud vendor may dynamically release some rented resources and lease some other resources, we model the set of resources as dynamic. In order to minimize the overhead for the cloud vendor, the resources should ideally be able to manage their connections and storage themselves by running appropriate distributed protocols. For that approach to be scalable and sufficiently robust it is important to maintain an *overlay network* between the resources that has a low diameter and a high expansion. In Section 3, we give an overview of the state of the art of algorithmic research in the area of distributed storage systems and propose some future work.

2   Online Leasing Problems

Typical infrastructure problems consider scenarios where one has to buy certain goods or resources (e.g., facilities, network nodes, or network connections) in order to generate or improve a given infrastructure (e.g., a supply network). Such problems have not only been widely studied in the offline- but also in the online-setting [26–30], where the decision which goods to buy when must be taken without knowledge of future demands. One of the best known online problems is the ski rental problem, where we either rent skies (on a daily basis) or buy them. Seen as an infrastructure problem, it gives us the additional possibility to rent (lease) a good to satisfy our "skiing demand" (instead of only buying). Meyerson [21] generalized this idea with the *Parking Permit Problem*. If we have two lease types, one for a single day and one of infinite length, we get a variant of the ski rental problem.

Parking Permit Problem. The Parking Permit Problem, as stated earlier, was the first problem to introduce leasing [21]. Meyersen [21] gave an $O(K)$-deterministic and an $O(\log K)$-randomized algorithm for the Parking Permit Problem. He showed that these bounds are optimum and gave a randomized $O(\log n \log K)$-algorithm for the leasing variant of Online Steiner Forest as well. The $O(K)$-deterministic algorithm for the Parking Permit Problem goes as follows. For each rainy day, we buy a 1-day permit, until there is some $(k \in K)$-interval where the optimum offline solution for the sequence of days seen so far, would buy a k-day permit. In this case, we also buy a k-day permit. The $O(\log K)$-randomized algorithm for the Parking Permit Problem first computes an $O(\log K)$-competitive fractional solution and then converts it, by randomized rounding, into a randomized integer solution which maintains the $O(\log K)$-competitive factor. A fractional algorithm is allowed to purchase fractional permits, such that on each rainy day, the total fractional permits sums up to at least one. The fractional algorithm is described as follows. We initially set all permits to fraction zero. As long as the sum of the total fractional permits for a rainy day is not one, we do the following. For each $1 \leq i \leq K$, we multiply the fraction by which the currently valid permit of type i is purchased by $1 + 1/C_i$, and then we add to it $1/K \cdot C_i$, where C_i is the cost of permit i.

Facility Leasing. Anthony and Gupta [20] presented the leasing model more generally in many infrastructure problems: Facility Location, Steiner Tree, and Set Cover. As a typical infrastructure problem, *Facility Leasing* was the first leasing variant which was studied. In this leasing variant of Facility Location (presented by Anthony and Gupta [20]), we are given a set F of m facilities and a set D of n clients. The goal is to minimize the costs of serving the clients by opening facilities and connecting each client to an open facility. Connecting a facility i to client j incurs a connection cost c_{ij}. However, in contrast to the classical Facility Location model, there is a notion of (discrete) time. Clients do not arrive all at once. Instead, at time $t \in N$, a subset $D_t \subseteq D$ of clients appear and these clients need to be served for the current time t only. Moreover, opening a facility is not merely a binary decision. Instead, in order to open a facility, we are asked to determine the point in time at which the facility is going to be opened and one of the K different lease types that is to be used to open it. A facility i opened at time t using lease type k incurs a cost c_i^k and can serve clients arriving during the interval $[t, t + l_k]$. The problem is known as *metric* when connection costs are assumed to satisfy the triangle inequality. The leasing model was studied in both *offline* and *online* settings. For the offline setting, Anthony and Gupta [20] gave an $O(K)$-approximation for Facility Leasing, which was a result of an interesting relationship between deterministic leasing problems and problems in multistage stochastic optimization. The $O(K)$-approximation for Facility Leasing was later improved to a 3-approximation by Nagarajan and Williamson [22]. In the online setting, Nagarajan and Williamson [22] gave an $O(K \log n)$-competitive algorithm for metric Facility Leasing. Their algorithm is an extension of the algorithm in [25] for Online Facility Location. We extended the results by Nagarajan and Williamson [22] for Facility Leasing by removing the dependency on n (and thereby on time) [24]. We gave an $O(l_{max} \log(l_{max}))$-competitive algorithm where l_{max} is the maximum lease length. Moreover, we showed that our algorithm has an $O(\log^2(l_{max}))$-competitive factor for many natural client arrival patterns. Such patterns include, for example, situations where the number of clients arriving in each time step does not vary too much, or is non-increasing, or is polynomially bounded in l_{max}.

Set Cover Leasing. In the leasing variant of Set Cover (presented by Anthony and Gupta [20]), elements in U, $|U| = n$, arrive over time and must be covered by sets in a family F of subsets of U, $|F| = m$. Each set can be leased for K different periods of time. Leasing a set S for a period k incurs a cost of c_S^k and allows S to cover its elements for the next l_k time steps. The objective is to minimize the total cost of the sets leased, such that the elements arriving at any time t are covered by sets which contain them and are leased during time t. Anthony and Gupta [20] gave an $O(\log n)$-approximation for the problem in the offline setting. This is the best possible unless $P = NP$ [31]. Currently, we have randomized algorithms for Set Cover Leasing and many of its variants [23]. Our algorithm for Set Cover Leasing has an expected $O(\log(Km) \log n)$-competitive factor and draws ideas from the Parking Permit Problem algorithm [21] and the randomized algorithm for non-metric Online Facility Location [29].

Future Research Directions. It is easy to see that the aspect of managing the renting of cloud resources may contain many online leasing problems as subproblems. Therefore, with the ongoing advances in the cloud computing technologies, many attempts to capture these advances in the form of online leasing problems are expected. These include new online leasing problems, in addition to variants of existing online leasing problems, which we stated above. From the technical point of view, the research in online leasing problems seems to be in its infancy. All the algorithms for the online leasing variants of the problems studied so far build upon the algorithms for the non-leasing variants of the corresponding problems and the Parking Permit Problem algorithm. Since all these problems generalize the Parking Permit Problem, the only lower bound we have for these problems is $\Omega(K + f())$ where K is the lower bound for the Parking Permit Problem and $f()$ is the lower bound for the underlying non-leasing variant of the problem. It is still not known whether we can prove stronger lower bounds for these problems. This means, we do not know yet the price we have to pay for leasing. In other words, we do not know whether there is a common difficulty among these problems in their leasing variant or the same techniques used for the non-leasing variants could be extended to solve the leasing variants.

3 Distributed Storage Systems

DHTs have been a popular choice for maintaining distributed storage systems. The widely used distributed database management system Apache Cassandra [5], for example, has been realized as a DHT-based distributed storage system [40] that, like many other DHTs, uses the concept of consistent hashing [42] to support scalability. A similar concept is introduced in [41], a predecessor of the current AmazonDynamDB offered by Amazon web services [6]. We will give a high-level overview of different implementations of DHTs and different concepts of load balancing in DHTs.

Distributed Hash Tables. DHTs were introduced as structured peer-to-peer networks. The most basic and popular concepts are Chord [46], Pastry [45] and Tapestry [47]. All of these DHTs map the set of hosts (randomly) to a virtual address space so that each host is responsible for a part or region of the address space. The data is usually mapped to addresses in that space using some pseudo-random hash function, and each data item is stored in that host responsible for the address of that data item. Whenever a host enters or leaves the system, the regions of some hosts may change, which may then cause the relocation of data. The simplest variant of this approach is the concept of consistent hashing [42]. In consistent hashing, the data items are hashed to points in the $[0, 1)$-interval, and the hosts are mapped to disjoint intervals in the same $[0, 1)$-interval, and a host stores all data items that are hashed to points in its interval. An alternative strategy is to hash data items and hosts to pseudo-random bit strings and to store (indexing information about) a data item at the host with the longest prefix match [43]. This concept is used by Pastry and Tapestry [47]. Chord, Pastry and Tapestry provide the following features for a set of n hosts:

- logarithmic diameter with high probability,
- (expected) logarithmic degree,
- (poly-)logarithmic structural changes in case of joining or leaving hosts, and
- the maximum load exceeds the average by at most a logarithmic factor with high probability.

Several other DHTs have been proposed since then that optimize the degree and the diameter. For example, Koorde [48] achieves a diameter of $O(\log n)$ with a degree of just $O(1)$ and a routing distance of $O(\log n/ \log \log n)$ with a degree of $O(\log n)$.

Load Balancing for DHTs. In order to improve the load balancing of DHTs like Chord, different approaches have been proposed in the last years. One concept already mentioned in the work introducing Chord [46] is the concept of virtual hosts. Instead of being responsible for a single region, a host simulates

$\Theta(\log n)$ virtual hosts and is responsible for all the regions assigned to these virtual hosts. With this strategy it can be shown that the maximal load exceeds the average load by a factor of just $O(1)$ instead of $O(\log n)$ with high probability. However, this increases the degree at each host by a factor of $O(\log n)$. The concept of virtual hosts has been further studied in [49, 50]. Another approach is based on the paradigm of many choices [52], i.e., a host is given set of possible positions in the address space from which it chooses the most balancing one. For example, in case of the Chord network a host would choose the position that cuts the longest of all intervals a host is currently responsible for. This concept has been used in [51, 53, 54] and can achieve a constant ratio of the maximal load and the minimal load provided that the data items are evenly distributed. However, the approaches above have a certain drawback. They assume that hosts join the network sequentially and they do not say how to balance the load in case of leaving hosts. This problem is attacked in [55, 56]. The authors propose a scheme in which each host chooses $O(\log n)$ places in the network and takes responsibility for only one of them. This place can change if some hosts leave or join, but each host migrates only among its $O(\log n)$ alternative places, and after each operation only $O(\log \log n)$ hosts have to migrate on expectation. Another solution in which hosts change their places over time is proposed in [57]. Each host estimates the number n of hosts currently in the system and then decides whether the interval it is responsible for is small, medium, or large. Hosts with small intervals are migrated to places where a long interval is cut to reach a constant ratio of the maximal interval size (resp. load) to the minimal interval size (resp. load). The number of hosts migrating is asymptotically optimal, but to find the right places for the migration can incur high communication costs depending on the diameter of the network and $\log n$. Some drawbacks of all the presented approaches is that they assume that data is evenly distributed and that the hosts are homogeneous. Furthermore, to calculate or estimate $O(\log n)$ the hosts need some kind of global knowledge. Another issue is that although DHTs are considered to be self-organizing, rarely mechanisms are proposed to maintain the system in cases of failing communication links or if hosts fail. Mostly, only join and leave routines are presented.

Heterogeneous DHTs. In a heterogeneous setting, each host u has its specific capacity $c(u)$, and the goal is to balance the load among the hosts so that each host u stores a fraction of $\frac{c(u)}{\sum_{\forall v} c(v)}$ of the data. The simplest solution would be to reduce the heterogeneous case to the homogeneous case by splitting a host of k times the base capacity (e.g., the minimum capacity of a host) into k many virtual hosts like already seen for homogeneous hosts. In [62] the main idea is not to place the virtual hosts belonging to a real host randomly in the identifier space but in a restricted range to achieve a low degree in the overlay network. However, they still need an estimation of the network size and a classification of nodes with high, average, and low capacity. A similar approach is presented in [63]. Rao et al. [50] proposed some schemes also based on virtual servers, where the data is moved from heavy nodes to light nodes to balance the load after the data assignment. However, the concept of virtual hosts has some major drawbacks. As already mentioned, an estimation of the network size n is needed. Furthermore, knowledge about the overall capacity of the DHT is needed to determine whether a host has a low, medium or high capacity. So far, the only solution independent of the overall capacity and the number of hosts is the weighted distributed hash table by Schindelhauer and Schomaker [65]. Their basic idea is to assign a distance function to each host that scales with the capacity of the host. A data element is then assigned to the host of minimum distance with respect to these distance functions. The authors show that when using a specific logarithmic function, a fair load balancing can be achieved. However, this approach lacks a distributed implementation, as the authors do not state how the hosts have to be interconnected to easily determine which data is stored on which host. Instead, they only present a centralized algorithm to calculate the mapping of the data to the hosts.

Self-Stabilizing DHTs. For a distributed system to be self-stabilizing, two properties have to be fulfilled:

- Convergence: The system has to be able to eventually reach a legal state from any initial state.
- Closure: Whenever the system starts in a legal state, it stays in a legal state.

Since the introduction of the concept of self-stabilizing in the seminal paper of Dijkstra [66], a huge volume of protocols for self-stabilizing systems has been proposed. However, until recently most of these protocols just dealt with the case that the interconnection network is static. Self-stabilizing overlay networks were only proposed in the past few years. Some prominent examples are a self-stabilizing hypertree [59], a self-stabilizing skip graph [60], and a universal protocol for self-stabilizing overlay networks [58]. Recently, we also proposed a self-stabilizing overlay network for hosts of different bandwidths [61].

In [64] we introduced the concept of self-stabilization [66] to the area of DHTs. In this work a distributed protocol is presented that recovers a variant of the Chord network from any state, as long as the set of hosts remains connected. Recently, we also presented a self-stabilizing overlay network that allows an efficient distributed implementation of the weighted distributed hash table by Schindelhauer and Schomaker [65]. We showed that in the overlay network each host has a degree of at most $O(\log n)$ with high probability and also the routing distance is $O(\log n)$ with high probability. Applying the analysis of [65], we were able to show that the load of each host is balanced in a fair way. A future challenge will be to obtain a self-stabilizing DHT that does not just handle arbitrary capacities but also arbitrary bandwidths.

Future Research Directions. As can be seen from the results above, a lot of work has already addressed the problem of load balancing in distributed storage systems. However, not much rigorous work is known if the hosts in that storage system show several degrees of heterogeneity like heterogeneous bandwidths and capacities. Also, for most approaches a uniform distribution of the data items is assumed, which cannot be ensured if related data is supposed to be stored close to each other or there are some geographic or security constraints. Cloud provides also offer different degrees of control. E.g., in Amazon Web Services a customer can specify and control the cloud service (a virtual machine (VM)) in a detailed way while in the Google App Engine the cloud service is automatically adapted to the application running on it, giving the customer less control. Also, the concept of self-stabilizing deserves further attention because the fact that a system is self-stabilizing does not necessarily mean that it is highly available. Here, it is important that the system is able to fix its faulty parts while all other parts that are still functional remain functional. In other words, a more local form of convergence and closure is needed, where closure is to be preserved for the legal parts while convergence is ensured for the illegal parts of the system.

References

1. Amazon Elastic Compute Cloud, http://aws.amazon.com/ec2/
2. Salesforces Force.com Cloud Computing Architecture,
   http://www.salesforce.com/platform/
3. Google App Engine, https://appengine.google.com
4. Windows Azur Platform, http://www.microsoft.com/windowsazure/
5. Apache Cassandra, http://cassandra.apache.org/
6. Amazon Web Services, http://aws.amazon.com/
7. Assuncao, M.D., Costanzo, A., Buyya, R.: Evaluating the Cost-Benefit of Using Cloud Computing to Extend the Capacity of Clusters. In: Proceedings of the 18th International Symposium on High Performance Distributed Computing (HPDC) (2009)
8. Malik, S., Huet, F.: Virtual Cloud: Rent Out the Rented Resources. In: Proceedings of the 6th IEEE International Conference for Internet Technology and Secured Transactions (ICITST), pp. 536–541 (2011)
9. Ben-Yehuda, O., Ben-Yehuda, M., Schuster, A., Tsafrir, D.: Deconstructing Amazon EC2 Spot Instance Pricing. In: Proceedings of the 3rd IEEE International Conference on Cloud Computing Technology and Science (Cloud-Com) (2011)
10. Ben-Yehuda, O., Ben-Yehuda, M., Schuster, A., Tsafrir, D.: The Resource-as-a-Service (RaaS) Cloud. In: Proceedings of the 4th USENIX Conference on Hot Topics in Cloud Computing (HotCloud) (2012)
11. Buyya, R., Ranjan, R., Calheiros, R.N.: InterCloud: Utility-Oriented Federation of Cloud Computing Environments for Scaling of Application Services. In: Hsu, C.-H., Yang, L.T., Park, J.H., Yeo, S.-S. (eds.) ICA3PP 2010, Part I. LNCS, vol. 6081, pp. 13–31. Springer, Heidelberg (2010)
12. Assuno, M., Buyya, R., Venugopal, S.: InterGrid: A Case for Internetworking Islands of Grids. Journal of Concurrency and Computation: Practice and Experience Archive 20(8) (2008)
13. Assuncao, M., Buyya, R.: Performance Analysis of Allocation Policies for Inter-Grid Resource Provisioning. Information and Software Technology 51(1), 42–55 (2009)
14. Bernstein, D., Ludvigson, E., Sankar, K., Diamond, S., Morrow, M.: Blueprint for the Inter-cloud Protocols and Formats for Cloud Computing Interoperability. In: Proceedings of the 4th International Conference on Internet and Web Applications and Services (2009)

15. Campbell, R., Gupta, I., Heath, M., Ko, S., Kozuch, M., Kunze, M., Kwan, T., Lai, K., Lee, H., Lyons, M., Milojicic, D., O'Hallaron, D., Soh, Y.: Open CirrusTM Cloud Computing Testbed: Federated Data Centers for Open Source Systems and Services Research. In: Proceedings of the Conference on Hot Topics in Cloud Computing, HotCloud (2009)
16. Celesti, A., Tusa, F., Villari, M., Puliafito, A.: How to Enhance Cloud Architectures to Enable Cross-Federation. In: Proceedings of the IEEE 3rd International Conference on Cloud Computing, CLOUD (2010)
17. Celesti, A., Tusa, F., Villari, M., Puliafito, A.: Three-Phase Cross-Cloud Federation Model: The Cloud SSO Authentication. In: Proceedings of the 2nd International Conference on Advances in Future Internet (2010)
18. Keahey, K., Tsugawa, M., Matsunaga, A., Fortes, J.: Sky Computing. Proceedings of the IEEE Journal of Internet Computing 13(5), 43–51 (2009)
19. Vinothina, V., Sridaran, R., Ganapathi, P.: A Survey on Resource Allocation Strategies in Cloud Computing. International Journal of Advanced Computer Science and Applications 3(6), 97–104 (2012)
20. Anthony, B.M., Gupta, A.: Infrastructure leasing problems. In: Fischetti, M., Williamson, D.P. (eds.) IPCO 2007. LNCS, vol. 4513, pp. 424–438. Springer, Heidelberg (2007)
21. Meyerson, A.: The parking permit problem. In: Proceedings of the 46th Annual IEEE Symposium on Foundations of Computer Science (FOCS), pp. 274–284 (2005)
22. Nagarajan, C., Williamson, D.P.: Offline and online facility leasing. In: Lodi, A., Panconesi, A., Rinaldi, G. (eds.) IPCO 2008. LNCS, vol. 5035, pp. 303–315. Springer, Heidelberg (2008)
23. Abshoff, S., Markarian, C., Meyer auf der Heide, F.: Online Algorithms for Set Cover Leasing Problems (in preparation)
24. Kling, P., Meyer auf der Heide, F., Pietrzyk, P.: An algorithm for online facility leasing. In: Even, G., Halldórsson, M.M. (eds.) SIROCCO 2012. LNCS, vol. 7355, pp. 61–72. Springer, Heidelberg (2012)
25. Fotakis, D.: A primal-dual algorithm for online non-uniform facility location. Journal of Discrete Algorithms 5, 141–148 (2007)
26. Alon, N., Awerbuch, B., Azar, Y., Buchbinder, N., Naor, J.: The online set cover problem. In: Proceedings of the 35th Annual ACM Symposium on the Theory of Computation (STOC), pp. 100–105 (2003)
27. Meyerson, A.: Online Facility Location. In: Proceedings of the 42nd Annual IEEE Symposium on Foundations of Computer Science, pp. 426–431 (2001)
28. Fotakis, D.: On the Competitive Ratio for Online Facility Location. In: Baeten, J.C.M., Lenstra, J.K., Parrow, J., Woeginger, G.J. (eds.) ICALP 2003. LNCS, vol. 2719, pp. 637–652. Springer, Heidelberg (2003)
29. Alon, N., Awerbuch, B., Azar, Y., Buchbinder, N.: A General Approach to Online Network Optimization Problems. In: Proceedings of the 15th Annual ACM-SIAM Symposium on Discrete Algorithms (SODA), pp. 577–586 (2004)
30. Buchbinder, N., Naor, J.: Online primal-dual algorithms for covering and packing problems. In: Brodal, G.S., Leonardi, S. (eds.) ESA 2005. LNCS, vol. 3669, pp. 689–701. Springer, Heidelberg (2005)
31. Alon, N., Moshkovitz, D., Safra, S.: Algorithmic construction of sets for k-restrictions. ACM Transactions on Algorithms 2, 153–177 (2006)
32. Armbrust, M., Fox, A., Griffith, R., Joseph, A.D., Katz, R., Konwinski, A., Lee, G., Patterson, D., Rabkin, A., Stoica, I.O.N., Zaharia, M.: A view of cloud computing. Communications of the ACM 53(4), 50–58 (2010)

33. Armbrust, M., Fox, A., Griffith, R., Joseph, A.D., Katz, R., Konwinski, A., Lee, G., Patterson, D., Rabkin, A., Stoica, I.O.N., Zaharia, M.: Above the clouds: A berkeley view of cloud computing. Technical Report (2009)

34. Kurze, T., Klems, M., Bermbach, D., Lenk, A., Tai, S., Kunze, M.: Cloud Federation. In: Proceedings of the 2nd International Conference on Cloud Computing, GRIDs, and Virtualization (CLOUD COMPUTING 2011) (2011)

35. Villegas, D., Bobroff, N., Rodero, I., Delgado, J., Liu, Y., Devarakonda, A., Fong, L., Sadjadi, S.M., Parashar, M.: Cloud federation in a layered service model. Journal of Computer and System Sciences 78(5), 1330–1344 (2012)

36. Zhang, Z., Zhang, X.: Realization of open cloud computing federation based on mobile agent. In: Proceedings of IEEE International Conference on Intelligent Computing and Intelligent Systems (ICIS 2009), pp. 642–646 (2009)

37. Ranjan, R., Buyya, R.: Decentralized overlay for federation of enterprise clouds. CoRR abs/0811.2563 (2008)

38. Rochwerger, B., Breitgand, D., Levy, E., Galis, A., Nagin, K., Llorente, I.M., Montero, R., Wolfsthal, Y., Elmroth, E., Cceres, J., Ben-Yehuda, M., Emmerich, W., Galn, F.: The reservoir model and architecture for open federated cloud computing. IBM Journal of Research and Development 53, 535–545 (2009)

39. Bernstein, D., Vij, D., Diamond, S.: An Intercloud Cloud Computing Economy - Technology, Governance, and Market Blueprints. In: SRII Global Conference (SRII 2011), pp. 293–299 (2011)

40. Lakshman, A., Malik, P.: Cassandra: a decentralized structured storage system. ACM SIGOPS Operating Systems Review 44(2), 35–40 (2010)

41. DeCandia, G., Hastorun, D., Jampani, M., Kakulapati, G., Lakshman, A., Pilchin, A., Sivasubramanian, S., Vosshall, P., Vogels, W.: Dynamo: amazon's highly available key-value store. In: Proceedings of twenty-first ACM SIGOPS Symposium on Operating Systems Principles (SOSP 2007), pp. 205–220 (2007)

42. Karger, D., Lehman, E., Leighton, T., Levine, M., Lewin, D., Panigrahy, R.: Consistent hashing and random trees: Distributed caching protocols for relieving hot spots on the World Wide Web. In: STOC 1997, pp. 654–663 (1997)

43. Plaxton, C.G., Rajaraman, R., Richa, A.W.: Accessing nearby copies of replicated objects in a distributed environment. In: SPAA 1997, pp. 311–320 (1997)

44. Ratnasamy, S., Francis, P., Handley, M., Karp, R., Shenker, S.: A scalable content-addressable network. In: SIGCOMM, pp. 161–172 (2001)

45. Rowstron, A., Druschel, P.: Pastry: Scalable, decentralized object location, and routing for large-scale peer-to-peer systems. In: Guerraoui, R. (ed.) Middleware 2001. LNCS, vol. 2218, pp. 329–350. Springer, Heidelberg (2001)

46. Stoica, I., Morris, R., Karger, D., Frans Kaashoek, M., Balakrishnan, H.: Chord: A scalable peer-to-peer lookup service for internet applications. In: SIGCOMM, pp. 149–160 (2001)

47. Zhao, B.Y., Huang, L., Stribling, J., Rhea, S.C., Joseph, A.D., Kubiatowicz, J.D.: Tapestry: a resilient global-scale overlay for service deployment. IEEE Journal on Selected Areas in Communications 22(1), 41–53 (2006)

48. Kaashoek, F., Karger, D.R.: Koorde: A Simple Degree-optimal Hash Table. In: Kaashoek, M.F., Stoica, I. (eds.) IPTPS 2003. LNCS, vol. 2735, pp. 98–107. Springer, Heidelberg (2003)

49. Godfrey, B., Lakshminarayanan, K., Surana, S., Karp, R., Stoica, I.: Load balancing in dynamic structured p2p systems. In: 23rd Conference of the IEEE Communications Society, INFOCOM (2004)

50. Rao, A., Lakshminarayanan, K., Surana, S., Karp, R., Stoica, I.: Load balancing in structured P2P systems. In: Kaashoek, M.F., Stoica, I. (eds.) IPTPS 2003. LNCS, vol. 2735, pp. 68–79. Springer, Heidelberg (2003)
51. Byers, J., Considine, J., Mitzenmacher, M.: Simple Load Balancing for DHTs. In: Kaashoek, M.F., Stoica, I. (eds.) IPTPS 2003. LNCS, vol. 2735, pp. 80–87. Springer, Heidelberg (2003)
52. Mitzenmacher, M., Richa, A.W., Sitaraman, R.: The power of two random choices: A survey of techniques and results. In: Handbook of Randomized Computing (2000)
53. Naor, M., Wieder, U.: Novel architectures for P2P applications: the continuous discrete approach. In: Proc. of the 15th ACM Symp. on Parallel Algorithms and Architectures (SPAA), pp. 50–59 (2003)
54. Naor, M., Wieder, U.: A simple fault tolerant distributed hash table. In: Kaashoek, M.F., Stoica, I. (eds.) IPTPS 2003. LNCS, vol. 2735, pp. 88–97. Springer, Heidelberg (2003)
55. Karger, D.R., Ruhl, M.: Simple efficient load balancing algorithms for peer-to-peer systems. In: Voelker, G.M., Shenker, S. (eds.) IPTPS 2004. LNCS, vol. 3279, pp. 131–140. Springer, Heidelberg (2005)
56. Karger, D.R., Ruhl, M.: Simple efficient load balancing algorithms for peer-to-peer systems. In: Proc. of the 16th ACM Symp. on Parallelism in Algorithms and Architectures (SPAA), pp. 36–43 (2004)
57. Bienkowski, M., Korzeniowski, M., Meyer auf der Heide, F.: Dynamic load balancing in distributed hash tables. In: van Renesse, R. (ed.) IPTPS 2005. LNCS, vol. 3640, pp. 217–225. Springer, Heidelberg (2005)
58. Berns, A., Ghosh, S., Pemmaraju, S.V.: Building self-stabilizing overlay networks with the transitive closure framework. In: Défago, X., Petit, F., Villain, V. (eds.) SSS 2011. LNCS, vol. 6976, pp. 62–76. Springer, Heidelberg (2011)
59. Dolev, S., Kat, R.: Hypertree for self-stabilizing peer-to-peer systems. In: NCA, pp. 25–32 (2004)
60. Jacob, R., Richa, A., Scheideler, C., Schmid, S., Täubig, H.: A distributed polylogarithmic time algorithm for self-stabilizing skip graphs. In: PODC, pp. 131–140 (2009)
61. Feldotto, M., Graffi, K., Scheideler, C.: HSkip+: A self-stabilizing overlay network for nodes with heterogeneous bandwidths. Technical report, University of Paderborn (2014)
62. Godfrey, P.B., Stoica, I.: Heterogeneity and Load Balance in Distributed Hash Tables. In: IEEE INFOCOM (2005)
63. Bienkowski, M., Brinkmann, A., Klonowski, M., Korzeniowski, M.: SkewCCC+: A heterogeneous distributed hash table. In: Lu, C., Masuzawa, T., Mosbah, M. (eds.) OPODIS 2010. LNCS, vol. 6490, pp. 219–234. Springer, Heidelberg (2010)
64. Kniesburges, S., Koutsopoulos, A., Scheideler, C.: Re-chord: a self-stabilizing chord overlay network. In: SPAA 2011, pp. 235–244 (2011)
65. Schindelhauer, C., Schomaker, G.: Weighted distributed hash tables. In: SPAA 2005, pp. 218–227 (2005)
66. Dijkstra, E.W.: Self-stabilizing systems in spite of distributed control. Commun. ACM 17, 643–644 (1974)
67. Kniesburges, S., Koutsopoulos, A., Scheideler, C.: CONE-DHT: A Distributed Self-Stabilizing Algorithm for a Heterogeneous Storage System. In: Afek, Y. (ed.) DISC 2013. LNCS, vol. 8205, pp. 537–549. Springer, Heidelberg (2013)

Communication Complexity Lower Bounds in Distributed Message-Passing

Rotem Oshman*

Center for Computational Intractability, Princeton University

Most theoretical models of distributed systems neglect the cost of local computation, and charge only for communication between the participants in the computation. For example, in shared memory models, we charge only for steps where processes interact with the shared memory, and in message-passing systems we charge only for messages sent between network nodes. For message-passing systems, traditional cost measures (with many exceptions) include the total number of messages and the number of rounds, but of late there has been much interest in characterizing the number of *bits* that need to be sent to solve various tasks, and also the round-complexity of solving problems under bandwidth restrictions.

To prove lower bounds in settings where local computation is "free" and communication is our main concern, it is natural to use techniques and results from the field of *communication complexity*, introduced by Yao in his seminal paper [17]. Two-player communication games involve two players, Alice and Bob, who receive private inputs X, Y (respectively) and wish to compute a joint function $f(X, Y)$ of their inputs (or accomplish some other task, e.g., produce outputs A, B satisfying some joint constraint $R(X, Y, A, B)$, or sample a value from some distribution specified by the inputs). Perhaps the most widely-applied lower bound in this setting is the lower bound on set disjointness [6,13]: in the set disjointness problem, Alice and Bob receive sets $X, Y \subseteq \{0, 1\}^n$, and their goal is to determine whether or not $X \cap Y = \emptyset$. In [6,13] it is shown that even using randomness, $\Omega(n)$ bits of communication are required to solve set disjointness with constant error probability; this bound was later strengthened to show that the players need to learn $\Omega(n)$ bits of *information* (in the information-theoretic sense) about each others' inputs [2]. The set disjointness lower bound has found applications in streaming algorithms, data structures, sublinear-time property testing, and other fields, and it has many interesting applications in distributed computing. We will discuss some of them.

Message-passing models that charge for bits of communication can be broadly classified into three classes.

1. CONGEST with point-to-point communication [11]: the most popular model, the CONGEST model assumes synchronous computation. In each round, each node has a budget of B bits that it may send on each of its communication links (possibly a different B-bit message on each link). We are interested in the number of rounds required to solve problems under this constraint.

* The author is supported by NSF grants CCF-0832797 and NSF CCF-1149888.

M. Halldórsson (Ed.): SIROCCO 2014, LNCS 8576, pp. 14–17, 2014.
© Springer International Publishing Switzerland 2014

For example, in [14] it is shown that even in networks with low diameter, finding a spanning tree (as well as various other tasks) requires $\Omega(\sqrt{n/B})$ rounds of computation, where n is the number of nodes in the network.

One instance of the CONGEST model with point-to-point communication is the *congested clique*, where the communication network is assumed to be complete. In this model there are several surprisingly fast algorithms for tasks such as routing, sorting, and subgraph detection [9,10,4], and lower bounds have been elusive. Some explanation for this is provided in [1], where it is shown that even slightly super-constant lower bounds for the congested clique would imply better lower bounds for bounded-depth circuits than are currently known.

A popular technique introduced in the textbook [8] for proving lower bounds in the CONGEST model is to reduce from two-player communication games by finding a *sparse cut* in the network, and then having the players simulate the distributed algorithm's run on the network, with Alice simulating the nodes on one side of the cut and Bob simulating the nodes on the other side.

2. CONGEST with communication by broadcast: this model is similar to the one above, but instead of communicating over point-to-point links, nodes communicate by broadcasting messages that are received by all nodes in their neighborhood. This more closely matches communication in wireless networks. The most widely-studied instance is the one where the network is the complete graph; this is the classical *shared blackboard* model of multi-party communication complexity [8], since broadcasting a message to all other players is equivalent to writing it on a blackboard that everyone can read. However, there are also lower bounds for the case where the network is not complete [7,5]. Lower bounds in this setting usually focus on round-complexity under restricted bandwidth, as in the previous model.

To prove lower bounds in this model, one also frequently uses reductions from two-player communication games, using a similar technique to the one outlined above.

3. Private-channel models of communication complexity: recently there has been much interest in studying the *total number of bits* required to solve various problems. In contrast to the models above, here we allow players to send an unbounded number of bits in each round, and we are generally not interested in the round complexity. For technical convenience, it is usually assumed that the communication network is a star, with a special player, the *coordinator*, at the center. In [12,15,16], lower bounds on *distributed streaming* are shown, including bounds on approximating the frequency moments of input distributed among the players and testing various graph properties. In [3], a tight lower bound of $\Omega(nk)$ bits is given for k-party set disjointness, where the players must determine whether or not their input sets have some global intersection; several of the lower bounds from [12,16] can also be obtained by reduction from set disjointness.

Lower bounds in this class of models are not as easy to obtain by reduction from two-player communication complexity, but in some cases they can be: the *symmetrization* technique, introduced in [12] and applied to harder problems in [15,16], allows one to reduce from two-player games to multiple players, by randomly choosing one player for Alice to simulate, and having Bob simulate all the other players as well as the coordinator. The input distribution needs to be chosen quite carefully for such reductions to go through. Symmetrization has its limits, and it cannot be used to show a tight lower bound on set disjointness; the bound from [3] was shown "from scratch" using *information complexity*, where we charge players for the bits that they learn about the other players' inputs.

Conclusion. Communication complexity lower bounds have many applications in distributed computing. In many cases, known lower bounds (most frequently, the set disjointness lower bound) can be applied as a black-box, using reductions from two-party communication games. However, in other cases, existing bounds do not quite suffice, for several reasons: first, even if a two-player reduction is possible, the particular two-player problems that are natural to reduce from may not have been studied before (as in [7]). Fortunately, recent developments in the field of information complexity have made two-party lower bounds much more approachable, even for those who are not experts in communication complexity. In other cases, a two-player reduction cannot yield a tight lower bound, and multi-player problems must be considered. This area of communication complexity is still somewhat in its infancy: although the communication complexity community has long been interested in multi-player communication complexity, most attention has until recently been focused on the *number-on-forehead* model, where each player can see the *other* players' inputs, but not its own. In contrast, distributed systems are more naturally modeled as *number-in-hand* games, where each player sees its own input. Although much progress has been made in the last few years, many fascinating problems remain open, and number-in-hand multi-party models are in general much less understood than their two-party counterpart.

References

1. Kuhn, F., Drucker, A., Oshman, R.: The computational power of the congested clique. To appear in PODC 2014 (2014)
2. Bar-Yossef, Z., Jayram, T.S., Kumar, R., Sivakumar, D.: An information statistics approach to data stream and communication complexity. J. Comput. Syst. Sci. 68(4), 702–732 (2004)
3. Braverman, M., Ellen, F., Oshman, R., Pitassi, T., Vaikuntanathan, V.: A tight bound for set disjointness in the message-passing model. In: Proc. 54th Symp. on Found. of Comp. Science (FOCS), pp. 668–677 (2013)
4. Dolev, D., Lenzen, C., Peled, S.: "Tri, tri again": Finding triangles and small subgraphs in a distributed setting (extended abstract). In: Aguilera, M.K. (ed.) DISC 2012. LNCS, vol. 7611, pp. 195–209. Springer, Heidelberg (2012)
5. Drucker, A., Kuhn, F., Oshman, R.: The communication complexity of distributed task allocation. In: PODC, pp. 67–76 (2012)

6. Kalyanasundaram, B., Schnitger, G.: The probabilistic communication complexity of set intersection. SIAM Journal on Discrete Mathematics 5(4), 545–557 (1992)
7. Kuhn, F., Oshman, R.: The complexity of data aggregation in directed networks. In: Peleg, D. (ed.) DISC 2011. LNCS, vol. 6950, pp. 416–431. Springer, Heidelberg (2011)
8. Kushilevitz, E., Nisan, N.: Communication Complexity. Cambridge University Press (2006)
9. Lenzen, C.: Optimal deterministic routing and sorting on the congested clique. In: Proc. 32nd Symp. on Principles of Distr. Comp. (PODC), pp. 42–50 (2013)
10. Patt-Shamir, B., Teplitsky, M.: The round complexity of distributed sorting: extended abstract. In: Proc. 30th Symp. on Principles of Distr. Comp. (PODC), pp. 249–256 (2011)
11. Peleg, D.: Distributed Computing: A Locality-sensitive Approach. Society for Industrial and Applied Mathematics (2000)
12. Phillips, J.M., Verbin, E., Zhang, Q.: Lower bounds for number-in-hand multiparty communication complexity, made easy. In: Proc. 23rd Symp. on Discrete Algorithms (SODA), pp. 486–501 (2012)
13. Razborov, A.: On the distributed complexity of disjointness. TCS: Theoretical Computer Science 106 (1992)
14. Das Sarma, A., Holzer, S., Kor, L., Korman, A., Nanongkai, D., Pandurangan, G., Peleg, D., Wattenhofer, R.: Distributed verification and hardness of distributed approximation. SIAM Journal on Computing 41(5), 1235–1265 (2012)
15. Woodruff, D.P., Zhang, Q.: Tight bounds for distributed functional monitoring. In: Proc. 44th Symp. on Theory of Comp. (STOC), pp. 941–960 (2012)
16. Woodruff, D.P., Zhang, Q.: When distributed computation is communication expensive. In: Afek, Y. (ed.) DISC 2013. LNCS, vol. 8205, pp. 16–30. Springer, Heidelberg (2013)
17. Yao, A.C.-C.: Some complexity questions related to distributive computing (preliminary report). In: STOC, pp. 209–213 (1979)

Distributed Algorithmic Foundations of Dynamic Networks

Gopal Pandurangan*

Division of Mathematical Sciences, Nanyang Technological University,
Singapore 637371 and Department of Computer Science and ICERM,
Brown University, Providence, RI 02912
gopalpandurangan@gmail.com

Introduction. Much of the well-established theory of distributed algorithms focuses on static networks, where nodes do not crash and edges maintain operational status forever. On the other hand, large real-world networks are inherently dynamic: the participants in peer-to-peer networks and social networks change over time, mobile nodes in wireless networks move in and out of each other's transmission range, and, in distributed data center networks, faulty machines need to be replaced by new machines without interrupting the operation of the remaining network. Dynamic network models, where the communication topology varies over time but where the set of nodes is fixed, have been studied extensively in literature. Recently, dynamic networks have been considered in the context of distributed computation, mostly assuming the model of [15,11,12]. In this setting, algorithms and complexity bounds for problems like distributed consensus [13], random walks [17,5], token dissemination [7,17], aggregation [6], and counting [1] have been developed. Most of this work assumes that the set of participating nodes is fixed. In contrast, much less is known about dynamic networks with *churn*, i.e., where the set of participants can change over time. Measurement studies of real-world dynamic networks [9,10,18] show that the churn rate (i.e. the number of nodes that can join/leave the network at the same time) is quite high: nearly 50% of the nodes can be replaced within an hour. Thus it is important to develop a rich theory of dynamic networks with churn.

Dynamic networks pose non-trivial challenges in solving even basic distributed problems. Algorithmic techniques developed for static networks are not readily applicable to dynamic networks[4]. We need rigorous models and algorithmic techniques for dynamic networks. In highly dynamic networks, where nodes can join and leave continuously and substantially change over time, doing non-trivial distributed computing tasks is particularly challenging. In particular, it is important to design algorithms that work continuously over time (not assuming any eventual quiescence or stabilization). In this talk, we will focus on the following fundamental distributed computing problems in the context of dynamic networks with churn:

* Supported in part by Nanyang Technological University grant M58110000, Singapore Ministry of Education (MOE) Academic Research Fund (AcRF) Tier 2 grant MOE2010-T2-2-082, Singapore MOE AcRF Tier 1 grant MOE2012-T1-001-094, and by a grant from the United States-Israel Binational Science Foundation (BSF).

M. Halldórsson (Ed.): SIROCCO 2014, LNCS 8576, pp. 18–22, 2014.

- Agreement: Nodes have to agree on a common value.
- Search and Storage: Find data and resources in the network; store data reliably and securely.
- Byzantine agreement: Solve agreement problem under the presence of malicious nodes.

We will give an overview of our recent results that make progress towards developing an algorithmic theory of dynamic networks. First, we will present a rigorous theoretical framework for studying dynamic networks. Then we will present efficient techniques and algorithms for solving agreement, Byzantine agreement, and search/storage. The techniques we use are efficient information spreading, support estimation (estimating aggregate functions) and random-walk based methods and show that these can be adapted to work even under highly dynamic networks with provable guarantees.

Model. Apart from nodes independently joining and leaving the network, the evolution of the network might also be triggered by coordinated attacks (e.g. denial-of-service attacks) with the goal of corrupting the communication in the network. In large-scale networks of millions of nodes, it is unrealistic to assume that all nodes faithfully execute the given protocol. Thus we must design scalable algorithms that can handle *adversarial* dynamic changes and are robust against malicious (i.e. *Byzantine* nodes [16]) that might collude and send corrupted messages. In particular, we assume that the churn is controlled by an all-powerful adversary that comes in two flavors: (1) *oblivious* — has complete knowledge and control of what nodes join and leave and at what time and has unlimited computational power — but is oblivious to the random choices made by the algorithm and (2) *adaptive* — in addition to the above it has knowledge of past random choices (but not current and future ones). Byzantine nodes, on the other hand, have knowledge about the entire state of network at every round (including random choices made by all the nodes) and can behave arbitrarily. However, they can communicate only through the (current) edges of the dynamic network and cannot send messages to nodes which are not neighbors at that point in time. Key parameters of the model are the amount of churn per round (called the *churn rate*) that the adversary can control and the number of Byzantine nodes (per round) that are allowed.

Our model is a synchronous dynamic network (with Byzantine nodes) represented by a *bounded-degree* graph with a dynamically changing topology (both nodes and edges change from round to round) whose nodes execute a distributed algorithm and whose edges represent connectivity in the network. To rule out trivial impossibility results, we assume that in each round the network is an *expander*. We note that although the churn can be quite high, the network size is *stable*, in particular, the results below assume that the size is n in every round (this can be relaxed somewhat). We refer to [4,3,2] for a detailed description of our model.

Main Results. We develop fast randomized algorithms that run in polylogarithmic number of rounds and prove lower bounds for several fundamental problems in this setting: First, we consider the distributed agreement problem in a network with εn churn per round (but without Byzantine nodes) where each node starts with an input value and eventually, we require nodes to decide on the same value. It is easy to see that in our adversarial setting, achieving global consensus among *all* nodes is impossible—the adversary can effectively isolate some nodes by subjecting all of their neighbors to churn in every round, since we assume bounded node degrees. Thus we consider *almost-everywhere (AE) agreement* (cf. [8]), where we allow a small number of nodes to decide on the wrong value. By leveraging properties of the exponential distribution and the expansion of the network, we obtain an algorithm (cf. [4]) that sends messages of only *polylogarithmic* (in n) size and achieves AE agreement in *polylogarithmic* number of rounds, despite the fact that, within $\lceil 1/\varepsilon \rceil = \Theta(1)$ rounds, the entire network can be subjected to churn, in other words, the amount of churn per round can be up to εn. Thus this result shows that agreement in possible even under linear (in the size of n) amount of churn. The above result works under an oblivious adversary. Under an adaptive adversary we show that the amount of churn that can be tolerated is less, up to $\varepsilon\sqrt{n}$. (Moreover, this algorithm requires polynomial sized messages per round.) In follow up paper [3], we show that this is essentially the best possible amount of churn that can be tolerated under an adaptive adversary, if one requires fast (i.e., polylogarithmic) number of rounds.

Next, we consider *Byzantine almost-everywhere (BAE) agreement* where malicious nodes can deviate arbitrarily from the protocol, which has been studied extensively throughout the past decades. In this setting, we prove that there is no polylogarithmic-time algorithm that achieves BAE agreement if the number of Byzantine nodes is $\omega(\sqrt{n}\log n)$ [3]. On the positive side, we present an algorithm that achieves BAE agreement by leveraging the fast mixing of random walks on expander graphs and employing a majority rule-based agreement algorithm. This algorithm tolerates up to $O(\sqrt{n}/\text{polylog}(n))$ Byzantine nodes and (adaptive) churn, thus matching our lower bound (up to logarithmic factors) [3].

Finally, we focus on the fundamental problem of storing, maintaining, and searching data in P2P networks [2]. Search in P2P networks is a well-studied fundamental application with a large body of work in the last decade or so, both in theory and practice (e.g., see the survey [14]). While many P2P systems/protocols have been proposed for efficient search and storage of data, a major drawback of almost all these is the lack of algorithms that work with provable guarantees under a large amount of churn per round. The problem is especially challenging since the goal is to guarantee that almost all nodes are able to efficiently store, maintain, and retrieve data, even under high churn rate. In such a highly dynamic setting, it is non-trivial to even just store data in a persistent manner; the churn can simply remove a large fraction of nodes in just one time step. On the other hand, it is costly to replicate too many copies of a data item to guarantee persistence. Thus the challenge is to use as little storage as possible and maintain the data for a long time, while at the same time designing efficient search algorithms that find the data quickly, despite high churn rate. We develop algorithms that, in the presence of $O(n/\text{polylog}(n))$ (oblivious) churn per round, (1) enable almost all (except $o(n)$) nodes to store data among their peers, and (2) guarantee that these data items can be successfully searched by almost all nodes. To satisfy the above requirements, it is sufficient if we store each data item at only $\Theta(n^{1/2+\epsilon})$ peers, for any small constant $\epsilon > 0$. Our algorithms are scalable, as they only require polylogarithmic number of rounds and messages of polylogarithmic size.

Conclusion. Large-scale, highly dynamic networks are increasingly dominant in the real world. Distributed algorithms that are robust, efficient, and secure are required. An important goal is to solve fundamental distributed computing problems with provable guarantees, under strong models. We have developed models, algorithms, and techniques that work even under a high amount of dynamism — however it requires properties, such as good expansion. A agenda for future work is to build on current framework to design even stronger algorithms that work with minimal assumptions.

References

1. Abshoff, S., Benter, M., Malatyali, M., Meyer auf der Heide, F.: On two-party communication through dynamic networks. In: Baldoni, R., Nisse, N., van Steen, M. (eds.) OPODIS 2013. LNCS, vol. 8304, pp. 11–22. Springer, Heidelberg (2013)
2. Augustine, J., Molla, A.R., Morsy, E., Pandurangan, G., Robinson, P., Upfal, E.: Storage and search in dynamic peer-to-peer networks. In: SPAA, pp. 53–62 (2013)
3. Augustine, J., Pandurangan, G., Robinson, P.: Fast byzantine agreement in dynamic networks. In: Fatourou, P., Taubenfeld, G. (eds.) PODC, pp. 74–83. ACM (2013)
4. Augustine, J., Pandurangan, G., Robinson, P., Upfal, E.: Towards robust and efficient computation in dynamic peer-to-peer networks. In: ACM-SIAM, SODA 2012, pp. 551–569. SIAM (2012)
5. Avin, C., Koucký, M., Lotker, Z.: How to explore a fast-changing world (Cover time of a simple random walk on evolving graphs). In: Aceto, L., Damgård, I., Goldberg, L.A., Halldórsson, M.M., Ingólfsdóttir, A., Walukiewicz, I. (eds.) ICALP 2008, Part I. LNCS, vol. 5125, pp. 121–132. Springer, Heidelberg (2008)
6. Cornejo, A., Gilbert, S., Newport, C.C.: Aggregation in dynamic networks. In: PODC, pp. 195–204 (2012)
7. Dutta, C., Pandurangan, G., Rajaraman, R., Sun, Z., Viola, E.: On the complexity of information spreading in dynamic networks. In: Khanna, S. (ed.) SODA, pp. 717–736. SIAM (2013)
8. Dwork, C., Peleg, D., Pippenger, N., Upfal, E.: Fault tolerance in networks of bounded degree. SIAM J. Comput. 17(5), 975–988 (1988)
9. Falkner, J., Piatek, M., John, J.P., Krishnamurthy, A., Anderson, T.E.: Profiling a million user dht. In: Internet Measurement Comference, pp. 129–134 (2007)
10. Krishna Gummadi, P., Saroiu, S., Gribble, S.D.: A measurement study of napster and gnutella as examples of peer-to-peer file sharing systems. Computer Communication Review 32(1), 82 (2002)
11. Kuhn, F., Oshman, R.: Dynamic networks: Models and algorithms. SIGACT News 42(1), 82–96 (2011)

12. Kuhn, F., Lynch, N., Oshman, R.: Distributed computation in dynamic networks. In: ACM STOC, pp. 513–522 (2010)
13. Kuhn, F., Oshman, R., Moses, Y.: Coordinated consensus in dynamic networks. In: Proceedings of the 30th Annual ACM SIGACT-SIGOPS Symposium on Principles of Distributed Computing, PODC 2011, pp. 1–10. ACM (2011)
14. Lua, E.K., Crowcroft, J., Pias, M., Sharma, R., Lim, S.: A survey and comparison of peer-to-peer overlay network schemes. IEEE Communications Survey and Tutorial (2004)
15. O'Dell, R., Wattenhofer, R.: Information dissemination in highly dynamic graphs. In: DIALM-POMC, pp. 104–110 (2005)
16. Pease, M.C., Shostak, R.E., Lamport, L.: Reaching agreement in the presence of faults. J. ACM 27(2), 228–234 (1980)
17. Das Sarma, A., Molla, A.R., Pandurangan, G.: Fast distributed computation in dynamic networks via random walks. In: Aguilera, M.K. (ed.) DISC 2012. LNCS, vol. 7611, pp. 136–150. Springer, Heidelberg (2012)
18. Sen, S., Wang, J.: Analyzing peer-to-peer traffic across large networks. In: Proceedings of the 2nd ACM SIGCOMM Workshop on Internet Measurment, IMW 2002, pp. 137–150. ACM, New York (2002)

The Beachcombers' Problem:
Walking and Searching with Mobile Robots

Jurek Czyzowicz[1], Leszek Gąsieniec[2], Konstantinos Georgiou[3],
Evangelos Kranakis[4], and Fraser MacQuarrie[4]

[1] Université du Québec en Outaouais, Department d'Informatique,
Gatineau, Québec, Canada
[2] University of Liverpool, Department of Computer Science, Liverpool, UK
[3] University of Waterloo, Dept. of Combinatorics & Optimization,
Waterloo, Ontario, Canada
[4] Carleton University, School of Computer Science, Ottawa, Ontario, Canada

Abstract. We introduce and study a new problem concerning the exploration of a geometric domain by mobile robots. Consider a line segment $[0, I]$ and a set of n mobile robots $r_1, r_2, \ldots, r_n$ placed at one of its endpoints. Each robot has a *searching speed* s_i and a *walking speed* w_i, where $s_i < w_i$. We assume that every robot is aware of the number of robots of the collection and their corresponding speeds.

At each time moment a robot r_i either walks along a portion of the segment not exceeding its walking speed w_i, or it searches a portion of the segment with speed not exceeding s_i. A search of segment $[0, I]$ is completed at the time when each of its points have been searched by at least one of the n robots. We want to develop efficient *mobility schedules* (algorithms) for the robots which complete the search of the segment as fast as possible. More exactly we want to maximize the *speed* of the mobility schedule (equal to the ratio of the segment length versus the time of the completion of the schedule).

We analyze first the offline scenario when the robots know the length of the segment that is to be searched. We give an algorithm producing a mobility schedule for arbitrary walking and searching speeds and prove its optimality. Then we propose an online algorithm, when the robots do not know in advance the actual length of the segment to be searched. The speed S of such algorithm is defined as $S = \inf_{I_L} S(I_L)$ where $S(I_L)$ denotes the speed of searching of segment $I_L = [0, L]$. We prove that the proposed online algorithm is 2-competitive. The competitive ratio is shown to be better in the case when the robots' walking speeds are all the same, approaching 1.29843 as n goes to infinity.

Keywords: Algorithm, Mobile Robots, On-line, Schedule, Searching, Segment, Speed, Walking.

1 Introduction

A line segment has to be explored collectively by n mobile robots initially placed at a segment endpoint. At every time moment a robot may perform either of

M. Halldórsson (Ed.): SIROCCO 2014, LNCS 8576, pp. 23–36, 2014.
© Springer International Publishing Switzerland 2014

the two different activities of *walking* and *searching*. While walking, each robot may *traverse* the domain with a speed not exceeding its maximal *walking speed*. During searching, the robot performs a more *elaborate* task on the domain. The bounds on the walking and searching speeds may be distinct for different robots, but we always assume that each robot can walk with greater maximal speed than it can search. Our goal is to design the movement of all robots so that each point of the domain is being searched by at least one robot and the time when the process is completed is minimized (i.e. its speed is maximized).

In many situations *two-speed* searching is a convenient way to approach exploration of various domains. For example *foraging* or *harvesting* a field may take longer than walking across. Intruder searching activity takes more time than uninvolved territory traversal. In computer science, problems such as *web pages indexing, forensic search, code inspection,* and *packet sniffing* require a more involved inspection process. Similar problems arise in many other domains. We use the analogy of the robots as *beachcombers* to emphasize that when searching a domain (e.g. a beach looking for things of value), robots would need to move slower than if they were simply traversing the domain.

In our problem, the searchers collaborate in order to terminate the searching process as quickly as possible. Our algorithms generate *mobility schedules* i.e. sequences of moves of the robots, which assure that every point of the environment is inspected by at least one robot while this robot was performing the searching activity.

1.1 Preliminaries

Let I_L denote the interval $[0, L]$ for any positive integer L. Consider n robots $r_1, r_2, \ldots, r_n$, each robot r_i having searching speed s_i and walking speed w_i, such that $s_i < w_i$. A *searching schedule* $\mathcal{A}$ of I_L is defined by an increasing sequence of moments $t_0 = 0, t_1, \ldots, t_z$, such that in each interval $[t_j, t_{j+1}]$, for $0 \le j < z$, every robot r_i either walks along some sub-segment of I_L not exceeding its walking speed w_i, or searches some sub-segment of I_L not exceeding its searching speed s_i. The searching schedule is *correct* if for each point $p \in I_L$ there is some $j \ge 1$ and some robot r_i, such that during the interval $[t_j, t_{j+1}]$ robot r_i searches the sub-segment of I_L containing point p. We assume that there is no communication between robots during the execution of the algorithm.

By the speed $S_{\mathcal{A}}(I_L)$ of schedule $\mathcal{A}$ searching interval I_L we mean the value of $S_{\mathcal{A}}(I_L) = L/t_z$. We call t_z the *finishing time* of the searching schedule. The searching schedule is optimal if there does not exist any other correct searching schedule having a speed larger than S.

It is easy to see that the schedule speed maximization criterion is equivalent to its finishing time minimization when the segment length is given or to the searched segment length maximization when the time bound is set in advance. However the speed maximization criterion applies better to the online problem when the objective of the schedule is to perform searching of an unknown-length segment or a semi-line. Such a schedule successively searches the intervals I_L

for the increasing values of L. The speed of such a schedule is defined as $S_A = \inf_{I_L} S_A(I_L)$.

Observe that any searching schedule may be converted to another one, which has the property that all sub-segments which were being searched (during some time intervals $[t_j, t_{j+1}]$ by some robots) have pairwise disjoint interiors. Indeed, if some sub-segment is being searched by two different robots (or twice by the same robot), the second searching may be replaced by the walk through it by the robot involved. Since the walking speed of any robot is always larger than its searching speed, the speed of such converted schedule is not smaller than the original one. Therefore, when looking for the optimal searching schedule, it is sufficient to restrict consideration to schedules whose searched sub-segments may only intersect at their endpoints. In the sequel, all searching schedules in our paper will have this property.

Notice as well, that, when looking for the most efficient schedule, we may restrict our consideration to schedules such that at any time moment a robot r_i is either searching using its maximal searching speed s_i, or walking with maximal allowed speed w_i. Indeed, whenever r_i searches (or walks) during a time interval $[t_j, t_{j+1}]$ using a non-maximal and not necessarily constant searching speed (resp. walking speed) we may replace it with a search (resp. walk) using the maximal allowed speed. It is easy to see that the search time of any point is never longer for the modified schedule, so the speed of such a schedule is not decreased.

We assume that all the robots start their exploration at the same time and that they are allowed to cross over each other. We study the offline and online versions of the problem.

Definition 1 (Beachcombers' Problem). *Consider an interval $I_L = [0, L]$ and n robots $r_1, r_2, \ldots, r_n$, initially placed at its endpoint 0, each robot r_i having searching speed s_i and walking speed w_i, such that $s_i < w_i$. The Beachcombers' Problem consists of finding an efficient correct searching schedule A of I_L. The speed S_A of the solution to the Beachcombers' Problem equals $S_A = I_L / t_z$, where t_z is the finishing time of A.*

Definition 2 (Online Beachcombers' Problem). *Consider n robots $r_1, r_2, \ldots, r_n$, initially placed at the origin of a semi-line I, each robot r_i having searching speed s_i and walking speed w_i, such that $s_i < w_i$. The Online Beachcombers' Problem consists of finding an efficient correct searching schedule A of I. The cost S_A of the solution to the Online Beachcombers' Problem, called the speed of A, equals $S_A = \inf_{I_L} S_A(I_L) = \inf_L \frac{|I_L|}{t_z(I_L)}$ where $I_L = [0, L]$ for any positive integer L and $t_z(I_L)$ denotes the time when the search of the segment $I_L = [0, L]$ is completed.*

1.2   Related Work

The original text on graph searching started with the work of Koopman [1]. Many papers followed studying searching and exploration of graphs (e.g. [2, 3]) or geometric environments, (e.g.[4–8]). The purpose of these studies was usually either to learn (map) an unknown environment (e.g.[2]) or to search it, looking for a target (motionless or mobile) (cf. [3]).

Many searching problems were studied from a game-theoretic viewpoint (see [5]). [5] presented an approach to rendezvous and searching, when two mobile players either collaborate in order to find each other, or they compete against each other - one willing to meet and the second one attempting to avoid the meeting. Searching one-dimensional environments (segments, lines, semi-lines), similarly to the present paper, despite the simplicity of the environment, often led to interesting results in this model (cf. [9–11]).

The efficiency of the searching or exploration algorithm is usually measured by the time used by the mobile agent, often proportional to the distance travelled. Many searching and especially exploration algorithms are *online*, i.e. they concern a priori unknown environments, cf. [12, 13]. Performance of such algorithms is usually expressed by the *competitive ratio*, i.e. the proportion of the time spent by the online algorithm versus the time of the optimal offline algorithm, which assumes the knowledge of the environment (cf. [14, 15]). Most exploration algorithms (e.g. [7, 8, 16] and several search algorithms (cf. [11–13]) use the competitive ratio to measure their performance.

Most of the above research concerned single robots. Collections of mobile robots, collaborating in order to reduce the exploration time, were used, e.g., in [17–20]. Most recently [16] studied tradeoffs between the number of robots and the time of exploration showing how a polynomial number of robots may search the graph optimally.

Some research studying mobile robots assumes distinct robot speeds. Varying mobile sensor speed was used in [21] for the purpose of sensor energy efficiency. [22] was utilizing distinct robot speeds to design fast converging protocols, e.g. for gathering. [23, 24] considered distinct speeds for robots performing boundary patrolling. However to the best of our knowledge, the present paper is the first one assuming two-speed robots for the problem of searching or exploration.

1.3   Outline and Results of the Paper

In Section 2 we first study the properties of optimal mobility schedules. We then propose Comb, an optimal algorithm for the Beachcombers' Problem, which requires $O(n \log n)$ computational steps, and prove its correctness. Section 3 is devoted to online searching, where the length of the segment to be searched is not known in advance. In this section we propose the online searching algorithm LeapFrog, prove its correctness and analyze its efficiency. We prove that the LeapFrog algorithm is 2-competitive. The competitive ratio is shown to be reduced to 1.29843 in the case when all robots' walking speeds are the same. Section 5 concludes the paper and proposes problems for further research. Due to space limitations all missing proofs can be found in the full paper.

2 Searching a Known Segment

We proceed by first identifying in Subsection 2.1 a number of structural properties exhibited by every optimal solution to the Beachcombers' Problem. This will allow us to conclude in Subsection 2.2 that the Beachcombers' Problem can be solved efficiently.

2.1 Properties of Optimal Schedules

In the following lemmas we identify some properties of optimal schedules. By the observation made in the preliminaries we suppose that the segment may be divided into a sequence of sub-intervals $\sigma_1, \sigma_2, \ldots, \sigma_k$, each sub-interval σ_i being solely and entirely searched by the same robot.

Lemma 1. *In every optimal schedule all robots terminate their work simultaneously and each robot completes its work by searching some non-empty part of the segment.*

Proof. Suppose, to the contrary, that the finishing time of some optimal schedule $\mathcal{A}$ is T and some robot r_e finishes its work at an earlier time $T - \epsilon$, for $\epsilon > 0$. We show that it is possible to rearrange the schedule (i.e. the times of walking and searching of all robots) so that 1° the rearranged schedule $\mathcal{A}'$ is correct, 2° it completes within time T, 3° robot r_e is the one which searches the very first sub-interval of the segment, and 4° r_e finishes its work at some time $T - \epsilon_1$, for $\epsilon_1 > 0$.

Let $[0, A]$ be the first sub-interval being searched by a particular robot in $\mathcal{A}$ and $[A, B]$ be the second such interval, with $0 < A < B < 1$. If it is r_e which searches $[0, A]$ in $\mathcal{A}$, we make r_e also search $X = [A, \min(\frac{\epsilon}{2}(s_e), \frac{B}{2})]$ - the portion of $[A, B]$, while the robot searching $[A, B]$ in $\mathcal{A}$ is switched to walking in X. Otherwise, if it is robot $r_f \neq r_e$ which searches $[A, B]$ in $\mathcal{A}$ we make robot r_e search interval $Y = [0, \min(\frac{\epsilon}{2}(w_e - s_e), \frac{A}{2})]$ while robot r_f in $\mathcal{A}$ is switched to walking in Y. Observe that, we made r_e search more than in $\mathcal{A}$, but still finishing within time $T - \frac{\epsilon}{2} < T$. However this permits some other robot r_f to switch from searching to walking in some non-empty sub-interval. As r_f was completing its work in schedule $\mathcal{A}$ within at most time T and $w_{r_f} > s_{r_f}$ robot r_f finishes now its work in time $T - \epsilon_1$ for some $\epsilon_1 > 0$.

As a result of the above rearrangement we have a correct schedule in which a sequence of subsequent sub-intervals $\sigma_1, \sigma_2, \ldots, \sigma_k$ are searched by the robots, such that the robots searching σ_1 and σ_2 both finish their work before time T (i.e. σ_1 is searched by r_e finishing not later than $T - \frac{\epsilon}{2}$ and σ_2 searched by r_f finishing not later than $T - \epsilon_2$). Similarly as above, we can make robot r_f search into sub-interval σ_3 for time $\frac{\epsilon_2}{2}$ permitting the robot responsible for searching σ_3 switch to walking for some positive amount of time and finish searching strictly before time T. By induction, we can thus reduce the finishing times of all robots contradicting that schedule $\mathcal{A}$ was optimal.

In case when some robot does not participate in searching or if it finishes its work by walking, we can reduce its makespan by cutting off the last walk without affecting the correctness of the schedule. We then repeat the same argument as above. □

Lemma 2. *In every optimal schedule, each robot searches a continuous interval.*

Proof. Suppose, to the contrary, that in an optimal schedule $\mathcal{A}$ some robot searches more than one interval $\sigma_1, \sigma_2, \ldots, \sigma_k$. Let σ_q be the last interval of this sequence, such that some robot, say r_j, searching σ_q searches also σ_p, for $1 \leq p < q - 1$. Denote by L the length of the segments between σ_q and σ_p, i.e. $L = \sum_{i=p+1}^{q-1} |\sigma_i|$. Consider moving interval σ_p distance L to the right and each σ_i, for $i = p+1, \ldots, q-1$ distance L to the left. By movement of an interval we mean that the robot searching this interval changes to the searching mode in the new position of the interval and to the walking mode in the old position (except in the intersection of the intervals). Observe that such rearranged schedule $\mathcal{A}'$ is correct and no robot searches or walks longer than in $\mathcal{A}$. However, the robot which was searching interval σ_{q-1} in $\mathcal{A}$ finishes its work earlier in $\mathcal{A}'$ (it finishes by walking in $\mathcal{A}'$). By Lemma 1 $\mathcal{A}'$ is not optimal, hence $\mathcal{A}$ as well. □

By Lemmas 1 and 2 every optimal schedule is defined by a sequence of intervals $\sigma_1, \sigma_2, \ldots, \sigma_n$, such that robot r_i walks at maximal speed through intervals $\sigma_1, \ldots, \sigma_{i-1}$ and finishes its work by searching σ_i (we suppose w.l.o.g. that we rename the robots with respect to the order of the intervals). The next lemma shows that such robots must be arranged in this sequence by the non-decreasing order of their walking speeds.

Lemma 3. *In every optimal schedule, for any two robots r_i, r_j with $w_i < w_j$, robot r_i searches a sub-interval closer to the starting point than the sub-interval of robot r_j.*

Proof. Suppose, to the contrary, that the robots are not arranged by the non-decreasing walking speed. Then there must exist a pair of consecutive robots r_i, r_{i+1} with $w_i > w_{i+1}$, searching, respectively, the consecutive intervals σ_i, σ_{i+1}. In what follows, we investigate the effect of switching the order of the search intervals of two robots r_i, r_{i+1}, so that the union of the intervals remains unchanged. In particular we will redistribute the portions of the union of the two intervals that each robot will search. Since we will only redistribute the union of intervals σ_i, σ_{i+1}, the remaining sub-intervals will remain the same, and so will the finishing search times of the remaining robots.

For the sake of notational convenience, let $|\sigma_i| = c_i$ and $|\sigma_{i+1}| = c_{i+1}$. We may also assume that $c_i + c_{i+1} = 1$, and therefore that $c_i = \lambda$ and that $c_{i+1} = 1 - \lambda$, after proper scaling of the intervals.

Note that robot r_i searches σ_i while robot r_{i+1} walks within σ_i and searches σ_{i+1}. Each robot will have to walk the distance $x \geq 0$ between the start point and the leftmost point of σ_i. By Lemma 1 all robots have the same finishing time, so we have $\frac{x}{w_i} + \frac{\lambda}{s_i} = \frac{x}{w_{i+1}} + \frac{\lambda}{w_{i+1}} + \frac{1-\lambda}{s_{i+1}}$. Solving for λ gives

$\lambda = \left(\frac{x}{w_{i+1}} - \frac{x}{w_i} + \frac{1}{s_{i+1}}\right) / \left(\frac{1}{s_i} + \frac{1}{s_{i+1}} - \frac{1}{w_{i+1}}\right)$. Therefore, as $\lambda \geq 0$ (by Lemmas 1, 2), the finishing time

$$T = \frac{x}{w_i} + \frac{\lambda}{s_i} = \frac{x}{w_i} + \frac{\frac{x}{s_i}\left(\frac{1}{w_{i+1}} - \frac{1}{w_i}\right)}{\frac{1}{s_i} + \frac{1}{s_{i+1}} - \frac{1}{w_{i+1}}} + \frac{1}{\left(\frac{1}{s_i} + \frac{1}{s_{i+1}} - \frac{1}{w_{i+1}}\right)s_i s_{i+1}}.$$

We now reschedule the robots so that robot r_{i+1} searches first, say a μ portion of $c_i + c_{i+1} = 1$, and robot r_i searches the remaining sub-interval of length $(1 - \mu)(c_i + c_{i+1}) = 1 - \mu$. This means that robot r_i will now walk the interval of length μ. Since by Lemma 1 the two robots must finish simultaneously, the same reasoning shows that $\frac{x}{w_i} + \frac{\mu}{w_i} + \frac{1-\mu}{s_i} = \frac{x}{w_{i+1}} + \frac{\mu}{s_{i+1}}$. As before, we can eliminate μ, so as to conclude that

$$T' = \frac{x}{w_{i+1}} + \frac{\mu}{s_{i+1}} = \frac{x}{w_{i+1}} + \frac{\frac{x}{s_{i+1}}\left(\frac{1}{w_i} - \frac{1}{w_{i+1}}\right)}{\frac{1}{s_i} + \frac{1}{s_{i+1}} - \frac{1}{w_i}} + \frac{1}{\left(\frac{1}{s_i} + \frac{1}{s_{i+1}} - \frac{1}{w_i}\right)s_i s_{i+1}}.$$

We show below that $T > T'$. Observe first that, since $w_i > w_{i+1}$ we have

$$K = \left(\frac{1}{\left(\frac{1}{s_i} + \frac{1}{s_{i+1}} - \frac{1}{w_{i+1}}\right)s_i s_{i+1}} - \frac{1}{\left(\frac{1}{s_i} + \frac{1}{s_{i+1}} - \frac{1}{w_i}\right)s_i s_{i+1}}\right) > 0$$

Hence

$$T - T' = \left(\frac{x}{w_i} + \frac{\frac{x}{s_i}\left(\frac{1}{w_{i+1}} - \frac{1}{w_i}\right)}{\frac{1}{s_i} + \frac{1}{s_{i+1}} - \frac{1}{w_{i+1}}} + \frac{1}{\left(\frac{1}{s_i} + \frac{1}{s_{i+1}} - \frac{1}{w_{i+1}}\right)s_i s_{i+1}}\right)$$
$$- \left(\frac{x}{w_{i+1}} + \frac{\frac{x}{s_{i+1}}\left(\frac{1}{w_i} - \frac{1}{w_{i+1}}\right)}{\frac{1}{s_i} + \frac{1}{s_{i+1}} - \frac{1}{w_i}} + \frac{1}{\left(\frac{1}{s_i} + \frac{1}{s_{i+1}} - \frac{1}{w_i}\right)s_i s_{i+1}}\right)$$
$$= x\left(\left(\frac{1}{w_i} - \frac{1}{w_{i+1}}\right) + \frac{\frac{1}{s_i}\left(\frac{1}{w_{i+1}} - \frac{1}{w_i}\right)}{\frac{1}{s_i} + \frac{1}{s_{i+1}} - \frac{1}{w_{i+1}}} + \frac{\frac{1}{s_{i+1}}\left(\frac{1}{w_{i+1}} - \frac{1}{w_i}\right)}{\frac{1}{s_i} + \frac{1}{s_{i+1}} - \frac{1}{w_i}}\right) + K$$
$$> x\left(-\left(\frac{1}{w_{i+1}} - \frac{1}{w_i}\right) + \frac{\frac{1}{s_i}\left(\frac{1}{w_{i+1}} - \frac{1}{w_i}\right)}{\frac{1}{s_i} + \frac{1}{s_{i+1}}} + \frac{\frac{1}{s_{i+1}}\left(\frac{1}{w_{i+1}} - \frac{1}{w_i}\right)}{\frac{1}{s_i} + \frac{1}{s_{i+1}}}\right) + K$$

where the last term above equals K which is > 0. Therefore in schedule $\mathcal{A}'$ the finishing time of both robots r_i, r_j is smaller than in $\mathcal{A}$, which by Lemma 1 contradicts the optimality of $\mathcal{A}$. $\square$

By using the properties obtained in Lemmas 1, 2 and 3, we determine a useful recurrence for the sub-intervals searched by robots in an optimal schedule.

Lemma 4. *Let the robots* $r_1, r_2, \ldots, r_n$ *be ordered in non-decreasing walking speed, and suppose that* T_{opt} *is the time of the optimal schedule. Then, the segment to be searched may be partitioned into successive sub-segments of lengths* $c_1, c_2, \ldots, c_n$ *and the optimal schedule assigns to robot* r_i *the* i^{th} *interval of length* c_i, *where the length* c_i *satisfies the following recursive formula, and where we assume, without loss of generality, that* $w_0 = 0$ *and* $w_1 = 1$.[1]

$$c_0 = 0; \qquad c_k = \frac{s_k}{w_k} \left(\left(\frac{w_{k-1}}{s_{k-1}} - 1 \right) c_{k-1} + T_{opt}(w_k - w_{k-1}) \right), \quad k \geq 1 \qquad (1)$$

Proof. From Lemma 2 we know that all robots search contiguous intervals. Since by Lemma 1 we need to utilize all robots, it follows that the optimal schedule defines a partition of the unit domain into n sub-intervals. Finally by Lemma 3, if we order the robots in non-decreasing walking speed, then robot r_i will search the ith in a row interval, showing the first claim of the lemma.

Now, from Lemma 1, we know that all robots finish at the same time, T_{opt}. Since all robots start processing the domain at the same time, robot k will walk its initial sub-interval of length $\sum_{i=1}^{k-1} c_i$ in time proportional to $1/w_k$, and in the remaining time it will search the interval of length c_k. Hence $c_k = s_k \left(T_{opt} - \frac{\sum_{i=1}^{k-1} c_i}{w_k} \right)$, from which we easily derive the desired recursion. □

2.2   The Optimal Schedule for the Beachcombers' Problem

As a consequence of Lemmas 1, 2 and 3 we have the following offline algorithm Comb producing an optimal schedule. The algorithm is parametrized by the real values c_i equal to the sizes of intervals to be searched by each robot r_i.

Figure 1(a) illustrates the schedule produced by algorithm Comb for a set of five robots. The trajectory of each robot (except the first one) is formed by a segment of walking (thin line) followed by a segment of searching (bold). The dashed line corresponds to the slope representing the speed S_{opt} of the schedule. Each sub-segment C_i in Fig. 1(a) has length c_i stated in Lemma 4.

We can now prove the following theorem:

Theorem 1. *The Beachcombers' Problem can be solved optimally in* $O(n \log n)$ *many steps.*

Proof. By Lemma 4 we need to order the robots by non-decreasing walking speed, which requires $O(n \log n)$ many steps. We then show how to compute all c_i in linear number of steps, modulo the arithmetic operations that depend on the encoding sizes of w_i, s_i.

---

[1] We set $w_0 = 0$ and $w_1 = 1$ for notational convenience, so that (1) holds. Note that w_0 does not correspond to any robot, while w_1 is the walking speed of the robot that will search the first sub-interval, and so will never enter walking mode, hence, w_1 does not affect our solution.

Algorithm Comb;
1. Sort the robots in non-decreasing walking speeds;
2. **for** $i \leftarrow 1$ **to** n **do**
3.   Robot r_i first walks the interval of length $\sum_{j=1}^{i-1} c_j$,
        and then searches interval of length c_i

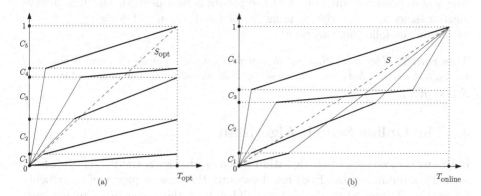

Fig. 1. Schedules produced by (a) Algorithm Comb and (b) Algorithm LeapFrog

Consider an imaginary unit time period. Starting with the slowest, for each robot, we use (1) to compute (in constant time) the sub-interval y_i it would search if it were to remain active for the unit time period. Consequently, we can compute in n steps the total length $\sum_{i=1}^{n} y_i$ of the interval that the collection of robots can search within a unit time period. This schedule, scaled to a unit domain, will have finishing time $T = 1/\sum_{i=1}^{n} y_i$. The length of the interval that robot r_k will search is then $c_k = y_k/\sum_{i=1}^{n} y_i$.   □

2.3   Closed Formulas for the Optimal Schedule of the Beachcombers' Problem

From the proof of Theorem 1 we can implicitly derive the time (and the speed) of an optimal solution to the Beachcombers' Problem. In what follows, we assume that $w_i = 0$, that the robots are ordered in non-decreasing walking speeds, and that $w_1 = 1$ (see Lemma 4).

Lemma 5. *Consider a set of robots such that in the optimal schedule each robot finishes searching in time T_{opt}. Robot r_k will search a sub-interval of length c_k, such that*

$$\frac{c_k}{T_{opt}} = s_k - \frac{s_k}{w_k} \sum_{r=1}^{k-1} s_r \prod_{j=r+1}^{k-1} \left(1 - \frac{s_j}{w_j}\right) \qquad (2)$$

Definition 3 (Search Power). *Consider a set of n robots $r_1, r_2, \ldots, r_n$, with $s_i < w_i$, $i = 1, \ldots, n$. We define the search power of any subset A of robots using* *a real function $g : 2^{[n]} \mapsto \mathbb{R}^+$ as follows: For any subset A, first sort the items in non-decreasing walking speeds w_i, and let $w_1^A, \ldots, w_{|A|}^A$ be that ordering (the superscripts just indicate membership in A). We define the evaluation function (search power of set A) as $g(A) := \sum_{k=1}^{|A|} s_k^A \prod_{j=k+1}^{|A|} \left(1 - \frac{s_j^A}{w_j^A}\right).$*

The search power of any subset of the robots is well defined, and it is always positive (since $s_i < w_i$). By summing (2) for $k = 1, \ldots, n$ and using $\sum_{k=1}^{n} c_k = 1$, we obtain the following theorem:

Theorem 2. *The speed S_{opt} of the optimal schedule equals the search power of the collection of robots. In other words, if N denotes the set of all robots, then $S_{opt} = g(N)$.*

3   The Online Search Algorithm

In this section we present an algorithm which produces a searching schedule for a segment of unknown size. Each robot executes the same sequence of movements for each unit interval of the segment. Therefore, the speed of searching each integer segment is the same. We call this the *swarm speed*. Contrary to the offline case, where all robots finish searching at the same time (at different positions), in the online algorithm the robots arrive at each integer point of the segment at the same time. Robots which cannot contribute to increasing the swarm speed are not used in the schedule. Each utilized robot r_i (called a *swarm robot*) searches a sub-segment of the unit segment of size c_i and walks along the remainder. The lengths of the sub-segments c_i are chosen to synchronize the arrival of all robots at every integer point. The sub-segments are pairwise interior disjoint and their union covers the entire unit segment, i.e. $\sum_{i=1}^{k} c_i = 1$.

Theorem 3. *Consider a partition of the unit interval into k consecutive non-overlapping segments $C_1, C_2, \ldots, C_k$, from left to right, of lengths $c_1, c_2, \ldots, c_k$, respectively. Assume that all the robots start (at endpoint 0) and finish (at endpoint 1) simultaneously. Further assume that the i_{th} robot r_i searches the segment C_i with speed s_i and walks the rest of the interval $I \setminus C_i$ with speed w_i such that $w_i > s_i$. Then the swarm speed satisfies $S = \frac{\sum_{i=1}^{k} \frac{1}{\delta_i}}{1 + \sum_{i=1}^{k} \frac{1}{w_i \delta_i}}$, where $\delta_i := \frac{1}{s_i} - \frac{1}{w_i}$, for $i = 1, 2, \ldots, k$.*

Proof. The partition of the interval $[0, 1]$ into segments as prescribed in the statement of the theorem gives rise to the equation

$$c_1 + c_2 + \cdots + c_k = 1. \qquad (3)$$

Let s be the speed of the swarm of n robots. Since all the robots must reach the other endpoint 1 of the interval at the same time, we have the following identities.

$$\frac{c_i}{s_i} + \frac{1 - c_i}{w_i} = \frac{1}{S}, \text{ for } 1 \leq i \leq k, \qquad (4)$$

where $\frac{c_i}{s_i}$ is the time spent searching and $\frac{1-c_i}{w_i}$ the time spent walking by robot r_i. Using the notation $\delta_i := \frac{1}{s_i} - \frac{1}{w_i}$, and substituting into Equation (4), after simplifications we get $c_i = \frac{1}{S\delta_i} - \frac{1}{w_i\delta_i}$, for $1 \leq i \leq k$. Using Equation (3) we see that $1 = \sum_{i=1}^{k} c_i = \sum_{i=1}^{k} \frac{1}{S\delta_i} - \sum_{i=1}^{k} \frac{1}{w_i\delta_i} = \frac{1}{S}\sum_{i=1}^{k} \frac{1}{\delta_i} - \sum_{i=1}^{k} \frac{1}{w_i\delta_i}$, which implies the theorem, as desired. □

We define the procedure SwarmSpeed which finds the speed of a swarm in linear time and algorithm OnlineSearch which defines the swarm. Algorithm OnlineSearch, defines the schedule for a swarm of k robots $r_1, r_2, \ldots, r_k$ out of the original n robots such that $w_1 \geq w_2 \geq \cdots \geq w_k$.

Once the swarm speed has been computed, we can compute the sub-segments' lengths c_i, that we call the *contribution* of robot r_i - the fraction of the unit interval that r_i is allotted to search.

---

real procedure SwarmSpeed();
1.   **var** $S \leftarrow 0$, $S_{num} \leftarrow 0$, $S_{den} \leftarrow 1$, δ : **real**; $i \leftarrow 1$: **integer**;
2.   **while** $i \leq n$ **and** $S < w_i$ **do**
3. $\delta \leftarrow 1/(\frac{1}{s_i} - \frac{1}{w_i})$;
4. $S_{num} \leftarrow S_{num} + \delta$; $S_{den} \leftarrow S_{den} + \delta/w_i$; $S = \frac{S_{num}}{S_{den}}$;
5. $i \leftarrow i + 1$;
6.   **return** S;

Algorithm LeapFrog(robot r_j);
1.   **var** $S \leftarrow$ SwarmSpeed();
2.   **if** $w_j \leq S$ **then**
3.   EXIT; {robot r_j stays motionless}
4.   **else**
5.   **for** $i \leftarrow 1$ **to** $j - 1$ **do**
6.   WALK$((\frac{1}{s} - \frac{1}{w_i})/(\frac{1}{s_i} - \frac{1}{w_i}))$;
7.   **while** not at line end **do**
8.   SEARCH$((\frac{1}{s} - \frac{1}{w_j})/(\frac{1}{s_j} - \frac{1}{w_j}))$;
9.   WALK$(1 - (\frac{1}{s} - \frac{1}{w_j})/(\frac{1}{s_j} - \frac{1}{w_j}))$;

The schedule produced by algorithm LeapFrog is illustrated at Fig. 1(b). We used the same set of robots as in the offline case, i.e. robots having identical walking and searching speeds as those in Fig. 1(a). Observe that robot r_1, while useful in the Comb algorithm, was not used by LeapFrog algorithm in Fig. 1(b), since its walking speed is smaller than the swarm speed. The swarm speed is represented by the slope of the dashed line in Fig. 1(b).

Lemma 6. *Algorithm* OnlineSearch *is correct (i.e. every point of $[0, +\infty)$ is searched).*

Proof. Let $C_j(i)$ denote the sub-segment of $[i, i+1]$ of length c_j which is searched by robot r_j. The lemma follows from the observation that $\bigcup_{j=1}^{k} C_j(i) = [i, i+1]$, for all $i \geq 0$ and $j = 1, \ldots, k$. □

4   Competitiveness of the Online Searching

In this section we discuss the competitiveness of the LeapFrog algorithm. Since competitive ratio is naturally discussed more often for cost optimization (minimization) problems, we assume in this section that we compare the finishing time (rather than speed) of the online versus offline solution. We show first that in the general case the LeapFrog Algorithm is 2-competitive.

Theorem 4. *Consider any set of robots $r_1, r_2, \ldots, r_n$, ordered by non-decreasing walking speed. If the completion time of the optimal schedule produced by Comb equals T_{opt} then the completion time T_{online} of the searching schedule produced by LeapFrog is such that $T_{online} < 2 \cdot T_{opt}$.*

Proof. As LeapFrog outputs schedules of the same speed for all integer-length segments it is sufficient to analyze its competitiveness for a unit segment. Assume, to the contrary, that the time T_{online} of the schedule output by LeapFrog is such that $T_{online} \geq 2 \cdot T_{opt}$. The swarm speed S of LeapFrog is then at most $S \leq 1/(2 \cdot T_{opt})$. Consider $C_1, C_2, \ldots, C_n$ - the sub-segments searched by robots $r_1, r_2, \ldots, r_n$, respectively. Recall that each robot r_i of Comb walks along segments $C_1, C_2, \ldots, C_{i-1}$ and searches C_i arriving at its right endpoint at time T_{opt}. Let i^* be the index such that the midpoint $1/2 \in C_{i^*}$ (or point $1/2$ is a common endpoint of C_{i^*} and C_{i^*+1}). Observe, that in time $2 \cdot T_{opt}$ each robot r_i, such that $i \geq i^*$ could reach the right endpoint of the unit segment, while searching its portion of length $2 \cdot |C_i|$. For each robot r_i, such that $i \geq i^*$, we have $w_i > 1/(2 \cdot T_{online}) \geq S$, each such robot is used by LeapFrog in lines 5-9. However, since $\sum_{i=i^*}^{n} 2 \cdot |C_i| > 1$ all robots r_i, for $i \geq i^*$ search a segment longer than 1, arriving at its right endpoint within time $2 \cdot T_{opt}$, or $T_{online} < 2 \cdot T_{opt}$ for the unit segment. This contradicts the earlier assumption.   □

Observe that, the competitive ratio of 2 may be approached as close as we want.

Proposition 1. *For any sufficiently small $\epsilon > 0$ there is a set of two robots for which the LeapFrog algorithm produces a schedule of completion time T_{online} such that $T_{online} = (2 - \epsilon)T_{opt}$.*

Proof. Let the speeds of the two robots be $s_1 = 1 - \epsilon/2, w_1 = 1, s_2 = 1, w_2 = (2 - \epsilon)/\epsilon$. As the swarm speed S computed in SwarmSpeed procedure equals 1, the line 2 of the LeapFrog algorithm excludes r_1 from the swarm, so the search is performed uniquely by r_1 with $T_{online} = 1$. Using Theorem 2 we get $S_{opt} = \sum_{k=1}^{2} s_k \prod_{j=k+1}^{2} \left(1 - \frac{s_j}{w_j}\right) = \left(1 - \frac{\epsilon}{2}\right)\left(1 - \frac{\epsilon}{2-\epsilon}\right) + 1 = 2 - \epsilon$. Hence $T_{opt} = 1/(2 - \epsilon)$ and $T_{online} = 1 = (2 - \epsilon)T_{opt}$ □

The following theorem concerns the competitiveness of the LeapFrog algorithm in the special case when all robot walking speeds are the same.

Theorem 5. *Let be given the collection of robots* $r_1, r_2, \ldots, r_n$ *with the same walking speed* $w = w_1 = \ldots = w_n$. *The* LeapFrog *algorithm has the competitive ratio* α_n *which is increasing in* n. *In particular,* $\alpha_2 = 1.115$, $\alpha_3 \approx 1.17605$, $\alpha_4 \approx 1.20386$ *and* $\lim_{n \to \infty} \alpha_n \approx 1.29843$.

Our strategy for proving Theorem 5 is to show that the competitive ratio of LeapFrog -among all problem instances when walking speeds are the same - is maximized when all robots' searching speeds are also the same. Because of lack of space, a section related to the proof of Theorem 5 is entirely deferred to the full paper.

5 Conclusion and Open Problems

We proposed and analyzed offline and online algorithms for addressing the beachcombers' problem. The offline algorithm, when the size of the segment to search is known in advance is shown to produce the optimal schedule. An interested reader may observe that, if all walking speeds are different, then the Comb algorithm is the only one achieving the optimal speed. The online searching algorithm is shown to be 2-competitive in general case and 1.29843-competitive when the robots' walking speeds are identical. We conjecture that there is no online algorithm with the competitive ratio of $(2 - \epsilon)$ for any $\epsilon > 0$. Observe that, instead of repeating the same search pattern for each unit-length segment, we could scale down such schedule and repeat it for a segment of arbitrary small value $\varepsilon > 0$. This would permit to keep the swarm speed S within an arbitrarily small range also for an interval of length being any real value (greater than 1) thus obtaining a competitive ratio of $(2 - \epsilon)$ for any $\epsilon > 0$. This conjecture is made only for the class of algorithms where the robots cannot communicate. A model which allows communication could possibly beat this ratio, and would be an interesting problem to study. Other possible open questions concern different domain topologies, robots starting from different initial positions or the case of faulty robots.

References

1. Koopman, B.O.: Search and screening. Operations Evaluation Group, Office of the Chief of Naval Operations, Navy Department (1946)
2. Deng, X., Papadimitriou, C.H.: Exploring an unknown graph. In: Proceedings of the 31st Annual Symposium on Foundations of Computer Science, pp. 355–361. IEEE (1990)
3. Fomin, F.V., Thilikos, D.M.: An annotated bibliography on guaranteed graph searching. Theor. Comput. Sci. 399(3), 236–245 (2008)
4. Albers, S., Henzinger, M.R.: Exploring unknown environments. SIAM J. Comput. 29(4), 1164–1188 (2000)
5. Alpern, S., Gal, S.: The theory of search games and rendezvous, vol. 55. Kluwer Academic Publishers (2002)

6. Baeza-Yates, R.A., Culberson, J.C., Rawlins, G.J.E.: Searching in the plane. Information and Computation 106, 234 (1993)
7. Czyzowicz, J., Ilcinkas, D., Labourel, A., Pelc, A.: Worst-case optimal exploration of terrains with obstacles. Inf. Comput. 225, 16–28 (2013)
8. Deng, X., Kameda, T., Papadimitriou, C.H.: How to learn an unknown environment (extended abstract). In: FOCS, pp. 298–303 (1991)
9. Bellman, R.: An optimal search problem. Bull. Am. Math. Soc., 270 (1963)
10. Beck, A.: On the linear search problem. Israel Journal of Mathematics 2(4), 221–228 (1964)
11. Demaine, E.D., Fekete, S.P., Gal, S.: Online searching with turn cost. Theoretical Computer Science 361(2), 342–355 (2006)
12. Albers, S.: Online algorithms: a survey. Math. Program. 97(1-2), 3–26 (2003)
13. Albers, S., Schmelzer, S.: Online algorithms - what is it worth to know the future? In: Algorithms Unplugged, pp. 361–366 (2011)
14. Berman, P.: On-line searching and navigation. In: Fiat, A., Woeginger, G.J. (eds.) Online Algorithms 1996. LNCS, vol. 1442, pp. 232–241. Springer, Heidelberg (1998)
15. Fleischer, R., Kamphans, T., Klein, R., Langetepe, E., Trippen, G.: Competitive online approximation of the optimal search ratio. SIAM J. Comput. 38(3), 881–898 (2008)
16. Dereniowski, D., Disser, Y., Kosowski, A., Pająk, D., Uznański, P.: Fast collaborative graph exploration. In: Fomin, F.V., Freivalds, R., Kwiatkowska, M., Peleg, D. (eds.) ICALP 2013, Part II. LNCS, vol. 7966, pp. 520–532. Springer, Heidelberg (2013)
17. Chalopin, J., Flocchini, P., Mans, B., Santoro, N.: Network exploration by silent and oblivious robots. In: Thilikos, D.M. (ed.) WG 2010. LNCS, vol. 6410, pp. 208–219. Springer, Heidelberg (2010)
18. Das, S., Flocchini, P., Kutten, S., Nayak, A., Santoro, N.: Map construction of unknown graphs by multiple agents. Theor. Comput. Sci. 385(1-3), 34–48 (2007)
19. Fraigniaud, P., Gasieniec, L., Kowalski, D.R., Pelc, A.: Collective tree exploration. Networks 48(3), 166–177 (2006)
20. Higashikawa, Y., Katoh, N., Langerman, S., Tanigawa, S.: Online graph exploration algorithms for cycles and trees by multiple searchers. J. Comb. Optim. (2012)
21. Wang, G., Irwin, M.J., Fu, H., Berman, P., Zhang, W., Porta, T.L.: Optimizing sensor movement planning for energy efficiency. ACM Transactions on Sensor Networks 7(4), 33 (2011)
22. Beauquier, J., Burman, J., Clement, J., Kutten, S.: On utilizing speed in networks of mobile agents. In: Proceeding of the 29th ACM SIGACT-SIGOPS Symposium on Principles of Distributed Computing, pp. 305–314. ACM (2010)
23. Czyzowicz, J., Gąsieniec, L., Kosowski, A., Kranakis, E.: Boundary patrolling by mobile agents with distinct maximal speeds. In: Demetrescu, C., Halldórsson, M.M. (eds.) ESA 2011. LNCS, vol. 6942, pp. 701–712. Springer, Heidelberg (2011)
24. Kawamura, A., Kobayashi, Y.: Fence patrolling by mobile agents with distinct speeds. In: Chao, K.-M., Hsu, T.-S., Lee, D.-T. (eds.) ISAAC 2012. LNCS, vol. 7676, pp. 598–608. Springer, Heidelberg (2012)

Reliable Shared Memory Abstraction on Top of Asynchronous Byzantine Message-Passing Systems

Damien Imbs[1], Sergio Rajsbaum[1], Michel Raynal[2,3], and Julien Stainer[3]

[1] Instituto de Mathematicas, UNAM, D.F. 04510, Mexico
[2] Institut Universitaire de France
[3] IRISA, Campus de Beaulieu, 35042 Rennes Cedex, France

Abstract. This paper is on the construction and the use of a shared memory abstraction on top of an asynchronous message-passing system in which up to t processes may commit Byzantine failures. This abstraction consists of arrays of n single-writer/multi-reader atomic registers, where n is the number of processes. Differently from usual atomic registers which record a single value, each of these atomic registers records the whole history of values written to it. A distributed algorithm building such a shared memory abstraction it first presented. This algorithm assumes $t < n/3$, which is shown to be a necessary and sufficient condition for such a construction. Hence, the algorithm is resilient-optimal. Then the paper presents distributed algorithms built on top of this shared memory abstraction, which cope with up to t Byzantine processes. The simplicity of these algorithms constitutes a strong motivation for such a shared memory abstraction in the presence of Byzantine processes.

For a lot of problems, algorithms are more difficult to design and prove correct in a message-passing system than in a shared memory system. Using a protocol stacking methodology, the aim of the proposed abstraction is to allow an easier design (and proof) of distributed algorithms, when the underlying system is an asynchronous message-passing system prone to Byzantine failures.

Keywords: APppproximate agreement, Asynchronous message-passing system, Atomic read/write register, Broadcast abstraction, Byzantine process, Distributed computing, Message-passing system, Quorum, Reliable broadcast, Reliable shared memory, Single-writer/multi-reader register, t-Resilience.

1 Introduction

Distributed computing. Distributed computing occurs when one has to solve a problem in terms of physically distinct entities (usually called nodes, processors, processes, agents, sensors, etc.) such that each entity has only a partial knowledge of the many parameters involved in the problem. In the following, we use the term *process* to denote a computing entity. From an operational point of view this means that the processes of a distributed system need to exchange information, and agree in some way or another, in order to cooperate to a common goal. If processes do not cooperate, the system is no longer a distributed system. Hence, a distributed system has to provide the processes with communication and agreement abstractions.

Designing distributed applications is not an easy task (e.g., see [2,13,20,21,22]). This is due to the fact that, due its very nature, no process can capture instantaneously the

M. Halldórsson (Ed.): SIROCCO 2014, LNCS 8576, pp. 37–53, 2014.

global state of the application it is part of. More precisely, as processes are geograph-
ically localized at distinct places, distributed applications have to cope with the uncertainty created by asynchrony and failures. As a simple example, it is impossible to
distinguish a crashed process from a very slow process in an asynchronous system prone
to process crashes.

As in sequential computing, a simple approach to facilitate the design of distributed
applications consists in designing appropriate abstractions. With such abstractions, the
application designer can think about solutions to solve problems at a higher conceptual
level than the one offered by the basic send/receive communication layer.

Byzantine behavior. This failure type has first been introduced in the context of synchronous distributed systems (e.g., [11,19], see also the monography [21]), and then
investigated in the context of asynchronous ones (e.g., see the textbooks [2,13,20]). A
process has a *Byzantine* behavior when it arbitrarily deviates from its intended behavior;
it then commits a Byzantine failure. Otherwise it is non-faulty (or non-Byzantine). This
bad behavior can be intentional (malicious) or simply the result of a transient fault that
altered the local state of a process, thereby modifying its behavior in an unpredictable
way. Let us notice that process crashes (unexpected halting) and communication omis-
sions, define a strict subset of Byzantine failures.

The major part of the papers on Byzantine failures considers (synchronous or asynchronous) message-passing systems, and mainly addresses agreement problems, such
as consensus and total order broadcast, or the construction of a Byzantine-tolerant disk
storage (e.g., [5,10,14,15]). Many of these papers consider registers built on top of
duplicated disks (servers), which are accessed by clients, and where disks and clients
may exhibit different type of failures. Moreover, in these client/server models, clients
communicate only with servers and vice versa (the communication graph is bipartite).

Content of the paper: Construction of an atomic read/write memory. Differently, this
paper is on the construction of a shared memory (atomic registers) on top of an asynchronous message-passing system where processes may exhibit a Byzantine behavior.
Its first contribution is the definition of a shared memory (atomic registers) in the context of Byzantine processes, and the design of an algorithm that builds such a shared
memory on top of an asynchronous message-passing systems where up to t processes
may be Byzantine. These registers differ from classical registers in that each register
contains the whole history of the values written to it. This allows preventing a malicious process from overwriting a previously written value letting correct processes
believe it wrote it only once. Hence, such a register is called a *h-register* (h standing for
"history"), and a read of it returns the complete history of writes to the registers. This
t-resilient shared memory is made up of n single-writer/multi-reader (SWMR) atomic
h-registers (one per process). The paper also shows that $t < n/3$ is a necessary and
sufficient requirement for such a construction.

This construction and the associated upper bound $t < n/3$ complement the previous
result known on the construction of an atomic shared memory in asynchronous crash-
prone message-passing systems, where it has been shown that $t < n/2$ is an upper
bound on the number of faulty processes [1]. Interestingly, the upper bound $t < n/3$ is
the same as the one for solving consensus in both Byzantine synchronous systems [11]

and Byzantine asynchronous systems (enriched with an appropriate oracle such as a common coin, e.g., [4,17,18]).

Content of the paper: From read/write h-registers to higher level abstractions. The second contribution of the paper consists of algorithms that solve distributed computing problems on top of the previous t-resilient shared memory abstraction. These algorithms illustrate the versatility of Byzantine-tolerant atomic h-registers. The first algorithm, which is pretty simple, solves the "correct-only" agreement problem (a weakened version of the consensus problem). Then the paper describes an algorithm that solves the multidimensional approximate agreement on top of the t-resilient shared memory abstraction. This algorithm can be seen as an adaptation to a Byzantine read/write shared memory system of Mendes-Herlihy's algorithm [16], which solves the same problem "directly" on top of an asynchronous Byzantine message-passing system.

As shown by these examples, the important feature of the proposed shared memory abstraction lies in the fact that it prevents Byzantine processes from corrupting synchronization among the correct processes. A Byzantine process can create inconsistency only on the values it writes, but any two correct processes see the same sequence of written values.

Roadmap. The paper is composed of 6 sections. Section 2 introduces the underlying Byzantine asynchronous message-passing model. Section 3 defines the notion of Byzantine-tolerant atomic read/write h-registers, and presents an algorithm that builds such h-registers on top of the basic Byzantine asynchronous message-passing model. This section shows also that $t < n/3$ is a necessary and sufficient requirement for such a construction. Then, Section 4 and Section 5 present algorithms that solve distributed computing problems on top of Byzantine-tolerant atomic h-registers. Finally, Section 6 concludes the paper. Due to page limitations, missing proofs and additional developments will be found in [9].

2   Computation Model, Reliable Broadcast and Two Properties

2.1   Computation Model

Computing entities. The system is made up of a set Π of n sequential processes, denoted p_1, p_2, ..., p_n. These processes are asynchronous in the sense that each process progresses at its own speed, which can be arbitrary and remains always unknown to the other processes.

Communication model. The processes cooperate by sending and receiving messages through bi-directional channels. The communication network is a complete network, which means that each process p_i can directly send a message to any process p_j (including itself). It is assumed that, when a process receives a message, it can unambiguously identify its sender. Each channel is reliable (no loss, corruption, or creation of messages), not necessarily first-in/first-out, and asynchronous (while the transit time of each message is finite, there is no upper bound on message transit times). Moreover, Byzantine processes are not prevented from reading messages and reordering them.

Byzantine failures. The model parameter t is an upper bound on the number of processes that can exhibit a Byzantine behavior [11,19]. A Byzantine process is a process that behaves arbitrarily: it may crash, fail to send or receive messages, send arbitrary messages, start in an arbitrary state, perform arbitrary state transitions, etc. Hence, a Byzantine process, which is assumed to send the same message m to all the processes, can send a message m_1 to some processes, a different message m_2 to another subset of processes, and no message at all to the other processes. Moreover, Byzantine processes can collude to "pollute" the computation.

Terminology and notation. A Byzantine process is also called a *faulty* process. A process that never commits a failure is called a *correct* (or *non-faulty*) process. Given an execution, let $\mathcal{C}$ and $\mathcal{F}$ denote the sets of correct and faulty processes, respectively.

This process model is denoted $\mathcal{BAMP}_{n,t}[c(n,t)]$, where $c(n,t)$ is a constraint imposed on the model parameter t, e.g., $(t < n/3)$.

2.2   Reliable Broadcast Abstraction

This section presents a reliable broadcast abstraction (denoted r-broadcast), that will be used to build atomic read/write h-registers. (Section 3). This abstraction is a simple generalization of a reliable broadcast due to Bracha [3]. While Bracha's abstraction is for a single broadcast, the proposed abstraction considers that each process can issue a sequence of broadcasts. It is shown in [3] that $t < n/3$ is a necessary requirement when one has to build such an abstraction in the presence of asynchrony and Byzantine failures.

Specification. The reliable broadcast abstraction provides each process with the operations R_broadcast() and R_deliver(). When a process p_i invokes R_broadcast() we say that "p_i r-broadcasts a value". Similarly, when p_i returns from an invocation of R_deliver() and obtains a value, we say "p_i r-delivers a value".

The operation R_broadcast() has two input parameters: a broadcast value v, and an integer sn, which is a local sequence number used to identify the successive r-broadcasts issued by the invoking process p_i. The sequence of numbers used by each (correct) process is the increasing sequence of consecutive integers.

- RB-Validity. If a correct process r-delivers a pair (v, sn) from a correct process p_x, then p_x invoked the operation R_broadcast(v, sn).
- RB-Integrity. Given any process p_i and any sequence number sn, a correct process r-delivers at most once a pair $(-, sn)$ from p_i.
- RB-Uniformity. If a correct process r-delivers a pair (v, sn) from p_i (possibly faulty), then all correct processes eventually r-deliver the same pair (v, sn) from p_i.
- RB-Termination. If the process that invokes R_broadcast(v, sn) is correct, all the correct processes eventually r-deliver the pair (v, sn).

RB-Validity relates the outputs to the inputs, namely what is r-delivered was r-broadcast. RB-Integrity states that there is no r-broadcast duplication. RB-Uniformity is an "all or none" property (it is not possible for a pair to be delivered by a correct process and to be never delivered by the other correct processes). RB-Termination is a liveness property: at least all the pairs r-broadcast by correct processes are r-delivered by them.

2.3 Two Preliminary Quorum-Related Properties

The following properties are central in the understanding and the proof of the construction of atomic SWMR h-registers described in the next section (Proofs in [9]).

Property 1. Let m, n, and t be positive integers. $(m > \frac{n+t}{2}) \Leftrightarrow (m \geq \lfloor \frac{n+t}{2} \rfloor + 1)$.

Property 2. Any two sets of processes Q_1 and Q_2 of size at least $\lfloor \frac{n+t}{2} \rfloor + 1$ have at least one correct process in their intersection.

3 Construction of Single-Writer/Multi-Reader Atomic h-Registers

3.1 Atomic Read/Write h-Registers in the Presence of Byzantine Processes

Single-writer/multi-reader (SWMR) h-registers. The fault-tolerant shared memory supplied to the upper abstraction layer is an array denoted $REG[1..n]$. For each i, $REG[i]$ is a single-writer/multi-reader (SWMR) h-register, i.e., $REG[i]$ can be written only by p_i. To that end, p_i invokes the operation $REG[i].\text{write}(v)$ where v is the value it wants to write. Any process p_j can read $REG[i]$. It invokes then the operation $REG[i].\text{read}()$.

Let us notice that the "single-writer" requirement is natural in the presence of Byzantine processes. If h-registers could be written by any process, it would be possible for the Byzantine processes to pollute the whole memory, and no non-trivial computation could be possible.

The value returned by a read operation. A h-register $REG[i]$ contains the sequence of values (also called a *history*) that have been written into it, and such a sequence is returned by the invocations of the operation $REG[i].\text{read}()$. Each h-register $REG[i]$ is initialized to the empty sequence (denoted ϵ), which corresponds to the default value $\perp$. It is assumed that no process can write $\perp$ into its h-register. Let us remark that returning a sequence of values is not a restriction, as, when a process obtains such a sequence h, it can always consider only its last value (or $\perp$ if $h = \epsilon$). $|h|$ denotes the length of h.

Notations. Let p_i and p_j be correct processes. We use the following notations.

- op_read$[j, i, h]$: execution by a correct process p_j of $REG[i].\text{read}()$, returning the history h.
- op_write$[i, wsn]$: wsn^{th} execution by a correct process p_i of $REG[i].\text{write}()$.

Specification. The correct behavior of an SWMR h-register is defined as follows.

- R-Termination (liveness). Let p_i be a correct process.
  - Each invocation of $REG[i].\text{write}()$ issued by p_i terminates.
  - $\forall j$, each invocation of $REG[j].\text{read}()$ issued by p_i terminates.
- R-Consistency (safety).
  - Single history. Let p_i be a (correct or faulty) process. There exists exactly one history H_i such that, for any correct process p_j, any op_read$[j, i, h]$ is such that h is a prefix of H_i.
  - Read followed by write. Let p_i and p_j be two correct processes. If op_read $[j, i, h]$ terminates before op_write$[i, wsn]$ starts, then $|h| < wsn$.

- Write followed by read. Let p_i and p_j be two correct processes. If op_write $[i, wsn]$ terminates before op_read$[j, i, h]$ starts, then $wsn \leq |h|$.
- No read inversion. Let p_i and p_j be two correct processes. If op_read$[j, i, h]$ terminates before op_read$[k, i, h']$ starts, then $|h'| \geq |h|$.

As, whatever i, the invocations of $REG[i]$.read() by a faulty process p_j can return any value, the previous specification do not need to take them into account. Moreover, it is possible that, while executing a code different from the code of the write operation, a faulty process modifies the content of its h-register $REG[j]$ (at the operational level, this happens when the messages it generates could have been sent by a correct implementation of the operation write()). The specification of the consistency of such a h-register $REG[j]$ takes this into account in the "no read inversion" property.

The previous properties state that each h-register is linearizable [8]. This means that it is possible to totally order the executions of the operations in such a way that (a) the execution of each operation appears as if it has been executed at a single point of the time line between its start event and its end event, (b) no two operations have been executed at the same time, and (c) each read by a correct process returns the sequence of values written before it in the sequence (when considering read/write h-registers, linearizability is atomicity [12,22]).

An important theorem associated with linearizability is the following [8]: If each object (h-register) is linearizable, then the set of all the objects, considered as a single object, is linearizable. This means that linearizable objects compose for free.

3.2   The Construction

An algorithm constructing an SWMR atomic (linearizable) h-register in the presence of up to t Byzantine processes, is described in Figure 1. This algorithm requires $t < n/3$, hence it is suited to the computing model $\mathcal{BAMP}_{n,t}[t < n/3]$. This algorithm uses a **wait**(*condition*) statement. The corresponding process is blocked until *condition* becomes satisfied. While a process is blocked, it can process the messages it receives.

Local variables. Each process p_i manages four local variables whose scope is the full computation (local variables are denoted with lower case letters, and subscripted by the process index i).

- $reg_i[1..n]$ is the local representation of the array $REG[1..n]$ of atomic SWMR h-registers. Each local register $reg_i[j]$ is initialized to the empty sequence ϵ whose size is 0 (the corresponding value being the default value $\perp$). The content of $reg_i[j]$ is called the local *history* of $REG[j]$, as known by p_i.
- wsn_i is a sequence number generator (initialized to 0) for the writes of $REG[i]$ (issued by p_i).
- $rsn_i[1..n]$ is a local array such that $rsn_i[j]$ is used to associate sequence numbers to the invocations of $REG[j]$.read() issued by p_i. Each $rsn_i[j]$ is initialized to 0.
- $approx\_rsn_i[1..n, 1..n]$ is a matrix of sequence numbers, each initialized to 0; $approx\_rsn_i[k, j]$ (initialized to 0) is the last read sequence number ($rsn_k[j]$) used by p_k to to read $REG[j]$, as known by p_i.

The operation $REG[i]$.write(v). This operation is implemented by the client lines 01-04 and the server lines 09-12. Process p_i first increases wsn_i and r-broadcasts the message WRITE(v, wsn_i). Let us remark that this is the only place where the algorithm uses the underlying reliable broadcast abstraction. The process p_i then waits for acknowledgments (message WRITE_DONE(wsn_i)) from a quorum including strictly more than $\frac{n+t}{2}$ processes, and finally terminates the write operation. As we have seen (Lemmas 1 and 2), the intersection of any two quorums of such a size contains at least one correct process. This property will be used to prove the consistency of the h-register $REG[i]$.

When p_i is r-delivered a message WRITE(v, wsn) from a process p_j, it first waits until the previous write of $REG[j]$ by p_j has locally terminated (line 09). Hence, all the writes of $REG[j]$ are locally processed by p_i in the order they have been issued by p_j. When $|reg_i[j]| + 1 = wsn$, p_i adds v at the tail of $reg_i[j]$ (line 10), and sends back an acknowledgment to p_j (line 11).

Finally, p_i sends to each p_k the message READ_VALUE(j, $approx\_rsn_i[k,j]$, $reg_i[j]$) to inform p_k that this write from p_j has locally been taken into account at p_i (line 12). As we will see, this is to help terminate the invocations of $REG[j]$.read() issued by correct processes.

---

operation $REG[i]$.write(v) **is**
(01) $wsn_i \leftarrow wsn_i + 1$;
(02) R_broadcast WRITE(v, wsn_i);
(03) **wait** $\big($WRITE_DONE(wsn_i) received from $> \frac{n+t}{2}$ different processes$\big)$;
(04) **return** ().

operation $REG[j]$.read() **is**
(05) $rsn_i[j] \leftarrow rsn_i[j] + 1$;
(06) broadcast READ(j, $rsn_i[j]$);
(07) **wait** $\big(\exists h : $READ_VALUE($j$, $rsn_i[j]$, h) received from $> \frac{n+t}{2}$ different processes$\big)$;
(08) **return** (h). % last value in h can be returned if not interested in history of $REG[j]$ %
%——

when a message WRITE(v, wsn) **from** p_j **is** R_delivered:
(09) **wait** ($|reg_i[j]| + 1 = wsn$);
(10) $reg_i[j] \leftarrow reg_i[j] \cdot v$;
(11) send WRITE_DONE(wsn) to p_j;
(12) **for** $k \in [1..n]$ **do** send READ_VALUE(j, $approx\_rsn_i[k,j]$, $reg_i[j]$) to p_k **end for**.

when a message READ(j, rsn) **from** p_k **is** received:
(13) **if** ($approx\_rsn_i[k,j] < rsn$) **then**
(14) $approx\_rsn_i[k,j] \leftarrow rsn$;
(15)   send READ_VALUE(j, rsn, $reg_i[j]$) to p_k
(16) **end if.**

Fig. 1. Array of SWMR atomic h-registers in $\mathcal{BAMP}_{n,t}[t < n/3]$ (code for p_i)

The operation $REG[j]$.read(). This operation is implemented by the client lines 05-08 and the server lines 12-16. When p_i wants to read $REG[j]$, it first broadcasts a read request appropriately identified (message READ(j, $rsn_i[j]$), lines 05-06), and waits for acknowledgment messages carrying the same history h (in READ_VALUE(j, $rsn_i[j]$, h)), from a quorum of strictly more than $\frac{n+t}{2}$ distinct processes (lines 07). When p_i stops waiting, it knows that, when they sent their acknowledgments, the history of $REG[j]$ was equal to h for strictly more than $\frac{n+t}{2}$ processes. When this occurs, p_i returns this history h as the result of the read operation (line 08).

On its server side, when a process p_i receives a read request message READ(j, rsn) from a process p_k, it first checks if its view of the read of $REG[j]$ by p_k is "late", i.e., if $approx\_rsn_i[k, j] < rsn$ (line 13). If it is the case, p_i updates $approx\_rsn_i[k, j]$ (line 14), and sends by return to p_k the message READ_VALUE(j, rsn, $reg_i[j]$), to inform it of its current value of $REG[j]$ (line 15). If $approx\_rsn_i[k, j] \geq rsn$, the read request is an old message and p_i ignores it.

Comparing with the crash failure model. It is known that the algorithms implementing an atomic register on top of an asynchronous message-passing system prone to process crashes, require that "read have to write" [1]. More precisely, before returning a value, a process must write this value to ensure atomicity. Doing so, it is not possible that two sequential read invocations, both concurrent with a write invocation, are such that the first read obtains the new value while the second read obtains the old value (demanding the reader to write the value it is about to return guarantees that there is no new/old inversion [20]).

As Byzantine failures are more severe than crash failures, the algorithm of Figure 1 needs to use a mechanism analogous to the "read have to write" to prevent new/old inversion from occurring. This is done by sending to each process p_k the customized message READ_VALUE(j, $approx\_rsn_i[j, k]$, $reg_i[j]$) issued at line 12. These messages play the role of writes, that allow the wait statement of line 07 to always terminate with a correct history value for $REG[j]$.

Message complexity. It follows from the previous discussion that, in addition to a reliable broadcast, each write generates $O(n)$ messages WRITE_DONE() and $O(n^2)$ messages READ_VALUE()), and each read generates at most $2n$ messages. Hence, while the algorithm presented in [1] requires the assumption $t < n/2$ (which is a necessary and sufficient requirement on the model parameter t) and $O(n)$ messages to implement an SWMR atomic register in the crash failure model, the proposed algorithm requires the assumption $t < n/3$ (which is shown to be necessary and sufficient, see below), and $O(n^2)$ messages plus a reliable broadcast, to implement SWMR atomic h-registers in the Byzantine asynchronous message-passing model.

3.3   Proof of the Construction and Upper Bound

Lemma 1. *Let $n \geq 3t + 1$. If p_i is correct and invokes $REG[i]$.write(), its invocation terminates* (Proof in [9]).

Lemma 2. (Single history). *Let p_i be a (correct or faulty) process. It exists a history H_i such that any invocation of the operation $REG[i]$.read() by a correct process returns a prefix of H_i* (Proof in [9]).

Lemma 3. *Let $n \geq 3t + 1$. If p_j is correct and invokes $REG[i]$.read(), its invocation terminates.*

Proof. The proof is by contradiction. let us assume that a correct process p_j invokes $REG[i]$.read() and this invocation never terminates. This means that the predicate associated with the wait statement of line 07 remains false forever, namely, $\nexists h$ such that the message READ_VALUE($i, rsn_j[i], h$) is received from strictly more than $\frac{n+t}{2}$ different processes.

As p_j is correct, it broadcasts the request message READ(i, sn) where $sn = rsn_j[i]$ (line 06), and this message is received by all correct processes. Moreover, sn is the greatest sequence number ever used by p_j to read $REG[i]$, and, due to the contradiction assumption, $rsn_j[i]$ keeps forever the value sn.

Let p_k be any correct process. When p_k receives READ(i, sn) from p_j, the predicate $approx\_rsn_k[j, i] < sn$ is satisfied (line 13). This is because sn is greater than all previous sequence numbers used by p_j to read $REG[i]$. It follows that p_k updates $approx\_rsn_k[j, i]$ to $sn = rsn_j[i]$, and sends READ_VALUE($i, sn, reg_k[i]$) to p_j (lines 14-15). Moreover, as the read of $REG[i]$ by p_j never terminates, $approx\_rsn_k[j, i]$ remains forever equal to $sn = rsn_j[i]$.

As the predicate of line 07 remains forever false at p_j, and p_j receives at least $(n - t)$ messages READ_VALUE($i, sn, -$) (one from each correct process), it follows that p_j receives at least two messages READ_VALUE(i, sn, h) and READ_VALUE(i, sn, h') such that h is a strict prefix of h or h' is a strict prefix of h (this is because, due to Lemma 2, both h and h' are prefixes of H_i). Without loss of generality, let h' be the shortest history received, and h be the longest.

Due to the RB-uniformity property of the underlying broadcast abstraction, it follows that all the correct processes r-delivers the same messages WRITE() from p_i, and process them in the same order (line 09). Let p_k be a correct process. It follows directly from the code of the algorithm that, each time p_k adds a value to $reg_k[i]$ (line 10), it sends a message READ_VALUE($i, sn, reg_k[i]$) to p_j (line 12). It follows that there is a finite time after which p_j has received the very same message READ_VALUE(j, sn, h) from strictly more than $\frac{n+t}{2}$ different processes. The predicate of line 07 becomes then satisfied. This contradicts the initial assumption, and the lemma follows. $\square_{Lemma\ 3}$

Lemma 4. (Read followed by write). *Let p_i and p_j be two correct processes. If the execution of op_read$[j, i, h]$ terminates before op_write$[i, wsn]$ starts, then the returned history h is such that $|h| < wsn$.*

Proof. As p_i is correct and has not yet invoked op_write$[i, wsn]$ when p_j terminates op_read$[j, i, h]$, it follows that no correct process r-delivers a message WRITE(v, wsn) from p_i before op_read$[j, i, h]$ terminates. Hence, when they received from p_j the message READ(i, sn) generated by op_read$[j, i, -]$, strictly more than $\frac{n+t}{2}$ different processes p_x (i.e., strictly more than $\frac{n-t}{2}$ correct processes) returned the same message READ_VALUE(i, sn, h) where $h = reg_x[i]$ and $|h| < wsn$. Consequently, the history h returned by p_j is smaller than wsn. $\square_{Lemma\ 4}$

Lemma 5. (Write followed by read). *Let $n > 3t$. Let p_i and p_j be two correct processes. If op_write$[i, wsn]$ terminates before op_read$[j, i, h]$ starts, then the returned history h is such that $|h| \geq wsn$.*

Proof. It follows from line 03 and lines 10-11 that, when op_write$[i, wsn]$ terminates (which implies before op_read$[j, i, h]$ starts), there is a quorum Q_W of strictly more than $\frac{n+t}{2}$ processes p_x such that $|reg_x[i]| \geq wsn$. Moreover, the invocation op_read$[j, i, h]$ obtains the same message READ_VALUES(i, sn, h) from a quorum Q_R including strictly more than $\frac{n+t}{2}$ correct processes. According to Lemmas 1 and 2, it follows that $Q_W \cap Q_R$ contains at least one correct process p_y. As this process is such that $|reg_y[i]| \geq wsn$, it follows that $|h| \geq wsn$. $\square_{Lemma\ 5}$

Lemma 6. (No read inversion). *Let $n > 3t$. Let p_j and p_k be two correct processes. If op_read$[j, i, h]$ terminates before op_read$[k, i, h']$ starts, we have then $|h| \leq |h'|$.*

Proof. To terminate, op_read$[j, i, h]$ received the same message READ_VALUE(j, rsn, h) from a quorum Q_{R1} of strictly more than $\frac{n+t}{2}$ different processes. Similarly, let Q_{R2} be the quorum of strictly more than $\frac{n+t}{2}$ processes that allowed op_read$[k, i, h']$ to terminate.

According to Lemmas 1 and 2, there is a correct process p_x in $Q_{R1} \cap Q_{R2}$. As, (a) p_x is correct and sent the message READ_VALUE(j, rsn, h) to p_j, and later sent the message READ_VALUE(j, rsn, h') to p_k, and (b) $reg_x[i]$ can only increase, we necessarily have $|h| \leq |h'|$, which concludes the proof of the lemma. $\square_{Lemma\ 6}$

The following theorem follows from the previous lemmas 1-6.

Theorem 1. *The algorithm described in Figure 1 implements an array of n SWMR atomic h-registers (one per process) in the system model $\mathcal{BAMP}_{n,t}[t < n/3]$.*

Theorem 2. *The condition $t < n/3$ is necessary to built an SWMR atomic h-register in $\mathcal{BAMP}_{n,t}[\emptyset]$.*

Proof. The algorithm presented in the previous section has shown that the condition $t < n/3$ is sufficient to built an SWMR atomic register. So, the rest of the proof addresses the necessity part of the condition.

The proof is by contradiction. Let us assume that there is an algorithm A that builds an atomic register in $\mathcal{BAMP}_{n,t}[n \leq 3t]$, which means that it satisfies the R-consistency and R-termination properties stated in Section 3.1. For simplicity, and without loss of generality, we consider the largest possible value of t, i.e., $n = 3t$. Let us first observe that to guarantee R-termination, a process cannot wait for messages from more than $n - t = 2t$ processes.

Let us partition the set of processes into three sets Q_1, Q_2 and Q_3, each of size (at most) t. Moreover, let us consider two processes $p_1 \in Q_1$ and $p_2 \in Q_2$.

Let us consider a first execution E_1 defined as follows.

- The set of Byzantine processes is Q_1; these processes do not send messages and appear as crashed.
- The process $p_2 \in Q_2$ writes a value v in $REG[2]$. Due to the R-termination property of the algorithm A, the invocation of $REG[2]$.write(v) by p_2 terminates. Let τ_w be the time instant at which this write terminates.

Let E_2 be a second execution defined as follows.

- All processes are correct, but the processes of Q_2 execute no step before τ_r (defined below).
- After τ_w, the process $p_1 \in Q_1$ reads the register $REG[2]$. Due to the R-termination property of the algorithm A, and because the processes of Q_2 could be Byzantine, the invocation of $REG[2]$.read() by p_1 terminates. Let τ_r be the time instant at which this read terminates. As, no process of Q_2 executes a step before the read terminates, p_2 does not write $REG[2]$ before τ_r. Consequently, according to the R-consistency property *read followed by write*, $REG[2]$ has still its initial value ϵ. It follows that the read operation by p_1 returns this initial value.

Let us finally consider E_3, a third execution defined as follows.

- The set of Byzantine processes is Q_3, and these processes behave exactly as in E_1 with respect to the processes of Q_2, and exactly as in E_2 with those of Q_1.
- The messages sent by the processes of Q_1 to the processes of Q_2 and by the processes of Q_2 to the processes of Q_1 are delayed until after τ_r.
- The messages exchanged between themselves by the processes of $Q_2 \cup Q_3$ are received at exactly the same time instants as in E_1. Similarly, the messages exchanged between themselves by the processes of $Q_1 \cup Q_3$ are received at exactly the same time instants as in E_2.
- As the very same time instants as in E_1, process $p_2 \in Q_2$ writes a value v in $REG[2]$. Since, from the point of view of the processes of Q_2, the executions E_1 and E_3 are indistinguishable, the invocation of $REG[2]$.write(v) by p_2 terminates at τ_w too.
- As in execution E_2, after τ_w the process $p_1 \in Q_1$ reads the register $REG[2]$. Since, from the point of view of the processes of Q_1, the executions E_2 and E_3 are indistinguishable, the invocation of $REG[2]$.read() by p_1 terminates at τ_r and returns ϵ. But this violates the R-Consistency property *write followed by read*, which contradicts the existence of Algorithm A.

$$\square_{Theorem\ 2}$$

4   A Simple Abstraction on Top of SWMR Atomic h-Registers

This section presents a simple use of the previous construction of an array of n SWMR atomic h-registers. Other examples are given in [9]. The simplicity of this algorithm comes from the high abstraction level provided by SWMR atomic h-registers built on top of a message-passing system.

The classical notations for atomic registers are used in the following, namely, given an atomic h-register XX, $XX \leftarrow v$ stands for XX.write(v), and $x \leftarrow XX$ stands for $x \leftarrow XX$.read(v). Moreover, only the last value of the sequence returned by XX.read(v) is considered.

Correct-Only Agreement. A correct-only agreement object is a one-shot object that provides the processes with a single operation denoted correct_only_agreement(). This operation is used by each process to propose a value and decide a set of values. A decided set contains only values proposed by correct processes and the decided sets satisfy the containment property. According to the topological bounds stated in [7], the problem captured by the correct-only agreement object can be solved if and only if $n > (\dim(I) + 2)t$, where $\dim(I)$ is the dimension of the colorless input complex I with which the problem is instantiated. In our context, this necessary and sufficient condition boils down to $n > (w + 1)t$, where $w > 1$ is the maximal number of distinct values that can be proposed by the correct processes in any execution (in the topology parlance, $w - 1$ is the greatest dimension of a simplex of the colorless input complex I).

A correct-only agreement object is defined by the following properties, where $output_i$ denotes the set of values output by a correct process p_i.

- Termination. The invocation of correct_only_agreement() by a correct process p_i terminates.
- Containment. If both p_i and p_j are correct and invoke correct_only_agreement(), then $output_i \subseteq output_j$ or $output_j \subseteq output_i$.
- Validity. The set $output_i$ of a correct process p_i is not empty and contains only values proposed by at least one correct process.

---

operation correct_only_agreement(v_i) **is**
(01) $IN[i] \leftarrow v_i$;
(02) **for** $x \in \{1, ..., n\}$ **do** $aux1[x] \leftarrow IN[x]$ **end for**;
(03) **for** $x \in \{1, ..., n\}$ **do** $aux2[x] \leftarrow IN[x]$ **end for**;
(04) **while** $[(aux1 \neq aux2) \vee (\nexists v : |\{j : aux1[j] = v\}| > t)]$ **do**
(05) $aux1 \leftarrow aux2$;
(06)   **for** $x \in \{1, ..., n\}$ **do** $aux2[x] \leftarrow IN[x]$ **end for**
(07) **end while**;
(08) $output_i \leftarrow \{v : |\{j : aux1[j] = v\}| > t\}$;
(09) return($output_i$).

Fig. 2. Correct-only agreement on top of $\mathcal{BAMP}_{n,t}[t < n/(w+1)]$ (code for p_i)

Algorithm. The algorithm implementing the operation correct_only_agreement(), is described in Figure 2. It uses an array $IN[1..n]$ of SWMR h-registers. A successful double scan is necessary but not sufficient to exit the while loop. Namely, a process p_i must additionally observe that at least one value has been proposed by $(t+1)$ processes (i.e., by at least one correct process). Finally, the output $output_i$ contains all the values that, from p_i's point of view, have been proposed by at least $(t + 1)$ processes.

The containment property is a consequence of the fact that the writes in the array $IN[1..n]$ are atomic, and the number of non-$\perp$ entries can only increase. The termination property is a consequence of the following observations: (a) there is a bounded number of processes, (b) the atomic h-registers are one-write registers, and (c) the condition $n > (w+1)t$. The validity follows from the condition $n > (w+1)t$ (hence there is at least one value that appears $(t+1)$ times), and the predicate of line 04.

5 Solving Multidimensional Approximate Agreement

This section shows how an algorithm designed for the Byzantine asynchronous message-passing system model can be easily adapted to the Byzantine asynchronous shared memory model introduced in Section 3. The shared memory abstraction being at a higher abstraction level than message-passing, the shared memory version of the algorithm seems much easier to understand.

5.1 The Multidimensional Approximate Agreement Problem

Approximate agreement. The ϵ-approximate agreement problem has been introduced in the context of synchronous and asynchronous message-passing systems where processes can commit Byzantine failures [6]. Each process proposes a value in $\mathbb{R}$, and each correct process has to decide a value such that: (a) any decided value is in the range of the values proposed by the correct processes (validity), and (b) the distance between any two values decided by correct processes is at most ϵ (agreement).

The condition $t < n/3$ is necessary and sufficient to solve ϵ-approximate agreement in both synchronous and asynchronous systems [6].

Multidimensional approximate agreement. The ϵ-approximate agreement problem has been generalized in [16] to the case where each input value is a point in $\mathbb{R}^d$ (space of dimension d). Such a point is defined by a size d vector (one entry per coordinate). The problem is then defined by the following properties. Let multi_approx_agreement() be the associated operation invoked by processes.

- Termination. The invocation of multi_approx_agreement() by a correct process p_i terminates.
- Validity. The value decided by any correct process is a point of $\mathbb{R}^d$ within the convex hull of the points proposed by the correct processes.
- Agreement. The Euclidean distance between any two points decided by correct processes is at most ϵ.

It is shown in [16] that $n > (d+2)t$ is a sufficient and necessary requirement for the problem to be solved.

5.2 Solving Multidimensional Approximate Agreement

An algorithm solving the multidimensional approximate agreement problem on top of the shared memory abstraction is presented in Figure 3. This algorithm is an adaptation of Mendes-Herlihy's algorithm designed to asynchronous message-passing [16].

Shared memory: arrays of n SWMR atomic h-registers. The processes cooperate through the following arrays of SWMR atomic h-registers. Each h-register is written at most once by its writer. An input value is a d-dimensional vector (coordinates of the input of p_i in $\mathbb{R}^d$).

- $VAL[1..n]$: array such that $VAL[i]$ is written by p_i to publish its input.
- $VIEW[1..n]$: array such that $VIEW[i]$ contains p_i's view of the inputs.
- $EST[1..][1..n]$: array such that $EST[r][i]$ contains p_i's estimate of its decision value at round r.
- $MAX[1..n]$: array such that $MAX[i] = r$ means that p_i has estimated that r rounds are enough to reach approximate agreement. Initially, $MAX[i] = +\infty$.

The value of any of these arrays is obtained with a collect() operation, which (differently from a snapshot) is an asynchronous and unordered read of the corresponding entries of the array [2,22].

Procedures used in the algorithm. Let X denote a multiset of values of $\mathbb{R}^d$ and $\mathcal{CH}(X)$ the convex hull of such a multiset. Algorithm 3 uses the notion of *safe area* [16]. The safe area of X is defined by $\mathsf{Safe}_t(X) = \bigcap_{X' \subseteq X, |X'| = |X| - t} \mathcal{CH}(X')$.Informally, this captures the region of $\mathbb{R}^d$ that is contained in all the convex hulls of the subsets of cardinal $n - t$ of $|X|$. If the values of $|X|$ are proposed by distinct processes, then this region is consequently contained in the convex hull of the subset of the values of X proposed by correct processes. It is shown in [16] that for any X such that $|X| > t(d + 1)$, the safe area $\mathsf{Safe}_t(X)$ is non-empty [16, Lemma 3.6].

The procedures bary(S), MD_Midpoint(S) and SingleDimMaxDist(S), where S is a convex polytope of $\mathbb{R}^d$, are called locally by the processes; bary(S) denotes the barycenter of S, while the two later are defined as follows, where $s[x]$ designate the xth coordinate of $s \in \mathbb{R}^d$:

$$\forall x \in \{1, \ldots, d\} \ : \ \mathsf{MD\_Midpoint}(S)[x] = (\max\{s[x], s \in S\} + \min\{s[x], s \in S\})/2,$$

$$\text{and} \quad \mathsf{SingleDimMaxDist}(S) = \max_{x \in \{1, \ldots, d\}} (\max\{s[x], s \in S\} - \min\{s[x], s \in S\}).$$

Local variables at each process p_i. Each process p_i manages the following local data structures: two arrays $my\_view_i[1..n]$ and $views_i[1..n]$, both initialized to $[\bot, \ldots, \bot]$; a set of points $safe\_init_i$; a local estimate est_i of the point that will be locally decided; an array $vals_i[1..n]$ of estimates values; an array $max\_r_i[1..n]$ containing round number upper bounds; and r_i which contains p_i's current round number.

Behavior of a process p_i. A process p_i first writes its input value v_i (point in $\in \mathbb{R}^d$) in $VAL[i]$, and collects a local view ($my\_view_i[1..n]$) including at least $(n - t)$ inputs (lines 01-02). Then, to make it public, p_i writes its view in $VIEW[i]$, and collects in $views_i$ the views of at least $(n - t)$ processes (lines 03-04). The process p_i then calculates in $safe\_init_i$ the barycenter of the safe area of these views (line 05). The parameters d and ϵ of the problem, and the set of barycenters $safe\_inits_i$ of its current instance, allow p_i to locally compute an upper bound on the number of rounds to be executed, which is written into the atomic h-register $MAX[i]$ (line 06). The set $safe\_inits_i$ is also used to compute p_i's first estimate (est_i) of its decision value (line 07).

---

operation multi_approx_agreement(v_i) **is**
(01) $VAL[i] \leftarrow v_i; r_i \leftarrow 0;$
(02) **repeat** $my\_view_i[1..n] \leftarrow VAL.$collect()
          until $(|\{x \mid my\_view_i[x] \neq \perp\}| \geq n - t)$ **end repeat**;
(03) $my\_view_i[1..n] \leftarrow VAL.$collect(); $VIEW[i] \leftarrow my\_view_i[1..n];$
(04) **repeat** $views_i[1..n] \leftarrow VIEW.$collect() **until** $(|\{x \mid views_i[x] \neq \perp\}| \geq n - t)$ **end**;
(05) $views_i[1..n] \leftarrow VIEW.$collect(); $safe\_inits_i \leftarrow \{$bary($Safe_t(X)) : X \in views_i\};$
(06) $MAX[i] \leftarrow \lceil \log_2 \left(\sqrt{d}/\epsilon \cdot \mathsf{SingleDimMaxDist}(safe\_inits_i) \right) \rceil;$
(07) $est_i \leftarrow$ bary($Safe_t(safe\_inits_i));$
(08) **repeat** $r_i \leftarrow r_i + 1;$
(09) $EST[r_i][i] \leftarrow est_i;$
(10)   **repeat** $vals_i[1..n] \leftarrow EST[r_i].$collect(); $max\_r_i[1..n] \leftarrow MAX.$collect()
(11)   **until** $(|\{x : vals_i[x] \neq \perp\}| \geq n - t) \vee (|\{x : max\_r_i[x] < r_i\}| > t)$ **end repeat**;
(12)   **if** $(|\{x : vals_i[x] \neq \perp\}| \geq n - t)$ **then** $vals_i[1..n] \leftarrow EST[r_i].$collect();
(13) $est_i \leftarrow$ MD_Midpoint($Safe_t(vals_i))$ **end if**
(14) **until** $(|\{x : max\_r_i[x] < r_i\}| > t)$ **end repeat**;
(15) return(est_i).

Fig. 3. Multidimensional approximate agreement on top of $\mathcal{BAMP}_{n,t}[t < n/(d+2)]$ (code p_i)

Then, process p_i starts a sequence of asynchronous rounds whose aim is to refine its current estimate est_i (lines 09-13) until it returns its last estimate value. This occurs when p_i attains a round r_i during which it sees a set of more than t processes –i.e., at least one correct process– such that each process p_j of this set posted in its atomic h-register $MAX[j]$ a last round upper bound smaller than r_i. This is captured by the outer termination predicate $(|\{x : max\_r_i[x] < r_i\}| > t)$ used at lines 12 and 13.

During a loop iteration r_i, p_i first writes its current decision value estimate in $EST[r_i][i]$ (line 09), and repeatedly collects both current estimates of decision values written in $EST[r_i][1..n]$, and upper bounds of the last round number from $MAX[1..n]$, until either the outer termination predicate is satisfied, or it sees at least $(n - t)$ estimates computed at round r_i (first sub-predicate of line 11). The use of the outer termination predicate in the inner repeat loop (line 11) allows p_i not to remain blocked forever waiting for a correct process that has already terminated. Then, if it knows enough values deposited in $EST[r_i]$, p_i collects again current estimates of decision values written in $EST[r_i]$ (line 12), and computes a new estimate est_i (line 13). Finally, once a large enough number of rounds is reached (line 14), p_i returns (decides) its current estimate value (line 15). The proof of the correctness of this algorithm is given in [9]. It is similar to, and follows the structure of, the one of Mendes-Herlihy's algorithm [16]

Theorem 3. *The algorithm presented in Figure 3 is a correct implementation of multi-dimensional approximate agreement in the $\mathcal{BAMP}_{n,t}[t < n/(d+2)]$ model.*

6 Conclusion

This paper has first proposed a clean notion of atomic h-registers in the presence of Byzantine failures, and has then shown how to build it on top of a Byzantine asynchronous message-passing system. More precisely, an algorithm building a shared memory abstraction, made up of n single-writer/multi-reader atomic h-registers, has been presented and proved correct. The paper has also shown that $t < n/3$ is a necessary and sufficient condition for such an algorithm.

The paper has then presented distributed algorithms suited to such a shared memory abstraction, which can cope with up to t Byzantine processes. The simplicity of these algorithms constitutes a strong motivation for the use of such a shared memory abstraction in the presence of Byzantine processes. As, for a lot of problems, algorithms are more difficult to design and prove correct in a message-passing system than in a shared memory system, the proposed abstraction should allow easier designs and proofs for other algorithms that have to cope with Byzantine failures.

Acknowledgments. This work has been partially supported by the French ANR projects DISPLEXITY (devoted to computability and complexity in distributed computing) and CO_2Dim, the Mexican projects UNAM PAPIIT IN107714 and ECOS-ANUIES, and the CONACYT project LAISLA.

References

1. Attiya, H., Bar-Noy, A., Dolev, D.: Sharing memory robustly in message passing systems. Journal of the ACM 42(1), 121–132 (1995)
2. Attiya, H., Welch, J.: Distributed computing: fundamentals, simulations and advanced topics, 2nd edn., 414 p. Wiley-Interscience (2004)
3. Bracha, G.: Asynchronous Byzantine agreement protocols. I&C 75(2), 130–143 (1987)
4. Cachin, C., Kursawe, K., Shoup, V.: Random oracles in Constantinople: practical asynchronous Byzantine agreement using cryptography. In: Proc. 19th Annual ACM Symposium on Principles of Distributed Computing (PODC 2000), pp. 123–132. ACM Press (2000)
5. Chockler, G., Malkhi, D.: Active disk Paxos with infinitely many processes. Distributed Computing 18(1), 73–84 (2005)
6. Dolev, D., Lynch, N.A., Pinter, S.S., Stark, E.W., Weihl, W.E.: Reaching approximate agreement in the presence of faults. Journal of the ACM 33(3), 499–516 (1986)
7. Herlihy, M.P., Kozlov, D., Rajsbaum, S.: Distributed computing through combinatorial topology, 336p. Morgan Kaufmann/Elsevier (2014) ISBN 9780124045781
8. Herlihy, M.P., Wing, J.M.: Linearizability: a correctness condition for concurrent objects. ACM Transactions on Programming Languages and Systems 12(3), 463–492 (1990)
9. Imbs, D., Rajsbaum, S., Raynal, M., Stainer, J.: Reliable shared memory abstractions on top of asynchronous t-resilient Byzantine message-passing systems. Tech Report 2018, 18p., IRISA, Université de Rennes, France (2014)
10. Ittai, A., Chockler, G., Keidar, I., Malkhi, D.: Byzantine disk paxos: optimal resilience with byzantine shared memory. Distributed Computing 18(5), 387–408 (2006)
11. Lamport, L., Shostack, R., Pease, M.: The Byzantine generals problem. ACM Transactions on Programming Languages and Systems 4(3), 382–401 (1982)
12. Lamport, L.: On interprocess communication, Part I: basic formalism. Distributed Computing 1(2), 77–85 (1986)
13. Lynch, N.A.: Distributed algorithms, 872p. Morgan Kaufmann Pub., San Francisco (1996) ISBN 1-55860-384-4
14. Malkhi, D., Merritt, M., Reiter, M.K., Taubenfeld, G.: Objects shared by Byzantine processes. Distributed Computing 16(1), 37–48 (2003)

15. Martin, J.-P., Alvisi, L.: A framework for dynamic Byzantine storage. In: Proc. Int'l Conference on Dependable Systems and Networks (DSN 2004), pp. 325–334. IEEE Press (2004)
16. Mendes, H., Herlihy, M.: Multidimensional approximate agreement in Byzantine asynchronous systems. In: Proc. 46th ACM STOC, pp. 391–400. ACM Press (2013)
17. Mostéfaoui, A., Moumem, H., Raynal, M.: Signature-free asynchronous Byzantine consensus with $t < n/3$ and $O(n^2)$ messages. In: Proc. 33rd ACM PODC. ACM Press (2014)
18. Mostéfaoui, A., Raynal, M.: Communication and agreement abstractions in the presence of Byzantine processes. Tech Report 2015, 24 p., IRISA, Univ. de Rennes, France (2014)
19. Pease, M., Shostak, R., Lamport, L.: Reaching agreement in the presence of faults. Journal of the ACM 27, 228–234 (1980)
20. Raynal, M.: Communication and agreement abstractions for fault-tolerant asynchronous distributed systems, 251p. Morgan & Claypool Pub. (2010) ISBN 978-1-60845-293-4
21. Raynal, M.: Fault-tolerant agreement in synchronous message-passing systems, 165p. Morgan & Claypool Pub. (2010) ISBN 978-1-60845-525-6
22. Raynal, M.: Concurrent programming: algorithms, principles and foundations, 515p. Springer (2013) ISBN 978-3-642-32026-2

Distributed Transactional Contention Management as the Traveling Salesman Problem

Bo Zhang, Binoy Ravindran, and Roberto Palmieri

Virginia Tech, Blacksburg VA 24060, USA
{alexzbzb,binoy,robertop}@vt.edu

Abstract. In this paper we consider designing contention managers for distributed software transactional memory (DTM), given an input of n transactions sharing s objects in a network of m nodes. We first construct a dynamic ordering conflict graph $G_c^*(\phi(\kappa))$ for an offline algorithm (κ, ϕ_κ). We show that finding an optimal schedule is equivalent to finding the offline algorithm for which the weight of the longest weighted path in $G_c^*(\phi(\kappa))$ is minimized. We further illustrate that when the set of transactions are dynamically generated, processing transactions according to a $\chi(G_c)$-coloring of G_c does not lead to an optimal schedule, where $\chi(G_c)$ is the chromatic number of G_c. We prove that, for DTM, any online work conserving deterministic contention manager provides an $\Omega(\max[s, \frac{s^2}{\overline{D}}])$ competitive ratio in a network with normalized diameter $\overline{D}$. Compared with the $\Omega(s)$ competitive ratio for multiprocessor STM, the performance guarantee for DTM degrades by a factor proportional to $\frac{s}{\overline{D}}$. To break this lower bound, we present a randomized algorithm CUTTING, which needs partial information of transactions and an approximate algorithm A for the traveling salesman problem with approximation ratio ϕ_A. We show that the average case competitive ratio of CUTTING is $O(s \cdot \phi_A \cdot \log^2 m \log^2 n)$, which is close to $O(s)$.

Keywords: Synchronization, Distributed Transactional Memory.

Transactional Memory [18] is an alternative synchronization model for shared memory objects that promises to alleviate the difficulties of manual implementation of lock-based concurrent programs, including composability. The recent integration of TM in hardware by major chip vendors (e.g., Intel, IBM), together with the development of dedicated GCC extensions for TM (i.e., GCC-4.7) has significantly increased TM's traction, in particular its software version (STM). Similar STM, distributed STM (or DTM) [12,20,7,19,14,13] is motivated by the difficulties of lock-based distributed synchronization.

In this paper we consider the data-flow DTM model [6], where transactions are immobile, and objects are migrated to invoking transactions. In a realization of this model [15], when a node initiates a transaction that requests a read/write operation on object o, it first checks whether o is in the local cache; if not, it invokes a *cache-coherence* protocol to locate o in the network. If two transactions access the same object at the same time, a contention manager is required to

M. Halldórsson (Ed.): SIROCCO 2014, LNCS 8576, pp. 54–67, 2014.
© Springer International Publishing Switzerland 2014

handle the concurrent request. The performance of a contention manager is often evaluated quantitatively by measuring its *makespan* — the total time needed to complete a finite set of transactions [1]. The goal in the design of a contention manager is often to minimize the makespan, i.e., maximize the throughput.

The first theoretical analysis of contention management in multiprocessors is due to Guerraoui *et. al.* [5], where an $O(s^2)$ upper bound is given for the Greedy manager for s shared objects, compared with the makespan produced by an optimal clairvoyant offline algorithm. Attiya *et. al.* [1] formulated the contention management problem as a *non-clairvoyant* job scheduling problem and improved this bound to $O(s)$. Furthermore, a matching lower bound of $\Omega(s)$ is given for any deterministic contention manager in [1]. To obtain alternative and improved formal bounds, recent works have focused on randomized contention managers [16,17]. Schneider and Wattenhofer [16] presented a deterministic algorithm called CommitRounds with a competitive ratio $\Theta(s)$ and a randomized algorithm called RandomizedRounds with a makespan $O(C \log M)$ for M concurrent transactions in separate threads with at most C conflicts with high probability. In [17], Sharma *et. al.* consider a set of M transactions and N transactions per thread, and present two randomized contention managers: Offline-Greedy and Online-Greedy. By knowing the conflict graph, Offline-Greedy gives a schedule with makespan $O(\tau \cdot (C + N \log MN))$ with high probability, where each transaction has the equal length τ. Online-Greedy is only $O(\log MN)$ factor worse, but does not need to know the conflict graph. While these works have studied contention management in multiprocessors, no past work has studied it for DTM, which is our focus.

Alternative solutions for reducing the abort rate in STM and DTM can be found in [4] and [3,11], respectively.

In this paper we study contention management in DTM. Similar to [1], we model contention management as a non-clairvoyant scheduling problem. To find an optimal scheduling algorithm, we construct a dynamic ordering conflict graph $G_c^*(\phi(\kappa))$ for an offline algorithm (κ, ϕ_κ), which computes a k-coloring instance κ of the dynamic conflict graph G_c and processes the set of transactions in the order of ϕ_κ. We show that the makespan of (ϕ, κ) is equivalent to the weight of the longest weighted path in $G_c^*(\phi(\kappa))$. Therefore, finding the optimal schedule is equivalent to finding the offline algorithm (ϕ, κ) for which the weight of the longest weighted path in $G_c^*(\phi(\kappa))$ is minimized. We illustrate that, unlike the one-shot scheduling problem (where each node only issues one transaction), when the set of transactions are dynamically generated, processing transactions according to a $\chi(G_c)$-coloring of G_c does not lead to an optimal schedule, where $\chi(G_c)$ is G_c's chromatic number.

We prove that for DTM, an online, work conserving deterministic contention manager provides an $\Omega(\max[s, \frac{s^2}{\overline{D}}])$ competitive ratio for s shared objects in a network with normalized diameter $\overline{D}$. Compared with the $\Omega(s)$ competitive ratio for multiprocessor STM, the performance guarantee for DTM degrades by a factor proportional to $\frac{s}{\overline{D}}$. This motivates us to design a randomized contention manager that has partial knowledge about the transactions in advance.

We thus develop an algorithm called CUTTING, a randomized algorithm based on an approximate TSP algorithm A with an approximation ratio ϕ_A. CUTTING divides the nodes into $O(C)$ partitions, where C is the maximum degree in the conflict graph G_c. The cost of moving an object inside each partition is at most $\frac{\text{ATSP}_A}{C}$, where ATSP_A is the total cost of moving an object along the approximate TSP path to visit each node exactly once. CUTTING resolves conflicts in two phases. In the first phase, a binary tree is constructed inside each partition, and a transaction always aborts when it conflicts with its ancestor in the binary tree. In the second phase, CUTTING uses a randomized priority policy to resolve conflicts. We show that the average case competitive ratio of CUTTING is $O(s \cdot \phi_A \cdot \log^2 m \log^2 n)$ for s objects shared by n transactions invoked by m nodes, which is close to the multiprocessor bound of $O(s)$ [1].

CUTTING is the first ever contention manager for DTM with an average-case competitive ratio bound, and constitutes the paper's contribution.

1   Preliminaries

DTM model. We consider a set of *distributed transactions* $\mathcal{T} := \{T_1, T_2, \ldots, T_n\}$ sharing up to s objects $\mathcal{O} := \{o_1, o_2, \ldots, o_s\}$ distributed on a network of m nodes $\{v_1, v_2, \ldots, v_m\}$, which communicate by message-passing links. For simplicity of the analysis, we assume that each node runs only a single thread, i.e., in total, there are at most m threads running concurrently.[1] A node's thread issues transactions sequentially. Specifically, node v_i issues a sequence of transactions $\{T_1^i, T_2^i, \ldots\}$ one after another, where transaction T_j^i is issued once after T_{j-1}^i has committed.

An execution of a transaction is a sequence of timed operations. There are four operation types that a transaction may take: *write, read, commit,* and *abort*. An execution ends by either a commit (success) or an abort (failure). When a transaction aborts, it is restarted from its beginning immediately and may access a different set of shared objects. Each transaction T_i has a *local execution duration* τ_i, which is the time T_i executes locally without contention (or interruption). Note that τ_i does not include the time T_i acquires remote objects. In our analysis, we assume a fixed τ_i for each transaction T_i. Such a general assumption is unrealistic if the local execution duration depends on the properties of specific objects. In that case, when a transaction alters the set of requested objects after it restarts, the local execution duration also varies. Therefore, if the local execution duration varies by a factor of c, then the performance of our algorithms would worsen by the same factor c.

A transaction performs a read/write operation by first sending a read/write access request through CC. For a read operation, the CC protocol returns a read-only copy of the object. An object can thus be read by an arbitrary number

---

[1] When a node runs multiple threads, our analysis can still be adopted by treating each thread as an individual node. This strategy overlooks the possible local optimization for the same threads issued by the same node. Therefore, multiprocessor contention management strategy can be used to improve performance.

of transactions simultaneously. For a write operation, the CC protocol returns the (writable) object itself. At any time, only one transaction can hold the object exclusively. A contention manager is responsible for resolving the conflict, and does so by aborting or delaying (i.e., postponing) one of the conflicting transactions.

A CC protocol moves objects via a specific path (e.g., the shortest path for Ballistic [6], or a path in a spanning tree for Relay [21]). We assume a fixed CC protocol with a *moving cost* d_{ij}, where d_{ij} is the communication latency to move an object from node v_i to v_j under that protocol. We can build a complete *cost graph* $G_d = (V_d, E_d)$, where $|V_d| = m$ and for each edge $(v_i, v_j) \in E_d$, the weight is d_{ij}. We assume that the moving cost is bounded: we can find a constant D such that for any d_{ij}, $D \geq d_{ij}$.

Conflict graph. We build the *conflict graph* $G_c = (\mathcal{T}_c, E_c)$ for the transaction subset $\mathcal{T}_c \subseteq \mathcal{T}$, which runs concurrently. An edge $(T_i, T_j) \in E_c$ exists if and only if T_i and T_j conflict. Two transactions conflict if they both access the same object and at least one of the accesses is a write. Let N_T denote the set of neighbors of T in G_c. The degree $\delta(T) := |N_T|$ of a transaction T corresponds to the number of neighbors of T in G_c. We denote $C = \max_i \delta(T_i)$, i.e., the maximum degree of a transaction. The graph G_c is dynamic and only consists of live transactions. A transaction joins $\mathcal{T}_c$ after it (re)starts, and leaves $\mathcal{T}_c$ after it commits/aborts. Therefore, N_T, $\delta(T)$, and C only capture a "snapshot" of G_c at a certain time. More precisely, they should be represented as functions of time. When there is no ambiguity, we use the simplified notations. We have $|\mathcal{T}_c| \leq \min\{m, n\}$, since there are at most n transactions, and at most m transactions can run in parallel. Then we have $\delta(T) \leq C \leq \min\{m, n\}$.

Let $o(T_i) := \{o_1(T_i), o_2(T_i), \ldots\}$ denote the sequence of objects requested by transaction T_i. Let $\gamma(o_j)$ denote the number of transactions in $\mathcal{T}_c$ that concurrently writes o_i and $\gamma_{max} = \max_j \gamma(o_j)$. Let $\lambda(T_i) = \{o : o \in o(T_i) \wedge (\gamma(o) \geq 1)\}$ denote the number of transactions in $\mathcal{T}_c$ that conflict with transaction T_i and $\lambda_{max} = \max_{T_i \subset \mathcal{T}_c} \lambda(T_i)$. We have $C \leq \lambda_{max} \cdot \gamma_{max}$ and $C \geq \gamma_{max}$.

2 The DTM Contention Management Problem

2.1 Problem Measure and Complexity

A contention manager determines when a particular transaction executes in case of a conflict. To quantitatively evaluate the performance of a contention manager, we measure the *makespan*, which is the total time needed to complete a set of transactions $\mathcal{T}$. Formally, given a contention manager A, makespan_A denotes the time needed to complete all transactions in $\mathcal{T}$ under A.

We measure the contention manager's quality, by assuming OPT, the optimal, centralized, clairvoyant scheduler which has the complete knowledge of each transaction (requested objects, locations, released time, local execution time). The quality of a contention manager A is measured by the ratio $\frac{\text{makespan}_A}{\text{makespan}_{\text{OPT}}}$, called the *competitive ratio* of A on $\mathcal{T}$. The competitive ratio of A is

$\max_{\mathcal{T}} \frac{\text{makespan}_A}{\text{makespan}_{\text{OPT}}}$, i.e., the maximum competitive ratio of A over all possible workloads.

An ideal contention manager aims to provide an optimal schedule for any given set of transactions. However, it is shown in [1] (for STM) that if there exists an adversary to change the set of shared objects requested by any transaction arbitrarily, no algorithm can do better than a simple sequential execution. Furthermore, even if the adversary can only choose the initial conflict graph and does not influence it afterwards, it is NP-hard to get a reasonable approximation of an optimal schedule [16].

We can consider the transaction scheduling problem for multiprocessor STM as a subset of the transaction scheduling problem for DTM. The two problems are equivalent as long as the communication cost (d_{ij}) can be ignored, compared with the local execution time duration (τ_i). Therefore, extending the problem space into distributed systems only increases the problem complexity.

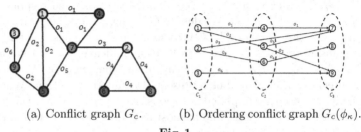

(a) Conflict graph G_c.   (b) Ordering conflict graph $G_c(\phi_\kappa)$.

Fig. 1.

We depict an example of a conflict graph G_c in Figure 1(a), which consists of 9 write-only transactions. Each transaction is represented as a numbered node in G_c. Each edge (T_i, T_j) is marked with the object which causes T_i and T_j to conflict (e.g., T_1 and T_4 conflict on o_1). We can construct a coloring of the conflict graph $G_c = (\mathcal{T}_c, E)$. A 3-coloring scenario is illustrated in Figure 1(a). Transactions are partitioned into 3 sets: $C_1 = \{T_1, T_2, T_3\}, C_2 = \{T_4, T_5, T_6\}, C_3 = \{T_7, T_8, T_9\}$. Since transactions with the same color are not connected, every set $C_i \subset \mathcal{T}_c$ forms an independent set and can be executed in parallel without facing any conflicts. With the same argument of [1], we have the following lemma.

Lemma 1. *An optimal offline schedule* OPT *determines a k-coloring κ of the conflict graph G_c and an execution order ϕ_κ such that for any two sets $C_{\phi_\kappa(i)}$ and $C_{\phi_\kappa(j)}$, where $i < j$, if (1) $T_1 \in C_{\phi_\kappa(i)}$, $T_2 \in C_{\phi_\kappa(j)}$, and (2) T_1 and T_2 conflict, then T_2 is postponed until T_1 commits.*

In other words, OPT determines the order in which an independent set C_i is executed. Generally, for a k-coloring of G_c, there are $k!$ different choices to order the independent sets. Assume that for the 3-coloring example in Figure 1(a), an execution order $\phi_\kappa = \{C_1, C_2, C_3\}$ is selected. We can construct an *ordering conflict graph* $G_c(\phi_\kappa)$, as shown in Figure 1(b).

Definition 1 (Ordering conflict graph). *For the conflict graph G_c, given a k-coloring instance κ and an execution order $\{C_{\phi_\kappa(1)}, C_{\phi_\kappa(2)}, \ldots, C_{\phi_\kappa(k)}\}$, the ordering conflict graph $G_c(\phi_\kappa) = (\mathcal{T}_c, E(\phi_\kappa), w)$ is constructed. $G_c(\phi_\kappa)$ has the following properties:*

1. *$G_c(\phi_\kappa)$ is a weighted directed graph.*
2. *For two transactions $T_1 \in C_{\phi_\kappa(i)}$ and $T_2 \in C_{\phi_\kappa(j)}$, a directed edge (or an arc) $(T_1, T_2) \in E(\phi_\kappa)$ (from T_1 to T_2) exists if: (i) T_1 and T_2 conflict over object o; (ii) $i < j$; and (iii) $\nexists T_3 \in C_{\phi_\kappa(j')}$, where $i < j' < j$, such that T_1 and T_3 also conflict over o.*
3. *The weight $w(T_i)$ of a transaction T_i is τ_i; the weight $w(T_i, T_j)$ of an arc (T_i, T_j) is d_{ij}.*

For example, the edge (T_1, T_4) in Figure 1(a) is also an arc in Figure 1(b). However, the edge (T_1, T_7) in Figure 1(a) no longer exists in Figure 1(b), because C_2 is ordered between C_1 and C_3, and T_1 and T_4 also conflict on o_1.

Hence, any offline algorithm can be described by the pair (κ, ϕ_κ), and the ordering conflict graph $G_c(\phi_\kappa)$ can be constructed. Given $G_c(\phi_\kappa)$, the execution time of each transaction can be determined.

Theorem 2. *For the ordering conflict graph $G_c(\phi_\kappa)$, given a directed path $P = \{T_{P(1)}, T_{P(2)}, \ldots, T_{P(L)}\}$ of L hops, the weight of P is defined as $w(P) = \sum_{1 \leq i \leq L} w(T_{P(i)}) + \sum_{1 \leq j \leq L-1} w(T_{P(j)}, T_{P(j+1)})$. Then transaction $T_0 \in \mathcal{T}_c$ commits at time: $\max_{P=\{T_{P(1)}, \ldots, T_0\}} t_{P(1)} + w(P)$, where $T_{P(1)}$ starts at time $t_{P(1)}$.*

Proof. We prove the theorem by induction. Assume $T_0 \in C_{\phi_\kappa(j)}$. When $j=1$, T_0 executes immediately after it starts. At time $t_0 + \tau_0$, T_0 commits. There is only one path that ends at T_0 in $G_c(\phi_\kappa)$ (which only contains T_0). The theorem holds.

Assume that when $j = 2, 3, \ldots, q - 1$, the theorem holds. Let $j = q$. For each object $o_i \in o(T_0)$, find the transaction $T_{0(i)}$ such that $T_{0(i)}$ and T_0 conflict over o_i, and $(T_{0(i)}, T_0) \in E(\phi_\kappa)$. If no such transaction exists for all objects, the analysis falls into the case when $j = 1$. Otherwise, for each transaction $T_{0(i)}$, from Definition 1, no transaction which requests access to o_i is scheduled between $T_{0(i)}$ and T_0. The offline algorithm (κ, ϕ_κ) moves o_i from $T_{0(i)}$ to T_0 immediately after $T_{0(i)}$ commits. Assume that $T_{0(i)}$ commits at $t_{0(i)}^c$. Then T_0 commits at time: $\max_{o_i \in o(T_0)} t_{0(i)}^c + w(T_{0(i)}, T_0) + w(T_0)$. Since $(T_{0(i)}, T_0) \in E(\phi_\kappa)$, then from the induction step, we know that $t_{0(i)}^c = \max_{P=\{T_{P(1)}, \ldots, T_{0(i)}\}} t_{P(1)} + w(P)$. Hence, by replacing $t_{0(i)}^c$ with $\max_{P=\{T_{P(1)}, \ldots, T_{0(i)}\}} t_{P(1)} + w(P)$, the theorem follows.

Theorem 2 illustrates that the commit time of transaction T_0 is determined by one of the *weighted paths* in $G_c(\phi_\kappa)$ which ends at T_0. Specifically, if every node issues its first transaction at the same time, the commit time of T_0 is solely determined by the longest weighted path in $G_c(\phi_\kappa)$ which ends at T_0. However, when transactions are dynamically generated over time, the commit time of a transaction also relies on the starting time of other transactions. To accommodate the dynamic features of transactions, we construct the *dynamic ordering conflict graph* $G_c^*(\phi_\kappa)$ based on $G_c(\phi_\kappa)$.

Definition 2 (Dynamic ordering conflict graph). *Given an ordering conflict graph $G_c(\phi_\kappa)$, the dynamic ordering conflict graph $G_c^*(\phi_\kappa)$ is constructed by making the following modifications on $G_c(\phi_\kappa)$:*

1. *For the sequence of transactions $\{T_1^i, T_2^i, \ldots, T_L^i\}$ issued by each node v_i, an arc (T_{j-1}^i, T_j^i) is added to $G_c^*(\phi_\kappa)$ for $2 \le j \le L$ and $w(T_{j-1}^i, T_j^i) = 0$.*
2. *If transaction T_j which starts at t_j does not have any incoming arcs in $G_c^*(\phi_\kappa)$, then $w(T_j) = t_j + \tau_j$.*

Theorem 3. *The makespan of algorithm (κ, ϕ_κ) is the weight of the longest weighted path in $G_c^*(\phi_\kappa)$: $makespan_{(\kappa,\phi_\kappa)} = \max_{P \in G_c^*(\phi_\kappa)} w(P)$*

Proof. We start the proof with special cases, and then extend the analysis to the general case. Assume that (i) each node issues only one transaction, and (ii) all transactions start at the same time. Then the makespan of (κ, ϕ_κ) is equivalent to the execution time of the last committed transaction: $makespan_{(\kappa,\phi_\kappa)} = \max_{T_0 \in \mathcal{T}_c, P \in G_c(\phi_\kappa), P=\{\ldots, T_0\}} w(P) = \max_{P \in G_c(\phi_\kappa)} w(P) = \max_{P \in G_c^*(\phi_\kappa)} w(P)$. Then, we can progressively relax the assumptions and use Theorem 2 to prove this theorem. Now, we relax the second assumption: each node issues a single transaction at arbitrary time points. Let P be the path which maximizes $makespan_{(\kappa,\phi_\kappa)}$. Therefore, T_{P_1} (the head of P) has no incoming arcs in $G_c^*(\phi_\kappa)$ (since each node only issues a single transaction). From the construction of $G_c^*(\phi_\kappa)$, $w(T_{P_1}) = t(P_1) + \tau_{P_1}$. We can find a path P^* in $G_c^*(\phi_\kappa)$ which contains the same elements as P with weight $w(P^*) = t(P_1) + w(P)$, which is the longest path in $G_c^*(\phi_\kappa)$.

Now, we relax the first assumption: each node issues a sequence of transactions, and all nodes start their first transactions at the same time. Similar to the first case, we have: $makespan_{(\kappa,\phi_\kappa)} = \max_{P \in G_c(\phi_\kappa), P=\{T_{P_1}, \ldots, T_0\}} t(P_1) + w(P)$.

Let P be the path which maximizes $makespan_{(\kappa,\phi_\kappa)}$. If T_{P_1} (the head of P) is the first transaction issued by a node, the theorem follows. Otherwise, $\forall o_i \in o(T_{P_1})$, T_{P_1} is the first transaction scheduled to access o_i by (κ, ϕ_κ), because there is no incoming arc to T_{P_1} in $G_c(\phi_\kappa)$. If T_{P_1} is the lth transaction issued by node v_j, when we convert from $G_c(\phi_\kappa)$ to $G_c^*(\phi_\kappa)$, the longest path P^* that ends at T_0 is a path starting from T_1^j to T_{l-1}^j, followed by an arc (T_{l-1}^j, T_{P_1}), and then followed by P. Note that T_{l-1}^j commits at t_{P_1} (the starting time of T_{P_1}). Hence, we have $w(P^*) = t(P_1) + w(T_{l-1}^j, T_{P_1}) + w(P)$. Since $w(T_{l-1}^j, T_{P_1}) = 0$ (from the construction of $G_c^*(\phi_\kappa)$), we have $t(P_1) + w(P) = w(P^*)$. We conclude that the path in $G_c(\phi_\kappa)$ corresponding to the commit time of transaction T_0 is equivalent to the longest path which ends at T_0 in $G_c^*(\phi_\kappa)$. The theorem follows.

Theorem 3 shows that, given an offline algorithm (κ, ϕ_κ), finding its makespan is equivalent to finding the longest weighted path in the dynamic ordering conflict graph $G_c^*(\phi_\kappa)$. Therefore, the optimal schedule OPT is the offline algorithm which minimizes the makespan.

Corollary 4. $makespan_{\text{OPT}} = \min_{\kappa, \phi_\kappa} \max_{P \in G_c^*(\phi_\kappa)} w(P)$

It is easy to show that finding the optimal schedule is NP-hard. For the *one-shot scheduling problem*, where each node issues a single transaction, if $\tau_0 = \tau$ for all transactions $T_0 \in \mathcal{T}$ and $D \ll \tau$, the problem becomes the classical node coloring problem. Finding the optimal schedule is equivalent to finding the chromatic number $\chi(G_c)$. As [10] shows, computing an optimal coloring, given complete knowledge of the graph, is NP-hard, and computing an approximation within the factor of $\chi(G_c)^{\frac{\log \chi(G_c)}{25}}$ is also NP-hard.

If $s = 1$, i.e., there is only one object shared by all transactions, finding the optimal schedule is equivalent to finding the traveling salesman problem (TSP) path in G_d, i.e., the shortest hamiltonian path in G_d. When the cost metric d_{ij} satisfies the triangle inequality, the resulting TSP is called the metric TSP, and has been shown to be NP-complete by Karp [9]. If the cost metric is symmetric, Christofides [2] presented an algorithm approximating the metric TSP within approximation ratio $3/2$. If the cost metric is asymmetric, the best known algorithm approximates the solution within approximation ratio $O(\log m)$ [8].

When each node generates a sequence of transactions dynamically, it is not always optimal to schedule transactions according to a $\chi(G_c)$-coloring. Since the conflict graph evolves over time, an optimal schedule based on a static conflict graph may lose potential parallelism in the future. In the dynamic ordering conflict graph, a temporarily-optimal scheduling does not imply that the resulting longest weighted path is optimal.

2.2   Lower Bound

Our analysis shows that to compute an optimal schedule, even knowing all information about the transactions in advance, is NP-hard. Thus, we design an online algorithm which guarantees better performance than that can be obtained by simple serialization of all transactions. Before designing the contention manager, we need to know what performance bound an online contention manager could provide in the best case. We first introduce the *work conserving* property [1]:

Definition 3. *A scheduling algorithm is work conserving if it always runs a maximal set of non-conflicting transactions.*

In [1], Attiya *et al.* showed that, for multiprocessor STM, a deterministic work conserving contention manager is $\Omega(s)$-competitive, if the set of objects requested by a transaction changes when the transaction restarts. We prove that for DTM, the performance guarantee is even worse.

Theorem 5. *For DTM, any online, work conserving deterministic contention manager is $\Omega(\max[s, \frac{s^2}{\overline{D}}])$-competitive, where $\overline{D} := \frac{D}{\min_{G_d} d_{ij}}$ is the normalized diameter of the cost graph G_d.*

Proof. The proof uses s^2 transactions with the same local execution duration τ. A transaction is denoted by T_{ij}, where $1 \leq i, j \leq s$. Each transaction T_{ij} contains a sequence of two operations $\{R_i, W_i\}$, which first reads from object o_i and then writes to o_i. Each transaction T_{ij} is issued by node v_{ij} at the same time, and object o_i is held by node v_{i1} when the system starts. For each i, we select a set of nodes $V_i := \{v_{i1}, v_{i2}, \ldots, v_{is}\}$ within the range of the diameter $D_i \leq \frac{D}{s}$.

Consider the optimal schedule OPT. Note that all transactions form an $s \times s$ matrix, and transactions from the same row ($\{T_{i1}, T_{i2}, \ldots, T_{is}\}$ for $1 \leq i \leq s$) have the same operations. Therefore, at the start of the execution, OPT selects one transaction from each row, thus s transactions start to execute. Whenever T_{ij} commits, OPT selects one transaction from the rest of the transactions in row i to execute. Hence, at any time, there are s transactions that run in parallel.

The order that OPT selects transactions from each row is crucial: OPT should select transactions in the order such that the weight of the longest weighted path in $G_c^*(\text{OPT})$ is optimal. Since transactions from different rows run in parallel, we have: makespan$_{\text{OPT}} = s \cdot \tau + \max_{1 \leq i \leq s} \text{TSP}(G_d(o_i))$, where $G_d(o_i)$ denotes the subgraph of G_d induced by s transactions requesting o_i, and $\text{TSP}(G_d(o_i))$ denotes the length of the TSP path of $G_d(o_i)$, i.e., the shortest path that visits each node exactly once in $\text{TSP}(G_d(o_i))$.

Now consider an online, work conserving deterministic contention manager A. Being work conserving, it must select to execute a maximal independent set of non-conflicting transactions. Since the first access of all transactions is a read, the contention manager starts to execute all s^2 transactions.

After the first read operation, for each row i, all transactions in row i attempt to write o_i, but only one of them can commit and the others will abort. Otherwise, atomicity is violated, since inconsistent states of some transactions may be accessed. When a transaction restarts, the adversary determines that all transactions change to write to the same object, e.g., $\{R_i, W_1\}$. Therefore, the rest $s^2 - s$ transactions can only be executed sequentially after the first s transactions execute in parallel and commit. Then we have: makespan$_A \geq (s^2 - s + 1) \cdot \tau + \min_{G_d} \text{TSP}(G_d(s^2 - s + 1))$, where $G_d(s^2 - s + 1)$ denotes the subgraph of G_d induced by a subset of $s^2 - s + 1$ transactions.

Now, we can compute A's competitive ratio. We have: $\frac{\text{makespan}_A}{\text{makespan}_{\text{OPT}}} \geq$

$$\max\left[\frac{(s^2-s+1)\cdot\tau}{s\cdot\tau}, \frac{\min_{G_d} \text{TSP}(G_d(s^2-s+1))}{\max_{1\leq i\leq s} \text{TSP}(G_d(o_i))}\right] \geq \max\left[\frac{s^2-s+1}{s}, \frac{(s^2-s+1)\cdot\min_{G_d} d_{ij}}{(s-1)\cdot\frac{D}{s}}\right] =$$

$\Omega(\max[s, \frac{s^2}{\overline{D}}])$. The theorem follows.

Theorem 5 shows that for DTM, an online, work conserving deterministic contention manager cannot provide a similar performance guarantee compared with multiprocessor STM. When the normalized network diameter is bounded (i.e., $\overline{D}$ is a constant, where new nodes join the system without expanding the diameter of the network), it can only provide an $\Omega(s^2)$-competitive ratio. In the next section, we present an online randomized contention manager, which needs partial information of transactions in advance, in order to provide a better performance guarantee.

3 Algorithm: Cutting

3.1 Description

We present the algorithm CUTTING, a randomized scheduling algorithm based on a partitioning constructed on the cost graph G_d. To partition the cost graph, we first construct an approximate TSP path (ATSP path) in G_d $\mathrm{ATSP}_A(G_d)$ by selecting an approximate TSP algorithm A. Specifically, A provides the approximation ratio ϕ_A, such that for any graph G, $\frac{\mathrm{ATSP}_A(G)}{\mathrm{TSP}(G)} = O(\phi_A)$. Note that if d_{ij} satisfies the triangle inequality, the best known algorithm provides an $O(\log m)$ approximation [8]; if d_{ij} is symmetric as well, a constant ϕ_A is achievable [2]. We assume that a transaction has partial knowledge in advance: a transaction T_i knows its required set of objects o_i after it starts. Therefore, a transaction can send all its object requests immediately after it starts.

Based on the constructed ATSP path ATSP_A, we define the (C, A) *partitioning* on G_d, which divides G_d into $O(C)$ partitions. A constructed partition P is a subset of nodes, which satisfies either: 1) $|P| = 1$; or 2) for any pair of nodes $(v_i, v_j) \in P$, $d_{ij} \leq \frac{\mathrm{ATSP}_A}{C}$.

Definition 4 ((C, A) partitioning). *In the cost graph G_d, the (C, A) partitioning $\mathcal{P}(C, A, v)$ divides m nodes into $O(C)$ partitions in two phases.*

Phase I. *Randomly select a node v, and let node v^j be the jth node (excluding v) on the ATSP path $\mathrm{ATSP}_A(G_d)$ starting from v. Hence, $\mathrm{ATSP}_A(G_d)$ can be represented by a sequence of nodes $\{v^0, v^1, \ldots, v^{m-1}\}$.*

Phase II. *Inside each partition $P_t = \{v^k, v^{k+1}, \ldots\}$, each node v^k is assigned a* partition index $\psi(v^j) = (j \mod k)$, *i.e., its index inside the partition.*
1. *Starting from v^0, add v^0 to P_1, and set P_1 as the current partition.*
2. *Check v^1. If $\mathrm{ATSP}_A(G_d)[v^0, v^1] \leq \frac{\mathrm{ATSP}_A(G_d)}{C}$, where $\mathrm{ATSP}_A(G_d)[v^1, v^2]$ is the length of the part of $\mathrm{ATSP}_A(G_d)$ from v^0 to v^1, add v^1 to P_1. Else, add v^1 to P_2, and set P_2 as the current partition.*
3. *Repeat Step 2 until all nodes are partitioned. For each node v^k and the current partition P_t, this process checks the length of $\mathrm{ATSP}_A(G_d)[v^j, v^k]$, where v^j is the first element added to P_t. If $\mathrm{ATSP}_A(G_d)[v^j, v^k] \leq \frac{\mathrm{ATSP}_A(G_d)}{C}$, v^k is added to P_t. Else, v^k is added to P_{t+1}, and P_{t+1} is set as the current partition.*

The conflict resolution also has two phases. In the first phase, CUTTING assigns each transaction a partition index. When two transactions T_1 (invoked by node v^{j_1}) and T_2 (invoked by node v^{j_2}) conflict, the algorithm checks: 1) whether they are from the same partition P_t; 2) If so, whether $\exists$ integer $\nu \geq 1$ such that $\lfloor \frac{\max\{\psi(v^{j_1}), \psi(v^{j_2})\}}{2^\nu} \rfloor = \min\{\psi(v^{j_1}), \psi(v^{j_2})\}$. Note that by checking these two conditions, an underlying binary tree $\mathrm{BT}(P_t)$ is constructed in P_t as follows:

1. Set v^{j_0} as the root of $\mathrm{BT}(P_t)$ (level 1), where $\psi(v^{j_0} = 0)$, i.e., the first node added to P_t.
2. Node v^{j_0}'s left pointer points to v^{j_0+1} and right pointer points to v^{j_0+2}. Nodes v^{j_0+1} and v^{j_0+2} belong to level 2.
3. Repeat Step 2 by adding nodes sequentially to each level from left to right. In the end, $O(\log_2 m)$ levels are constructed.

Note that by satisfying these two conditions, the transaction with the smaller partition index must be an *ancestor* of the other transaction in $\mathrm{BT}(P_t)$. Therefore, a transaction may conflict with at most $O(\log_2 m)$ ancestors in this case. CUTTING resolves the conflict greedily so that the transaction with the smaller partition index always aborts the other transaction.

In the second phase, each transaction selects an integer $\pi \in [1, m]$ randomly when it starts or restarts. If one transaction is not an ancestor of another transaction, the transaction with the lower π proceeds and the other transaction aborts. Whenever a transaction is aborted by a remote transaction, the requested object is moved to the remote transaction immediately.

3.2   Analysis

We now study two efficiency measures of CUTTING from the average-case perspective: the average response time (how long it takes for a transaction to commit on average) and the average makespan (i.e., the expected value produced by the randomization in the algorithm).

Lemma 6. *A transaction T needs $O\big(C \log^2 m \log n\big)$ trials from the moment it is invoked until it commits, on average.*

Proof. We start from a transaction T invoked by the root node $v^\psi \in \mathrm{BT}(P_t)$. Since v^ψ is the root, T cannot be aborted by another ancestor in $\mathrm{BT}(P_t)$. Hence, T can only be aborted when it chooses a larger π than π', which is the integer chosen by a conflicting transaction T' invoked by node $v^{\psi'} \in P_{t'}$. The probability that for transaction T, no transaction $T' \in N_T$ selects the same random number $\pi' = \pi$ is: $\mathbf{Pr}(\nexists T' \in N_T | \pi' = \pi) = \prod_{T' \in N_T}(1 - \frac{1}{m}) \geq (1 - \frac{1}{m})^{\delta(T)} \geq (1 - \frac{1}{m})^m \geq \frac{1}{e}$. Note that $\delta(T) \leq C \leq m$. On the other hand, the probability that π is at least as small as π' for any conflicting transaction T' is at least $\frac{1}{(C+1)}$. Thus, the probability that π is the smallest among all its neighbors is at least $\frac{1}{e(C+1)}$.

We use the following Chernoff bound:

Lemma 7. *Let $X_1, X_2, \ldots, X_n$ be independent Poisson trials such that, for $1 \leq i \leq n$, $\mathbf{Pr}(X_i = 1) = p_i$, where $0 \leq p_i \leq 1$. Then, for $X = \sum_{i=1}^n X_i$, $\mu = \mathbf{E}[X] = \sum_{i=1}^n p_i$, and any $\delta \in (0, 1]$, $\mathbf{Pr}(X < (1-\delta)\mu) < e^{-\delta^2 \mu/2}$.*

By Lemma 7, if we conduct $16e(C + 1) \ln n$ trials, each having a success probability $\frac{1}{e(C+1)}$, then the probability that the number of successes X is less than $8 \ln n$ becomes: $\mathbf{Pr}(X < 8 \ln n) < e^{-2 \ln n} = \frac{1}{n^2}$.

Now we examine the transaction T^l invoked by node $v^{\psi^l} \in P_t$, where v^{ψ^l} is the left child of the root node v^ψ in $\text{BT}(P_t)$. When T^l conflicts with T, it aborts and holds off until T commits or aborts. Hence, T^l can be aborted by T at most $16e(C+1)\ln n$ times with probability $1 - \frac{1}{n^2}$. On the other hand, T^l needs at most $16e(C+1)\ln n$ to choose the smallest integer among all conflicting transactions with probability $1 - \frac{1}{n^2}$. Hence, in total, T^l needs at most $32e(C+1)\ln n$ trials with probability $(1 - \frac{1}{n^2})^2 > (1 - \frac{2}{n^2})$.

Therefore, by induction, the transaction T^L invoked by a level-L node v^{ψ^L} of $\text{BT}(P_t)$ needs at most $(1 + \log_2 L)\log_2 L \cdot 8e(C+1)\ln n$ trials with probability at least $1 - \frac{(1+\log_2 L)\log_2 L}{2n^2}$. Now, we can calculate the average number of trials: $\mathbf{E}[\# \text{ of trials a transaction needs to commit}] = O(C\log^2 m \log n)$.

Since when the starting point of the ATSP path is randomly selected, the probability that a transaction is located at level L is $1/2^{L_{max}-L+1}$. The lemma follows.

Lemma 8. *The average response time of a transaction is $O\big(C\log^2 m \log n \cdot (\tau + \frac{\text{ATSP}_A}{C})\big)$.*

Proof. From Lemma 6, each transaction needs $O(C\log^2 m \log n)$ trials, on average. We now study the duration of a trial, i.e., the time until a transaction can select a new random number. Note that a transaction can only select a new random number after it is aborted (locally or remotely). Hence, if a transaction conflicts with a transaction in the same partition, the duration is at most $\tau + \frac{\text{ATSP}_A}{C}$; if it conflicts with a transaction in another partition, the duration is at most $\tau + D$. Note that a transaction sends its requests of objects simultaneously once after it (re)starts. If a transaction conflicts with multiple transactions, the first conflicting transaction it knows is the transaction closest to it. From Lemma 6, a transaction can be aborted by transactions from other partitions by at most $16e(C+1)\ln n$ times. Hence, the expected commit time of a transaction is $O(C\log^2 m \log n \cdot (\tau + \frac{\text{ATSP}_A}{C}))$. The lemma follows.

Theorem 9. *The average-case competitive ratio of* CUTTING *is $O(s \cdot \phi_A \cdot \log^2 m \log^2 n)$.*

Proof. By following the Chernoff bound provided by Lemma 7 and Lemma 8, we can prove that CUTTING produces a schedule with average-case makespan $O(C\log^2 m \log n \cdot (\tau + \frac{\text{ATSP}_A}{C}) + (N \cdot \log^2 m \log^2 n \cdot \tau + \text{ATSP}_A))$, where N is the maximum number of transactions issued by the same node. We then find that $\text{makespan}_{\text{OPT}} \geq \max_{1 \leq i \leq s}(\tau \cdot \max[\gamma_i, N] + \text{TSP}(G_d(o_i)))$, since γ_i transactions concurrently conflict on object o_i. Hence, at any given time, only one of them can commit, and the object moves along a certain path to visit γ_i transactions one after another. Then we have: $\text{makespan}_{\text{OPT}} \geq \max_{1 \leq i \leq s}(\tau \cdot \max[\gamma_i, N] + \text{TSP}(G_d(o_i))) \geq \tau \cdot \max[\frac{\sum_{1 \leq i \leq s} \gamma_i}{s}, N] + \frac{\sum_{1 \leq i \leq s} \text{TSP}(G_d(o_i))}{s}$. Therefore, the competitive ratio of CUTTING is: $\frac{\text{makespan}_{\text{CUTTING}}}{\text{makespan}_{\text{OPT}}} = s \cdot \log^2 m \log^2 n \cdot \frac{\tau \cdot C + \text{ATSP}_A}{\tau \cdot \sum_{1 \leq i \leq s} \gamma_i + \sum_{1 \leq i \leq s} \text{TSP}(G_d(o_i))}$. Note that $C \leq \sum_{1 \leq i \leq s} \gamma_i$ and $\sum_{1 \leq i \leq s} \text{TSP}(G_d(o_i)) \geq \text{TSP}(G_d)$. The theorem follows.

4   Conclusions

CUTTING provides an efficient average-case competitive ratio. This is the first such result for the design of contention management algorithms for DTM. The algorithm requires that each transaction be aware of its requested set of objects when it starts. This is essential in our algorithms, since each transaction can send requests to objects simultaneously after it starts. If we remove this restriction, the original results do not hold, since a transaction can only send the request of an object once after the previous operation is done. This increases the resulting makespan by a factor of $\Omega(s)$.

Acknowledgement. This work is supported in part by US National Science Foundation under grants CNS-1116190.

References

1. Attiya, H., Epstein, L., Shachnai, H., Tamir, T.: Transactional contention management as a non-clairvoyant scheduling problem. In: PODC, pp. 308–315 (2006)
2. Christofides, N.: Worst case analysis of a new heuristic for the traveling salesman problem. Technical Report CS-93-13, G.S.I.A., Carnegie Mellon University, Pittsburgh, USA (1976)
3. Diegues, N.L., Romano, P.: Bumper: Sheltering Transactions from Conflicts. In: SRDS, pp. 185–194 (2013)
4. Diegues, N.L., Romano, P.: Time-warp: lightweight abort minimization in transactional memory. In: PPOPP, pp. 167–178 (2014)
5. Guerraoui, R., Herlihy, M., Pochon, B.: Toward a theory of transactional contention managers. In: PODC, pp. 258–264 (2005)
6. Herlihy, M., Sun, Y.: Distributed transactional memory for metric-space networks. Distributed Computing 20(3), 195–208 (2007)
7. Hirve, S., Palmieri, R., Ravindran, B.: HiperTM: High Performance, Fault-Tolerant Transactional Memory. In: Chatterjee, M., Cao, J.-N., Kothapalli, K., Rajsbaum, S. (eds.) ICDCN 2014. LNCS, vol. 8314, pp. 181–196. Springer, Heidelberg (2014)
8. Kaplan, H., Lewenstein, M., Shafrir, N., Sviridenko, M.: Approximation algorithms for asymmetric TSP by decomposing directed regular multigraphs. J. ACM 52, 602–626 (2005)
9. Karp, R.M.: Reducibility Among Combinatorial Problems. In: Miller, R.E., Thatcher, J.W. (eds.) Complexity of Computer Computations, pp. 85–103 (1972)
10. Khot, S.: Improved Inaproximability Results for MaxClique, Chromatic Number and Approximate Graph Coloring. In: FOCS, pp. 600–609 (2001)
11. Kim, J., Palmieri, R., Ravindran, B.: Enhancing Concurrency in Distributed Transactional Memory through Commutativity. In: Wolf, F., Mohr, B., an Mey, D. (eds.) Euro-Par 2013. LNCS, vol. 8097, pp. 150–161. Springer, Heidelberg (2013)
12. Palmieri, R., Quaglia, F., Romano, P.: OSARE: Opportunistic Speculation in Actively REplicated Transactional Systems. In: SRDS, pp. 59–64 (2011)
13. Romano, P., Palmieri, R., Quaglia, F., Carvalho, N., Rodrigues, L.: An Optimal Speculative Transactional Replication Protocol. In: ISPA, pp. 449–457 (2010)

14. Romano, P., Palmieri, R., Quaglia, F., Carvalho, N., Rodrigues, L.: Brief announcement: on speculative replication of transactional systems. In: SPAA, pp. 69–71 (2010)

15. Saad, M.M., Ravindran, B.: HyFlow: a high performance distributed software transactional memory framework. In: HPDC, pp. 265–266 (2011)

16. Schneider, J., Wattenhofer, R.: Bounds on Contention Management Algorithms. In: Dong, Y., Du, D.-Z., Ibarra, O. (eds.) ISAAC 2009. LNCS, vol. 5878, pp. 441–451. Springer, Heidelberg (2009)

17. Sharma, G., Estrade, B., Busch, C.: Window-Based Greedy Contention Management for Transactional Memory. In: Lynch, N.A., Shvartsman, A.A. (eds.) DISC 2010. LNCS, vol. 6343, pp. 64–78. Springer, Heidelberg (2010)

18. Shavit, N., Touitou, D.: Software Transactional Memory. In: PODC, pp. 204–213 (1995)

19. Siek, K., Wojciechowski, P.T.: Brief announcement: towards a fully-articulated pessimistic distributed transactional memory. In: SPAA, pp. 111–114 (2013)

20. Turcu, A., Ravindran, B., Palmieri, R.: Hyflow2: a high performance distributed transactional memory framework in scala. In: PPPJ, pp. 79–88 (2013)

21. Zhang, B., Ravindran, B.: Dynamic analysis of the relay cache-coherence protocol for distributed transactional memory. In: IPDPS, pp. 1–11 (2010)

The Complexity Gap between Consensus and Safe-Consensus

(Extended Abstract)*

Rodolfo Conde and Sergio Rajsbaum

Instituto de Matemáticas, Universidad Nacional Autónoma de México
Ciudad Universitaria, México D.F. 04510, México
aragorn@ciencias.unam.mx,
rajsbaum@im.unam.mx

Abstract. In the *consensus* task each process proposes a value, and all correct processes have to decide the same value. In addition, *validity* requires that the decided value is a proposed value. Afek, Gafni and Lieber (DISC'09) introduced the *safe-consensus* task, by weakening the validity requirement: if the first process to invoke the task returns before any other process invokes it, then it outputs its input; otherwise, when there is concurrency, the consensus output can be arbitrary, not even the input of any process. Surprisingly, they showed that safe-consensus is equivalent to consensus, in a system where any number of processes can crash (e.g., wait-free).

We show that safe-consensus is nevertheless a much weaker communication primitive, in the sense that any wait-free implementation of consensus requires $\binom{n}{2}$ safe-consensus black-boxes, and this bound is tight. The lower bound proof uses connectivity arguments based on subgraphs of *Johnson graphs*. For the upper bound protocol that we present, we introduce the *g-2coalitions-consensus* task, which may be of independent interest. We work in an iterated model of computation, where the processes repeatedly: write their information to a (fresh) shared array, invoke safe-consensus boxes and snapshot the contents of the shared array.

Keywords: Consensus, safe-consensus, coalition, Johnson graph, connectivity, distributed algorithms, lower bounds, wait-free computing, iterated models.

1 Introduction

The ability to agree on a common decision is key to distributed computing. The most widely studied agreement abstraction is *consensus*. In the consensus task each process proposes a value, and all correct processes have to decide the same value. In addition, *validity* requires that the decided value is a proposed value.

Herlihy's seminal paper [21] examined the power of different synchronization primitives for *wait-free computation*, e.g., when computation completes in a finite

---

* Partially supported by PAPIIT-UNAM IN107714.

number of steps by a process, regardless of how fast or slow other processes run, and even if some of them halt permanently. He showed that consensus is a universal primitive, in the sense that a solution to consensus (with read/write registers) can be used to implement any synchronization primitive in a wait-free manner. Also, consensus cannot be wait-free implemented from read/write registers alone [17,26]; indeed, all modern shared-memory multiprocessors provide some form of universal primitive.

Afek, Gafni and Lieber [2] introduced *safe-consensus*, which seemed to be a synchronization primitive much weaker than consensus. The validity requirement becomes: if the first process to invoke the task returns before any other process invokes it, then it outputs its input; otherwise, when there is concurrency, the consensus output can be arbitrary, not even the input of any process. In any case, all processes must agree on the same output value. Trivially, consensus implements safe-consensus. Surprisingly, they proved that the converse is also true, by presenting a wait-free implementation of consensus using safe-consensus black-boxes and read/write registers. Why is it then, that safe-consensus seems a much weaker synchronization primitive?

Our Results. We show that while consensus and safe-consensus are wait-free equivalent, any wait-free implementation of consensus for n processes requires $\binom{n}{2}$ safe-consensus black-boxes, and this bound is tight.

Our main result is the lower bound. It uses connectivity arguments based on subgraphs of *Johnson graphs*, and an intricate combinatorial and bivalency argument, that yields a detailed bound on how many safe-consensus objects of each type (fan-in) are used by the implementation protocol. For the upper bound, we present a simple protocol, based on the new *g-2coalitions-consensus* task, which may be of independent interest.

We work in an iterated model of computation [28], where the processes repeatedly: write their information to a (fresh) shared array, invoke (fresh) safe-consensus boxes and snapshot the contents of the shared array.

Related Work. Distributed computing theory has been concerned from early on with understanding the relative power of synchronization primitives. The wait-free context is the basis to study other failure models e.g. [7], and there is a characterization of the wait-free, read/write solvable tasks [25]. For instance, the weakening of consensus, *set agreement*, where n processes may agree on at most $n - 1$ different input values, is still not wait-free solvable [8,25,31] with read/write registers only. The *renaming* task where n processes have to agree on at most $2n - 1$ names has also been studied in detail e.g. [4,11,12,13,14].

Iterated models e.g. [10,23,24,28,29,30] facilitate impossibility results, and (although more restrictive) facilitate the analysis of protocols [19]. We follow in this paper the approach of [20] that used an iterated model to prove the separation result that set agreement can implement renaming but not vice-versa, and expect our result can be extended to a general model using simulations, as was done in [18] for that separation result. For an overview of the use of topology to study computability, including the use of iterated models and simulations see [22].

Afek, Gafni and Lieber [2] presented a wait-free protocol that implements consensus using $\binom{n}{2}$ safe-consensus black-boxes (and read/write registers). Since our implementation uses the weaker, iterated form of shared-memory, it is easier to prove correct. Safe-consensus was used in [2] to show that the g-tight-group-renaming task [3] is as powerful as g-consensus.

The idea of the classical consensus impossibility result [17,26] is (roughly speaking) that the executions of a protocol in such a system can be represented by a graph which is always connected. The connectivity invariance has been proved in many papers using the critical state argument introduced in [17], or sometimes using a layered analysis as in [27]. Connectivity can be used also to prove time lower bounds e.g. [5,16,27]. We extend here the layered analysis to prove a lower bound on the number of objects needed to implement consensus.

In a previous work [15] we had already studied an iterated model extended with the power of safe-consensus. However, that model had the restriction that in each iteration, all processes invoke the same safe-consensus object. We showed that set agreement can be implemented, but not consensus. The imposibility proof uses much simpler connectivity arguments than those of this paper.

The paper is organized as follows. Section 2 describes the model of computation. Section 3 contains the protocol that solves n-process consensus using $\binom{n}{2}$ safe-consensus boxes. In Section 4, we present the lower bound proof for three processes. It illustrates some of the main ideas of the general case. For lack of space the general case, as well as some proofs have been deferred to the full version of the paper.

2   Model and Task Definitions

Our model is an extension of the standard iterated version [10] of the usual read/write shared memory model e.g. [6]. There are $n \geqslant 2$ processes $\Pi = \{p_1, \ldots, p_n\}$, which execute in an asynchronous *wait-free* setting, i.e., any number of processes may crash.

A one-shot snapshot object S is a shared memory array which provides two atomic operations:

- S.update(v): when called by process p_j, it writes the value v to the register $S[j]$.
- S.scan(): returns a copy of the whole shared memory array S.

Each operation of S can be used by a process at most once. It is proven in [1,9] that snapshot objects can be wait-free implemented using only read/write shared memory registers.

The Iterated Model. In this model, the processes can use two kinds of communication media. The first is a shared memory that is structured as an infinite array $SM[i]$ $(i \geqslant 0)$ of snapshot objects; the second medium is an infinite array $T[i]$ of shared objects. The processes communicate between them through the snapshot objects and the shared objects of T, in an asynchronous and round-based pattern.

The general form of the protocols in the iterated model is given in the pseudocode of Figure 1. All the variables r, sm, val, $input$ and dec are local to process p_i and only when we analyze a protocol, we add a subindex i to a variable to specify it is local to p_i. Initially, r is zero and sm is assigned the contents of the readonly variable $input$, which contains the input value for process p_i; all other variables are initialized to $\perp$. In each round, p_i increments by one the loop counter r, accesses the current shared memory array $SM[r]$, writing all the information it has stored in sm and val (full information) and then p_i decides which shared object it is going to invoke by executing a deterministic function h that returns an index l, then p_i invokes the shared object $T[l]$ with some value v. Then, p_i takes a snapshot of the shared array, and finally, p_i checks if dec is equal to $\perp$, if so, it executes a deterministic function δ to determine if it may $decide$ a valid output value or $\perp$. Notice that in each round of a protocol, each process invokes at most one shared object of the array T.

Definition of Consensus and Safe-Consensus Tasks. The tasks of interest in this paper are the *consensus* and *safe-consensus* [2] tasks.

Consensus. Every process starts with some initial input value taken from the set $\{0, 1\}$ and must output a value such that:

 - Termination: Each process must eventually output some value.
 - Agreement: All processes output the same value.
 - Validity: If some process outputs v, then v is the initial input of some process.

Safe-consensus. Every process starts with some initial input value taken from a set I and must output a value such that Termination and Agreement are satisfied, and:

 - Safe-Validity: If a process p_i starts executing the task and outputs before any other process starts executing the task, then its decision is its own proposed input value. Otherwise, if two or more processes access the safe-consensus task concurrently, then any decision value is valid.

The safe-consensus task [2] is the result of weakening the validity condition of consensus.

From now on, we work exclusively in the iterated model, where the shared objects invoked by the processes solve safe-consensus. We assume that the input values that the processes feed to the safe-consensus objects are their own ids (without loss of generality).

3 Solving Consensus with Safe-Consensus

In this section, we argue that there exists an iterated protocol that solves the consensus task using precisely $\binom{n}{2}$ safe-consensus objects. The complete specification of such protocol will be included in the full version of the paper.

```
(1)  init r ← 0; sm ← input; dec ← ⊥; val ← ⊥;

(2)  loop forever
(3)        r ← r + 1;
(4)        SM[r].update(sm, val);
(5)        val ← T[h(⟨r, id, sm, val⟩)].exec(v);
(6)        sm ← SM[r].scan();

(7)        if (dec = ⊥) then
(8)            dec ← δ(sm, val);
(9)        end if
(10) end loop
```

Fig. 1. General form of a protocol in the iterated model

A simple way to describe the protocol that solves consensus is by seeing it as a protocol in which the processes use a set of $\binom{n}{2}$ shared objects which represent an intermediate task which can be implemented using one snapshot object and one safe-consensus object. This task is our new g-2coalitions-consensus task. It can be defined (roughly) as follows:

2Coalitions-consensus. We have g processes $p_1, \ldots, p_g$ and each one starts with some initial input value of the form $x = \langle v_1, v_2 \rangle$, where $v_i \in I \cup \{\bot\}$ such that $v_1 \neq \bot$ or $v_2 \neq \bot$. Let $x.left$ denote the value v_1 and $x.right$ the value v_2. if $x_1, \ldots, x_g$ are the input values of all processes, then it must hold that for all i, j such that $x_i.left \neq \bot$ and $x_j.left \neq \bot$, then $x_i.left = x_j.left$. A similar rule must hold if $x_i.right \neq \bot$ and $x_j.right \neq \bot$. Also, there must exists a unique process with input value $\langle v, \bot \rangle$ with $v \neq \bot$ and process p_g must be the only process with input value $x_g = \langle \bot, v' \rangle$, where $v' \neq \bot$. Each process must output a value such that Termination and Agreement are satisfied, and:

- 2coalitions-Validity: If some process outputs v, then there must exists a process p_j with input x_j such that $x_j = \langle v, u \rangle$ or $x_j = \langle u, v \rangle$ with $v \in I$.

Using the task g-2coalitions-consensus, the protocol can be described graphically as shown in Figure 2, for the case of $n = 4$. In each round of the protocol, some processes invoke a 2coalitions-consensus object, represented by the symbol $2CC_i$. In round one, p_1 and p_2 invoke the object $2CC_1$ with input values $\langle v_1, \bot \rangle$ and $\langle \bot, v_2 \rangle$ respectively, (where v_i is the initial input value of process p_i) and the consensus output u_1 of $2CC_1$ is stored by p_1 and p_2 in some local variables. In round two, p_2 and p_3 invoke the $2CC_2$ object with inputs $\langle v_2, \bot \rangle$ and $\langle \bot, v_3 \rangle$ respectively and they keep the output value u_2 in local variables. Round three is executed by p_3 and p_4 in a similar way, to obtain the consensus value u_3 from the 2coalition-consensus object $2CC_3$. At the beginning of round four, p_1, p_2 and p_3 gather the values u_1, u_2 obtained from the objects $2CC_1$ and $2CC_2$ to invoke the $2CC_4$ 2coalition-consensus object with the input values $\langle u_1, \bot \rangle, \langle u_1, u_2 \rangle$ and

$\langle \perp, u_2 \rangle$ respectively (Notice that p_2 uses a tuple with both values u_1 and u_2) and they obtain a consensus value u_4. Similar actions are taken by the processes p_2, p_3 and p_4 in round five with the shared object $2CC_5$ and the values u_2, u_3 to compute an unique value u_5. Finally, in round six, all processes invoke the last shared object $2CC_6$, with the respective input tuples

$$\langle u_4, \perp \rangle, \langle u_4, u_5 \rangle, \langle u_4, u_5 \rangle, \langle \perp, u_5 \rangle,$$

and the shared object returns to all processes a unique output value u, which is the decided output value of all processes, thus this is the final consensus of the processes.

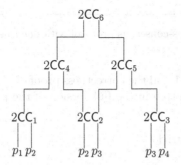

Fig. 2. The structure of the 4-consensus protocol using 2coalitions-consensus tasks

The protocol of Figure 3 implements g-2coalitions-consensus. Each process p_i receives as input a tuple with values satisfying the properties of the 2coalitions-consensus task and then in lines 3-5, p_i writes its input tuple in shared memory using the snapshot object SM; invokes the safe-consensus object with its id as input, storing the unique output value u of the shared object in the local variable val and finally, p_i takes a snapshot of the memory. Later, what happens in Lines 6-10 depends on the output value u of the safe-consensus object. If $u = g$, then by the Safe-Validity property, either p_g invoked the object or at least two processes invoked the safe-consensus object concurrently and as there is only one process with input tuple $\langle v, \perp \rangle$, p_i will find an index j with $sm[j].right \neq \perp$ in line 7, assign this value to dec and in line 11 p_i decides. On the other hand, if $u \neq g$, then again by the Safe-Validity condition of the safe-consensus task, either process p_u is running and invoked the safe-consensus object or two or more processes invoked concurrently the shared object and because all processes with id not equal to g have input tuple $\langle z, y \rangle$ with $z \neq \perp$, it is guaranteed that p_i can find an index j with $sm[j].left \neq \perp$ and assign this value to dec to finally execute line 11 to decide its output value. All processes decide the same value because of the properties of the input tuples of the 2coalitions-consensus task and the Agreement property of the safe-consensus task.

Theorem 1. *There exists an iterated protocol that solves the consensus task for n processes using $\binom{n}{2}$ safe-consensus objects.*

```
(1)   procedure g-2coalitions-consensus(v₁, v₂)
(2)   begin
(3)       SM.update(⟨v₁, v₂⟩);
(4)       val ← safe-consensus.exec(id);
(5)       sm   ← SM.scan();
(6)       if val = g then
(7)           dec ← choose any sm[j].right ≠ ⊥;
(8)       else
(9)           dec ← choose any sm[j].left ≠ ⊥;
(10)      end if
(11)      decide dec;
(12) end
```

Fig. 3. A g-2coalitions-consensus protocol with one safe-consensus object

The proof of Theorem 1 and the correctness proof of the 2coalitions-consensus protocol of Figure 3 are given in the full version of the paper.

4   The Lower Bound

Our main result is a matching lower bound on the number of safe-consensus objects needed to solve consensus. For lack of space, we present here only the case of three processes, and defer the general case to the full version of the paper.

Further Model Terminology. We need some additional definitions. Let Π be a set of n processes, and $\mathcal{A}$ a protocol in the iterated model with safe-consensus objects (see Figure 1). A *local state* s_i of a process $p_i \in \Pi$ is defined by the contents of its local variables. An *initial local state* of p_i is a local state in which all its local variables are set to $\bot$, except the *input$_i$* variable, which contains the input of p_i. An *output local state* is a local state in which the local variable dec contains a non-$\bot$ value.

For any $n \geqslant 1$, define $\bar{n} = \{1, \ldots, n\}$ and for $m \in \bar{n}$, let $V_{n,m} = \{c \subseteq \bar{n} \mid |c| = m\}$. Given the protocol $\mathcal{A}$ and an execution, we define for each $m \leqslant n$ the set $\Gamma_{\mathcal{A}}(n, m) \subseteq 2^{\bar{n}}$ as follows: $b = \{i_1, \ldots, i_m\} \in \Gamma_{\mathcal{A}}(n, m)$ if and only if in some iteration of the protocol $\mathcal{A}$, the processes $p_{i_1}, \ldots, p_{i_m}$ invoke the same safe-consensus object. For example, if $m = 3$ and $c = \{i, j, k\} \in \Gamma_{\mathcal{A}}(n, 3)$, then in at least one round, processes p_i, p_j and p_k invoke a safe-consensus object. If in other iteration these processes invoke together another safe-consensus object, then these two invocations are represented by the same set $c \in \Gamma_{\mathcal{A}}(n, 3)$. On the other hand, if $d = \{i, j, l\} \notin \Gamma_{\mathcal{A}}(n, 3)$, then there does not exist an execution of $\mathcal{A}$ in which only the three processes p_i, p_j and p_l invoke a safe-consensus shared object. For the consensus protocol of Section 3, we have that for $n = 4$, $\Gamma_{\mathcal{A}}(4, 2) = \{\{1, 2\}, \{2, 3\}, \{3, 4\}\}$, $\Gamma_{\mathcal{A}}(4, 3) = \{\{1, 2, 3\}, \{2, 3, 4\}\}$ and $\Gamma_{\mathcal{A}}(4, 4) = \{\{1, 2, 3, 4\}\}$. A set $b \in \Gamma_{\mathcal{A}}(n, m)$ is called a m-*box* or simply a *box*. For notation consistency an element $d \in \Gamma_A(n,1)$ is called a *trivial box*, it represents a safe-consensus object invoked only by one process, which of course does not give any additional information to the process. A trivial box is used to model a process that does not invoke a safe-consensus object in a round. Let $\Gamma_A(n) = \bigcup_{m=2}^{n} \Gamma_A(n,m)$, $\nu_A(n,m) = |\Gamma_A(n,m)|$ and $\nu_A(n) = \sum_{m=2}^{n} \nu_A(n,m)$.

Our lower bound says that any iterated consensus protocol A using safe-consensus, must satisfy the inequality $\nu_A(n) \geqslant \binom{n}{2}$ for all $n \geqslant 2$. Moreover, A also satisfies the inequalities $\nu_A(n,m) > n - m$ for all $n \geqslant 2$ and $2 \leqslant m \leqslant n$.

An *event* is performed by a process p_i, which applies one of the following actions: an update (W), a scan (R), or an invocation to a safe-consensus (S) object. Any of these operations may be preceded/followed by some local computation. It is convenient to consider events performed concurrently. If $p_{i_1}, \ldots, p_{i_k}$ are processes, then we denote the fact that $p_{i_1}, \ldots, p_{i_k}$ execute concurrently the event E by $E(X)$ where $X = \{i_1, \ldots, i_k\}$, and E may be W, R or S. A *round schedule* is a finite sequence of the form

$$E_1(X_1), \ldots, E_r(X_r),$$

that encodes the way in which processes with ids in the set $X_1 \cup \cdots \cup X_r$ take the steps represented by the events $E_1, \ldots, E_r$. For example, in the round schedule

$$W(1,3), S(1,3), R(1,3), W(2), S(2), R(2),$$

processes p_1, p_3 perform an update, invoke safe-consensus (not necessarily the same object) and execute scan, each one concurrently and in the given order; after that, p_2 executes solo the same events in the same order.

A *global state (at the end of an iteration)* is a vector[1] $S = \langle s_1, \ldots, s_n; SM; b_1, \ldots, b_q; o_1, \ldots, o_q \rangle$, where s_i is the local state of process p_i, SM is the state of the shared memory, $b_1, \ldots, b_q \in \Gamma_A(n) \cup \Gamma_A(n,1)$ are the boxes that specify the way in which the processes invoked the safe-consensus shared objects to enter the local states s_i's and for $j = 1, \ldots, q$, o_j is the output value of the safe-consensus shared object represented by the box b_j. An *initial state* is a global state in which every local state is an initial local state, all registers in the shared memory are set to $\perp$, the set of boxes is empty and the set of output values of shared objects is empty. A *decision state* is a global state in which all local states are output states. A global state Q is a *successor* of the global state S if and only if there is an execution of A starting from the state S, and after executing A a finite number of rounds, the global state is Q. If π is any round schedule and S is a global state, the successor of S obtained by running A (starting in the state S) one iteration with the round schedule π is denoted by $S \cdot \pi$. When referring to a global state S, we usually omit the word global and simply refer to S as a state.

Two states S, P are said to be *adjacent* if there exists a non-empty subset $X \subseteq \overline{n}$ such that all processes with ids in X have the same local state in both S and P. That is, for each $i \in X$, p_i cannot *distinguish* between S and P. We

---

[1] Although the elements $b_1, \ldots, b_q, o_1, \ldots, o_q$ can be obtained from the local states $s_1, \ldots, s_n$, it is convenient to write them explicitly in the definition of S.

denote this by $S \overset{X}{\sim} P$. Clearly, two adjacent states belong to the same iteration. S and P are *connected*, if we can find a *path* of states $\mathfrak{p} \colon S = P_1 \sim \cdots \sim P_r = P$, such that for all j with $1 \leqslant j \leqslant r - 1$, P_j and P_{j+1} are adjacent.

For disjoint sets $A_1, \ldots, A_q \subset \bar{n}$, the round schedule $\xi(A_1, \ldots, A_q, Y)$ is:

$$W(A_1), S(A_1), R(A_1), \ldots, W(A_q), S(A_q), R(A_q), W(Y), S(Y), R(Y), \quad (1)$$

where $Y = \bar{n} - (\bigcup_{i=1}^{q} A_i)$. Sometimes, if there is no confusion, we omit the set Y and just write $\xi(A_1, \ldots, A_q)$. For any state S and $u \geqslant 0$, define

$$S \cdot \xi^u(A_1, \ldots, A_q) = \begin{cases} S & \text{if } u = 0, \\ (S \cdot \xi^{u-1}(A_1, \ldots, A_q)) \cdot \xi(A_1, \ldots, A_q) & \text{otherwise.} \end{cases}$$

I.e. $S \cdot \xi^u(A_1, \ldots, A_q)$ is the state that we obtain after we run the protocol $\mathcal{A}$ (starting from S) u rounds with the round schedule $\xi(A_1, \ldots, A_q)$ in each iteration.

The Connectivity of Iterated Protocols with Safe-Consensus. We recall some classical definitions regarding consensus protocols: If S is a state, we say that S is *v-valent* if there is an execution starting from S, where a process outputs v. S is *univalent* if in every execution starting from S, processes output the same value. If S is not univalent, then S is *bivalent*. By definition, any state where all processes have the same input v is v-univalent.

Roughly, a typical consensus impossibility proof shows that a protocol $\mathcal{A}$ cannot solve consensus because there exist one execution in which processes decide a consensus value v, and a second execution where the consensus output of the processes is v', with $v \neq v'$, such that the global states of these executions are connected [15,17,25,26]. Any protocol that solves consensus, must be able to prevent the existence of such paths of connected states.

Lemma 1. *Consider a protocol that satisfies the agreement and termination properties of consensus. Let I, J be any two initial states. If for all rounds $r \geqslant 0$, I^r, J^r are connected successor states of I and J respectively, then I and J are v-valent for the same value v.*

Thus, any protocol that has two initial states, one 0-univalent and one 1-univalent, and satisfies Lemma 1, cannot solve consensus.

Our main result is the following.

Theorem 2. *If $\mathcal{A}$ is an iterated protocol for n-consensus using safe-consensus objects, then for every $m \in \{2, \ldots, n\}$, $\nu_{\mathcal{A}}(n, m) > n - m$.*

To prove the theorem by contradiction, suppose that $\mathcal{A}$ solves consensus and for some $m_0 \in \{2, \ldots, n\}$, it is true that $\nu_{\mathcal{A}}(n, m_0) \leqslant n - m_0$, i.e., at most $n - m_0$ subsets of processes of size m_0 can invoke safe-consensus shared objects in the protocol $\mathcal{A}$.

The Lower Bound for 3-process Consensus Protocols. We show here that if $\mathcal{A}$ is a protocol which solves consensus for three processes, then $\mathcal{A}$ must satisfy the inequalities $\nu_{\mathcal{A}}(3, m) > 3 - m$ for $m \in \{2, 3\}$. We investigate what happens if $\mathcal{A}$ does not satisfy the given inequalities for some m_0. There are two cases to consider.

Case $m_0 = 2$. Assume that $\mathcal{A}$ is such that $\nu_{\mathcal{A}}(3, 2) \leqslant 3 - 2 = 1$, that is, at most two fixed processes can invoke together safe-consensus shared objects. We use a simple combinatorial result.

Lemma 2. *Let $U \subset V_{n,2}$ with $|U| \leqslant n - 2$. Then there exists a partition $\overline{n} = A \cup B$ such that*

$$(\forall b \in U)(b \subseteq A \text{ or } b \subseteq B). \tag{2}$$

The previous result can be proven using subgraphs of *Johnson graphs*. Lemma 2 will be used to prove Lemma 5, which is a structural result that we use to construct a bivalency argument to show the lower bound in the case of $m_0 = 2$ in the proof of Theorem 2.

Lemma 3. *Let S be a state of $\mathcal{A}$ in some round $r \geqslant 0$ and $\overline{n} = A \cup B$ a partition of $\overline{n}$ such that*

$$(\forall b \in \Gamma_{\mathcal{A}}(3, 2))(b \subseteq A \text{ or } b \subseteq B). \tag{3}$$

Then there exists a path $\mathfrak{p} \colon S \cdot \xi(A) \overset{B}{\sim} S \cdot \xi(\overline{3}) \overset{A}{\sim} S \cdot \xi(B)$ of connected states in round $r + 1$ of $\mathcal{A}$.

Proof. To build the path $\mathfrak{p}$, it is enough to show that there exists the possibility that the output values of the safe-consensus shared objects invoked by the processes are the same in the three states $S \cdot \xi(A), S \cdot \xi(\overline{3})$ and $S \cdot \xi(B)$. We have cases, according to the way in which the processes invoke the safe-consensus objects in round $r + 1$.

> *Case a).* If each process invokes solo a safe-consensus object, then by the Safe-Validity property of safe-consensus, each process receives its own id as output value from the shared object it invokes, so that all processes see the same output values from the safe-consensus objects in the states $S \cdot \xi(A), S \cdot \xi(\overline{3})$ and $S \cdot \xi(B)$.
>
> *Case b).* Suppose that processes p_i, p_j invoke a safe-consensus object and p_k invokes solo another safe-consensus object. This fact is represented by the 2-box $b_1 = \{i, j\}$ and the trivial box $b_2 = \{k\}$. By the Safe-Validity property of safe-consensus, p_k always receives as output value from the shared object represented by b_2 its own id, thus p_k sees the same safe-consensus value in the three states. Now, as $b_1 \in \Gamma_{\mathcal{A}}(3, 2)$, by Equation (3), we know that $b_1 \subseteq A$ or $b_1 \subseteq B$, so that in each state of $\mathfrak{p}$, processes p_i, p_j invoke concurrently the safe-consensus object represented by b_1 and by the Safe-Validity property, the returned value of the safe-consensus can be arbitrary. Thus there exists executions of $\mathcal{A}$ in which we can make the safe-consensus object represented by b_1 output the same value in the three states $S \cdot \xi(A), S \cdot \xi(\overline{3})$ and $S \cdot \xi(B)$.

Case c). Now suppose that all three processes invoke the same safe-consensus object, which is represented by the 3-box $b = \overline{3}$. Because $\overline{3} = A \cup B$ and $A \cap B = \varnothing$, it must be true that $|b \cap A| = 2$ or $|b \cap B| = 2$. Without loss of generality, assume that $|b \cap A| = 2$, then $|b \cap B| = 1$ and by the Safe-Validity property, the output value of the shared object represented by b must be k in the state $S \cdot \xi(B)$, and in the states $S \cdot \xi(A), S \cdot \xi(\overline{3})$, the output value can be arbitrary (because in these two states, at least two processes are invoking concurrently the safe-consensus object represented by b). Therefore there exists executions of $\mathcal{A}$ in which the output value of the safe-consensus object is k in the three states $S \cdot \xi(A), S \cdot \xi(\overline{3})$ and $S \cdot \xi(B)$. It follows that the path $\mathfrak{p}$ exists.

Lemma 4. *Suppose that for the protocol $\mathcal{A}$ there exist a partition $\overline{n} = A \cup B$ satisfying Equation 3 and a sequence $\mathfrak{p} \colon S_0 \overset{X_1}{\sim} \cdots \overset{X_l}{\sim} S_l$ of connected states in round $r \geqslant 0$ of $\mathcal{A}$, such that $X_i = A$ or $X_i = B$ for all $i \in \{1, \ldots, l\}$. Then in round $r + 1$ of $\mathcal{A}$ there exists a path $\mathfrak{q} \colon Q_0 \overset{Y_1}{\sim} \cdots \overset{Y_s}{\sim} Q_s$ of connected states and the following properties hold:*

I) Each state Q_k is of the form $Q_k = S_j \cdot \xi(X)$, where $X = A$ or $X = B$;
II) $(\forall j \in \{1, \ldots, s\})(Y_j = A$ or $Y_j = B)$.

Proof. To find the path $\mathfrak{q}$ satisfying I) and II), we use induction on l. In the base case $l = 1$, $\mathfrak{p} \colon S_0 \overset{X_1}{\sim} S_1$ with $X_1 = A$ or $X_1 = B$. It is easy to see that the path $S_0 \cdot \xi(X_1) \overset{X_1}{\sim} S_1 \cdot \xi(X_1)$ fulfills conditions I), II). For the induction hypothesis, suppose that for the path $S_0 \overset{X_1}{\sim} \cdots \overset{X_{l'}}{\sim} S_{l'}$ with $1 \leqslant l' < l$, we have build the path $\mathfrak{q}' \colon Q_1 \overset{Y_1}{\sim} \cdots \overset{Y_{s'}}{\sim} Q_{s'}$ satisfying I) and II) of the conclusion of the Lemma. We now show how to connect $Q_{s'}$ with a successor state of $S_{l'+1}$. Let $X_{l'+1}$ be the set of processes that cannot distinguish between $S_{l'}$ and $S_{l'+1}$. By the induction hypothesis, $Q_{s'} = S_{l'} \cdot \xi(X)$, where $X = A$ or $X = B$. We have cases.

Case $X = X_{l'+1}$. In this case we use the small path $S_{l'} \cdot \xi(X) \overset{X}{\sim} S_{l'+1} \cdot \xi(X)$.
Case $X \neq X_{l'+1}$. Without loss of generality, assume that $X = A$ and $X_{l'+1} = B$. We apply Lemma 3 to obtain the path $S_{l'} \cdot \xi(A) \overset{B}{\sim} S_{l'} \cdot \xi(\overline{3}) \overset{A}{\sim} S_{l'} \cdot \xi(B)$. Combining this path with the path $S_{l'} \cdot \xi(B) \overset{B}{\sim} S_{l'+1} \cdot \xi(B)$, we are done.

We have by induction the sequence of connected states $Q_1, \ldots, Q_s$ from the sequence $S_1, \ldots, S_l$ satisfying the required properties, and the result follows.

Lemma 5. *If $\mathcal{A}$ is an iterated protocol for three processes using safe-consensus objects with $\nu_A(3, 2) \leqslant 1$ and I is an initial state in $\mathcal{A}$, then there exists a partition of the set $\overline{3} = A \cup B$ such that for all $u \geqslant 1$, the states $I \cdot \xi^u(A)$ and $I \cdot \xi^u(B)$ are connected.*

Proof. We can apply Lemma 2 to the set $\Gamma_A(3, 2) \subset V_{3,2}$ to find the partition of $\overline{3}$ and then we use induction combined with Lemma 4. We omit the details.

Case $m_0 = 3$. The last case to consider is when the three processes cannot invoke the same safe-consensus shared object together. To prove this case, we need one structural result, regarding paths of connected states in an iterated protocol. With this result, we can build a bivalency argument to prove the lower bound for the case of $m_0 = 3$ in the proof of Theorem 2.

Lemma 6. *Suppose that $\mathcal{A}$ is a protocol with safe-consensus objects for three processes such that $\nu_{\mathcal{A}}(3,3) = 0$. If S, Q are two initial states connected by a path $\mathsf{q}_0 \colon S \sim \cdots \sim Q$, then for all $u \geqslant 0$, there exist successor states S^u, Q^u of S and Q respectively, in round u of $\mathcal{A}$, such that S^u and Q^u are connected.*

Proof. Let $\mathcal{A}$ be a protocol for three processes with $\nu_{\mathcal{A}}(3,3) = 0$. We use induction on the round number u. For the base case $u = 0$, by hypothesis we have the path q_0 which fulfills the conclusion of the lemma.

For the induction hypothesis, assume that for $u \geqslant 0$, we have the path $\mathsf{q}_u \colon S_0 \overset{X_1}{\sim} \cdots \overset{X_q}{\sim} S_q$, where $S^u = S_0$ and $Q^u = S_q$. To build the path $\mathsf{q}_{u+1} \colon S^{u+1} \sim \cdots \sim Q^{u+1}$, connecting S^{u+1} and Q^{u+1}, successor states of S^u and Q^u respectively, we proceed by induction on q. In the base case $q = 1$, we have that q_u is the path $S_0 \overset{X_1}{\sim} S_1$, here we easily build the path $S_0 \cdot \xi(X_1) \overset{X_1}{\sim} S_1 \cdot \xi(X_1)$. Suppose that for $1 \leqslant l < q$, we have build the path $\mathsf{q}' \colon R_1 \sim \cdots \sim R_s$, where R_1 is a successor state of S_1 and $R_s = S_l \cdot \xi(X)$ is a successor states of S_l. We now wish to connect R_s (a successor state of S_l) with a successor state of S_{l+1}. Let X_{l+1} be the set of processes which cannot distinguish between S_l and S_{l+1}. As $\nu_{\mathcal{A}}(3,3) = 0$, In any execution of $\mathcal{A}$, the three processes cannot invoke the same safe-consensus shared object, thus in round $u + 1$ of $\mathcal{A}$, they must invoke the safe-consensus objects in one of the following two possibilities:

- Each process invokes a safe-consensus object solo.
- Two processes invoke a safe-consensus object and the third process invokes solo another shared object.

If the processes invoke three separate safe-consensus objects, then we build the following path from $R_s = S_l \cdot \xi(X)$ to $S_{l+1} \cdot \xi(X_{l+1})$

$$S_l \cdot \xi(X) \overset{\overline{3}-X}{\sim} S_l \cdot \xi(\overline{3}) \overset{\overline{3}-X_{l+1}}{\sim} S_l \cdot \xi(X_{l+1}) \overset{X_{l+1}}{\sim} S_{l+1} \cdot \xi(X_{l+1}). \qquad (4)$$

As each process sees its own id as the output value of the safe-consensus object it invokes, then the only way that a process can distinguish between two states, is by means of the contents of the shared memory. Therefore the path given in Equation (4) exists.

In case that two processes invoke a safe-consensus object, represented by the 2-box $b \in \Gamma_{\mathcal{A}}(3,2)$ and the other process invokes solo an object represented by the trivial box c, then we build a path from $R_s = S_l \cdot \xi(X)$ to $S_{l+1} \cdot \xi(X_{l+1})$ as follows: First notice that the two states $S_l \cdot \xi(X)$ and $S_l \cdot \xi(\overline{3})$ are indistinguishable for the processes with ids in the set $\overline{3} - X$, this is because we can find executions of $\mathcal{A}$ in which the safe-consensus values of the objects represented by b and c are the same in the two previous states (Safe-Validity), proving this is an easy case analysis. Now, we need to connect the state $S_l \cdot \xi(\overline{3})$ with the state $S_l \cdot \xi(X_{l+1})$. We have subcases on the size of the set X_{l+1}.

Case $|X_{l+1}| = 1$. If $X_{l+1} = c$, then we have that $S_l \cdot \xi(\overline{3}) \overset{b}{\sim} S_l \cdot \xi(X_{l+1})$, because by the Safe-Validity property of safe-consensus, we can find executions of $\mathcal{A}$ in which the output value of the safe-consensus object represented by b is the same in the states $S_l \cdot \xi(\overline{3})$ and $S_l \cdot \xi(X_{l+1})$, thus all processes with ids in b cannot distinguish between these two states. On the other hand, if $X_{l+1} \neq c$, then $X_{l+1} \subset b$ and we have the path $S_l \cdot \xi(\overline{3}) \overset{c}{\sim} S_l \cdot \xi(X_{l+1})$.

Case $|X_{l+1}| = 2$. If $X_{l+1} = b$, then we claim that, as in the last part of the previous case, $S_l \cdot \xi(\overline{3}) \overset{c}{\sim} S_l \cdot \xi(X_{l+1})$. When $X_{l+1} \neq b$, it must be true that $X_{l+1} = \{j\} \cup c$, where $j \in b$. The path that we need to build here is

$$S_l \cdot \xi(\overline{3}) \overset{c}{\sim} S_l \cdot \xi(\{j\}) \overset{b}{\sim} S_l \cdot \xi(\{j\}, c) \overset{c}{\sim} S_l \cdot \xi(\{j\} \cup c).$$

The arguments to prove that this path exists, are very similar to previous arguments, using the Safe-Validity property of the safe-consensus task. This finishes the cases to connect the states $S_l \cdot \xi(\overline{3})$ and $S_l \cdot \xi(X_{l+1})$.

Finally, we connect the states $S_l \cdot \xi(X_{l+1})$ and $S_{l+1} \cdot \xi(X_{l+1})$ with the small path $S_l \cdot \xi(X_{l+1}) \overset{X_{l+1}}{\sim} S_{l+1} \cdot \xi(X_{l+1})$. Thus, we have connected the state $R_s = S_l \cdot \xi(X)$ with a successor state of S_{l+1} and this completes the proof of the induction step. By induction on q, we have build the path q_{u+1} from the path q_u, satisfying the conclusion of the lemma.

Therefore we have proven that given the path q_u, connecting the states S^u and Q^u, we can build a new path q_{u+1}, connecting successor states of S^u and Q^u respectively, so that by induction on u, the result is valid for all $u \geqslant 0$. This finishes the proof.

To complete the proof of Theorem 2, we use all the previous results as follows.

Proof of Theorem 2. (Case $n = 3$) Assume that there is some $m_0 \in \{2, 3\}$ such that $\nu_{\mathcal{A}}(3, m_0) \leqslant 3 - m_0$. Let O, U be the initial states in which all processes have as input values 0s and 1s respectively. We now find successor states of O and U in each round $r \geqslant 0$, which are connected. We have two cases:

Case $m_0 = 2$. By Lemma 5, there exists a partition of $\overline{3} = A \cup B$ such that for any state S and any $r \geqslant 0$, $S \cdot \xi^r(A)$ and $S \cdot \xi^r(B)$ are connected. Let OU be the initial state in which all processes with ids in A have as input value 0s and all processes with ids in B have as input values 1s. Then for all $r \geqslant 0$

$$O \cdot \xi^r(A) \overset{A}{\sim} OU \cdot \xi^r(A) \quad \text{and} \quad OU \cdot \xi^r(B) \overset{B}{\sim} U \cdot \xi^r(B)$$

and by Lemma 5, the states $OU \cdot \xi^r(A)$ and $OU \cdot \xi^r(B)$ are connected. Thus, for any r, we can connect the states $O^r = O \cdot \xi^r(A)$ and $U^r = U \cdot \xi^r(B)$.

Case $m_0 = 3$. It is known that any two initial states for consensus are connected [17], so that we can connect O and U with a sequence q of initial states of $\mathcal{A}$, By Lemma 6, for each round $r \geqslant 0$ of $\mathcal{A}$, there exist successor states O^r, U^r of O and U respectively, such that O^r and U^r are connected.

In this way, we have connected successor states of O and U in each round of the protocol $\mathcal{A}$. Now, O is a 0-valent, initial state, which is connected to the initial state U, so that we can apply Lemma 1 to conclude that U is 0-valent. But this contradicts the fact that U is a 1-valent state, so we have reached a contradiction. Therefore $\nu_{\mathcal{A}}(3, m) > 3 - m$ for $m = 2, 3$.

References

1. Afek, Y., Attiya, H., Dolev, D., Gafni, E., Merritt, M., Shavit, N.: Atomic snapshots of shared memory. J. ACM 40(4), 873–890 (1993)
2. Afek, Y., Gafni, E., Lieber, O.: Tight group renaming on groups of size g is equivalent to g-consensus. In: Keidar, I. (ed.) DISC 2009. LNCS, vol. 5805, pp. 111–126. Springer, Heidelberg (2009)
3. Afek, Y., Gamzu, I., Levy, I., Merritt, M., Taubenfeld, G.: Group renaming. In: Baker, T.P., Bui, A., Tixeuil, S. (eds.) OPODIS 2008. LNCS, vol. 5401, pp. 58–72. Springer, Heidelberg (2008)
4. Attiya, H., Bar-Noy, A., Dolev, D., Peleg, D., Reischuk, R.: Renaming in an Asynchronous Environment. Journal of the ACM (July 1990)
5. Attiya, H., Dwork, C., Lynch, N., Stockmeyer, L.: Bounds on the time to reach agreement in the presence of timing uncertainty. J. ACM 41, 122–152 (1994), http://dx.doi.org/10.1145/174644.174649
6. Attiya, H., Welch, J.: Distributed Computing: Fundamentals, Simulations and Advanced Topics. John Wiley & Sons (2004)
7. Borowsky, E., Gafni, E., Lynch, N., Rajsbaum, S.: The bg distributed simulation algorithm. Distrib. Comput. 14(3), 127–146 (2001)
8. Borowsky, E., Gafni, E.: Generalized flp impossibility result for t-resilient asynchronous computations. In: STOC 1993: Proceedings of the Twenty-fifth Annual ACM Symposium on Theory of Computing, pp. 91–100. ACM, New York (1993)
9. Borowsky, E., Gafni, E.: Immediate atomic snapshots and fast renaming. In: PODC 1993: Proceedings of the Twelfth Annual ACM Symposium on Principles of Distributed Computing, pp. 41–51. ACM, New York (1993)
10. Borowsky, E., Gafni, E.: A simple algorithmically reasoned characterization of wait-free computation (extended abstract). In: PODC 1997: Proceedings of the Sixteenth Annual ACM Symposium on Principles of Distributed Computing, pp. 189–198. ACM, New York (1997)
11. Castañeda, A., Herlihy, M., Rajsbaum, S.: An equivariance theorem with applications to renaming. In: Fernández-Baca, D. (ed.) LATIN 2012. LNCS, vol. 7256, pp. 133–144. Springer, Heidelberg (2012), http://dx.doi.org/10.1007/978-3-642-29344-3_12
12. Castañeda, A., Rajsbaum, S.: New combinatorial topology upper and lower bounds for renaming. In: Proceedings of the Twenty-seventh ACM Symposium on Principls of Distributed Computing, PODC 2008, pp. 295–304. ACM, New York (2008), http://doi.acm.org/10.1145/1400751.1400791
13. Castañeda, A., Rajsbaum, S.: New combinatorial topology bounds for renaming: The upper bound. J. ACM 59(1), 3:1–3:49 (2012), http://doi.acm.org/10.1145/2108242.2108245
14. Castañeda, A., Rajsbaum, S., Raynal, M.: The renaming problem in shared memory systems: An introduction. Computer Science Review 5(3), 229–251 (2011), http://www.sciencedirect.com/science/article/pii/S1574013711000116

15. Conde, R., Rajsbaum, S.: An introduction to the topological theory of distributed computing with safe-consensus. Electronic Notes in Theoretical Computer Science 283, 29–51 (2012), http://www.sciencedirect.com/science/article/pii/S1571066112000059, proceedings of the workshop on Geometric and Topological Methods in Computer Science (GETCO)

16. Dwork, C., Moses, Y.: Knowledge and common knowledge in a byzantine environment: Crash failures. Information and Computation 88(2), 156–186 (1990), http://www.sciencedirect.com/science/article/pii/0890540190900149

17. Fischer, M.J., Lynch, N.A., Paterson, M.S.: Impossibility of distributed consensus with one faulty process. J. ACM 32(2), 374–382 (1985)

18. Gafni, E., Rajsbaum, S.: Distributed programming with tasks. In: Lu, C., Masuzawa, T., Mosbah, M. (eds.) OPODIS 2010. LNCS, vol. 6490, pp. 205–218. Springer, Heidelberg (2010), http://dx.doi.org/10.1007/978-3-642-17653-1_17

19. Gafni, E., Rajsbaum, S.: Recursion in distributed computing. In: Dolev, S., Cobb, J., Fischer, M., Yung, M. (eds.) SSS 2010. LNCS, vol. 6366, pp. 362–376. Springer, Heidelberg (2010), http://dx.doi.org/10.1007/978-3-642-16023-3_30

20. Gafni, E., Rajsbaum, S., Herlihy, M.: Subconsensus tasks: Renaming is weaker than set agreement. In: Dolev, S. (ed.) DISC 2006. LNCS, vol. 4167, pp. 329–338. Springer, Heidelberg (2006)

21. Herlihy, M.: Wait-free synchronization. ACM Trans. Program. Lang. Syst. 13(1), 124–149 (1991), http://dx.doi.org/10.1145/114005.102808

22. Herlihy, M., Kozlov, D., Rajsbaum, S.: Distributed Computing Through Combinatorial Topology. Morgan Kaufmann (2013), http://store.elsevier.com/Distributed-Computing-Through-Combinatorial-Topology/Maurice-Herlihy/isbn-9780124045781/

23. Herlihy, M., Rajsbaum, S.: The topology of shared-memory adversaries. In: PODC 2010: Proceeding of the 29th ACM Symposium on Principles of Distributed Computing, pp. 105–113. ACM, New York (2010)

24. Herlihy, M., Rajsbaum, S.: The topology of distributed adversaries. Distributed Computing 26(3), 173–192 (2013), http://dx.doi.org/10.1007/s00446-013-0189-9

25. Herlihy, M., Shavit, N.: The topological structure of asynchronous computability. J. ACM 46(6), 858–923 (1999)

26. Loui, M.C., Abu-Amara, H.H.: Memory requirements for agreement among unreliable asynchronous processes. In: Preparata, F.P. (ed.) Parallel and Distributed Computing, Advances in Computing Research, pp. 163–183. JAI Press, Greenwich (1987)

27. Moses, Y., Rajsbaum, S.: A layered analysis of consensus. SIAM J. Comput. 31(4), 989–1021 (2002)

28. Rajsbaum, S.: Iterated shared memory models. In: López-Ortiz, A. (ed.) LATIN 2010. LNCS, vol. 6034, pp. 407–416. Springer, Heidelberg (2010)

29. Rajsbaum, S., Raynal, M., Travers, C.: An impossibility about failure detectors in the iterated immediate snapshot model. Inf. Process. Lett. 108(3), 160–164 (2008)

30. Rajsbaum, S., Raynal, M., Travers, C.: The iterated restricted immediate snapshot model. In: Hu, X., Wang, J. (eds.) COCOON 2008. LNCS, vol. 5092, pp. 487–497. Springer, Heidelberg (2008)

31. Saks, M., Zaharoglou, F.: Wait-free k-set agreement is impossible: The topology of public knowledge. SIAM J. Comput. 29(5), 1449–1483 (2000)

The Simultaneous Number-in-Hand Communication Model for Networks: Private Coins, Public Coins and Determinism[*]

Florent Becker[1], Pedro Montealegre[1], Ivan Rapaport[2,3], and Ioan Todinca[1]

[1] Univ. Orléans, INSA Centre Val de Loire, LIFO EA 4022, Orléans, France
[2] Departamento de Ingeniería Matemática, Univ. de Chile, Chile
[3] Centro de Modelamiento Matemático (UMI 2807 CNRS), Univ. de Chile, Chile

Abstract. We study the multiparty communication model where players are the nodes of a network and each of these players knows his/her own identifier together with the identifiers of his/her neighbors. The players simultaneously send a unique message to a referee who must decide a graph property. The goal of this article is to separate, from the point of view of message size complexity, three different settings: deterministic protocols, randomized protocols with private coins and randomized protocols with public coins. For this purpose we introduce the boolean function TWINS. This boolean function returns 1 if and only if there are two nodes with the same neighborhood.

1 Introduction

In the *number-in-hand* multiparty communication model there are k players. Each of these k players receives an n-bit input string x_i and they all need to collaborate in order to compute some function $f(x_1, \ldots, x_k)$. Despite its simplicity, the case $k > 2$ started to be studied very recently [1, 2, 4, 6–8, 13, 14].

There are different communication modes for the *number-in-hand* model. In this paper we focus on the *simultaneous message* communication mode, in which all players simultaneously send a unique message to a referee. The referee collects the messages and computes the function f. The computational power of both the players and the referee is unlimited. When designing a protocol for function f, the goal is to minimize the size of the longest message generated by the protocol. This minimum, usually depending on n, is called the *message size complexity* of f. Typical questions in communication complexity consist in designing protocols with small messages, and proving lower bounds on the size of such messages.

Several authors considered the case where the data distributed among the players is a graph [1, 4, 13, 14]. Informally, each player knows a set of edges of the graph and together they must decide a graph property, e.g., connectivity. Again

---

[*] This work has been partially supported by CONICYT via Basal in Applied Mathematics (I.R.), Núcleo Milenio Información y Coordinación en Redes ICM/FI P10-024F (I.R.) and Fondecyt 1130061 (I.R.)

M. Halldórsson (Ed.): SIROCCO 2014, LNCS 8576, pp. 83–95, 2014.
© Springer International Publishing Switzerland 2014

we can observe two different settings. In one of them, the edges are distributed among the players in an adversarial way [1, 14]. In this work, following [1, 4], we consider the setting where each player corresponds to a node of the graph, and thus each player knows the identifier of this node together with the identifiers of its neighbors, represented as an n-bits vector (in the vector x_i of player i, the bit number j is set to 1 if and only if the nodes i and j are adjacent). For the sake of simplicity we assume that the graph has n nodes numbered from 1 to n, hence there are $k = n$ players, and we call this model *number-in-hand for networks*.

For many natural functions the messages are much shorter when randomization is allowed [12]. In the randomized setting, there are significant differences between the communication complexities of protocols using *public coins* (shared by all players and the referee) and the more restrictive setting where each player has his own, *private coin*. We emphasize that in the *number-in-hand communication model for networks*, each edge is "known" by two players, thus we have some shared information. Not surprisingly, as pointed out in [14], this model is stronger than the one where edges are distributed in an adversarial way among players.

Related Work

The number-in-hand model with simultaneous messages and $k = 2$ players.
The case of two players is not new and it has been intensively studied. Clear separations have been proved between deterministic, private coins and public coins protocols in this case. For instance, the message size complexity of the EQ function, which simply tests whether the two n-bit inputs are equal, is $\Theta(n)$ for deterministic protocols [12], $\mathcal{O}(1)$ for randomized protocols with public coins with constant one-sided error [3], and $\Theta(\sqrt{n})$ for randomized protocols with private coins and constant one-sided error [3] (see Section 2 for details). More generally, Babai and Kimmel [3] proved that for any function f its randomized message size complexity, for private coins protocols, is at least the square root of its deterministic message size complexity. Chakrabarti *et al.* [5] proved that, for some family of functions, the gap between deterministic and randomized message size complexity with private coins is smaller that the square root.

The number-in-hand communication model for networks.
For deterministic protocols, Becker *et al.* [4] show that graphs of bounded degeneracy can be completely reconstructed by the referee using messages of size $\mathcal{O}(\log n)$, and several natural problems like deciding whether the graph has a triangle, or if its diameter is at most 3, have message size complexity of $\Theta(n)$. For randomized protocols with public coins, Ahn, Guha and McGregor [1, 2] introduced a beautiful and powerful technique for *graph sketching*. The technique works both for streaming models and for the *number-in-hand for networks*, and allows to solve CONNECTIVITY using messages of size $\mathcal{O}(\log^2 n)$. The protocols have two-sided, $\mathcal{O}(1/n^c)$ error, for any constant $c > 0$.

Our Results. In this paper we separate the deterministic, the randomized with private coins and the randomized with public coins settings of the *number-in-hand* *for networks* communication model. The separations are made using problem TWINS and some variants. The boolean function $\text{TWINS}(G)$ returns 1 if and only if graph G has two twins (that is, two nodes having the same neighborhood). We also consider function $\text{TWIN}_x(G)$, where x is the identifier of a node, and the result is 1 if and only if there is some other node having the same neighborhood as x.

We prove that the deterministic message size complexity of TWINS and TWIN_x is $\Theta(n)$. Also, both functions can be computed by randomized protocols with public coins and message size $\mathcal{O}(\log n)$. These protocols, based on the classical fingerprint technique, have one-sided error $\mathcal{O}(1/n^c)$ for any constant $c > 0$. Observe that the situation for private coins is very different from the case of the *number-in-hand* model with two players, where the gap between private coins and determinism is at most the square root.

In order to separate the private and public coins settings we use a boolean function called TRANSLATED-TWINS (see Section 2 for details). We prove that the message size complexity of this function in the private coins setting is $\Omega(\sqrt{n})$, while it is $\mathcal{O}(\log n)$ in the public coins setting. The main results of this paper are summarized in Table 1.

	TWINS	TWIN_x	TRANSLATED-TWINS
Deterministic	$\Theta(n)$	$\Theta(n)$	$\Theta(n)$
Randomized private-coins	$\mathcal{O}(\sqrt{n}\log n)$	$\mathcal{O}(\log n)$	$\Omega(\sqrt{n})$, $\mathcal{O}(\sqrt{n}\log n)$
Randomized public-coins	$\mathcal{O}(\log n)$	$\mathcal{O}(\log n)$	$\mathcal{O}(\log n)$

There are several natural problems that cannot be solved with randomized protocols using $o(n)$ bits. In the last part of this paper (Theorem 5) we sketch how the arguments of [4], for proving negative results on deterministic protocols, can be extended to the randomized setting. More precisely, we prove that the randomized public coin message size complexity of the boolean functions $\text{TRIANGLE}(G)$ (that outputs 1 if and only if G has a triangle) and $\text{DIAM3}(G)$ (that outputs 1 if and only if G has diameter at most 3) is $\Omega(n)$.

2 Preliminaries

Number-in-Hand. The *number-in-hand* communication model is defined as follows. Let f be a function having as input k boolean vectors of length n. There are k players $\{p_1, \ldots, p_k\}$ who wish to compute the value of f on input $(x_1, \ldots, x_k) \in (\{0,1\}^n)^k$. Player p_i only sees the input x_i, and also knows his own number i. We only consider here the *simultaneous messages* communication mode, in which all the k players simultaneously send a message to a *referee*. After that, the referee (another player who sees none of the inputs) announces the value $f(x_1, \ldots, x_k)$ using only the information contained in the k messages.

A *deterministic protocol* $\mathcal{P}$ for function f describes the algorithms of the players (for constructing the messages) and of the referee (for retrieving the final result) that correctly computes f on all inputs. An ϵ-*error randomized protocol*

$\mathcal{P}$ for f is a protocol in which every player and the referee are allowed to use a sequence of random bits, and for all $(x_1, \ldots, x_k) \in (\{0,1\}^n)^k$ the referee outputs $f(x_1, \ldots, x_k)$ with probability at least $1 - \epsilon$. For boolean functions f we define a *one-sided ϵ-error randomized protocol* in the same way, with exception that for all $(x_1, \ldots, x_k) \in (\{0,1\}^n)^k$ such that $f(x_1, \ldots, x_k) = 1$, the referee always outputs 1.

We distinguish between two sub-cases of randomized protocols: (i) the *private-coin* setting, in which each player, including the referee, flips private coins and (ii) the *public-coin* setting, where the coins are shared between players, but the referee can still have his own private coins.

The *cost* of a protocol $\mathcal{P}$, denoted $C(\mathcal{P})$, is the length of the longest message sent to the referee. The *deterministic message size complexity*, denoted $C^{\det}(f)$, is the minimum cost of any deterministic protocol computing f. Analogously, we denote $C_\epsilon^{\text{priv}}(f)$, $C_\epsilon^{\text{pub}}(f)$, as the message size complexity for ϵ-error public and private protocols, respectively.

Number-in-Hand for Networks. *Number-in-hand for networks* is a particular case of *number-in-hand* where each party is a node of an n-vertex graph with vertices numbered from 1 to n. Therefore, in this model, $k = n$, player p_i corresponds to the node i and the inputs $x_1, \ldots, x_n$ correspond to the rows of the adjacency matrix of some simple undirected graph G of size n. Hence, the input of player (node) i is the characteristic function of the neighborhood $N_G(i)$ (i.e. $j \in N_G(i)$ if and only if $ij \in E(G)$).

All our graphs are undirected, so for any pair i, j of nodes, the bit number i of player j equals the bit number j of player i. In full words, each edge of the graph is known by the two players corresponding to its end-nodes. All our protocols use $\Omega(\log n)$ bits. We assume, w.l.o.g., that each node sends its own number in the message transmitted to the referee.

Known Results. Let us recall some classical results of the *number-in-hand* model with two players. Babai and Kimmel [3] have shown that the order of magnitude of the private-coins randomized message size complexity of any function f is at least the square root of the deterministic message size complexity of f. They also characterize completely the function: $\text{EQ} : \{0,1\}^n \times \{0,1\}^n \to \{0,1\}$, where $\text{EQ}(x, y) = 1$ iff $x = y$.

Proposition 1 ([3]). *Consider the number-in-hand model with two players and a constant $\epsilon > 0$. The function* EQ *on two n-bit boolean vectors has the following message size complexities: $C^{\det}(\text{EQ}) = n$, $C_\epsilon^{\text{priv}}(\text{EQ}) = \Theta(\sqrt{n})$ and $C_\epsilon^{\text{pub}}(\text{EQ}) = \mathcal{O}(1)$. For any boolean function f, $C_\epsilon^{\text{priv}}(f) = \Omega(\sqrt{C^{\det}(f)})$.*

We also use the following result of Chakrabarti et al. [5] for private coins protocol; the deterministic part is a matter of exercise.

Proposition 2 ([5]). *Consider the boolean function* OREQ *that takes as input two boolean $n \times n$ matrices, and the output is 1 if and only if there is some $1 \le i \le n$ such that the i-th lines of the two matrices are equal. Then, for any $\epsilon < 1/2$, $C_\epsilon^{\text{priv}}(\text{OREQ}) = \Omega(n\sqrt{n})$. Also, $C^{\det}(\text{OREQ}) = \Theta(n^2)$.*

The Problems. We now come back to the *number-in-hand for networks* model. In this framework we shall study three boolean functions on graphs.

- TWINS(G) outputs 1 if and only if G has two vertices u and v with the same neighborhood, i.e., such that $N(u) = N(v)$.
- TWINS$_x(G)$ is a "pointed" version of previous function. Its output is 1 if and only if there is a vertex y such that $N(y) = N(x)$.
- TRANSLATED-TWINS is defined on input graphs G of size $2n$, labeled from 1 to $2n$. Its output is 1 if and only if G has a vertex i such that, for any vertex j, $j \in N(i) \iff j + n \in N(i + n)$. In other words, the output is 1 if and only if there exists i such that $N(i) + n = N(i + n)$.

For reductions we also use the function RECONSTRUCTION(G), whose output is G itself, i.e., the adjacency matrix of G. Note that if a deterministic protocol computes RECONSTRUCTION on the family of n-vertex graphs $\mathcal{G}_n$, then such protocol must generate messages of size at least $\log(|\mathcal{G}|)/n$ (see also [4]).

3 Deterministic Protocols

Theorem 1. *The deterministic message size complexity of functions* TWINS, TWIN$_x$ *and* TRANSLATED-TWINS *is* $\Theta(n)$.

The upper bounds of $\mathcal{O}(n)$ are trivial so we only need to prove the lower bounds. For the first two problems, we use the following reduction.

Lemma 1. *Assume that there is a deterministic protocol solving* TWINS *(resp.* TWINS$_{2n+1}$*) on* $2n + 1$*-node graphs using messages of size* $g(n)$*. Then one can solve* RECONSTRUCTION *on* n*-node graphs using messages of size* $2g(n)$*.*

Proof. Let G be an arbitrary n-nodes graph, i be an integer between 1 and n and S be a subset of $\{1, \ldots, n\}$ not containing i. Denote by $H(i, S)$ the graph on $2n + 1$ nodes obtained as follows (see Figure 1):

1. $H[\{1, \ldots, n\}] = G$.
2. For each $n + 1 \leq j \leq 2n$, its unique neighbor with identifier at most n is $j - n$.
3. Node $2n + 1$ is adjacent exactly to the nodes of S and to $i + n$.

Claim. We claim that TWINS$(H(i, S))$ (resp. TWINS$_{2n+1}(H(i, S))$ is true if and only if $N_G(i) = S$.

Clearly, if $N_G(i) = S$ then node i is a twin of $2n + 1$ in graph H. Conversely, we prove that if $H(i, S)$ has two twins u and v then one of them is $2n + 1$. This comes from the fact that the edges between $\{1, \ldots, n\}$ and $\{n + 1, \ldots, 2n\}$ in $H(i, S)$ form a matching, so no two nodes of $\{1, \ldots, 2n\}$ may be twins. Now assume that $2n+1$ has a twin u. Since $N_{H(i,S)}(2n+1) \cap \{n+1, \ldots, 2n\} = \{i+n\}$, the only possibility is that $u = i$. Eventually, i and $2n + 1$ are twins if and only if $N_G(i) = S$, which proves our claim.

Now assume that we have a distributed protocol for TWINS (or TWINS$_{2n+1}$) on graphs with $2n + 1$ nodes (actually it suffices to consider graphs from the family H described above). We construct an algorithm for RECONSTRUCTION on an arbitrary n-nodes graph G.

The players construct their messages as follows. Each node i sends the message m_i that it would send in the TWINS protocol if it had neighborhood $N_G(i) \cup \{i + n\}$ and the message m_i^+ that it would send in the same protocol with neighborhood $N_G(i) \cup \{i + n, 2n + 1\}$. That makes messages of size $2g(n)$.

The referee needs to retrieve the neighborhood $N_G(i)$ for each i, from the set of messages. For each i and each subset S of $\{1, \ldots, n\}$ not containing i, we simulate the behavior of the referee for TWINS on graph $H(i, S)$. For this purpose, for each $j \leq n$ we use message m_j if $j \notin S$ and message m_j^+ if $j \in S$. The messages for nodes $k > n$ can be constructed directly by the referee. Note that TWINS($H(i, S)$) is true iff $N_G(i) = S$, thus we can reconstruct $N_G(i)$. Eventually, this allows to solve RECONSTRUCTION on graph G. The same arguments work of we replace the TWINS protocol by TWINS$_{2n+1}$.   □

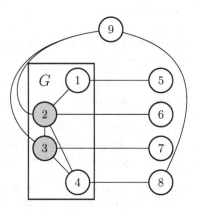

Fig. 1. $H(4, S)$, when $S = \{2, 3\}$

Remark 1. Since problem RECONSTRUCTION on n-node graphs requires messages of size $\Omega(n)$, we conclude that any deterministic protocol for either TWINS or TWINS$_{2n+1}$ also requires messages of size $\Omega(n)$.

For problem TRANSLATED-TWINS, we provide a reduction from OREQ (see Proposition 2 in Section 2). It will be used both for deterministic and randomized protocols with private coins.

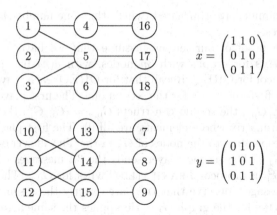

Fig. 2. Examples of graphs G_x^1 (top) and G_y^2 (bottom), for a given input (x, y). This is a *yes* instance since $x_3 = y_3$.

Lemma 2. *Assume that there is a protocol solving* TRANSLATED-TWINS *for* $6n$-node graphs using messages of size $g(n)$, in any of our three settings. Then there is a protocol for function OREQ, in the same setting, using messages of size $3ng(n)$.

Proof. Let x and y be two $n \times n$ boolean matrices. We construct a graph $G_{x,y}$ with $6n$ nodes such that TRANSLATED-TWINS$(G_{x,y}) =$ OREQ(x, y).

The graph G is formed by two connected components G_x^1 and G_y^2 of $3n$ nodes each, encoding the two matrices as follows (see Figure 2 for an example).

G_x^1 has $3n$ nodes numbered from 1 to $2n$ and from $5n + 1$ to $6n$. For any $i, j \in \{1, \ldots, n\}$ we put an edge between node i and node $j + n$ if and only if $x_{i,j} = 1$. Then for any $i \in \{1, \ldots, n\}$ we put an edge between node $i + n$ and node $i + 4n$. In other words, the node subsets $\{1, \ldots, n\}$ and $\{n + 1, \ldots, 2n\}$ induce a bipartite graph representing matrix x, and the node subsets $\{n + 1, \ldots, 2n\}$ and $\{5n + 1, \ldots, 6n\}$ induce a perfect matching.

The construction of G_y^2, with nodes numbered from $2n + 1$ to $5n$ is similar. For any $i, j \in \{1, \ldots, n\}$ we put an edge between node $i + 3n$ and node $j + 4n$ if and only if $y_{i,j} = 1$. Also, for any $i \in \{1, \ldots, n\}$, we put an edge between node $4n + i$ and node $2n + i$. Thus the node subsets $\{3n + 1, \ldots, 4n\}$ and $\{4n + 1, \ldots, 5n\}$ form a bipartite graph corresponding to matrix y. The subsets $\{4n + 1, \ldots, 5n\}$ and $\{2n + 1, \ldots, 3n\}$ induce a matching.

We claim that TRANSLATED-TWINS$(G_{x,y}) =$ OREQ(x, y). Assume that OREQ$(x, y) = 1$. There is an index i such that line number i in x equals line number i in y. Then, by construction, the neighborhood of node $i + 3n$ in $G_{x,y}$ is the neighborhood of node i, translated by an additive term $3n$.

Conversely, assume that there is some node $u \in \{1, \ldots, 3n\}$ such that the neighborhood of u is the translated neighborhood of $u + 3n$. By construction, the only possibility is that $u \leq n$ (because of the numberings of the matchings the other nodes cannot have translated twins), thus line number u is the same in the two matrices.

To achieve the proof of our lemma, assume that we have a protocol for TRANSLATED-TWINS for graphs with $3n$ nodes, with $g(n)$ bits per message. We design a protocol for OREQ. Recall that for OREQ, each player has a matrix, say x for the first one and y for the second one. The first player constructs graph $G_{x,0} = (G_x^1, G_0^2)$, the second constructs $G_{0,y} = (G_0^1, G_y^2)$ (here 0 denotes the $n \times n$ boolean matrix whose elements are all 0). The first player sends the $3n$ messages corresponding to the nodes of G_x^1 in the TRANSLATED-TWINS protocol for graph $G_{x,0}$. The second player sends the $3n$ messages corresponding to the nodes of G_y^2 in protocol TRANSLATED-TWINS for $G_{0,y}$. The referee collects these $6n$ messages; observe that they are exactly those sent by protocol TRANSLATED-TWINS for the graph $G_{x,y}$. He applies the same algorithm as the referee of TRANSLATED-TWINS on these messages. By the claim above, its output is TRANSLATED-TWINS$(G_{x,y})$, thus OREQ(x,y). Note that the messages used here are of size $\mathcal{O}(3ng(n))$ and that our arguments hold for any type of protocol. $\qquad\square$

This achieves the proof of Theorem 1.

4   Randomized Protocols

Theorem 2. *For any constant $c > 0$, TWINS, TWINS$_x$ and TRANSLATED-TWINS can be solved by randomized protocols with public coins using messages of size $O(\log n)$ and $1/n^c$ one-sided error. Problem TWINS$_x$ can also be solved by a randomized protocol with private coins using messages of size $O(\log n)$ and $1/n^c$ one-sided error.*

Proof. Let $n^{c+3} < p \leq 2n^{c+3}$ be a prime number. A random $t \in \mathbb{Z}_p$ is chosen uniformly at random using $\mathcal{O}(\log(n))$ public random bits. Given an n-bits vector $a = (a_1, \ldots, a_n)$, consider the polynomial $P_a = a_1 + a_2 X + a_3 X^2 + \ldots a_n X^{n-1}$ in $Z_p[X]$ and let $FP(a,t) = P_a(t)$. $FP(a,t)$ is sometimes called the "fingerprint" of vector a. Clearly two equal vectors have equal fingerprints, and, more important, for any two different vectors a and b, the probability that $FP(a,t) = FP(b,t)$ is at most $1/n^{c+2}$ (because the polynomial $P_a - P_b$ has at most n roots and t was chosen uniformly at random, thus the probability that t is a root of $P_a - P_b$ is at most $1/n^{c+2}$, see e.g., [11]).

Let x_i be the input vector of player (node) number i, i.e., the characteristic function of its neighborhood $N(i)$. A protocol for TWINS consists in each node sending the message $m_i = FP(x_i, t)$. The referee outputs 1 if and only if $m_i = m_j$ for some pair $i \neq j$. A protocol for TWINS$_x$ send the same messages, but this time the referee checks whether $m_x = m_i$ for some $i \neq x$. The protocol for TRANSLATED-TWINS on n-node graphs is slightly different. If a node $i \leq n/2$ has a neighbor $j > n/2$, it sends a special "no" message specifying that it cannot be a candidate for having a translated twin. Otherwise, let y_i^1 be the $n/2$-bits vector formed by the $n/2$ first bits of x_i. Thus y_i^1 is the characteristic vector of

$N(i) \cap \{1, \dots, n/2\}$. Player i sends the message $m_i = FP(y_i^1, t)$. Symmetrically, for nodes labelled $i > n/2$, if i has some neighbor $j \leq n/2$ it sends the "no" message. Otherwise, let y_i^2 be the $n/2$-bits vector formed by the last $n/2$ bits of x_i. Hence y_i^2 corresponds to $N(i) \cap \{n/2, \dots, n\}$, "translated" by $-n/2$. Player i sends the message $m_i = FP(y_i^2, t)$. Then the referee returns 1 if $m_i = m_{i+n/2}$ for some $i \leq n/2$.

Clearly, for protocol TWINS (resp. TWINS$_x$, TRANSLATED-TWINS), if the input graph is a yes-instance then the protocol outputs 1. The probability that TWINS answers 1 on a no-instance is the probability that $FP(m_i, t) = FP(m_j, t)$ for two nodes i and j with different neighborhoods. For each fixed pair of nodes this probability is at most $1/n^{c+2}$, so altogether the probability of a wrong answer is at most $1/n^c$. With similar arguments for TWINS$_x$ and TRANSLATED-TWINS the probability of a wrong answer is at most $1/n^{c+1}$, since the referee makes n tests and each may be a false positive with probability at most $1/n^{c+2}$.

For TWINS$_x$ with private coins, each node i sends a bit stating if it sees x, a number t_i chosen uniformly at random in the interval $n^{c+2} < p \leq 2n^{c+2}$ and also $FP(x_i, t_i)$. The referee retrieves the neighborhood of node x (which was sent bit by bit by all the others) and then, for each $i \neq x$, it constructs $FP(x_x, t_i)$ and compares it to $FP(x_i, t_i)$. If the values are equal for some i, the referee outputs 1, otherwise it outputs 0. Again any yes-instance will answer 1, and the probability that a no-instance (wrongly) answers 1 is at most $1/n^c$.   □

The fact that TRANSLATED-TWINS requires $\Omega(\sqrt{n})$ bits per node for any private coins, ϵ-error randomized protocol follows directly by Lemma 2 and Proposition 2.

Theorem 3. *For any $\epsilon < 1/2$, $C_\epsilon^{\mathrm{priv}}(\text{TRANSLATED-TWINS}) = \Omega(\sqrt{n})$.*

Theorems 2 and 3 show that problem TRANSLATED-TWINS separates the private coins and the public coins protocols.

In order to complete the table of the Introduction, we also observe that problems TWINS$_x$ and TRANSLATED-TWINS can be solved by randomized private coins protocols using $\mathcal{O}(\sqrt{n} \log n)$ bits.

Theorem 4. *For any $c > 0$, there is a randomized private coins protocol for TWINS and TRANSLATED-TWINS using messages of size $\mathcal{O}(\sqrt{n} \log n)$ and having $1/n^c$ one-sided error.*

Proof. Babai and Kimmel in [3] propose a private coins protocol with $1/3$ one sided error and $\mathcal{O}(\sqrt{n})$ communication cost for EQ_n, in the *number-in-hand* model with two players (see Proposition 1). Let us call this protocol $\mathcal{P}_0$. As the authors point out, this protocol is symmetrical, in the sense that both players compute the same function on their own input. We define the protocol $\mathcal{P}$ as one obtained by simulating $(c + 2) \log_3 n$ calls to protocol $\mathcal{P}_0$. More formally, in $\mathcal{P}$ each player creates $(c + 2) \log_3 n$ times the message that it would create in $\mathcal{P}_0$, using at each time independent tosses of private coins. The referee answers 1 if and only if the referee of $\mathcal{P}_0$ would have answered 1 on each of the $(c + 2) \log_3 n$

pairs of messages. Therefore $\mathcal{P}$ is a private coin randomized protocol for EQ_n with one sided error smaller than $1/n^{c+2}$, and cost $\mathcal{O}(\sqrt{n}\log n)$.

A one sided private coin randomized protocol $\mathcal{P}'$ for TWINS is one where each node plays the role of Alice in $\mathcal{P}$ taking as an input the characteristic function of its neighborhood, and then the referee simulates the role of the referee in $\mathcal{P}$ for each pair of messages. Similarly, a protocol $\mathcal{P}''$ for TRANSLATED-TWINS works as follows: each node i sends "no" in the same cases described in the proof of Theorem 2, and otherwise it simulates the role of Alice on input y_i^1 formed by the first $n/2$ bits of x_i, if $i \leq n/2$ or on input y_i^2 formed by the $n/2$ last bits of x_i if $i > n/2$, where x_i is the characteristic function of $N(i)$. The referee then simulates the referee of $\mathcal{P}$ on the messages of i and $i+n$ every time none of them say "no".

Since $\mathcal{P}$ has just one sided error, if TWINS (resp. TRANSLATED-TWINS) is *true*, $\mathcal{P}'$ (resp. $\mathcal{P}''$) will always accept. On the other hand, if TWINS (resp. TRANSLATED-TWINS) is *false*, then the probability that $\mathcal{P}'$ (resp. $\mathcal{P}''$) accepts is the probability that $\mathcal{P}$ accepts for at least one pair of vertices, and then the error of $\mathcal{P}'$ (resp. $\mathcal{P}''$) is at most n^2 times (resp. n times) the error of $\mathcal{P}$. We obtain that $\mathcal{P}'$ and $\mathcal{P}''$ have at most $1/n^c$ one sided error, and communication cost $\mathcal{O}(\sqrt{n}\log n)$.   □

Consider the boolean function $\mathrm{TRIANGLE}(G)$ that outputs 1 if and only if G has a triangle, and the function $\mathrm{DIAM3}(G)$, that outputs 1 if and only if G has diameter at most 3. In [4] is shown that the deterministic message sizes of these problems are lower-bounded by $\Omega(n)$, using a reduction from RECONSTRUCTION. However, as seen in Theorem 1, a reduction from RECONSTRUCTION does not imply lower-bounds on the message sizes of randomized protocols.

In the following theorem, we extend the techniques in [4] to reduce the problems $\mathrm{TRIANGLE}(G)$ and $\mathrm{DIAM3}(G)$ from INDEX, showing that the message sizes of randomized protocols for these problems are also of size $\Omega(n)$.

Theorem 5. *For any $\epsilon < 1/2$, any public coins randomized protocol computing* $\mathrm{TRIANGLE}(G)$ *(resp.* $\mathrm{DIAM3}(G)$*) with ϵ two-sided error uses messages of size* $\Omega(n)$.

Proof. Consider the INDEX function in the model *number-in-hand* with two players: the first player, say Alice, has as input an m-bits boolean vector x and the second player, Bob, has an integer $q, 1 \leq i \leq m$. Then $\mathrm{INDEX}(x,q) = x_q$, the qth coordinate of Alice's vector. We will use the fact that for any $\epsilon < 1/2$, any public coins randomized protocol for INDEX requires $\Omega(m)$ bits (see, e.g., [9, 10] for a proof). We may assume w.l.o.g. that $m = n^2$.

In [4], Becker *et al.* show that for the deterministic communication cost for TRIANGLE and DIAM3 is $\Theta(n)$, by showing that if there is a protocol $\mathcal{P}$ of cost c for TRIANGLE or DIAM3, then there is a protocol for RECONSTRUCTION in bipartite graphs of cost $2c$. We slightly modify their proof to obtain a reduction from INDEX.

Let $\epsilon < 1/2$, and $\mathcal{P}$ be a ϵ-error randomized public coins protocol for TRI-ANGLES on n-nodes graphs, using $c(n)$ bits. We give a protocol for INDEX using $2n \cdot c(2n + 1)$ bits.

Let x be an $m = n^2$-bits vector. Let H_x be the bipartite graph with vertex set $\{1, \ldots, 2n\}$, such that for any $1 \leq k, l \leq n$, if $x_{(k-1)n+l} = 1$ then H_x has an edge between nodes k and $l + n$. Consider the family of graphs $H_x(i, j)$ obtained from H_x by adding a node $2n + 1$ whose neighbors are nodes i and $j + n$ (for any $1 \leq i, j \leq n$). Observe that $H_x(i, j)$ has a triangle if and only if $x_{(i-1)n+j} = 1$, in which case the triangle is formed by the nodes $\{i, j + n, 2n + 1\}$. To simplify the notation we also define the graph $H_x(0, 0)$ obtained from H_x by adding an isolated node $2n + 1$.

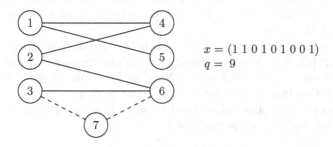

$x = (1\ 1\ 0\ 1\ 0\ 1\ 0\ 0\ 1)$
$q = 9$

Fig. 3. An illustration of $H_x(3, 6)$ when $x = (1, 1, 0, 1, 0, 1, 0, 0, 1)$ and $q = 9$

The protocol for INDEX is as follows. Bob sends its input q, which only costs $\mathcal{O}(\log n)$ bits. Alice constructs the family of graphs $H_x(i, j)$, for all pairs $1 \leq i, j \leq n$ and for $(i, j) = (0, 0)$. Any node $k \leq 2n$ has exactly two possible of neighborhoods, depending whether it is adjacent to $2n + 1$ or not. For each $k \leq 2n$, Alice creates the message $m^+(k)$ that the protocol for TRIANGLE would send for node k in the graph $H_x(k, 1)$ (if $k \leq n$) or in the graph $H(1, k - n)$ (if $k > n$). It also creates the message $m^-(k)$ that TRIANGLE would construct for node k in the graph $H_x(0, 0)$. In full words, $m^-(k)$ corresponds to the case when the neighborhood of k is the same as in H_x, and $m^+(k)$ to the case when this neighborhood is the neighborhood in H_x, plus node $2n + 1$. Then Alice sends, for each k, $1 \leq k \leq 2n$, the pair of messages $(m^-(k), m^+(k))$. Therefore Alice uses $2n \cdot c(2n + 1)$ bits. It remains to explain how the referee retrieves the bit x_q. Let i, j such that $q = (i - 1)n + j$. Observe that $x_q = 1$ if and only if graph $H_x(i, j)$ has a triangle, therefore the referee must simulate the behavior of the referee for TRIANGLE on $H_x(i, j)$. For this purpose, the referee computes the message that node $2n + 1$ would have sent on this graph (it only depends on i and j) and observes that protocol $\mathcal{P}$ on $H_x(i, j)$ would have sent message $m^+(i)$, $m^+(j + n)$ and $m^-(k)$ for any $k \leq 2n$ different from i and j. Therefore the referee can give the same output as $\mathcal{P}$ on $H_x(i, j)$, that is it outputs bit x_q. The protocol for INDEX will have ϵ error and will use $2n \cdot c(2n + 1)$ bits. Thus $\mathcal{P}$ requires $\Omega(n)$ bits.

The proof for DIAM3 is based on a similar reduction. Let $D_x(i,j)$ be the graph obtained from H_x by adding three nodes : node $2n+1$ seeing all nodes $k \leq 2n$, node $2n+2$ seeing i and node $2n+3$ seeing $j+n$. Graph $D_x(0,0)$ is similar with the difference that nodes $2n+2$ and $2n+3$. Observe (see also [4]) that $D_x(i,j)$ has diameter 3 if and only if $x_{(i-1)n+j} = 1$. The rest of the proof follows as before.   □

5   Open Problems

The first natural challenge is to determine the message size complexity of function TWINS for randomized protocols with private coins. Using the techniques of Babai and Kimmel [3] for EQ, one can prove that TWINS can be solved by a one-sided, bounded error protocol with private coins and messages of size $\mathcal{O}(\sqrt{n}\log n)$. We believe that the message size complexity of TWINS for private coins protocols is $\Omega(\sqrt{n})$.

More surprisingly, to the best of our knowledge, the message size complexity of CONNECTIVITY is wide open. Recall that, in the randomized, public coins setting, there exists a protocol using $\mathcal{O}(\log^2 n)$ bits, due to Ahn, Guha and McGregor [1]. Can this upper bound be improved to $\mathcal{O}(\log n)$? For randomized protocols with private coins and/or for deterministic protocols, can one prove a lower bound of $\Omega(n^c)$ for some constant $c < 1$?

References

1. Ahn, K.J., Guha, S., McGregor, A.: Analyzing graph structure via linear measurements. In: Proc. of the 23rd Annual ACM-SIAM Symposium on Discrete Algorithms, SODA 2012, pp. 459–467 (2012)
2. Ahn, K.J., Guha, S., McGregor, A.: Graph sketches: Sparsification, spanners, and subgraphs. In: Proc. of the 31st Symposium on Principles of Database Systems, PODS 2012, pp. 5–14 (2012)
3. Babai, L., Kimmel, P.G.: Randomized simultaneous messages: Solution of a problem of Yao in communication complexity. In: Proc. of the 12th Annual IEEE Conference on Computational Complexity, pp. 239–246 (1997)
4. Becker, F., Matamala, M., Nisse, N., Rapaport, I., Suchan, K., Todinca, I.: Adding a referee to an interconnection network: What can(not) be computed in one round. In: Proc. of the 25th IEEE International Parallel and Distributed Processing Symposium, IPDPS 2011, pp. 508–514 (2011)
5. Chakrabarti, A., Shi, Y., Wirth, A., Yao, A.: Informational complexity and the direct sum problem for simultaneous message complexity. In: Proc. of the 42nd IEEE Symposium on Foundations of Computer Science, FOCS 2001, pp. 270–278 (2001)
6. Drucker, A., Kuhn, F., Oshman, R.: The communication complexity of distributed task allocation. In: Proc. of the 2012 ACM Symposium on Principles of Distributed Computing, PODC 2012, pp. 67–76 (2012)
7. Gronemeier, A.: Asymptotically optimal lower bounds on the NIH-multi-party information complexity of the AND-function and disjointness. In: Proc. of the 26th International Symposium on Theoretical Aspects of Computer Science, STACS 2009, pp. 505–516 (2009)

8. Jayram, T.S.: Hellinger strikes back: A note on the multi-party information complexity of AND. In: Dinur, I., Jansen, K., Naor, J., Rolim, J. (eds.) APPROX and RANDOM 2009. LNCS, vol. 5687, pp. 562–573. Springer, Heidelberg (2009)
9. Kremer, I., Nisan, N., Ron, D.: On randomized one-round communication complexity. Computational Complexity 8, 21–49 (1999)
10. Kremer, I., Nisan, N., Ron, D.: Errata for: "on randomized one-round communication complexity". Computational Complexity 10, 314–315 (2001)
11. Kushilevitz, E.: Communication complexity. Advances in Computers 44, 331–360 (1997)
12. Kushilevitz, E., Nisan, N.: Communication Complexity. Cambridge University Press, New York (1997)
13. Phillips, J.M., Verbin, E., Zhang, Q.: Lower bounds for number-in-hand multiparty communication complexity, made easy. In: Proceedings of the 23rd Annual ACM-SIAM Symposium on Discrete Algorithms, SODA 2012, pp. 486–501 (2012)
14. Woodruff, D.P., Zhang, Q.: When distributed computation is communication expensive. In: Afek, Y. (ed.) DISC 2013. LNCS, vol. 8205, pp. 16–30. Springer, Heidelberg (2013)

Approximation of the Degree-Constrained Minimum Spanning Hierarchies

Miklós Molnár, Sylvain Durand, and Massinissa Merabet

University Montpellier 2, Laboratory LIRMM UMR 5506,
CC477, 161 rue Ada,
34095 Montpellier Cedex 5, France
{miklos.molnar,sylvain.durand,massinissa.merabet}@lirmm.fr

Abstract. Degree-constrained spanning problems are well known and are mainly used to solve capacity constrained routing problems. The degree-constrained spanning tree problems are NP-hard and computing the minimum cost spanning tree is not approximable. Often, applications (such as some degree-constrained communications) do not need trees as solutions. Recently, a more flexible, connected, graph related structure called hierarchy was proposed to span a set of vertices under constraints. This structure permits a new formulation of some degree-constrained spanning problems. In this paper we show that although the newly formulated problem is still NP-hard, it is approximable with a constant ratio. In the worst case, this ratio is bounded by 3/2. We provide a simple heuristic and prove its approximation ratio is the best possible for any algorithm based on a minimum spanning tree.

Keywords: Graph theory, Networks, Degree-Constrained Spanning Problem, Spanning Hierarchy, Approximation.

1   Introduction

Solving spanning problems in a cost efficient manner is important in several domains. For instance, implementing a minimum cost communication network or solving the routing in micro-circuits are classic examples for optimal spanning problems. Often in graphs, a given set of vertices should be spanned by the minimum cost structure. In the literature, solutions are mainly considered to be sub-graphs. For example, the structure which spans all the vertices in a graph with minimum cost is a minimum spanning tree (MST).

In some practical cases, different additional constraints are imposed. Various constrained spanning problems have been analyzed in graphs (cf. some examples in [1,2,3]). Here we are interested in the degree-constrained spanning problem. In this constrained spanning problem, a positive integer value $d(v)$ is assigned to each vertex $v \in V$ of an undirected graph $G = (V, E)$. This value represents the maximum degree of the vertex in the spanning structure (usually in a tree). This degree is potentially different from the degree $d_G(v)$ of v in G. Note that only values $0 < d(v) \le d_G(v)$ need to be considered for realistic cases. This degree bound can express two different facts:

M. Halldórsson (Ed.): SIROCCO 2014, LNCS 8576, pp. 96–107, 2014.
© Springer International Publishing Switzerland 2014

1. the vertex has a global "budget" to connect neighbor vertices (this budget approach can be found in [4])

2. because of its limited instantaneous "capacity", the vertex can perform a given action (a branching) for each of its visit only for a limited number of neighbor vertices.

The first case corresponds to the *degree-constrained spanning tree problem*. It has been formulated in [5] and has been extensively studied. For a long time, it is known that it is not always possible to span the vertices using trees with respect of the degree constraints. Moreover, negative results are also known on the approximability of the degree-constrained spanning tree problems [4].

In our paper we suppose that the degree bound expresses the limited capacity of the vertex for each visit (case 2). Moreover, we suppose that the limit is the same constant value, valid for all vertices in the graph.

For communications, the connectivity of the routes is inevitable but these routes can correspond to non-simple graph-related structures as walks, trails, etc. To span a set of vertices in a connected manner, a non-simple, tree-based structure has been proposed [6]. This structure, called hierarchy, is obtained by a homomorphic mapping of vertices between a tree and an arbitrary graph (cf. Section 3). A new formulation of degree-constrained problems is possible and profitable for some applications if the constraints concern each visit of the vertices. To solve these problems, it was demonstrated that

a) it is possible to span the vertices of the graph with respect of the degree bounds even if spanning trees satisfying the constraints do not exist

b) in some cases, a spanning hierarchy with lower cost can be found even if spanning trees respecting the constraints exist [7].

One possible application domain of the spanning hierarchies is the broadcast in all-optical WDM networks where the splitting capacity of the vertices is limited (for example in [8]). To solve the optical routing problem under the degree constraints, a set of light trees (abusively called light forest) is usually proposed. Let us notice that in the literature not only tree-based solutions can be found. In [9], a special walk (a light-trail) is computed to cover the vertices without branching. The spanning hierarchies give a good alternative to find efficient spanning structures generalizing walks when branching are allowed. In this paper we demonstrate that the optimum of the degree bounded spanning hierarchy problem can be approximated. We propose a simple and efficient algorithm providing a good approximation of the optimal value.

In Section 2, we propose a quick presentation of the well-known degree-constrained spanning tree problem. After the related definitions, the degree constrained minimum spanning hierarchy problem and its complexity are presented in Section 3. The algorithm proposed in Section 5 uses the result of Section 4 to span stars and computes polynomially a spanning hierarchy respecting a given degree bound. The proposed algorithm guarantees a constant approximation ratio. The presentation is closed by discussions on the performances of the algorithm and on some perspectives.

2   Related Works

The Degree-Constrained Minimum Spanning Tree DCMST problem was firstly introduced and investigated in [5] (it is also briefly mentioned in [10]). Let us suppose that the maximal degree of any vertices in the spanning tree must be at most $B \geq 2$. The authors justified the fact that this problem is NP-hard by stating that solving the DCMST problem with the degree bound B equal to two is equivalent to solve the minimum Hamiltonian path problem. Otherwise, by reducing the DCMST problem to an equivalent symmetric traveling salesman problem (TSP), Garey and Johnson [11] showed that this problem is NP-hard for any fixed constant $2 \leq B \leq |V - 1|$. Ravi showed that approximate the DCMST problem within a constant factor of the cost of the optimal tree is NP-hard [12]. In unweighted graphs, Furer and Raghavachari [13] gave an elegant algorithm that returns a spanning tree in which the degree of each vertex is at most $B + 1$, or returns a witness certifying that the degree bounds are infeasible. Goemans proved in [14] that this result can be generalized to weighted graphs. In polynomial time, we can find a spanning tree of maximum degree at most $B + 1$ whose cost is no more than the cost of a minimum cost tree with maximum degree at most B. Note that these results are formulated for homogeneous degree bound. When the degree bounds depend on the vertices, Goemans proved that one can find in polynomial time a spanning tree of maximum degree at most $B + 2$ whose cost is no more than the cost of a minimum cost tree with maximum degree at most B. The best result was presented by Singh and Lau in [15]. Their algorithm computes a spanning tree of minimum cost which violates the degree upper-bounds by at most one. Since it is not possible to obtain any approximation algorithm for the original problem, insisting on the satisfaction of all the degree upper bounds, this result is the best possible.

To solve spanning problems with different constraints, the hierarchy concept was proposed in [6].

3   Problem Definition

Our objective is to find a minimum cost spanning structure without the hypothesis that this structure must be a sub-graph. For instance, it may be an arbitrary route connecting vertices.

Let $G = (V, E)$ be an undirected connected graph with vertex set V and edge set E. The graph G is valuated by a strictly positive cost $c(e)$ associated to every edge $e \in E$. We are searching for routes in this graph. We suppose that the logical scheme of a route (the adjacency relation of nodes, vertices in the route, the succession of operations, etc.) is given by a connected graph $F = (W, D)$. For instance, F can be a path (a sequence of adjacent vertices and edges), it can be a cycle (if the vertices and operations have to be repeated in a cyclical manner), or an other graph. The association between the logical route F and the physical topology G can be given by a homomorphic mapping h and this "structure" can then be given by a triplet (F, h, G). Trivially, the resulting structure (route) is not necessarily a sub-graph in G. For instance, a *walk* or a *traversal* are connected routes, which may contain vertices and edges in G several times.

Definition 1 (Hierarchy). *If F is a tree, the connected structure defined by $H = (F, h, G)$ is called a* hierarchy *(cf. an example in Figure 1).*

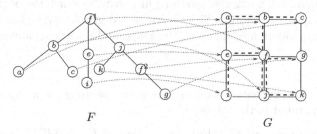

$$F \qquad\qquad G$$

Fig. 1. Homomorphic mapping of vertices to define a hierarchy

To formulate the optimal spanning problem under capacity-like constraints, some simple definitions are needed. The *cost* of a structure $H = (F, h, G)$ is the sum of the costs of the edges used in H: $c(H) = \sum_{e' \in D} c(e)$, where $e \in E$ is the edge associated with $e' \in D$.

If an edge in G is used several times (it is associated to several edges in F), its cost is summarized several times. Since a hierarchy H in a graph G is given by a triplet (F, h, G), and F is a tree, we talk about a leaf of the hierarchy when the concerned vertex is a leaf in F. Similarly, we talk about internal vertices concerning the non-leaf vertices in F. Several vertices of F may correspond to the same vertex v of G. These different occurrences will be labeled $v^1, v^2, \ldots$ if needed.

Our analysis deals with the minimum cost spanning problem of a graph G, where a positive integer B is given to bound the degree of vertices in the optimal route.[1] That is, the degree of each vertex in F (and not in G) is limited by B. Trivially, in interesting cases $2 \le B < \max_{v \in V} d_G(v)$. The minimum cost, connected structure spanning the vertex set of G and respecting the degree constraints is always a hierarchy. With these considerations, we define our spanning problem as follows.

Definition 2 (Degree Constrained Minimum Spanning Hierarchy problem). *Given a connected graph $G = (V, E)$, a cost $c(e)$ for each $e \in E$ and an integer $B \ge 2$, the problem consists in finding a hierarchy $H = (F, h, G)$ where h is a homomorphism from a tree $F = (W, D)$ to $G = (V, E)$ such that:*

- *Each vertex $v \in V$ is associated with at least one vertex $v' \in W$.*
- *The degree constraints are respected in F: $d_F(v') \le B, \forall v' \in W$.*
- *The cost $c(H)$ is minimal.*

---

[1] In DCMST problems, the degree bound expresses the overall capacity or budget of a vertex, but in our problem this bound corresponds to the maximal degree of each occurrence of the vertex in the spanning structure.

In the following, we will call the optimal solution "Degree Constrained Minimum Spanning Hierarchy" abbreviated by DCMSH.

Lemma 1. *For any degree bound $B \geq 2$, the DCMSH problem always has a solution.*

Proof. A traversal is a particular spanning hierarchy, in which the degree of each vertex occurrence is at most 2. Since a connected graph always has traversals, there are always hierarchies spanning the graph and respecting any degree constraint $B \geq 2$. ∎

The problem of the degree constrained minimum spanning hierarchy is NP-hard as it is demonstrated in the following.

Lemma 2. *If among all the Minimum Spanning Trees (MST) of a graph G there exists one satisfying the degree constraint, it is an optimal solution for the DCMSH problem and all the optimal solutions are trees in G.*

Proof. Obvious. The minimum cost spanning structure to connect all the vertices without any constraint is the MST, which is connected and does not contain any redundancy. So if one of the MSTs, for instance a tree T^* respects the degree constraint, it is optimal for the spanning problem and also for the DCMSH problem.

Now suppose that an optimal hierarchy $H = (T, h, G)$ exists and it is not a tree in G. Because the MST T^* is an optimal solution of our problem, the cost $c(H)$ of the optimal hierarchy must be the same that the cost $c(T^*)$ of the MST solution. Trivially, the cost of a hierarchy is greater than or equal to the cost of its image in G: $c(I) \leq c(H)$, where I is the image (the sub-graph generated by H in G). Then, it contains at least a cycle in G (a duplicated edge is considered as a cycle). I covers the vertex set V. Two possibilities can arise.
1. I is a tree and its cost is lower bounded by the cost of the MST: $c(T^*) \leq c(I)$. In this case, there is at least one duplicated edge in H (remember that H is not a simple tree) and $c(I) < c(H)$. Finally: $c(T^*) < c(H)$ and consequently H can not be optimal.
2. I is not a tree. By eliminating some redundancies with non-zero length, a tree T' spanning V is obtained. Trivially, $c(T') < c(I)$ and $c(I) < c(H)$. Finally: $c(T^*) \leq c(T') < c(H)$. ∎

Remark 1: The cost of the MST is therefore a lower bound for the DCMSH problem.
Remark 2: The result is not true if we only consider the *spanning trees* (and not the MSTs) satisfying the degree constraint.

Theorem 1. *The DCMSH problem is NP-hard for all $B \geq 2$.*

Proof. Let $G = (V, E)$ be a graph with $c(e) = 1, \forall e \in E$. Let $G' = (V', E')$ be the graph obtained by adding $B - 2$ leaves connected by edges of cost 1 to each vertex of V. In G', $|V'| = |V| + |V|(B - 2) = (B - 1)|V|$. Any spanning tree of G' has a cost equal to $(B - 1)|V| - 1$. There is a degree-constrained spanning hierarchy of cost $(B - 1)|V| - 1$ in G' if and only if there is a Hamiltonian path in G (remember, that the Hamiltonian path contains $|V| - 1$ edges).

Suppose that there is a degree-constrained spanning hierarchy $H = (T, h, G')$ of cost $(B - 1)|V| - 1$ in G'. Regarding its cost, H is a tree of G'. If we remove all the $(B - 2)|V|$ vertices of $V' \setminus V$ from H, we obtain a connected subgraph in which all vertices have a degree lower or equal to two, which is a Hamiltonian path of G.

Reciprocally, adding $B - 2$ leaves to each vertices of a Hamiltonian path of G gives a tree satisfying the degree constraint, which is a DCMSH in G' because of Lemma 2. ∎

Since the problem is NP-hard, guaranteed approximation algorithms are interesting to solve it in practical cases. To obtain an approximation of the DCMSH in an arbitrary connected graph, our approach is based on two elements:

- We consider the MST of the graph (which cost is a lower bound for every spanning hierarchy) as a start point.
- We decompose this tree into a set of connected stars. Each star is spanned by hierarchies with guarantee of cost and with respect to the degree constraint.

4   Degree Constrained Span of a Star with Hierarchies

Let S_k be a star with k edges, c its central vertex, and $c(S_k)$ the sum of its edges costs. Suppose that $B < k$. Then the minimum spanning hierarchy respecting the degree constraint contains several times the central vertex. Some leaves may also be duplicated. Since all edges of S_k must appear at least once in the hierarchy to ensure the spanning of all vertices, the computation of the DCMSH in a star is equivalent to the minimization of the length of the duplicated edges.

In the following, we propose a simple hierarchy computation to span stars with respect to the degree constraint B. The proposed algorithm does not guarantee the optimality of the hierarchy spanning the star, but it is enough to guarantee a good approximation ratio.

In our proposition, when edge duplications are needed, the less cost edges are used in an increasing order of edge costs. Moreover, these selected edges are duplicated at most once. Formally, let us make a partition of the edges of the star as follows. Let us create $\lfloor k/(B-1) \rfloor + 1$ sets in the partition. Each set, except one (the last), contains $B - 1$ edges (if $k \bmod (B - 1) = 0$, the last edge set is empty). The $\lfloor k/(B-1) \rfloor$ less cost edges are distributed in the partition: each of them is in a separated set (if the last edge set is empty, there is no less cost edge in this set). Each edge set corresponds to a "sub-star", which respects the degree constraint (with at most $B - 1$ edges). To obtain a connected hierarchy H_{S_k} spanning all the leaves, the sub-stars should be connected by the duplication of some edges. The less cost edge of each set is duplicated to make these connections. The central vertex is present in the final hierarchy as many times as there are sets in the partition. So, each central vertex occurrence respects the degree constraint B and the obtained structure is a hierarchy. Figure 2 illustrates the spanning hierarchy for $B = 4$ with $k \bmod (B-1) = 0$ (the interest of the last occurrence of vertex c will be justified in the following).

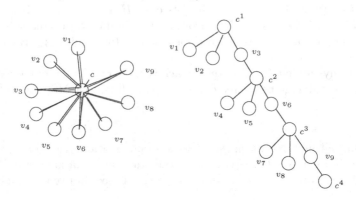

Fig. 2. Spanning hierarchy of a star computed by the proposed heuristic

Lemma 3. *The spanning hierarchy H_{S_k} computed by the proposed algorithm contains $N_c = \lfloor k/(B-1) \rfloor + 1$ times the central vertex c s.t. each occurrence respects the degree constraint. If $N_c \geq 2$, the first and the last occurrences have a degree strictly lower than B. The cost ratio $r = c(H_{S_k})/c(S)$ is bounded by $B/(B-1)$.*

Proof. By construction, each occurrence of c in H_{S_k} have a degree at most B. In each sub-star of the partition, there is an occurrence of c and the number of exclusively spanned leaves is at most equal to $B - 1^2$. It is $B - 1$ for all occurrences of c except eventually one (the last occurrence has not obligatory $B-1$ adjacent vertices). There are at most $\lfloor k/(B-1) \rfloor$ duplicated edges. Let D be the set of these duplicated edges. By choosing the less cost edges to duplicate, the cost of the duplicated part $c(D) = \sum_{e \in D}(c(e))$ of the star is limited by

$$c(D) \leq \frac{\lfloor k/(B-1) \rfloor}{k} c(S) \leq \frac{k/(B-1)}{k} c(S) = \frac{1}{B-1} c(S)$$

An upper bound of the cost ratio is given by:

$$r = \frac{c(H_S)}{c(S)} = \frac{c(S) + c(D)}{c(S)} \leq \frac{B}{B-1}$$

∎

Remark: If $N_c = 1$ (case of $k < B - 1$), the central vertex has a degree strictly lower than $B - 1$.

The spanning hierarchy H_{S_k} corresponds to a caterpillar (tree in which all the vertices are within distance 1 of a central path), each vertex in this central path

---

^{2} A leave is spanned exclusively, if it belong to only one sub-star of the partition.

has a degree at most B. Moreover, it ensures that the central vertex occurrences in the first and in the last sub-stars have a degree less than B (if there is only one star, $deg(c) < B - 1$, cf. Remark).

5 An Approximation Algorithm for the DCMSH Problem

Since the cost of an MST gives a lower bound for the DCMSH problem, upper bounds for approximation algorithm can be computed regarding the MST instead of the optimal spanning hierarchy. In the following, we propose an approximation algorithm based on a decomposition of the MST in the graph.

5.1 A Star Decomposition of the MST

The MST, can be decomposed into a set of stars in the following way. Let $T = (V_T, E_T)$ be an MST with $|V_T| > 2$ and v_1 an arbitrary vertex in T. Then v_1 can be considered as the central vertex of a star S_1. Some neighbor vertices of v_1 in S_1 are leaves in T while some others may be branching vertices. The branching vertices can be considered as central vertices of following stars. Recursively, the entire tree can be covered by stars which are edge disjoint. Figure 3 illustrates the decomposition.

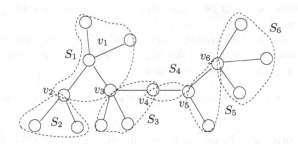

Fig. 3. A star decomposition of a tree

Since the stars are edge disjoint and cover all edges of T, trivially: $c(T) = \sum_{i=1}^{k} c(S_i)$, where $S_i, i = 1, \ldots, k$ indicate the stars in the decomposition.

5.2 The Proposed Algorithm to Approximate the DCMSH

To compute an approximation of the DCMSH in a given graph, we propose the following algorithm.

1. Compute an MST of the graph.
2. Decompose this MST using stars $S_1, S_2, \ldots, S_k$.
3. For each star S_i, compute a spanning hierarchy H_i as proposed in the previous section.

4. "Re-connect" the spanning sub-hierarchies H_i to form a connected spanning hierarchy H_A. A connection is needed, if a leaf in a star coincides with the central vertex of another one. For example, between two neighbor sub-hierarchies spanning stars S_i and S_j, a leaf of S_i corresponds to the central vertex in S_j. In H_i, the leaves of S_i are not duplicated and have a degree 1 or 2. Let us indicate by l_i a leaf in S_i, which corresponds to the central vertex c_j of S_j associated to a vertex v_k in the original graph. Remember that c_j can be repeated in H_j but in this case its first occurrence has a degree $B-1$.

   (a) If l_i has a degree 1 in H_i, it can be aggregated with the first occurrence of c_j in H_j and only one vertex can represent this vertex in the final hierarchy (this vertex in H_A corresponding to v_k respects the degree constraint B). It is the case of the vertex v_3 in our figure.

   (b) If l_i has a degree 2 in H_i (it is not a leaf), the connection can be made as follows.

   (b.a) If the corresponding central vertex c_j has only one occurrence in H_j, than this occurrence is of degree strictly less than $B-1$. Consequently, l_i and c_j can be aggregated in the final hierarchy and the aggregated vertex respects the degree constraint (cf. vertex v_2 in the figure).

   (b.b) If there are several occurrences of c_j in H_j, the first and the last occurrences have a degree at most $B-1$ and the two adjacent edges of l_i can be attached to these two occurrences without the violation of the degree constraint by the different vertices (l_i can be duplicated and each occurrence of l_i can be aggregated by one occurrence of c_j with degree less than B in H_j).

5. The hierarchy H_A can contain useless return edges (edges returning to a central vertex occurrence of a star s.t. the degree of this occurrence is equal to one). The useless edges can be deleted.

Theorem 2. *The previous algorithm offers an $R \le \frac{B}{B-1}$ approximation of the optimal solution.*

Proof. The algorithm is based on a decomposition of the MST T^* into a set of edge disjoint stars. Let $c(S_i)$ be the cost of the star $S_i, i = 1, \ldots, k$ in the decomposition. Trivially $c(T^*) = \sum_{i=1}^{k} c(S_i)$.

Using the result of Lemma 3, the obtained spanning hierarchy length is bounded by

$$c(H) = \sum_{i=1}^{k} c(H_{S_i}) \le \sum_{i=1}^{k} \frac{B}{B-1} c(S_i) = \frac{B}{B-1} c(T^*)$$

The approximation ratio is immediately.

$$R = \frac{c(H)}{c(H^*)} \le \frac{c(H)}{c(T^*)} \le \frac{B}{B-1}$$

■

Remark 1: If $deg(c) < B$ for all vertices $c \in V_{T^*}$, then the algorithm returns the MST, which is the optimum in this case.

Remark 2: If $B = 2$, the algorithm performs a deep-first search type traversal in the MST.

Moreover, we propose to discuss the fact that our computation is not directly related to the optimal spanning hierarchy but to the MST of the graph.

5.3 Discussion about the Heuristic

Since the proposed algorithm only uses the edges of an MST, the resulting hierarchy may be of poor quality for small values of B but the following theorem shows that its cost is the best which can be obtained when computing based on an MST.

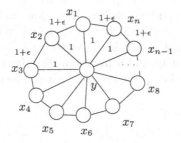

Fig. 4. A wheel graph used in Theorem 3

Theorem 3. *No constant approximation ratio lower than $B/(B-1)$ can be achieved for any heuristic only based on an MST.*

Proof. Let $G = (V, E)$ be a wheel graph with a central vertex y (see Figure 4). Suppose that $c(y, x_i) = 1$ for $i = 1, ..., n$, $c(x_i, x_{i+1}) = 1 + \epsilon$ for $i = 1, ..., n - 1$, and $c(x_n, x_1) = 1 + \epsilon$.

Trivially, the path $P = (y, x_1, x_2, x_3, ..., x_{n-1}, x_n)$ is a spanning hierarchy of G, which respects the degree constraint for any $B \geq 1$ and with a cost $c(P) = 1 + (n - 1)(1 + \epsilon)$.

The Minimum Spanning Tree of G is the star S of center y with n leaves. Let $H^* = (T^*, h^*, S)$ be an optimal hierarchy spanning the star S and respecting the degree constraint.

In T^*, there can be only one occurrence of every vertex corresponding to a leaf of S. If a leaf x_i of S has at least two occurrences in T^*

- If one of them is a leaf of T^*, it can be removed from T^* leading to a hierarchy spanning the same set of vertices with a smaller cost.
- Else, all occurrences are internal vertices of T^*. Let x_i^1 and x_i^2 be two occurrences. Let x_j be a leaf of T^* and $T^{*'}$ be the tree constructed from T^* by deleting the leaf x_j and replacing the label x_i^1 by x_j. Since all the neighbors of x_i^1 in T^* are occurrences of y, there still exists a homomorphism $h^{*'}$ between $T^{*'}$ and S leading to the same contradiction.

So, T^* is a particular bipartite graph where the partition of the vertices can be made as follows: one vertex set with the n_y occurrences of y and the other with the n vertices corresponding to the leaves of S. Since T^* is a tree, its number of edges is equal to its number of vertices minus 1. Consequently, $c(H^*) = n+n_y-1$. Any occurrence of y has at most B neighbors in T^*. So the number of edges of H^* is at most $n_y * B$ and we have $n + n_y - 1 \le n_y * B$ which implies $\frac{n-1}{B-1} \le n_y$. The cost of H^* is then at least $c(H^*) \ge n + \frac{n-1}{B-1} - 1 = \frac{B(n-1)}{B-1}$.

Hence, the approximation ratio of any heuristic only based on an MST is greater or equal to $\frac{c(H^*)}{c(P)} = \frac{\frac{B(n-1)}{B-1}}{1+(n-1)(1+\epsilon)}$ and $\frac{(n-1)}{1+(n-1)(1+\epsilon)}$ can be as close to 1 as wanted for n large enough and ϵ small enough.   ∎

When the computation of the spanning hierarchy is not based on the MST, more interesting results can be obtained. For example, let the minimum Hamiltonian walk problem (case of) rapidly be reviewed. When $B = 2$, our approximation ratio is equal to 2, which is the worth case. Nevertheless, in this case, the problem is equivalent to find a minimum hamiltonian path in the metrical closure of G. It can thus be approximated with a ratio of $3/2$ using for example the remarks of [16].

6   Conclusion

In this paper, we consider the problem of finding a minimum cost spanning structure when the degree of the vertices is bounded by an integer B. When this bound is due to a limited capacity each time the vertex is visited, the optimal structure is a hierarchy. We show that the problem is still NP-hard, but we provide an approximation algorithm to compute a degree constrained minimum spanning hierarchy with a ratio $B/(B-1)$. Since the problem is equivalent to find a minimum hamiltonian path when $B = 2$, a ratio of $3/2$ can always be assured. We also proved that the proposed approximation is the best possible with a heuristic based only on a minimum spanning tree. Future work will consist in an improvement of the ratio and showing that the problem is APX-complete (or to find a PTAS).

References

1. Papadimitriou, C.H., Yannakakis, M.: The Complexity of Restricted Minimum Spanning Tree Problems (Extended Abstract). In: Maurer, H.A. (ed.) ICALP 1979. LNCS, vol. 71, pp. 460–470. Springer, Heidelberg (1979)
2. Cieslik, D.: The vertex degrees of minimum spanning trees. European Journal of Operational Research 125, 278–282 (2000)
3. Ruzika, S., Hamacher, H.W.: A Survey on Multiple Objective Minimum Spanning Tree Problems. In: Lerner, J., Wagner, D., Zweig, K.A. (eds.) Algorithmics. LNCS, vol. 5515, pp. 104–116. Springer, Heidelberg (2009)

4. Ravi, R., Marathe, M.V., Ravi, S.S., Rosenkrantz, D.J., Hunt III, H.B.: Approximation algorithms for degree-constrained minimum-cost network-design problems. Algorithmica 31, 58–78 (2001)
5. Deo, N., Hakimi, S.: The shortest generalized Hamiltonian tree. In: Sixth Annual Allerton Conference, pp. 879–888 (1968)
6. Molnár, M.: Hierarchies to Solve Constrained Connected Spanning Problems. Technical Report 11029, LIRMM (2011)
7. Merabet, M., Durand, S., Molnar, M.: Exact solution for bounded degree connected spanning problems. Technical Report 12027, Laboratoire d'Informatique de Robotique et de Microélectronique de Montpellier - LIRMM (2012)
8. Zhou, Y., Poo, G.S.: Optical multicast over wavelength-routed wdm networks: A survey. Optical Switching and Networking 2, 176–197 (2005)
9. Ali, M., Deogun, J.: Cost-effective implementation of multicasting in wavelength-routed networks. IEEE J. Lightwave Technol., Special Issue on Optical Networks 18, 1628–1638 (2000)
10. Obruca, A.K.: Spanning tree manipulation and the travelling salesman problem. The Computer Journal 10, 374–377 (1968)
11. Garey, M.R., Johnson, D.S.: Computers and Intractability: A Guide to the Theory of NP-Completeness. W. H. Freeman & Co., New York (1979)
12. Ravi, R., Marathe, M.V., Ravi, S.S., Rosenkrantz, D.J., Hunt III, H.B.: Many birds with one stone: multi-objective approximation algorithms. In: Proceedings of the Twenty-fifth Annual ACM Symposium on Theory of Computing, STOC 1993, pp. 438–447. ACM, New York (1993)
13. Fürer, M., Raghavachari, B.: Approximating the minimum degree spanning tree to within one from the optimal degree. In: Proceedings of the Third Annual ACM-SIAM Symposium on Discrete Algorithms, SODA 1992, pp. 317–324. Society for Industrial and Applied Mathematics, Philadelphia (1992)
14. Goemans, M.: Minimum bounded degree spanning trees. In: 47th Annual IEEE Symposium on Foundations of Computer Science, FOCS 2006, pp. 273–282 (2006)
15. Singh, M., Lau, L.C.: Approximating minimum bounded degree spanning trees to within one of optimal. In: STOC 2007: Proceedings of the Thirty-ninth Annual ACM Symposium on Theory of Computing, pp. 661–670. ACM, New York (2007)
16. Hoogeveen, J.A.: Analysis of Christofides' heuristic: Some paths are more difficult than cycles. Oper. Res. Lett. 10, 291–295 (1991)

Secluded Path via Shortest Path[*]

Matthew P. Johnson[1,2], Ou Liu[2], and George Rabanca[2]

[1] Department of Math and Computer Science, Lehman College, Cuny
[2] PhD Program in Computer Science, The Graduate Center, Cuny

Abstract. We provide several new algorithmic results for the secluded path problem, specifically approximation and optimality results for the static algorithm of [3,5], and an extension (h-Memory) of it based on de Bruijn graphs, when applied to bounded degree graphs and some other special-graph classes which can model wireless communication and line-of-sight settings. Our primary result is that h-Memory is a PTAS for degree-Δ unweighted, undirected graphs, providing a $\left\lceil \sqrt{\frac{\Delta+1}{h+1}} \right\rceil$-approximation in time $O(n \log n)$; in particular, 0-Memory (i.e., static) provides a $\sqrt{\Delta + 1}$-approximation (i.e., $\epsilon = \sqrt{\Delta + 1} - 1$), tightening the previous analysis of this algorithm, and Δ-Memory is optimal (i.e., $\epsilon = 0$), and is faster than the known optimal algorithm for this setting [3].

We also show that 0-Memory and 1-Memory give constant approximations for unit-disk graphs and planar graphs, and that an extension of h-Memory solves many other tessellation graphs. Finally, we prove that the problem is NP-hard on node-weighted graphs of degree 3.

1 Introduction

Let the *neighborhood $N[P]$ of a path P* be the set of all nodes within distance at most 1 from some node on the path. Equivalently (for paths of length greater than 1), this is the union of the neighborhoods of all the nodes on the path. The *secluded path problem* be defined as follows: given a graph on n nodes and two specified nodes s and t, find a path from s to t of minimum size neighborhood.

Earlier work showed the problem is very hard to approximate in general and gave some approximation results where the factor typically was not constant but depended on parameters of the graph. Recently, an optimal dynamic programming algorithm [3] was given for the (unweighted, undirected) setting where the degree is bounded by a constant Δ, with a running time of $O(n^2)$. The DP sub-instances represent being at a node and implicitly "remembering" the Δ

---

[*] Research was sponsored by the Army Research Laboratory and was accomplished under Cooperative Agreement Number W911NF-09-2-0053. The views and conclusions contained in this document are those of the authors and should not be interpreted as representing the official policies, either expressed or implied, of the Army Research Laboratory or the U.S. Government. The U.S. Government is authorized to reproduce and distribute reprints for Government purposes notwithstanding any copyright notation here on.

M. Halldórsson (Ed.): SIROCCO 2014, LNCS 8576, pp. 108–120, 2014.
© Springer International Publishing Switzerland 2014

previous nodes you have visited. If Δ is not treated as a constant, however, then the $O(\Delta^\Delta)$ previous node sequences yield a running time of $O(\Delta^\Delta \cdot n^2)$.

Consider the situation where Δ is fixed, but is large enough that the $O(\Delta^\Delta \cdot n^2)$ running time is prohibitive. If the Δ^Δ factor comes from "remembering" Δ previous steps, then a natural question is whether remembering a smaller number of steps would give us some approximation guarantee. That is, despite the constant-Δ problem being polynomial solvable, because of the sort of polynomial it is, there may nonetheless be value in a PTAS. In this paper we provide such a PTAS, providing the usual tradeoff between approximation guarantee and running time, except in this case providing a $(1 + \epsilon)$-approximation for $\epsilon \geq 0$ rather than $\epsilon > 0$.

Our algorithm constructs de Bruijn subgraphs [4] based on the input graph G and on a memory parameter h. Each node in the constructed graph G_h corresponds to a length-h (or, for nodes close to the source, length$\leq h$) walk within G. That is, these fixed-length walks in G are interpreted as "k-mers" on symbols V generating an h-dimensional de Bruijn subgraph, or equivalently, as a graphic representation of the memory's configuration space.

Intuitively, many of our arguments in this paper take the following form. We wish to optimize for some function $cost(\cdot)$, but this is hard. Instead we optimize for some alternative function $cost'(\cdot)$, which is easy. Although $cost'(\cdot)$ may differ from $cost(\cdot)$ on many inputs (i.e., it may be inaccurate), it will match $cost(\cdot)$ on optimal solutions, or nearly matches it on optimal solutions, or nearly match it on *some* optimal solutions, and so by optimally solving according to $cost'(\cdot)$ we will find a good solution under $cost(\cdot)$.

We also apply this algorithm (and extensions) to certain special graph classes that correspond to two application settings. The first is wireless communication, where nodes of the graph correspond to unit disks located in the plane and two nodes share an edge if their disks intersect. For problem instances in which the nodes may occur at arbitrary locations in the plane the corresponding graph class is that of *unit disk graphs*. Alternatively, nodes may be placed in a regular arrangement, corresponding to a *grid graph*. A natural family of such graphs is yielded by considering tilings or tessellations of the plane, which partition the area of the plane into (regular) polygons. In this case, each tile can be interpreted as the unit disk kissing the tile's corners, in which case tiles share and edge iff the corresponding disks overlap. (This graph is the dual graph of the tessellation interpreted as a planar graph drawing.) Two fundamental types of tessellations are a) *regular* or *Platonic* and b) *semiregular* of *Archimedean* (see [7,2]). There are three Platonic tessellations: those whose tiles are triangles, squares, and hexagons. A relaxation of the technical geometric definition of such tilings yields the 8 Archimedean tilings, each of which consist of two types of tiles. The resulting disk graph will then have disks of two sizes.

A second application setting is the movement through city streets. Suppose we can travel along streets from intersection to intersection, there is an *observer* at each intersection, and that when we are at an intersection we are *visible to* the observers at the next intersection for each adjoining street. That is visibility is determined by line of sight between intersections, where the city blocks are obstacles and when there are multiple collinear intersections, each is visible only to its neighboring intersections. This setting is modeled by planar graphs, where intersections become nodes, streets become edges, and city blocks become faces. Note that the (non-dual) graph of a tessellation is a special case of planar graphs that can be interpreted in this way.

Related Work. The problem considered in this paper, recently introduced by [3], is a variation of the classical shortest path. In the standard version of this problem, a cost measure is associated with edges, e.g., representing length and the task is to identify a shortest path form s to t. The cost of the path is a *linear* sum of its constituent edges.

In this problem, the cost is on the nodes "seeing" the message rather than on the edges. Selecting an edge (u, v) means that u will transmit to v, but also to its entire neighborhood. The deeper difference, therefore, is that we pay not just for the nodes constituent to the path itself but for neighboring nodes as well, which results in the total cost being a nonlinear function of the chosen nodes' transmission costs.

Chechik et al. [3] gave a number of negative and positive results, including the following. They showed, by reduction from Red-Blue Set Cover [1,13], the problem is strongly inapproximable on unweighted undirected graphs with unbounded degree (more specifically, is hard to approximate with ratio $O(2^{\log^{1-\epsilon} n})$), where n is the number of nodes in the graph G, assuming $\mathbf{NP} \not\subseteq \mathbf{DTIME}\left(n^{\mathrm{poly}\log n}\right)$). Conversely, they showed that the static algorithm gives a $(\sqrt{\Delta} + 3)$-approximation.

They showed that the problem is NP-hard already on node-weighted graphs of degree 4 graphs and directed graphs of degree 3;[1] for the unweighted undirected setting with any constant maximum degree Δ, they gave a polynomial-time (albeit exponential in Δ) optimal dynamic programming algorithm. They also gave a result implying a 6-approximation for planar graphs as a special case, and noted without proof that a 3-approximation can be obtained. (We provide a proof here for completeness.)

A variant of the problem was also recently introduced as the Thinnest Path Problem [6], which involves hypergraphs and subsumes the secluded path problem. They gave a $\sqrt{\frac{n}{2}}$-approximation result for the static algorithm (see Sec. 2) applied to the hypergraph setting and a 19-approximation analysis of this algorithm for the case of unit-disk graphs.

Finally, turning to geometric settings, similarly motivated problems have been studied in the networking and sensor networks communities, where sensors are often modeled as unit disks. For example, the Maximal Breach Path problem [12] is defined in the context of traversing a region of the plane that contains sensor nodes at predetermined points, and its objective is to maximize the minimum distance between the points on the path and the the sensor nodes.

Similarly motivated problems have been studied in the context of path planning in AI "stealth" path planning problems, in which the task is to find a minimum "visibility" path from source to destination, have been considered in

---

[1] Degree 4 is stated but the construction in fact requires only degree 3.

[8,10,11]. Although the motivation is similar, such problems are technically quite different from the graph-based problems studied here; nonetheless, our unit-disk and planar settings can be viewed as simplified, stylized version of their geometric settings, where visibility may typically be defined in terms of line-of-sight. Finally, a dual problem studied extensively is *barrier coverage*, i.e., the (deterministic or stochastic) placement of sensors in order to make it difficult for an adversary to cross the region unseen (see [9] and the references therein).

Contributions. Our main result is a PTAS for Δ-degree unweighted undirected graphs, which provides a $\left\lceil \sqrt{\frac{\Delta+1}{h+1}} \right\rceil$-approximation, where h is the memory parameter, and its two special cases. First, when $\epsilon = 0$ it provides an optimal solution in $O(n \log n)$ time, faster than the known $O(n^2)$ optimal algorithm of [3] for this setting (treating h and Δ as constants for both), and with a much simpler analysis. Second, when $\epsilon = \sqrt{\Delta+1} - 1$, the algorithm collapses to the static algorithm, which we show provides a $\sqrt{\Delta+1}$ approximation, which is a tighter than the known analysis of [3] for this algorithm.

Our other positive results are show: 1) static is an 8-approximation for unit-disk graphs, improving on the 19-approximation analysis of [5] for this case, and 1-Memory is a 4-approximation; 2) 1-Memory is a 2-approximation algorithm for (directed) planar graphs (omitted from this version); and 3) (1,6)-Memory is optimal for the hexagon grid graph, and (2,12)-Memory is optimal for the square grid graph (both node-weighted); various other tessellation graphs can be solved similarly (omitted). Finally, we prove that the problem is NP-hard on node-weighted graphs of degree 3, improving on the known degree-4 hardness result [3].

2 Preliminaries

Definition 1. *For a path P in the given graph $G = (V, E)$, let $v \in P$ indicate that v is a node on the path. Let $N[P] = \bigcup_{v \in P} N[v_i]$, where $N[v]$ indicates the neighborhood of v. Let $cost(P) = |N[P]|$ denote the cost of P, P^* indicate an optimal path, and $cost_0(P) = \sum_{v \in P}(deg(v) - 1)$ be the static cost of P. Observe that $cost(P) \leq cost_0(P)$. Let the static graph $G_0 = (V_0, E_0)$ be a directed weighted graph where $V_0 = V$ and for each $\{u, v\} \in E$ there exist $(u, v), (v, u) \in E'$ with weights $deg(u) - 1$ and $deg(v) - 1$, respectively.*

We adopt the convention that a path P from s to t is treated as a path from s to some neighbor of t, so that every node on the path *transmits*. Thus $t \in N[P]$ but $t \notin P$. We call nodes that transmit (i.e., nodes on the path) *transmitters*, we call other nodes *nontransmitters* or *non-path* nodes; and we call nodes that receive, i.e., nodes in $N[P]$, *receivers*. Notice that $N[P]$ includes P itself (unless P is a single node), so all transmitters are also receivers. The input graph is assumed to be undirected with unweighted edges and nodes, unless otherwise stated. The extended graphs we construct will be directed and edge-weighted. We use OPT and ALG throughout to represent the cost of an optimal solution and the cost of the (current) algorithm's solution, respectively. Without loss of generality, we ignore the cost of the source node (trivially) receiving when convenient.

Definition 2. *A path Q is* shifted successor *of a path P if $P = (v_1, v_2, ..., v_h)$ and $Q = (v_2, v_3, ..., v_h, u)$, for some node u.*

To motivate the algorithm used in this paper, consider the incremental costs of each edge e_i added, sequentially, to a partial path P. Let $e_i = (v_i, v_{i+1})$, with vertex indices renumbered accordingly, and let P_i indicate the subpath $e_1, e_2, ..., e_i$ of P. Then the cost of e_1 is $1 + \deg(v_1)$ and for each subsequent edge e_i it is the number of v_i's neighbors being encountered for the first time, i.e., $|N[v_i] - N[P_{i-1}]|$.

Given a graph G, consider the graph G' constructed as follows: for each node v_i of G, introduced a "column" of $O(n!)$ nodes, corresponding to the different possible histories of v_i (i.e., simple paths from the source to v_i). For each edge (v_i, v_j) of G, we introduce in G' two collections of directed edges. First, we introduce edges of the form $v_i^a \to v_j^{a'}$ for all possible histories for v_i and v_j that are *consistent* in the sense that a indicates a simple path from s to a neighbor of v_i and a' indicates that extended to v_i, i.e., to a neighbor of v_j. Similarly, for edges $v_j^a \to v_i^{a'}$. The cost of edge $v_i^a \to v_j^b$ is set to $|N[v_i] - N[P_{i-1}^a]|$. Also, add to G' a new source node and destination node, connected with cost 1 (respectively, 0) edges to all the nodes of the source (respectively, destination) column.

By construction, we have that 1) each node v_i^a in G' corresponds to a simple path in G from source to node v_i, and 2) the cost of each edge $v_i^a \to v_j^b$ of G' corresponds to the the *secluded cost* of the transmission of node v_i in G *after reaching it by taking subpath a from source to v_i*. Thus we state:

Observation 1. *For each path P in G there will be a corresponding path P' in G' whose cost (i.e., sum of edge weights) equals P's secluded cost, and vice versa.*

Of course, the graph G' constructed above is larger than G by a factor of $O(n!)$. We will find that for several graph classes, we can restrict ourselves to a constant-size memory h, and hence ensure that our expanded graph is only a constant factor (specifically, $O(\Delta^h)$) larger than the original. That is, each node v in G now maps to a column of $O(\Delta^h)$ nodes in G_h. Each of these nodes corresponds to some possible length-h subpath reaching a neighbor of v (or shorter subpath reaching v_i in the case of v near to the source, or a null path if it *is* the source (see Fig. 3.2). Directed edges in G_h now correspond to "recent history" of length $\le h$ and a shifted successor of it, rather than a monotonic extension as above. The graph G_h can equivalently be viewed as a subgraph of a certain h-dimensional de Bruijn graph whose "symbols" are the nodes of G.

We now observe that computing the cost of $v_i^a \to v_j^{a'}$ on *only some of* the preceding nodes can only *increase* the cost:

Observation 2. *The cost of a path in G_h can be only greater than the cost of the corresponding path in G', and thus also the* secluded *cost of the corresponding path in G.*

Algorithm 1. Static (0-Memory)

1. **for** each edge (u, v) **do**
2. $e_{uv} \leftarrow \deg(u) - 1$
3. **end for**
4. run Dijkstra on the graph with edge weights $\{e_{uv}\}$

Algorithm 2. h-Memory

1. given the input graph, construct the h-dimensional De Bruijn graph G_h, as described in the text
2. run Dijkstra on G_h

We note that in both an optimal solution and in the path returned by Static and h-Memory, each transmitter will be adjacent to its successor and/or predecessor but to no other nodes on the path, since otherwise a shortcut would have produced a shorter path. Call a path with no potential shortcuts *minimal*. By definition of the edge costs, in G_h we are charged for a non-source node when it receives the transmission from its predecessor on the path. A path in G_h costs more than the corresponding path in G when a *nontransmitting* or *non-path* node receives multiple times.

3 Δ-Degree Graphs

3.1 Static

We begin by proving an approximation guarantee for the memoryless static algorithm, when run on graphs of maximum degree Δ. The proof in this section is a warmup for Sec. 3.3 and is omitted from this version.

Theorem 1. *0-Memory (i.e., static) gives a $\sqrt{\Delta + 1}$-approximation on Δ-degree unweighted undirected graphs.*

3.2 Δ-Memory

Now we consider another extreme, i.e., the h-Memory algorithm with $h = \Delta$. We will use $d_P(a, b)$ to indicate the *distance on the path* P between $a, b \in P$, which we define as the number of edges separating them on P, i.e., the length of subpath $P[a, b]$.

We will first prove a lemma, slightly strengthening a lemma of [3], which gave a similar result for $\Delta + 1$. It tells us there exist optimal solutions in which the length of the subpath between two nodes with a common neighbor is bounded by Δ.

Lemma 1. *There exists an optimal solution P^* in which, if $N[u] \cap N[v] \neq \emptyset$ for $u, v \in P^*$, we have $d_{P^*}(u, v) \leq \Delta$.*

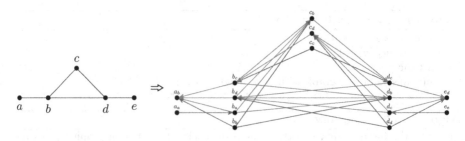

Fig. 1. Construction of the 1-Memory graph G_1 for a simple input graph G. In this example, all b_b's outgoing edges are given cost 3, b_a's are given cost 2, and b_c's and b_d's are given cost 1.

Proof. Suppose two nodes u_0, u_k on a (possibly optimal) path P share a neighbor $x \notin P$. Let $d_P(u_0, u_k) = k$, and name the intermediate nodes so that this subpath consists of the nodes $u_0, u_1, ..., u_{k-1}, u_k$. Let δ be the degree of x, and let P' a modification of P that replaces the subpath $P[u_0, u_k]$ with the subpath u_0, x, u_k.

Let us now consider the costs of these two paths. First we examine the cost of P'. When u_0 transmits, u_1 and x receive; when x transmits, u_k and x's other $\delta - 2$ neighbors (ignoring u_0) receive; when u_k transmits, u_{k-1} receives, for a total of $\delta + 2$ receivers. (We ignore without loss of generality any other neighbors of u_0 and u_k, which will receive from both P' and P.)

Now consider $P[u_0, u_k]$. Its cost is k for nodes $u_1, ..., u_k$ plus 1 for x, plus any other neighbors those nodes may have, for a total of $cost(P[u_0, u_k]) \geq k + 1$. If $k \geq \delta + 1$ then

$$cost(P[u_0, u_k]) \geq k + 1 \geq \delta + 2 = cost(u_0, x, u_k)$$

and so if P is optimal then P' is too. Otherwise, $k < \delta + 1$, and indeed $d_{P*}(u_0, u_k) = k \leq \delta \leq \Delta$.   □

We are now ready to prove the theorem. It shows that while Δ-Memory overestimates some paths' costs—even some optimal paths' costs—it optimally computes the costs of a least *some* optimal path, and so it will find one.

Theorem 2. Δ-*Memory (for any constant Δ) is an $O(n \log n)$-time optimal algorithm for unweighted undirected degree-Δ graphs.*

Proof. Let $cost(P)$ indicate the secluded cost of a path P in the input graph G, and let $cost_h(P)$ indicate the static cost of the corresponding path in G_h.

The lemma tells us that there is always an optimal solution P^* in which the cost of v_i's transmission is determined by (aside from v_i's neighborhood) not the entire subpath from source to v_i but only the last Δ nodes visited prior to reaching v_i. Thus the cost of the corresponding path in G_h will equal the secluded cost of P^*:

$$cost(OPT) = cost_h(OPT) \tag{1}$$

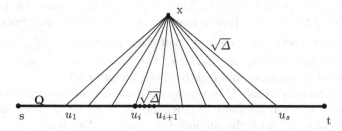

Fig. 2. Approximation guarantee of h-Memory

Obs. 2, however, told us that the cost of a path in G_h (for any h, and so in particular for G_Δ) can only be greater than the secluded cost of the corresponding path in G.

Then we have:

$$cost(ALG) \le cost_\Delta(ALG) \le cost_\Delta(OPT) = cost(OPT)$$

The first inequality follows from Obs. 2, the second inequality from the fact that the algorithm chooses a path of optimal $c_\Delta(\cdot)$ cost, and the equality from Eq. 1.

Since $\deg(G) = \Delta$, the number of length$\le\Delta$ subpaths to any node of G is $O(\Delta^\Delta)$. With Δ constant, the size of G_Δ only a constant larger than G, i.e., $|V_\Delta| = O(|V|)$ and $|E_\Delta| = O(|E|)$. Since $|E| = O(|V|)$ (because Δ is constant), Dijkstra can be run on G_Δ in time $O(n \log n)$. □

3.3 h-Memory for $0 \le h \le \Delta$

Finally, we generalize the two previous results for for values of the memory parameter h ranging from 0 to Δ, using an extension of the argument for Theorem 1.

Theorem 3. *h-Memory (for any constant Δ) gives a $\left\lceil \sqrt{\frac{\Delta+1}{h+1}} \right\rceil$-approximation on unweighted undirected Δ-degree graphs in $O(n \log n)$ time.*

Proof. Let $cost_h(P)$ indicate the cost of a path P from the point of view of h-Memory: the cost when node $v_i \in P$ transmits is the number of neighbors receiving who did not receive in the previous h transmissions, i.e., $|N(v_i) - \bigcup_{j=i-h}^{i-1} N(v_j)|$.

We show that an optimal path P^* can be converted into a path Q of $cost_h(Q) \le \left\lceil \sqrt{\frac{\Delta+1}{h+1}} \right\rceil \cdot cost(P^*)$. The existence of such a path Q establishes that $ALG \le \left\lceil \sqrt{\frac{\Delta+1}{h+1}} \right\rceil \cdot OPT$.

Initialize Q to P^* and let all nodes of the graph be *unmarked* (see Fig. 3.3). Consider a node x not on Q. Say that nodes $u_1, ..., u_s$ on Q are h-separated neighbors of x, i.e., they are all in $N(x)$ but for each $2 \le i \le s$, the h nodes preceding u_i on Q are not in $N(x)$. That is, between each successive pair of x's neighbors on Q there are at least h nodes on Q that are not neighbors of x.

Now say that a node x not on Q is *bad* if it has more than $\left\lceil \sqrt{\frac{\Delta+1}{h+1}} \right\rceil$ unmarked h-separated neighbors on Q, say, $u_1, ..., u_s$. Let the *length* of x be $d_Q(u_1, u_s)$. Say that x is *worst* if has maximum length among all bad nodes.

Now we iteratively modify Q by repeatedly doing the following: choose a worst node x with unmarked h-separated neighbors $u_1, ..., u_s$ on Q. Then we replace the subpath $Q[u_1, u_s]$ with the subpath (u_1, x, u_s), and we mark x.

Now we examine the final value of $cost_h(Q)$.

Consider a move that adds a node x to the path. This occurs when x has $s > \left\lceil \sqrt{\frac{\Delta+1}{h+1}} \right\rceil$ neighbors on Q. We now examine how this move changes the cost of Q. Prior to the move, the nodes $succ_Q(u_1), ..., u_s$ and x all receive (due to the transmission of $u_1, ..., pred_Q(u_s)$). Recall that $u_1, ..., u_s$ are h-separated, with at least h nodes in between each successive pair u_i, u_{i+1}, and so $||[u_1, u_s]|| \geq (s-1) \cdot (h+1) + 1$. Therefore the transmissions of $u_1, ..., pred_Q(u_s)$ contribute at least $(s-1) \cdot (h+1) \geq \sqrt{\frac{\Delta+1}{h+1}} \cdot (h+1) = \sqrt{(\Delta+1) \cdot (h+1)}$ to $cost(Q)$.

After the move, the transmission of x (to nodes $u_2, .., u_s$ and possibly others) is of $cost_h$ at most $\Delta - 1$. In addition, there is $cost_h$ 1 each for $succ_Q(u_1)$ and x to both receive from u_1. Altogether, the modified path incurs an h-memory cost of at most $\Delta + 1$ due to the transmitters u_1, x, replacing the true cost s of the nodes they replace.

Thus we charge $\Delta + 1$ to the substring $P^*[u_1 u_{s-1}]$, which is less than $\frac{\Delta+1}{\sqrt{(\Delta+1) \cdot (h+1)}} = \sqrt{\frac{\Delta+1}{h+1}}$ times the cost of that substring's transmissions.

Now consider neighbors y of Q that we never make such moves based on, i.e., nodes y that are neighbors of at most $\left\lceil \sqrt{\frac{\Delta+1}{h+1}} \right\rceil$ unmarked h-separated nodes on Q. We pay at most $\left\lceil \sqrt{\frac{\Delta+1}{h+1}} \right\rceil$ in h-memory cost for each such node *due to transmissions of unmarked nodes*, rather than the true cost of 1. Thus the h-memory cost of Q is at most $\left\lceil \sqrt{\frac{\Delta+1}{h+1}} \right\rceil$ times the true cost of P^*.

Since h is bounded by the constant Δ, the number of length-h subpaths is $O(\Delta^h)$, and so the running time is again $O(n \log n)$.   □

Finally, we observe that for any $\epsilon \geq 0$, setting[2] $e = \lfloor \epsilon \rfloor$ and $h = \frac{\Delta+1}{(1+e)^2} - 1$ yields $\left\lceil \sqrt{\frac{\Delta+1}{h+1}} \right\rceil = \sqrt{\frac{\Delta+1}{h+1}} = 1 + e \leq 1 + \epsilon$. The second equality is obtained by algebra, and the first equality holds because $1 + e$ is an integer. In terms of Δ and ϵ, the running time is $O(\Delta^{\Delta/(1+\epsilon)^2} n \cdot \log(\Delta^{\Delta/(1+\epsilon)^2} n))$. Thus we conclude:

Corollary 1. *h-Memory (for any constant Δ) is a PTAS for the the secluded path problem on unweighted undirected Δ-degree graphs.*

---

[2] $e = 0$ when $\epsilon < 1$.

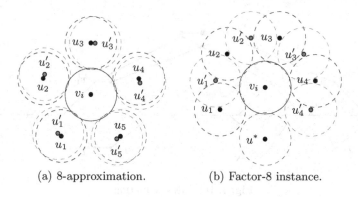

(a) 8-approximation.   (b) Factor-8 instance.

Fig. 3. 8-approximation of Static on unit disk graphs

4 Unit-Disk Graphs

We begin by proving a lemma showing that the cost of every (minimal) secluded path in G is inflated by a factor of a most 8 in G_0. Thus there will exist a path in G_0 of length at most 8 times the optimal secluded path cost, which therefore upper-bounds the cost of the solution returned by the algorithm.

Lemma 2. *In the case of a minimal solution, a nontransmitter receives at most 8 times.*

Proof. To obtain the bound, first recall that a unit disk can have at most 5 mutually disjoint neighbors, say $u_1, ..., u_5$ (see Fig. 3(a)). Therefore there can be at most 10 transmitting neighbors of v lying on a minimal subpath.

Consider the (minimal) subpaths $P_{10} = (u_1, u'_1, ..., u_2, u'_2, ..., u_3, u'_3, ..., u_4, u'_4,$ $..., u_5, u'_5)$ and P_9, with the latter the same except with u_1 omitted (assume u'_5 is not the destination), in the instance of Fig. 3(a). First v has no other neighbors. Then P_{10} costs more (at least 20: each transmission by one of v's 10 neighbors transmits to v and has cost at least 2) than the subpath $P' = (u_1, v, u'_5)$ does (as low as 2+9+2=13, with v being paid for only once), and so P_{10} cannot be a subpath of a shortest path. (Similarly for P_9, with costs 18 versus 12.) In order to be charged for v_i more than once (i.e., in order for it not to lie on the shortest path), having v_i transmit must be prohibitively expensive, i.e., it must have additional neighbors who do not receive the message when path P transmits, co-located at, say, node u^*.

If v *does* have any additional neighbors, they each would necessarily receive the transmission, at least once, from the nodes of P_{10} (or P_9), since the minimality property implies that these disks occupy an arc sweeping out greater than 5/6 of v's, and so there is no space for another disk to intersect v but not P_9.

Combined with the minimality property, the constraint that the path's disks not intersect with u^* limits its number to 8 (see Figs. 3(a) and 3(b)).   □

Proposition 1. *Static provides an 8-approximation when run on node-weighted UDGs.*

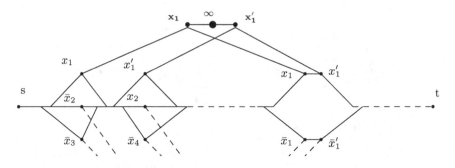

Fig. 4. Hardness reduction

Proof.

$$cost(ALG) \leq cost_0(ALG) \leq cost_0(OPT) \leq 8 \cdot cost(OPT)$$

The inequalities follow from 1) Obs. 2, 2) the fact that the algorithm returns a path of minimum $cost_0(\cdot)$, and 3) the lemma. □

An example in which the factor-8 approximation obtains is shown in Fig. 3(b). Here $u_1, u'_1, ..., u_4, u'_4$ are individual nodes; v_i indicates n co-located nodes (for some large n), and u^* indicates n^2 co-located nodes.

Proposition 2. *1-Memory provides a 4-approximation when run on node-weighted UDGs.*

Proof. The proof proceeds very similarly to the proof of Proposition 1, this time showing that we pay for each nontransmitter at most 4 times. Because of the 1-Memory property, in the instance of Figs. 3(a) and 3(b), path $P_{10} = (u_1, u'_1, ..., u_2, u'_2, ..., u_3, u'_3, ..., u_4, u'_4, ..., u_5, u'_5)$ would be charged for node v_i only once. Arrangements such the one shown in Fig. 3(a) can be extended to paths P' in which v_i receives from the transmissions of multiple *noncontiguous* subpaths. We will then be charged for v_i once for each such maximal subpath. Recall that a unit disk can intersect with at most 5 independent unit disks, meaning at most 5 such subpaths. As before, the fact that v_i is not chosen to transmit means there must be a neighbor of v_i to avoid, which reduces the worst-case total to 4. □

5   NP-Hardness

Proposition 3. *The secluded path problem is NP-hard on node-weighted of degree 3.*

Proof. We reduce from the special case of 3-SAT in which each literal appears at most twice. We create a graph in which the optimal secluded path cost equals $2n$ if the formula is satisfiable and is more otherwise (see Fig. 4). For each clause we add a *clause gadget*, in which the path splits into three length-2 subpaths, corresponding to the literals of a clause, which then merge back together. (The first clause gadget from the left in the example shown in the figure is for ($x_1 \vee \bar{x}_2 \vee \bar{x}_3$).) Each clause gadget's "literal nodes" (x_i) have a neighbor of weight 1 ($\mathbf{x_i}$ or $\mathbf{x'_i}$). Those neighbors have very heavy neighbors themselves, so that they cannot be part of any shortest path. The cost of all other vertices in the graph is 0. A clause gadget thus forces a good solution path to choose one of three alternatives corresponding to the clause's three literals.

To prevent an optimal solution path from visiting both x_i and $\bar{x}_i$, we add *variable gadgets* at the end of the construction (one variable gadget is shown in the figure). Each (for variable x_i) forces a solution path to make one of two choices for a variable, which means paying 2 nodes neighboring x_i, x'_i ($\mathbf{x_i}$ and $\mathbf{x'_i}$) or those neighboring $\bar{x}_i, \bar{x}'_i$ ($\mathbf{\bar{x}_i}$ and $\mathbf{\bar{x}'_i}$). Thus the variable gadgets by themselves therefore give a cost a total of $2n$ Thus a path will cost more than $2n$ iff it corresponds to a non-satisfying assignment.

6 Discussion

Several of the results in this paper used a lemma that on an optimal solution path two nodes with a common neighbor cannot be too far apart, which permitted the use of de Bruijn graphs to implement enough memory to correctly represent transmission costs using edge costs. In a similar spirit, the unit-disk approximation result depended on showing that there cannot be too many nodes on an optimal path sharing a neighbor. In all cases, we could then obtain good results computing shortest paths.

The itchiest open problem is proving the problem to be easy or hard on planar and/or unit-disk graphs. Note that the triangular grid graph, whose status is also open, is a special case of both the planar and unit-disk graphs. It appears that positive results for these settings will require fundamentally different techniques.

References

1. Carr, R.D., Doddi, S., Konjevod, G., Marathe, M.V.: On the red-blue set cover problem. In: SODA, pp. 345–353 (2000)
2. Chavey, D.: Tilings by regular polygons: A catalog of tilings. Computers & Mathematics with Applications 17(1-3), 147–165 (1989)
3. Chechik, S., Johnson, M.P., Parter, M., Peleg, D.: Secluded connectivity problems. In: Bodlaender, H.L., Italiano, G.F. (eds.) ESA 2013. LNCS, vol. 8125, pp. 301–312. Springer, Heidelberg (2013)
4. de Bruijn, N.G.: A combinatorial problem. Koninklijke Nederlandse Akademie v. Wetenschappen 49(49), 758–764 (1946)
5. Gao, J., Zhao, Q., Swami, A.: The thinnest path problem for secure communications: A directed hypergraph approach. In: Allerton Conference on Communications, Control, and Computing (2012)
6. Gao, J., Zhao, Q., Swami, A.: The thinnest path problem for secure communications: A directed hypergraph approach. In: Allerton Conference on Communication, Control, and Computing, pp. 847–852 (2012)

7. Grunbaum, B., Shephard, G.C.: Tilings and Patterns. W. H. Freeman and Company (1987)
8. Johansson, A., Dell'Acqua, P.: Knowledge-based probability maps for covert pathfinding. In: Boulic, R., Chrysanthou, Y., Komura, T. (eds.) MIG 2010. LNCS, vol. 6459, pp. 339–350. Springer, Heidelberg (2010)
9. Liu, B., Dousse, O., Wang, J., Saipulla, A.: Strong barrier coverage of wireless sensor networks. In: MobiHoc, pp. 411–420 (2008)
10. Marzouqi, M., Jarvis, R.: New visibility-based path-planning approach for covert robotic navigation. Robotica 24(6), 759–773 (2006)
11. Marzouqi, M.A., Jarvis, R.A.: Robotic covert path planning: A survey. In: RAM, pp. 77–82 (2011)
12. Meguerdichian, S., Koushanfar, F., Potkonjak, M., Srivastava, M.B.: Coverage problems in wireless ad-hoc sensor networks. In: INFOCOM, pp. 1380–1387 (2001)
13. Peleg, D.: Approximation algorithms for the label-cover$_{max}$ and red-blue set cover problems. J. Discrete Algorithms 5(1), 55–64 (2007)

Distributed Approximation
of Minimum Routing Cost Trees

Alexandra Hochuli[1], Stephan Holzer[2,*], and Roger Wattenhofer[1]

[1] ETH Zurich, 8092 Zurich, Switzerland
{hochulia,wattenhofer}@ethz.ch
[2] MIT, Cambridge, MA 02139, USA
holzer@mit.edu

Abstract. We study the NP-hard problem of approximating a Minimum Routing Cost Spanning Tree in the message passing model with limited bandwidth (CONGEST model). In this problem one tries to find a spanning tree of a graph G over n nodes that minimizes the sum of distances between all pairs of nodes. In the considered model every node can transmit a different (but short) message to each of its neighbors in each synchronous round. We provide a randomized $(2+\varepsilon)$-approximation with runtime $\mathcal{O}(D + \frac{\log n}{\varepsilon})$ for unweighted graphs. Here, D is the diameter of G. This improves over both, the (expected) approximation factor $\mathcal{O}(\log n)$ and the runtime $\mathcal{O}(D \log^2 n)$ stated in [13].

Due to stating our results in a very general way, we also derive an (optimal) runtime of $\mathcal{O}(D)$ when considering $\mathcal{O}(\log n)$-approximations as in [13]. In addition we derive a deterministic 2-approximation.

1 Introduction

A major goal in network design is to minimize the cost of communication between any two vertices in a network while maintaining only a substructure of the network. Despite the fact that a tree is the sparsest substructure of a network it can be surprisingly close to the optimal solution. Every network contains a tree whose total cost of communication between all pairs of nodes is only a factor two worse than the communication cost when all edges in the graph are allowed to be used!

The problem of finding trees that provide a low routing cost is studied since the early days of computing in the 1960s [18] and is known to be NP-hard [12] on weighted and unweighted graphs[1]. These days networks of computers and electric devices are omnipresent and trees offer easy and fast implementations for applications. In addition, trees serve as the basis for control structures as well

---

[*] Corresponding author. Part of this work was done at ETH Zurich. At MIT the author was supported by the following grants: AFOSR Contract Number FA9550-13-1-0042, NSF Award 0939370-CCF, NSF Award CCF-1217506, NSF Award number CCF-AF-0937274.
[1] Even for seemingly simpler versions than those which we study the problem remains NP-hard [23].

M. Halldórsson (Ed.): SIROCCO 2014, LNCS 8576, pp. 121–136, 2014.

as for information gathering/aggregation and information dissemination. This explains why routing trees are computed and used by wide spread protocols such as the IEEE 802.1D standard [3]. When bridging [19] is used in Local Area Networks (LAN) and Personal Area Networks (PAN), a spanning tree is computed to define the (overlay) network topology. Finding such a tree with low routing cost is crucial. As [3] demonstrates, current implementations do not perform well under the aspect of optimizing the routing costs and there is the need to find better and faster solutions. The nature of this problem and growth of wired and wireless networks calls for fast and good distributed implementations.

In this paper we present new approaches for distributed approximation of a Minimum Routing Cost Spanning Tree (MRCT) while extending previous work for approximation of those. By doing so we improve both, the round complexity and the approximation factor of the best known (randomized) result in a distributed setting for unweighted graphs. Our main contribution is an algorithm that computes a $\left(2 - \frac{2}{n} + \min\left\{\frac{\log n}{D}, \alpha(n, D)\right\}\right)$-approximation in time $\mathcal{O}\left(D + \frac{\log n}{\alpha(n,D)}\right)$ w.h.p.[2]. Previously, the best known distributed approximation for MRCT [13] (on weighted graphs) achieved an (expected) approximation-ratio of $\mathcal{O}(\log n)$ using randomness. The bound on the runtime of the algorithm of [13] is $\mathcal{O}(n \log^2 n)$ in the worst case – even when the network is fully connected (a clique). For unweighted graphs, the authors of [13] specify this runtime to be $\mathcal{O}(D \log^2 n)$. The distributed algorithms we present in this paper are for unweighted graphs as well[3] and compared to the (expected) approximation-ratio $\mathcal{O}(\log n)$ of [13] we essentially obtain a (guaranteed) approximation-ratio $2 + \varepsilon$ in time $\mathcal{O}(D + \frac{\log n}{\varepsilon})$ w.h.p.. This follows from choosing $\alpha(n, D) = \varepsilon$ for an arbitrary small $\varepsilon > 0$. When choosing $\alpha(n, D) = \log n$, we obtain the same approximation ratio as in [13] in time $\mathcal{O}(D)$. To be general, we leave the choice of $\alpha(n, D)$ to the reader depending on the application.

Besides this randomized solution we present a deterministic algorithm running in linear time $\mathcal{O}(n)$ achieving an approximation-ratio of 2.

2   Model and Basic Definitions

Our network is represented by an undirected graph $G = (V, E)$. Nodes V correspond to processors, computers or routers. Two nodes are connected by an edge from set E if they can communicate directly with each other. We denote the number of nodes of a graph by n, and the number of its edges by m. Furthermore we assume that each node has a unique ID in the range of $\{1, \ldots, 2^{\mathcal{O}(\log n)}\}$, i.e. each node can be represented by $\mathcal{O}(\log n)$ bits. Nodes initially have no knowledge of the graph G, other than their immediate neighborhood.

We consider a synchronous communication model, where every node can send B bits of information over all its adjacent edges in one synchronous round of

---

[2] A more precise statement can be found in Theorem 3. This Theorem also considers a generalized version of MRCT.

[3] They extend to graphs with certain realistic weight-functions.

communication. We also consider a modified model, where time is partitioned into synchronized slots, but a message might receive a delay when traversing an edge. This delay might not be uniform but fixed for each edge. In principle it is allowed that in each round a node can send different messages of size B to each of its neighbors and likewise receive different messages from each of its neighbors. Typically we use $B = \mathcal{O}(\log n)$ bits, which allows us to send a constant number of node or edge IDs per message. Since communication cost usually dominates the cost of local computation, local computation is considered to be negligible. For $B = \mathcal{O}(\log n)$ this message passing model is known as CONGEST model [15]. We are interested in the number of rounds that a distributed algorithm needs to solve some problem. This is the time complexity of the algorithm.

To be more formal, we are interested in evaluating a function $g : \mathbb{G}_n \to S$, where $\mathbb{G}_n$ is the set of all graphs over n vertices and S is e.g. $\{0,1\}$, $\mathbb{N}$ or $\mathbb{G}_n$, and define distributed round complexity as follows:

Definition 1 (Distributed round complexity). *Let $\mathcal{A}$ be the set of distributed deterministic algorithms that evaluate a function g on the underlying graph G over n nodes (representing the network). Denote by $R^{dc}(A(G))$ the distributed round complexity (indicated by dc) representing the number of rounds that an algorithm $A \in \mathcal{A}$ needs in order to compute $g(G)$. We define $R^{dc}(g) = \min_{A \in \mathcal{A}} \max_{G \in \mathbb{G}_n} R^{dc}(A(G))$ to be the smallest amount of rounds/time slots any algorithm needs in order to compute g.*

We denote by $R_\varepsilon^{dc-rand}(g)$ the randomized round complexity of g when the algorithms have access to randomness and compute the desired output with an error probability smaller than ε. By w.h.p. (with high probability) we denote a success probability larger than $1 - 1/n$.

The unweighted shortest path in G between two nodes u and v is a path with minimum number of edges among all (u,v)-paths. Denote by $d_G(u,v)$ the unweighted distance between two nodes u and v in G which is the length of an unweighted shortest (u,v)-path in G. We also say u and v are $d_G(u,v)$ hops apart. By $\omega_G : E \to \mathbb{N}$ we denote a graph's weight function and by $\omega_G(e)$ the weight of an edge in G. By $\omega_G(u,v) := \min_{\{P|P \text{ is } (u,v)\text{-path in } G\}} \sum_{e \text{ is edge in } P} \omega_G(e)$ we define the weighted distance between two nodes u and v, that is the weight of a shortest weighted path in a graph G connecting u and v[4].

The time-bounds of our algorithms as well as those of previous algorithms depend on the diameter of a graph. We also use the eccentricity of a node.

Definition 2 (Eccentricity, diameter). *The weighted eccentricity $ecc_{\omega_G}(u)$ in G of a node u is the largest weighted distance to any other node in the graph, that is $ecc_{\omega_G}(u) := \max_{v \in V} \omega_G(u,v)$. The weighted diameter $D_\omega(G) := \max_{u \in V} ecc_{\omega_G}(u) := \max_{u,v \in V} \omega_G(u,v)$ of a graph G is the maximum weighted distance between any two nodes of the graph. The unweighted diameter (or hop*

---

[4] Note that in the context of MRCT, ω often corresponds to the cost of an edge. In the literature the routing cost between any node u and v in a given spanning tree T of G is usually denoted by $c_T(u,v)$, while in generalized versions of MRCT, the weight of an edge can be different from the cost. In this paper we use $\omega_T(u,v) = c_T(u,v)$.

diameter) $D_h(G) := \max_{u,v \in V} \min_{\{P|P \text{ is } (u,v)\text{-path}\}} |P|$ *of a graph G is the max-imum number of hops between any two nodes of the graph. Here $|P|$ indicates the number of edges on path P.*

We often write D_ω and D_h instead of $D_\omega(G)$ and $D_h(G)$ when we refer to the diameter of a graph G in context. Observe that $D_h = D_\omega$ for unweighted graphs.

Finally, we define the problems that we study.

Definition 3 (S-Minimum Routing Cost Tree (S-MRCT)). *Let S be a subset of the vertices V in G. The S-routing cost of a subgraph H is defined as $RC_S(H) := \sum_{u,v \in S} \omega_H(u,v)$ and denotes the routing cost of H with respect to S. An S-MRCT is a subgraph T of G that is a tree, contains all nodes S and has minimum S-routing cost $RC_S(T)$ among all spanning trees of T.*

This is a generalization of the MRCT problem [22]. According to this definition V-MRCT (i.e. $S = V$) and MRCT of [22] are equivalent. Therefore all results are valid for the classical MRCT problem when choosing $S := V$.

In this paper we consider approximation algorithms for these problems. Given an optimization problem P, denote by OPT the cost of the optimal solution for P and by SOL_A the cost of the solution of an algorithm A for P. We say A is ρ-approximative for P if $OPT \leq SOL_A \leq \rho \cdot OPT$ for any input.

Fact 1 *The eccentricity of any node is a good approximation of the diameter. For any node $u \in V$ we know that $ecc_{\omega_G}(u) \leq D_\omega(G) \leq 2 \cdot ecc_{\omega_G}(u)$.*

3   Our Results

In Section 8 we prove the following two theorems.

Theorem 2. *In the CONGEST model, the deterministic algorithm proposed in Section 8 needs time $\mathcal{O}(|S| + D_\omega)$ to compute a $(2 - 2/|S|)$-approximation for S-MRCT when using either uniform weights for all edges or a weight function $\omega(e)$ that reflects the delay/edge traversal time of edge e.*

Theorem 3. *Let $\alpha(n, D_\omega)$ be some function in n and D_ω. The randomized algorithm proposed in Section 8 computes w.h.p. a $\left(2 - \frac{2}{|S|} + \min\left\{\frac{\log n}{D_\omega}, \alpha(n, D_\omega)\right\}\right)$-approximation for S-MRCT in the CONGEST model in time $\mathcal{O}\left(D_\omega + \frac{\log n}{\alpha(n, D_\omega)}\right)$ when using either uniform weights for all edges or a weight function $\omega(e)$ that reflects the delay/edge traversal time of edge e.*

We emphasize that the analysis of [20] yields a 2-approximation when compared to the routing cost in the original graph[5] and that we modify this analysis.

---

[5] Note that most other approximation algorithms are with respect to the routing cost of a minimal routing cost tree of the graph. In the full version of this paper [8] we provide an example that shows that sometimes even no subgraph with $o(n^2)$ edges exists that yields better approximations to the routing cost in the original graph than the trees presented here. From this we conclude that algorithms that compare their result only to the routing cost of the minimum routing cost tree do not always yield better results than those presented here.

4   Related Work

Minimum Routing Cost Trees are also known as uniform Minimum Communication Cost Spanning Trees [16,17] and shortest Total Path Length Spanning Trees [21]. Furthermore the MRCT problem is a special case of the Optimal Network Problem, first studied in the 1960s by [18] and later by [6]. In [20] Wong presented heuristics and approximations to the Optimal Network Problem with a restriction that makes the problem similar to the MRCT problem and obtained a 2-approximation. In [12] it is shown that this restricted version, which Wong studied on unweighted graphs, is NP-hard as well. It seems that earlier the authors of [11] formulated a similar problem under the name "Optimum communication spanning tree" where in addition to costs on edges, we are given a requirement-value $r_{u,v}$ for each pair of vertices that needs to be taken into account when computing the routing cost. In this setting one wants to find a tree T such that $\sum_{u,v \in V} r_{u,v} d_T(u, v)$ is minimized. In [22] it is argued that for metric graphs, the results by [1,2,4] yield a $\mathcal{O}(\log n \log \log n)$-approximation to this problem. Using a result presented in [7], this can be improved to be an $\mathcal{O}(\log n)$-approximation. In [13] it is shown how to implement this result in a distributed setting. They state their result depending on the shortest path diameter $D_{sp}(G) := \max_{u,v \in V}\{|P| \; |P$ is a shortest weighted (u, v)-path$\}$ of a graph. This diameter represents the maximum number of hops of any shortest weighted path between any two nodes of the graph. The authors of [13] obtain a randomized approximation of the MRCT with expected approximation-ratio $\mathcal{O}(\log n)$ in time $\mathcal{O}\left(D_{sp} \cdot \log^2(n)\right)$. Observe that this might be only a $\mathcal{O}(n \log^2 n)$-approximation even in a graph with $D_h = 1$ and $D_{sp} = n - 1$, such as a clique where all edges have weight n except $n - 1$ edges of weight 1 forming a line as a subgraph.[6] In our distributed setting we know that it is hard to approximate an MRCT due to Theorem 4.

Theorem 4 (Version of Theorem 5.1. of [5]). *For any polynomial function $\alpha(n)$, numbers p, $B \geq 1$, and $n \in \{2^{2p+1}pB, 3^{2p+1}pB, \ldots\}$, there exists a constant $\varepsilon > 0$ such that in the CONGEST model any distributed $\alpha(n)$-approximation algorithm for the MRCT problem whose error probability is smaller than ε requires $\Omega\left(\left(\frac{n}{pB}\right)^{\frac{1}{2} - \frac{1}{2(2p+1)}}\right)$ time on some $\Theta(n)$-vertex graph of diameter $2p + 2$.*

For certain realistic weight-functions our randomized algorithm breaks this $\Omega(\sqrt{n} + D)$-time lower bound. This is no contradiction, as the construction of [5] heavily relies on being able to choose highly different weights, which might not always appear in practice: in current LAN/PAN networks, weights (delays) usually differ only by a small factor. In case the weights are indeed the delay-times, the runtime of our algorithm just depends on the maximal delay that occurs between any two nodes in the network. Observe that also the runtime of the algorithm of [13] stated for arbitrary weight functions does not contradict this

---

[6] According to [22] it is NP-hard to find an MRCT in a clique.

approximation lower bound. The algorithm's runtime depends on the shortest path diameter D_{sp}, which is $\Theta(\sqrt{n}+D)$ in the worst case graphs provided in [5]. Finally we want to point out that for weighted graphs it might be possible to combine the recent result of [14] with the techniques developed in this paper. This might improve over the approximation factor of [13] for weighted graphs while getting a better runtime in some cases.

Related work in the non-distributed setting includes [22], where a PTAS to find the MRCT of a weighted undirected graph is presented. It is shown how to compute a $(1 + 2/(k + 1))$-approximation for any $k \geq 1$ in time $\mathcal{O}\left(n^{2k}\right)$. Details on the limits of transferring this PTAS into our distributed setting can be found in the full version of this paper [8]. In [8] we also summarize further related work in other models (non-distributed and parallel) that deal with the MRCT problem as well as the related problem of computing low stretch spanning trees.

5   Trees That 2-Approximate the Routing Cost

The main structure we need in this section are shortest path trees:

Definition 4 (Shortest path tree). *A shortest path tree (SP-tree) rooted in a node v, is a tree that connects any node u to the root v by a shortest path in G. In unweighted graphs, this is simply a breadth first-search tree.*

Previously it was known due to Wong [20], Theorem 3, that there is an SP-tree, which 2-approximates the routing cost of an MRCT. We restate this result by using an insight stated in Wong's analysis such that this tree not only 2-approximates the routing cost $RC_V(T)$ of an MRCT T of G (which is a V-MRCT) as Wong stated it, but even yields a 2-approximation of the routing cost $RC_V(G)$ when using shortest paths in the network G itself. Thus, on average the distances between two pairs in the tree are only a factor 2 worse than the distances in G.

The algorithm that corresponds to Wong's analysis computes and evaluates n SP-trees, one for each node in V. We show, that for the S-MRCT problem it is sufficient to consider only those shortest path trees rooted in nodes of S. At the same time, a slightly more careful analysis yields a slightly improved approximation factor of $2 - 2/|S|$, which is of interest for small sets S. Before we start, we define a useful measure for the analysis.

Definition 5 (Single source routing cost). *By $SSRC_S(v) := \sum_{u \in S} \omega_G(v, u)$ we denote the sum of the single source routing costs from node v to every other node in S by using edges in G.*

Note that for simplicity we defined an SP-tree to contain all nodes of V. However, one could also consider the subtree where all leaves are nodes in S. The measures RC_S and $SSRC_S$ would not change, as any additional edges are never used by any shortest paths and thus do not contribute to the S-routing cost of the tree. Such a tree can easily be obtained from the tree we compute.

Theorem 5. *Let $|S|$ be at least 2. In weighted graphs, the SP-tree T_v rooted in a node v with minimal single source routing cost $SSRC_S(v) = \min_{u \in S} SSRC_S(u)$ over all SP-trees rooted in nodes of S is a $(2 - 2/|S|)$-approximation to the S-routing cost $RC_S(G)$ in G.*

Corollary 1. *In weighted graphs, an SP-tree with minimum routing cost over all SP-trees rooted in nodes of S is a $(2 - 2/|S|)$-approximation to an S-MRCT.*

The proof of this theorem uses and modifies the ideas of the proof of Theorem 3 in [20]. The following proof is an adapted version of this proof.

Proof. Let v be the node for which the SP-tree T_v has minimal single source routing cost with respect to S among all SP-trees, that is $v := arg \min_{v \in V} SSRC_S(v)$.

The cost of connecting a node $u \neq v$ to all other nodes in S using edges in T_v is upper bounded by $(|S| - 2) \cdot w_G(v, u) + SSRC_S(v)$. This essentially describes the cost of connecting u to each other node by a path via the root v and using edges in T_v. Therefore the total routing cost $RC_S(T_v)$ for S using the network T_v can be bounded by

$$RC_S(T_v) \leq SSRC_S(v) + \sum_{v \neq u \in S} ((|S| - 2) \cdot w_G(v, u) + SSRC_S(v)).$$

As $|S| \geq 2$, this can be further transformed and bounded to be

$$= |S| \cdot SSRC_S(v) + (|S| - 2) \sum_{u \in S} w_G(v, u)$$

$$= |S| \cdot SSRC_S(v) + (|S| - 2) \cdot SSRC_S(v)$$

$$= (2 - 2/|S|) \cdot |S| \cdot SSRC_S(v)$$

$$\leq (2 - 2/|S|) \cdot \sum_{u \in S} SSRC_S(u).$$

Where the last bound follows, as $SSRC_S(v)$ is minimal among all $SSRC(u)$ for $u \in S$. Since $\sum_{u \in V} SSRC_S(u)$ is the same as $RC_S(G)$, we obtain that $RC_S(T_v) \leq 2RC_S(G)$. □

6   Considering few Randomly Chosen SP-Trees Is Almost as Good

We show that when investigating a small subset of all SP-trees chosen uniformly at random, with high probability one of these trees is a good approximation as well.

Lemma 1. *Let $\beta(n, D)$ be a positive function in n and D and define $\gamma := \left\lceil \frac{2 - 2/|S|}{\beta(n,D)} \right\rceil + 1$. Assume $S \subseteq V$ is of size at least $\gamma \ln n$. Let S' in turn be a subset of S chosen uniformly at random among all subsets of S of size $\gamma \ln n$. Let $v \in S'$ be a node such that $SSRC_S(v) = \min_{u \in S'} SSRC_S(u)$. Then $RC_S(T_v) \leq (2 - 2/|S| + \beta(n, D))RC_S(G)$.*

Proof. For simplicity, without loss of generality we assume that $|S|$ is a multiple of γ. Denote by $v_1, \ldots, v_{|S|}$ the nodes in S such that $SSRC_S(v_1) \leq SSRC_S(v_2) \leq \cdots \leq SSRC_S(v_{|S|})$. That is they are ordered corresponding to their single source routing costs. We say a node v is good, if the corresponding SP-tree T_v is among the $1/\gamma$-fraction of the SP-trees with lowest single source routing cost[7]. Therefore v is good if $SSRC_S(v) \leq SSRC_S(v_{|S|/\gamma})$ with respect to the above order of the trees.

First we prove that w.h.p. set S' contains a good node. Second we prove, that the corresponding SP-tree yields the desired approximation ratio.

1) Probability analysis: We know that $Pr_{v \in S}[v \text{ is good}] = 1/\gamma$. Furthermore each node $v \in S$ is included in set S' independent of the other nodes. Therefore we can conclude that the probability that at least one of the nodes v in S' is good is $1 - \left(1 - \frac{1}{\gamma}\right)^{|S'|} = 1 - \left(1 - \frac{1}{\gamma}\right)^{\gamma \ln n} > 1 - 1/n$ and thus high.

2) Approximation-ratio analysis: Let v_i be a good node. As in the proof of Theorem 5 we know that $RC_S(T_{v_i}) \leq (2 - 2/|S|) \cdot |S| \cdot SSRC_S(v_i)$..As $RC_S(G) = \sum_{u \in S} SSRC_S(u)$ and v_i is good, we can conclude that $SSRC_S(v_i) \leq \frac{1}{(1-1/\gamma) \cdot |S|} \cdot RC_S(G)$ as there are at most $(1-1/\gamma)|S|$ nodes v_j with $SSRC_S(v_j) \geq SSRC_S(v_i)$. Equality is approached in the worst case, where $j := |S|/\gamma$ and $SSRC_S(v_j) = 0$ for each $j < i$ and $SSRC_S(v_i) = SSRC_S(v_j)$ for all $j \geq i$.

Combined with Bound (6) it follows that $RC_S(T_{v_i}) \leq \frac{2-2/|S|}{1-1/\gamma} \cdot RC_S(G)$. Due to the choice of γ we conclude the statement of the Lemma.

7   How to Compute the Routing Cost of Many SP-Trees in Parallel

In Theorem 5 (and Lemma 1) we demonstrated that an SP-tree T_v with minimum single source routing cost yields a 2-approximation for $RC_S(G)$. The single source routing cost of a tree can be computed by computing distances between the root of a tree and nodes in S. However, instead of finding an SP-tree with smallest single source routing cost the literature usually considers finding an SP-tree with smallest routing cost. This is done e.g. in [20]. The reason for this is that the bound in the proof of Lemma 5 is not sharp when using the single source routing cost. To see this, we recall that while obtaining the bound, one approximates the distance between two nodes in the tree by adding up their distance to the root. Thus the bound considers the single source routing cost of an SP-tree. Compared to this, the routing cost takes the actual distance of the two nodes in an SP-tree into account. An explicit example for a graph that contains a node u such that $RC_S(T_u) < RC_S(T_v)$, where T_v has minimum single source routing cost is given in the full version of this paper [8]. Like in [20] we focus on this more powerful version of finding a tree of small routing cost.

---

[7] Due to the choice of $\gamma := \left\lceil \frac{2-2/|S|}{\beta(n,D)} \right\rceil + 1$ a good tree is among the $n\beta(n, D)$ cheapest trees.

Lemma 2. *Let $S := \{v_1, \ldots, v_{|S|}\}$ be a subset[8] $S \subseteq V$ of all nodes of a graph. Then we can compute the values $RC_S(T_{v_1}), \ldots, RC_S(T_{v_{|S|}})$ in time $\mathcal{O}(D_\omega + |S|)$ when using either uniform weights for all edges or a weight function implied by the delay/edge traversal time.*

The proof of this lemma can be found at the end of this section. First, we describe our algorithm that is used to prove this lemma. In Part 1 of this algorithm we start by computing SP-trees T_v for each $v \in S$. A pseudocode for this algorithm can be found as Algorithm 7.1. Part 2 deals with computing the routing cost of a single tree and is described later in this section.

We start by noting that for the weight functions we consider an SP-tree is just a Breath First Search tree (BFS-tree). This part is essentially the same as in the S-SP algorithm of [10] extended to edge-weights derived from the delays to send a message. We also store some additional data that is used later in Algorithm 7.2 to compute routing costs but was not needed for the S-SP computation in [10]. In Algorithm 7.2, for each node $v \in S$ an SP-tree T_v is constructed using what we call delayed breadth first search (DBFS). By DBFS we think of a breadth first search, where traversing edge (u, u') takes $\omega_G(u, u')$ time slots. In the end each node u in the graph knows $\omega_G(u, v)$. In addition each node u knows for each $v \in S$ its parent in the corresponding tree T_v. Furthermore node u knows at what time the DBFS, that computed T_v, sent its message to u via u's parent. During Algorithm 7.2, these timestamps are used to compute the routing cost of all these trees in time $\mathcal{O}(|S| + D_\omega)$.

Remark 1. Compared to Algorithm S-SP presented in [10] we added Lines 2, 6 and 26 in Algorithm 7.1 and extended the algorithm to certain delay functions as mentioned above (the proof in [10] can be naturally extended to those.) By doing so, we can store in $\tau[v]$ the time when a message of the computation of tree T_v was received the first time (via edge *parent_in_T_v*). In the end, $\omega_u[v]$ stores the distance $\omega_G(v, u)$ to v and *parent_in_T_v* indicates the first edge of a (u, v)-path witnessing this.

Despite its similarity to algorithm S-SP in [10], we describe Algorithm 7.1 in more detail for completeness. For the simplicity of the writeup, we refer to u not only as a node, we use u to refer to u's ID as well. Each node u stores $\delta(u)$ sets L_i, one for each of the $\delta(u)$ neighbors $u_1, \ldots, u_{\delta(u)}$ of u, and the sets L and L_{delay} to keep track of which messages were received, transmitted or need to be delayed. At the beginning, if $u \in S$, all these sets contain just u, else they are empty (Lines 1–7). Set L_{delay} is always initialized to be empty. Furthermore u maintains an array ω_u that eventually stores at position v (indicated by $\omega_u[v]$) the distance $\omega_G(u, v)$ to node v. Initially $\omega_u[v]$ is set to infinity for all v and is updated as soon as the distance is known (Line 27). In each node u, array τ stores at position v the time when a message of the computation of tree T_v was received the first time in u. At any time, set L contains all node IDs corresponding to

---

[8] Note that S used here can be e.g. S as in Section 5 or the smaller set S' as in Section 6.

Algorithm 7.1. Computing $SSRC_S(v)$ for each $v \in S$ Part 1 (executed by node u)

1: $L := \emptyset$; $\omega_u := \{0,0,\ldots,0\}$; $L_{delay} := \emptyset$;
2: $\tau := \{\infty,\infty,\ldots,\infty\}$ // **new**
3: **if** $u \in S$ **then**
4:     $L := \{u\}$;
5:     $\omega_u(u) := 0$;
6:     $\tau(u) := 0$; // **new**
7: **end if**
8: $L_1,\ldots,L_{\delta(u)} := L$;
9: **if** u equals 1 **then**
10:     **compute** $D'_\omega := ecc(u)$; //** According to Fact 1, D_ω is smaller than $2 \cdot D'_\omega$.
11:     **broadcast** D'_ω;
12: **else**
13:     **wait until** D'_ω was **received**;
14: **end if**
15: //** Compute S shortest path trees
16: **for** $t = 1,\ldots,|S| + 2 \cdot D'_\omega$ **do**
17:     **for** $i = 1,\ldots,\delta(u)$ **do**
18:         $$(l_i,\omega_i) := \begin{cases} \perp & : \text{if } L_i \setminus \cap L_{delay} = \emptyset \\ \arg\min\{v \in L_i \setminus L_{delay}| \\ \tau[v] + \omega_G(u,v) \geq t\} & : \text{else} \end{cases}$$
19:     **end for**
20:     within one time slot:
        if $l_1 \neq \perp$ **then send** $(l_1,\omega_u[l_1] + \omega_G(u,u_1))$ to neighbor u_1;
        receive (r_1,ω_1) from u_1;
          $\vdots$
        if $l_{\delta(u)} \neq \perp$ **then send** $\left(l_{\delta(u)},\omega_u[l_{\delta(u)}] + \omega_G\left(u,l_{\delta(u)}\right)\right)$ to neighbor $u_{\delta(u)}$;
        receive $\left(r_{\delta(u)},\omega_{\delta(u)}\right)$ from $u_{\delta(u)}$;
21:     $R := \{r_i | r_i < l_i \text{ and } i \in 1\ldots\delta(u)\} \setminus L$
22:     $s := \begin{cases} \infty & \text{if } L_{delay} = \emptyset \\ \min(L_{delay}) & \text{else} \end{cases}$
23:     **if** $s \leq \min(R)$ **and** $s < \infty$ **then**
24:         $L_{delay} := L_{delay} \setminus \{s\}$;
25:     **end if**
26:     **for** $i = 1,\ldots,\delta(u)$ **do**
27:         **if** $r_i < l_i$ **then**
28:             //** T_{l_i}'s message is delayed due to T_{r_i}.
29:             **if** $r_i \notin L$ **then**
30:                 $\tau[r_i] := t$; // **new**
31:                 $\omega_u[r_i] = \omega_i$;
32:                 $L := L \cup \{r_i\}, L_1 := L_1 \cup \{r_i\}, L_2 := L_2 \cup \{r_i\}$,
                    $\ldots L_{i-1} := L_{i-1} \cup \{r_i\}, L_{i+1} := L_{i+1} \cup \{r_i\}, \ldots L_{\delta(u)} := L_{\delta(u)} \cup \{r_i\}$;
33:                 **if** $\min(R) < r_i$ **or** $s < r_i$ **then**
34:                     $L_{delay} = L_{delay} \cup \{r_i\}$
35:                 **end if**
36:                 $parent_in_T_{r_i} := $ neighbor i;
37:             **end if**
38:         **else**
39:             $L_i := L_i \setminus \{l_i\}$; //** T_{l_i}'s message was successfully sent to neighbor i.
40:         **end if**
41:     **end for**
42: **end for**

the tree computations (where each node with a stored ID is the root initiating the computation of such a tree) that already reached u until now. The set L_{delay} contains all root IDs that reached v until time t but are marked to be delayed before forwarded. This ensures that we indeed compute BFS-trees.

Set L_i contains all IDs of L except those that could be forwarded successfully to neighbor u_i in the past. We say an ID l_i is forwarded successfully to neighbor u_i, if u_i is not sending a smaller ID r_i to u at the same time.

To compute the trees in Algorithm 7.1, the unique node with ID 1 computes D'_ω and thus a 2-approximation to the distance-diameter D_ω. This value is subsequently broadcast to the network (Lines 8–12). Then the computation of the $|S|$ trees starts and runs for $|S| + 2D'_\omega$ time steps. Lines 14–17 make sure that at any time the smallest ID, that is not marked to be delayed and was not already forwarded successfully to neighbor u_i is sent to u_i together with the length of the shortest (v, u_i)-path that contains u. In Line 18 we define the set R of all IDs that are received successfully in this time slot for the first time. This set is then used to decide whether to remove an ID s from L_{delay} in Lines 20 and 21, since all IDs that cause a delay to s are transmitted successfully by now. ID s is computed in Line 20. ID s is the smallest element of L_{delay} and is removed from L_{delay} if no other ID smaller than s was received successfully for the first time in this timeslot.

If a node ID r_i was received successful for the first time (verified in Lines 23 and 25), we update $\tau[r_i]$ and $\omega_u[r_i]$, add r_i to the according lists (Lines 28–30) and remember who u's parent is in T_{r_i} (Line 31). In case the ID v was received the first time from several neighbors, the algorithm as we stated it chooses the edge with lowest index i. On the other hand if we did not successfully receive a message from neighbor u_i but sent successfully a message to neighbor u_i, the transmitted ID is removed from L_i (Line 33).

Lemma 3. *Algorithm 7.1 computes an SP-tree T_v for each $v \in S$ in time* $\mathcal{O}(|S| + D_\omega)$.

Proof. This is essentially Theorem 6.1. in [9] stated for Algorithm 7.1 instead of Algorithm S-SP of [9]. Those parts of the two algorithms which contribute to the runtime and correctness are equivalent.

Now Part 2 of our algorithm calculates the routing cost of each tree T_v in parallel in time $\mathcal{O}(D_\omega + |S|)$. A pseudocode of this algorithm is stated in Algorithm 7.2.

To compute the routing cost of a tree, we look at each edge e in each tree T_v and compute the number of (v, w)-paths in T_v that contain the edge e, for $v, w \in S$. The sum of these numbers for each edge in a tree is the tree's routing cost. Given a tree T, for each edge e in T, the edge partitions the tree into two trees (when e was removed). To be more precise, denote by w_e, w'_e the two vertices to which e is incident. Edge e partitions the vertices of T into two subsets, which we call Z^1_e and Z^2_e defined by:

$$Z^1_e(T) := \{w \in S | e \text{ is contained in the unique } (w_e, w)\text{-path in } T\}$$
$$Z^2_e(T) := \{w \in S | e \text{ is contained in the unique } (w'_e, w)\text{-path in } T\}$$

We observe that edge e occurs in all $|Z_e^2(T)|$ paths from any node $v \in Z_e^1(T)$ to any node $w \in Z_e^2(T)$. Note that the total number of paths in which e occurs is $|Z_e^1(T)| \cdot |Z_e^2(T)|$. This fact is later used to compute $RC_S(T)$.

Algorithm 7.2. Computing $RC_S(T_v)$ for each $v \in S$ alternative Part 2 (executed by node u)

1: $rcs := \{\infty, \ldots, \infty\}$; $//^{**}$ is updated during the runtime of the algorithm.
2: **if** $u \in S$ **then**
3:    $z := \{1, \ldots, 1\}$; $//^{**}$ is updated during the runtime of the algorithm.
4: **else**
5:    $z := \{0, \ldots, 0\}$;
6: **end if**
7: **for** $t = 1, \ldots, |S| + 2D'_\omega$ **do**
8:    within one time slot:
     For each $v \in L$ such that $t = |S| + 2 \cdot D'_\omega - \tau[v]$ **send** $(v, rcs[v], z[v])$ to $parent\_in\_T_v$;
     receive (v_1, r_1, z_1) from neighbor u_1; $//^{**}$ r_1 equals $rcs(T_{v_1}, u_1)$,
                  $//^{**}$ z_1 equals $Z_{(u,u_1)}^1(T_{v_1})$
     receive (v_2, r_2, z_2) from neighbor u_2; $//^{**}$ r_2 equals $rcs(T_{v_2}, u_2)$,
                  $//^{**}$ z_2 equals $Z_{(u,u_2)}^1(T_{v_2})$

        $\vdots$

     receive $(v_{\delta(u)}, r_{\delta(u)}, z_{\delta(u)})$ from $u_{\delta(u)}$; $//^{**}$ $r_{\delta(u)}$ equals $rcs\left(T_{v_{\delta(u)}}, u_{\delta(u)}\right)$,
                $//^{**}$ $z_{\delta(u)}$ equals $Z_{(u,u_{\delta(u)})}^1\left(T_{v_{\delta(u)}}\right)$
9:    **for** $i = 1, \ldots, \delta(u)$ **do**
10:      **if** $v_i \neq \bot$ **then**
11:        $rcs[v_i] := rcs[v_i] + r_i + 2\omega_G(u,v) \cdot z_i \cdot (|S| - z_i)$;
12:        $z[v] := z[v] + z_i$;
13:      **end if**
14:    **end for**
15: **end for**
16: $//^{**}$ Now $rcs[u]$ equals $RC_S(T_u)$ in case that $u \in S$. Else it is ∞ and was never modified.

Lemma 4. *For a tree T, the routing cost $RC_S(T)$ can be restated as $RC_S(T) = 2 \cdot \sum_{e \in T} |Z_e^1(T)| \cdot |Z_e^2(T)| \cdot \omega_G(e)$.*

The proof of this lemma can be found in the full version of this paper [8].

To formulate the definition of $RC_S(T)$ in this way helps us to argue that we can compute $RC_S(T)$ recursively in a bottom-up fashion for any T. To do so, we consider trees to be oriented such that we use the notion of child/parent.

Definition 6 (Subtree, partial routing cost). *Given a tree T, for each node u in an oriented tree T, we define $T|_u$ to be the subtree of T rooted in u containing all descendants of u in T. Denote by V_v the vertices in $T|_v$. Given* node u, denote by $rc_S(T, u)$ the part of the routing cost $RC_S(T)$ that is due to the edges in $T|_u$. We define $rc_S(T, u)$ in a recursive way. In case that $T|_u$ consists of only one node, $T|_u$ contains no edges that could contribute to $rc_S(T, u)$ and we set $rc_S(T, u) := 0$. In case that $T|_u$ contains more than one node, we denote the children of u in T by $u_1, \ldots, u_{\delta(u)-1}$ and define $rc_S(T, u) :=$ $\sum_{i=1}^{\delta(u)-1} rc_S(T, u_i) + 2 \cdot \sum_{i=1}^{\delta(u)-1} w_G(u, u_i) \cdot |Z^1_{(u,u_i)}(T)| \cdot |Z^2_{(u,u_i)}(T)|$.

Note that $rc_S(T, u)$ is a measure with respect to the routing cost in T and thus different from $RC_S(T|_u)$. Besides $RC_S(T|_u)$ being undefined when $T|_u$ does not contain all nodes in S, $RC_S(T|_u)$ would take only routing cost within $T|_u$ into account.

We now formally prove that $rc_S(T, u)$ essentially describes the contribution of edges in subtree $T|_u$ to the total routing cost and conclude:

Lemma 5. *Let T be a tree rooted in node r. Then $RC_S(T) = rc_S(T, r)$.*

The proof of this lemma can be found in the full version of this paper [8].

Using this insight we are able to compute $RC_S(T_v)$ for all $v \in S$ in parallel recursively in a bottom-up fashion. This is by computing $rc_S(T_v, u)$ for each u based on aggregating $rc_S(T_v, u_j)$ for each of u's children. For each $v \in S$ these computations of $RC_S(T_v)$ run in parallel. A schedule on how to do these bottom-up computations in time $\mathcal{O}(|S| + D_\omega)$ is provided by using the inverted entries of τ.

In more detail each node u computes for each $v \in S$ the costs $rc_S(T_v, u)$ (stored in $rc_S[v]$) of its subtree of T_v as well as the number of nodes in $T_v|_u$ (stored in $z[v]$ and sends this information to its parent in T_v. When we computed T_v in Algorithm 7.1, we connected u via edge $parent\_in\_T_v$ to T_v at time $\tau[v]$. To avoid congestion we send information from u to its parent in T_v only at time $t = |S| + 2D'_\omega - \tau[v]$ (Line 7). Note that this schedule differs from the one that is implied by the computation of the trees in the sense that now only edges in the tree are used, while more edges were scheduled while building the trees. The edges used now in time slot $t = |S| + 2D'_\omega - \tau[v]$ are a subset of those scheduled at time $t = |S| + 2D'_\omega - \tau[v]$ while constructing the trees, such that there is no congestion from this modification.

At the same time as u sends, u receives messages from its neighbors. E.g. neighbor u_i might send $rc_S(T_{v'}, u_i)$ and $Z^1_{(u,u_i)}(T_{v'})$ for another node v'. In Lines 8 − 11 node u updates its memory depending on the received values. In the end the node with ID 1 computes $v := arg\min_{v \in V} RC_S(T_v)$ via aggregation using T_1. Node 1 informs the network that tree T_v is a 2-approximation to an S-MRCT.

Theorem 6. *The algorithm presented in this section computes all $|S|$ values $RC_S(T_v)$ for each node $v \in S$ in time $\mathcal{O}(|S| + D_\omega)$.*

Proof. **Runtime:** The construction of the $|S|$ trees in Algorithm 7.1 takes at most $\mathcal{O}(|S| + D_\omega)$ rounds as stated in Lemma 3. To forward/compute the costs from the leaves to the roots $v \in S$ in Algorithm 7.2 takes $|S| + 2D'_\omega$ since we just use the schedule τ of this length computed in Algorithm 7.1. Thus the total time used is $\mathcal{O}\left(|S| + D_\omega\right)$.

Correctness: We consider time slot $|S| + 2D'_\omega - \tau[v]$. If u is a leaf of T_v, it sends $(v, 0, 1)$ to its parent in T_v in case $u \in S$, else it sends $(v, 0, 0)$, which is correct. In case u is not a leaf, each child u_i has sent $rc_S\left(T_v, u_i\right)$ (stored in r_i) as well as $Z^1_{(u,u_i)}\left(T_v\right)$ (stored in z_i) to u at an earlier point in time. This is true as time-stamp $\tau[v]$ stored in u_i is always larger than time-stamp $\tau[v]$ stored in u, as u_i is a child of u. Each time u received some of these values from its children in T_v, it updated its memory according to Lemma 5 (Lines $8 - 11$ of Algorithm 7.2), leading to sending the correct values $rc_S\left(T_v, u\right)$ and $Z^1_{(parent\_in\_T_v,u)}\left(T_v\right)$ to its parent in T_v at time $|S| + 2D'_\omega - \tau[v]$. Thus in any case u sends the correct values.

We conclude that each node $v \in S$ has computed $rc_S(T_v, v) = RC_S(T_v)$ after Algorithm 7.2 has finished.

8   Proofs of Main Results

We put the tools of the previous sections together and prove the Theorems of Section 1.

Proof. (of Theorem 2). First, Algorithms 7.1 and 7.2 are used to compute $RC_S(v)$ for each $v \in S$. For each such node v, the value $RC_S(v)$ is stored in node v itself. A leader node (e.g. with lowest ID, which can be found in time $\mathcal{O}(D_\omega)$) computes $u := arg\min_{v \in V} RC_S(v)$ via aggregation using T_l, where l is the leader node. As stated in Theorem 5 the tree T_u is a $(2 - 2/|S|)$-approximation of a S-MRCT. The leader node informs the network that tree T_u is a $(2 - 2/|S|)$-approximation to an S-MRCT. The runtime follows from Lemma 2 and the fact, that to determine u by aggregating the corresponding minimum and to broadcast u can be done in time $\mathcal{O}(D_\omega)$.

Proof. (of Theorem 3). First we select a subset $S' \subseteq S$ of the size stated in Lemma 1. Each node joins a set S'' with probability $c \cdot s/n$, where s is the (desired) size of S' stated in Lemma 1 and c a constant depending on a Chernoff bound used now. Using such a Chernoff Bound, w.h.p. S'' is of size $c \cdot s$ or some constant $c \geq 1$. Now all IDs of nodes in S'' are sent to the leader who selects and broadcasts a subset S' of the desired size among the IDs of S''.

From now on the algorithm works exactly as in the proof of Theorem 2, except that the algorithm is run on S' instead of S (it computes and aggregates each $RC_S(v)$ for $v \in S'$ instead of S). As stated in Lemma 1, a tree T_u is found that is a $(2 - 2/|S| + \beta(n, D))$-approximation of an S-MRCT. The leader node informs the network that tree T_u is a $(2 - 2/|S| + \beta(n, D))$-approximation to an S-MRCT. Choosing $\beta(n, D) := \min\left\{\frac{\log n}{D}, \alpha(n, D)\right\}$ yields the desired approximation ratio of $2 - 2/|S| + \min\left\{\frac{\log n}{D}, \alpha(n, D)\right\}$, as stated in the Theorem.

Runtime analysis: As $s = \left(\left\lceil \frac{2-2/|S|}{\beta(n,D)} \right\rceil + 1 \right) \cdot \ln n$, selecting a set S'' and deriving S' can be done w.h.p. in time

$$\mathcal{O}(D + s) = \mathcal{O}\left(D + \left(\left\lceil \frac{2 - 2/|S|}{\beta(n, D)} \right\rceil + 1 \right) \cdot \ln n \right) = \mathcal{O}\left(D + \frac{\log n}{\beta(n, D)} \right),$$

which is $\mathcal{O}\left(D + \frac{\log n}{\alpha(n,D)} \right)$ due to the choice of β. The same runtime follows from Lemma 2 for computing the single source routing costs for all $v \in S'$. Combined with the fact that the aggregation and broadcast of u can be done in time $\mathcal{O}(D)$, the stated result is obtained.

Acknowledgment. We would like to thank Benjamin Dissler and Mohsen Ghaffari for helpful discussions and insights.

References

1. Bartal, Y.: Probabilistic approximation of metric spaces and its algorithmic applications. In: Proceedings of the 37th Annual IEEE Symposium on Foundations of Computer Science, FOCS 1996, Burlington, Vermont, USA, October 14-16, pp. 184–193 (1996)
2. Bartal, Y.: On approximating arbitrary metrices by tree metrics. In: Vitter, J.S. (ed.) Proceedings of the 30th Annual ACM Symposium on Theory of Computing, STOC 1998, Dallas, Texas, USA, May 23-26, pp. 161–168 (1998)
3. Campos, R., Ricardo, M.: A fast algorithm for computing minimum routing cost spanning trees. Computer Networks 52(17), 3229–3247 (2008)
4. Charikar, M., Chekuri, C., Goel, A., Guha, S.: Rounding via trees: deterministic approximation algorithms for group steiner trees and k-median. In: Vitter, J.S. (ed.) Proceedings of the 30th Annual ACM Symposium on Theory of Computing, STOC 1998, Dallas, Texas, USA, May 23-26, pp. 114–123 (1998)
5. Das Sarma, A., Holzer, S., Kor, L., Korman, A., Nanongkai, D., Pandurangan, G., Peleg, D., Wattenhofer, R.: Distributed verification and hardness of distributed approximation. SIAM Journal on Computing 41(5), 1235–1265 (2012)
6. Dionne, R., Florian, M.: Exact and approximate algorithms for optimal network design. Networks 9(1), 37–59 (1979)
7. Fakcharoenphol, J., Rao, S., Talwar, K.: A tight bound on approximating arbitrary metrics by tree metrics. In: Larmore, L.L., Goemans, M.X. (eds.) Proceedings of the 35th Annual ACM Symposium on Theory of Computing, STOC 2003, San Diego, California, USA, June 9-11, pp. 448–455 (2003)
8. Hochuli, A., Holzer, S., Wattenhofer, R.: Distributed approximation of minimum routing cost trees. Computing Research Repository CoRR, abs/1406.1244 (2014), http://arxiv.org/abs/1406.1244
9. Holzer, S., Peleg, D., Roditty, L., Tal, E., Wattenhofer, R.: Optimal distributed all pairs shortest paths and applications (2014), http://www.dcg.ethz.ch/~stholzer/APSP-full.pdf, preliminary full version of two merged papers to be submitted to a journal). New versions available on request

10. Holzer, S., Wattenhofer, R.: Optimal distributed all pairs shortest paths and applications. In: Kowalski, D., Panconesi, A. (eds.) Proceedings of the 31st Annual ACM SIGACT-SIGOPS Symposium on Principles of Distributed Computing, PODC 2012, Funchal, Madeira, Portugal, July 16-18, pp. 355–364 (2012)
11. Hu, T.C.: Optimum communication spanning trees. SIAM Journal on Computing 3(3), 188–195 (1974)
12. Johnson, D.S., Lenstra, J.K., Rinnooy Kan, A.H.G.: The complexity of the network design problem. Networks 8(4), 279–285 (1978)
13. Khan, M., Kuhn, F., Malkhi, D., Pandurangan, G., Talwar, K.: Efficient distributed approximation algorithms via probabilistic tree embeddings. In: Bazzi, R.A., Patt-Shamir, B. (eds.) Proceedings of the 27th Annual ACM SIGACT-SIGOPS Symposium on Principles of Distributed Computing, PODC 2008, Toronto, Ontario, August 18-21, pp. 263–272 (2008)
14. Nanongkai, D.: Distributed approximation algorithms for weighted shortest paths. To appear in: Proceedings of the 46th Annual ACM Symposium on Theory of Computing, STOC 2014, New York, USA, May 31-June 3 (2014)
15. Peleg, D.: Distributed computing: a locality-sensitive approach. Society for Industrial and Applied Mathematics, Philadelphia (2000)
16. Peleg, D.: Low stretch spanning trees. In: Diks, K., Rytter, W. (eds.) MFCS 2002. LNCS, vol. 2420, pp. 68–80. Springer, Heidelberg (2002)
17. Reshef, E.: Approximating minimum communication cost spanning trees and related problems. Master's thesis, Weizmann Institute of Science, Rehovot, Israel (1999)
18. Scott, A.J.: The optimal network problem: Some computational procedures. Transportation Research 3(2), 201–210 (1969)
19. Wikipedia. Bridging (networking) (April 28, 2014),
    http://en.wikipedia.org/wiki/Bridging_(networking)
20. Wong, R.T.: Worst-case analysis of network design problem heuristics. SIAM Journal of Algebraic Discrete Methods 1(1), 51–63 (1980)
21. Wu, B.Y., Chao, K.-M., Tang, C.Y.: Approximation algorithms for the shortest total path length spanning tree problem. Discrete Applied Mathematics 105(1), 273–289 (2000)
22. Wu, B.Y., Lancia, G., Bafna, V., Chao, K.-M., Ravi, R., Tang, C.Y.: A polynomial-time approximation scheme for minimum routing cost spanning trees. SIAM Journal on Computing 29(3), 761–778 (1999)
23. Wu, B.Y.: A polynomial time approximation scheme for the two-source minimum routing cost spanning trees. Journal of Algorithms 44(2), 359–378 (2002)

Randomized Lower Bound
for Distributed Spanning-Tree Verification*

Taisuke Izumi

Graduate School of Engineering, Nagoya Institute of Technology, Japan
t-izumi@nitech.ac.jp

Abstract. The *distributed verification* is the problem of deciding whether the subgraph induced by an input edge set L has a desired property (e.g., spanning trees, connectivity, cycle containment, and so on) or not. In this paper, we consider the lower bounds for the distributed verification of spanning trees and Hamiltonian paths. While the original work of the distributed verification by Das Sarma et al. [1] has shown their $\tilde{\Omega}(\sqrt{n})$-round lower bounds, that result is applied only for deterministic algorithms. Recently, their randomized lower bounds are proved by Elkin et al. [3], but the proof strategy is quite complicated. The primary contribution of this paper is that the same randomizied lower bounds are obtained by a simple and elementary reduction from the well-known two-party communication complexity of the set-disjointness function. We also show a tight lower bound for the verification problem of *low-diameter* spanning trees. By a simple modification of our proof, we can show that the randomized $\Omega(\min\{\sqrt{n}/\log n, h\})$-round lower bound holds for the verification of spanning trees with diameter h. This result implies that the naive approach (i.e., the breadth-first search along the edges in L) is the best possible for the verification of low-diameter spanning trees.

1   Introduction

The problem of the distributed verification is stated as follows: The distributed system is a (weighted or unweighted) network $G = (V, E)$, and we have a subset L of E as the input of the problem. A graph property P, such as spanning trees, connectivity, cycle containment, and so on, is also given. The distributed verification algorithm must decide whether the graph $G(L)$ induced by L has the property P or not (that is, the value of $P(G(L))$) as fast as possible. A trivial and universal solution for the problem is to aggregate all information about L and decide $P(G(L))$ in centralized ways. If the communication bandwidth of each link is not bounded, this approach gives an optimal-time algorithm with $O(D)$ rounds, where D is the diameter of G. However the assumption of so rich bandwidth is far from real systems, and thus the challenge of the verification problem is to solve it in the environment with limited bandwidth. Theoretically, such environments are called as the CONGEST model, where processes work under

---

* This work is supported in part by KAKENHI No.25106507 and No.25289114.

M. Halldórsson (Ed.): SIROCCO 2014, LNCS 8576, pp. 137–148, 2014.

the round-based synchrony, and each link can transfer $O(\log n)$-bit messages per one round. An importance of distributed verification problems is that they cleverly capture the difficulty of several graph problems in the CONGEST model, e.g., minimum spanning tree, st-shortest path, and diameter. A lot of hardness results for such global problems are presented as the corollaries of complexity analyses for the distributed verification. In addition, the verification problem itself is of interest to application sides. For example, the verification of spanning trees, which is the problem we consider in this paper, can be used for the failure detection of broadcast trees and routing tables.

In this paper, we focus on the distributed verification of spanning trees and Hamiltonian paths. More precisely, we consider the lower bound for Hamiltonian paths because its lower bound also deduces the bound for spanning trees by the existence of a simple reduction scheme. The paper initiating verification problems [1] gives a general framework to obtain the lower bounds for verification problems based on the reduction from the two-party communication complexity by Yao [19]. The two-party communication complexity is a theory to reveal the amount of communication to compute a global function whose inputs are distributed among two players. The reduction framework in [1] induces $\Omega(\sqrt{n}/\log n + D)$-round lower bounds for many verification problems, which includes both Hamiltonian paths and spanning trees. However, while most of them are obtained from the two-party communication complexity for the set-disjointness function, the verification of Hamiltonian paths and Spanning trees is reduced from that for the equality function. It is well-known that the communication complexity for the N-bit two-party equality is $\Omega(N)$ bits only for *deterministic* protocols, but $\Theta(\log N)$ bits for *randomized* protocols. Since $\Theta(\log N)$-bit complexity does not suffice to lead lower bounds for verification problems, the lower bounds for the spanning-tree and Hamiltonian-path verification derived in [1] hold only for deterministic algorithms. Recently, Elkin et al. [3] proved that the same lower bounds hold for the randomized cases in relation to the context of the quantum distributed computing. Since $O(\sqrt{n}\log^* n + D)$ rounds suffices to verify spanning trees deterministically (which is easily deduced from the result for the minimum-spanning tree construction by Garay et al. [5]), their result implies that randomization does not help so much to achieve the faster verification of spanning trees and Hamiltonian paths.

Our primary contribution is to give an elementary proof of the same randomized lower bound. As we mentioned above, the proof by Elkin et al. [3] is based on some unconventional computational models, and quite complicated. In contrast, our proof is built on the standard framework in [1]. The core of our proof is to provide a new reduction from the two-party set-disjointness to the verification of Hamiltonian paths and spanning trees. Compared with the proofs in prior work, our reduction is extremely simple and the gadget size is small, and thus the asymptotic bound derived by our reduction has a better coefficient. We also show a tight lower bound for the verification problem of *low-diameter* spanning trees, which verifies the property of "spanning tree with diameter less than h" for any given h. By a simple modification of our proof, we can show that

$\Omega(\min\{\sqrt{n}/\log n, h\})$-round lower bound for the verification of spanning trees with diameter h. This result implies that the naive algorithm, performing the breadth-first search along the edges in L, is the best possible for verifying the spanning trees of diameter $o(\sqrt{n}/\log n)$.

The paper is organized as follows: In Section 2 we state the related work. Section 3 provides the notations and definitions used in the paper. The main result is shown in Section 4. We present a discussion for the case of low-diameter spanning trees in Section 5. Finally the paper is concluded in Section 6.

2 Related Work

The paper by Das Sarma et al. [1] is the first one explicitly considering the distributed verification problem, which has given a general framework to lead lower bounds and approximation hardness for a vast class of problems. It is used in several following papers to obtain the complexity for a number of graph problems: Weighted/unweighted diameter and all-pair shortest paths [16,7,10,11], minimum cuts [6,13], distance sketches [10], weighted single-source shortest paths [10,13], fast random walks [14], and so on. For most of those problems $\tilde{\Omega}(\sqrt{n})$-round lower bounds are obtained, and some of them have nearly-tight upper-bound results, which are also considered in the papers cited above. The verification of minimum-spanning trees is also considered by Kor et al. [9].

While the framework by Das Sarma et al. [1] pointed out a general relationship interconnecting the communication complexity theory and distributed complexity theory, the construction of worst-case instances used in the framework is much inspired by the earlier papers leading the time lower bound for the distributed MST construction [17,12,2].

Since it is inherently difficult to lead $\omega(\sqrt{n})$-round lower bounds using the same strategy, an approach proving much powerful lower bound is also considered. The paper by Frischknecht et al. [4] provided a construction of hard instances which can potentially induce $\omega(\sqrt{n})$-round lower bounds, and proved that the time lower bound for the exact computation of the unweighted diameter is $\Omega(n/\log n)$.

3 Preliminaries

3.1 Round-Based Distributed Systems

A distributed system consists of n nodes interconnected with communication links. We model it by a undirected graph $G = (V, E)$, where $V = \{v_0, v_1, \cdots, v_{n-1}\}$ is the set of nodes, and $E \subseteq V \times V$ is the set of links (edges). The diameter of G is denoted by D, and the set of edges incident to v_i is denoted by I_i. Executions of the system proceed with a sequence of consecutive rounds. In each round, each process sends a (possibly different) message to each neighbor, and within the round, all messages are received. After receiving its messages, the process performs local computation. Throughout this paper, we restrict the number of bits transmittable through any communication link per one round to $O(\log n)$ bits. This is known as the CONGEST model.

We assume an initial knowledge for the value of n. This assumption is not essential because it is easily realized by a standard aggregation algorithm (e.g., see [15]), which incurs only $O(D)$ extra rounds.

3.2   Distributed Verification Problems

Let P be a predicate defined over all undirected graphs. We assume that an input label is assigned to each edge in G. Let $l : E \rightarrow \{0, 1\}$ be the labeling function. We define the set L of edges as $L = \{e \mid l(e) = 1\}$, and define $G(L)$ as the subgraph of G induced by L. The verification problem for predicate P is to determine whether $P(G(L))$ is true or not for any labeling function l. More precisely, each process v_i initially knows the set of edges $L \cap I_i$, as the input, and after the run of the algorithm it must output the value of $P(G(L))$.

A concrete problem of the distributed verification is specified by its predicate P. In this paper we consider the spanning-tree verification problem, (the corresponding predicate is denoted by P_{stree}) and the Hamiltonian-path verification problem (denoted by P_{Ham}). That is, $P_{stree}(G)$ (resp. $P_{Ham}(G)$) is true if and only if G is a tree (resp. path) of n nodes. In what follows, we often refer to each verification problem as the corresponding predicate.

It is known that P_{Ham} is not easier than P_{stree}. The following theorem has been proved:

Theorem 1 (Das Sarma et al.[1]). *If some (randomized or deterministic) algorithm can solve P_{stree} within t rounds, there exists an algorithm solving P_{Ham} within $t + O(D)$ rounds.*

This theorem implies that we only have to concentrate the lower bound for P_{Ham} and the upper bound for P_{stree}, which directly induces the lower bound for P_{stree} and the upper bound for P_{Ham}.

4   Randomized Lower Bounds for P_{Ham}

4.1   Two-Party Communication Complexity

The *communication complexity* theory is first introduced by Yao [19]. Roughly speaking, it is the theory to reveal the amount of communications to compute a global function whose inputs are distributed over the network. The most successful scenario in the communication complexity theory is the *two-party* communication complexity, where two players, called Alice and Bob, have N-bit strings $\mathbf{a} = (a_0, a_1, \cdots, a_{N-1})$ and $\mathbf{b} = (b_0, b_1, \cdots, b_{N-1})$ respectively and compute a global function $f : \{0, 1\}^N \times \{0, 1\}^N \rightarrow \{0, 1\}$. The communication complexity of a two-party protocol is the number of one-bit messages exchanged by the protocol for the worst case input (if the protocol is randomized, it is defined as the expected number at the worst-case input). One of the most useful problems in the communication complexity theory is the *set-disjointness*, which is defined as follows:

Definition 1. *The N-bit set-disjointness function $disj_N : \{0,1\}^N \times \{0,1\}^N \to \{0,1\}$ is defined as follows:*

$$disj_N(\mathbf{a}, \mathbf{b}) = \begin{cases} 1 & \text{if } \exists i \in [0, N-1] : a_i = b_i = 1, \\ 0 & \text{otherwise} \end{cases}$$

For this problem, the following theorem is known.

Theorem 2 (Kalyanasundaram and Schnitger, Razborov [8,18]). *The communication complexity of the N-bit set-disjointness problem is $\Omega(N)$.*

To obtain the lower bounds, this paper uses a variation of the two-party computation problem in distributed settings. We assume that Alice and Bob are placed at two nodes in a network of n nodes, and have N-bit strings $\mathbf{a}$ and $\mathbf{b}$, respectively. It is also assumed that each node in the network (including ones other than Alice and Bob) knows everything (i.e., the complete knowledge of the network topology) except for the N-bit strings held by Alice and Bob. Then all nodes must work cooperatively for outputting the value of $f(\mathbf{a}, \mathbf{b})$ as fast as possible. In what follows, we call this problem setting the *networked two-party computation* (and the networked set-disjointness problem if $f = disj_N$). Note that the measurement of the networked two-party computation is not the amount of communication, but the number of rounds.

Obviously the time complexity of networked two-party computation problems relies on the target function f and the topology of the network. The core of the reduction from the networked set-disjointness problem to the distributed verification is the existence of a class of graphs which well transforms the communication lower bound for the N-bit set-disjointness into the time lower bound for its networked version [1]. The next subsection we look at the construction of those graphs. For simplicity of the argument, throughout the paper, we assume that N is a power of 2, i.e., $N = 2^p$ for some nonnegative integer p. Note that the assumption is not essential and it is not difficult to remove it by considering a slightly larger instance of the original N-bit instance. That is, we consider 2^q-bit instances instead of N-bit ones, where q is the minimum integer satisfying $2^q \geq N$. this modification does not change the asymptotic complexity we show below.

4.2   Graph Construction

The construction shown in this subsection almost follows the result by Das Sarma et al. [1]. Let $\Gamma(N)$ be the graph we construct. It is built by the following steps:

1. We first prepare $4N$ paths of length N, each of which is denoted by P_i ($0 \leq i \leq 4N - 1$). The nodes constituting P_i are identified by $v_{(i,0)}, v_{(i,1)}, \cdots, v_{(i,N-1)}$ from left to right. We further prepare a node $v_{(4N,0)}$. Edges $(v_{(i,0)}, v_{(i+1,0)})$ and $(v_{(i,0)}, v_{(i+2,0)})$ are added for any $i \in [0, 4N - 2]$. Edges $(v_{(i,N-1)}, v_{(i+1,N-1)})$ and $(v_{(i,N-1)}, v_{(i+2,N-1)})$ are added for any $i \in [0, 4N - 3]$.

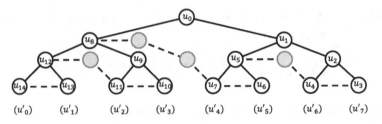

Fig. 1. An example of $AT(8)$

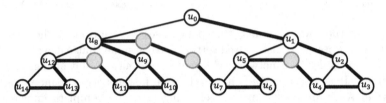

Fig. 2. The Hamiltonian path in $AT(8)$

2. Construct a spacial structure referred as $AT(N)$. The base structure of $AT(N)$ is a complete binary tree with N leaves. The nodes in the tree are labeled by $u_0, u_1, \cdots u_{2N-1}$ in the DFS order where right children always precede to the left. We further augment several paths to make the tree have the Hamiltonian path whose visiting order follows the node indices. Let h and h' be the height of u_i and u_{i+1} in the tree, a path of length $|h - h'| + 1$ is augmented between u_i and u_{i+1} if they are not adjacent to each other. Finally, we give an alias to each leaf node. We refer leaf nodes as $u'_0, u'_1, \cdots, u'_{N-1}$ from left to right. An example of $AT(8)$ is shown in Figure 1, where the dotted lines and gray nodes constitute the paths augmented to the complete binary tree. The Hamiltonian path from u_0 to u_{2N-1} is presented in Figure 2.

3. Add edges $(u'_i, v_{(j,i)})$ for any $i \in [0, N-1]$ and $j \in [0, 4N-1]$.

4. Put Alice and Bob at u'_0 and $u'_{(N-1)}$.

The whole construction is illustrated in Figure 3. Note that the number n of nodes in $\Gamma(N)$ is $\Theta(N^2)$, and its diameter is $D = O(\log n)$. For this graph, we can show the following theorem.

Theorem 3 (Das Sarma et al. [1]). *For any algorithm solving the networked N-bit set-disjointness in $\Gamma(N)$ with high probability, its worst-case running time is $\Omega(N/\log N + D)$ $(= \Omega(\sqrt{n}/\log n + D))$ rounds.*

While the graph used in this paper is a slightly modified version of the original construction in [1], the theorem above is proved in the almost same way. So we just quote it without the proof.

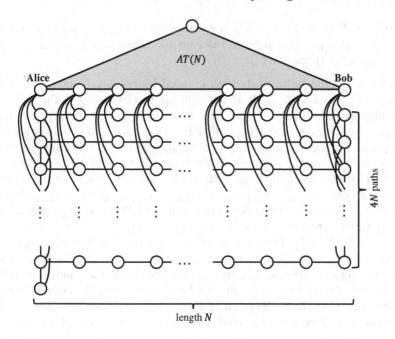

Fig. 3. Construction of $\Gamma(N)$ and a simulation of imaginary nodes

4.3 Reduction

In this subsection we show a reduction from the networked N-bit set disjointness to P_{Ham} incurring a constant number of extra rounds. Combining this reduction with Theorem 3, we obtain the $\Omega(\sqrt{n}/\log n)$-round lower bound for P_{Ham}. Precisely we prove the following lemma.

Lemma 1. *If P_{Ham} is solvable within r rounds with high probability, then the networked N-bit set-disjointness problem in $\Gamma(N)$ is solvable within $r+1$ rounds with high probability.*

Proof. Let $\mathcal{A}$ be an algorithm solving P_{Ham} within r expected rounds. The core of the proof is that any instance $(\mathbf{a},\mathbf{b})$ of the networked N-bit set-disjointness is encoded into some instance L of P_{Ham} such that L constitutes a Hamiltonian path if and only if $(\mathbf{a},\mathbf{b})$ is disjoint. Let $L_{(\mathbf{a},\mathbf{b})}$ is the instance for P_{Ham} corresponding to the set-disjointness instance $(\mathbf{a},\mathbf{b})$. The instance $L_{(\mathbf{a},\mathbf{b})}$ is constructed as follows:

1. $L_{(\mathbf{a},\mathbf{b})}$ contains all the edges constituting the Hamiltonian path of $AT(N)$ in $\Gamma(N)$ and edge $(u'_0, v_{(0,0)})$.
2. For any $i \in [0, 4N-1]$, all edges in P_i are included in $L_{(\mathbf{a},\mathbf{b})}$.
3. If $\mathbf{a}[i] = 0$ ($i \in [0, N-1]$), we add edges $(v_{(4i+1,0)}, v_{(4i+2,0)})$ and $(v_{(4i+3,0)}, v_{(4(i+1),0)})$. Otherwise, we add edges $(v_{(4i+1,0)}, v_{(4i+3,0)})$ and $(v_{(4i+2,0)}, v_{(4(i+1),0)})$ (Figure 4(a)).

4. If $\mathbf{b}[i] = 0$ ($i \in [0, N-1]$), we add edges $(v_{(4i,N-1)}, v_{(4i+1,N-1)})$ and $(v_{(4i+2,N-1)}, v_{(4i+3,N-1)})$. Otherwise, we add edges $(v_{(4i+1,0)}, v_{(4i+3,0)})$ and $(v_{(4i+2,4(i+1))})$ (Figure 4(b)).

Let $\Gamma_i(N) = (V_i, E_i)$ be the subgraph induced by $v_{(4(i+1),0)}$ and the nodes in P_{4i}, P_{4i+1}, P_{4i+2}, and P_{4i+3}. It is not difficult to check that $L_{(\mathbf{a},\mathbf{b})} \cap E_i$ is a Hamiltonian path of $\Gamma_i(N)$ if and only if either $\mathbf{a}[i] = 0$ or $\mathbf{b}[i] = 0$ holds (see Figure 5). Thus, if $\mathbf{a}$ and $\mathbf{b}$ are disjoint, then $L_{\mathbf{a},\mathbf{b}}$ is a Hamiltonian path (which passes through $u_0, u_0', v_{(0,0)}, v_{(4,0)}, v_{(8,0)}, \cdots v_{(4N,0)}$). The final step of the reduction is to solve the networked N-bit set disjointness problem using $\mathcal{A}$. At the first round, the node corresponding to Alice (i.e., u_0') sends the value $\mathbf{a}[i]$ to the processes $v_{(4i,0)}$, $v_{(4i+1,0)}$, $v_{(4i+2,0)}$, $v_{(4i+3,0)}$, and $v_{(4(i+1),0)}$ for each $i \in [0, N-1]$. Similarly, the node corresponding to Bob (i.e., u_{N-1}') sends the value $\mathbf{b}[i]$ to the processes $v_{(4i,N-1)}$, $v_{(4i+1,N-1)}$, $v_{(4i+2,N-1)}$, and $v_{(4i+3,N-1)}$ for each $i \in [0, N-1]$. Then all processes construct the instance $L_{\mathbf{a},\mathbf{b}}$: The processes $v_{0,0}, v_{1,0}, \cdots, v_{4N,0}$ and $v_{0,N-1}, v_{1,N-1}, \cdots, v_{4N-1,0}$ can respectively identify the incident edges in $L_{\mathbf{a},\mathbf{b}}$ by the messages from Alice and Bob, and all other processes can identify them locally. From round two, the system runs the algorithm $\mathcal{A}$ solving the verification of P_{Ham}. It terminates by round $r+1$ in expectation, and the verification result decides the disjointness of $\mathbf{a}$ and $\mathbf{b}$. □

Consequently we have the main theorem below:

Theorem 4. *For any (possibly randomized) algorithm solving the networked N-bit set-disjointness in $\Gamma(N)$ with high probability, its worst-case running time is $\Omega(N/\log n + D)$ $(= \Omega(\sqrt{n}/\log n + D))$ rounds.*

5   Verifying Low-Diameter Spanning Trees

The verification problem of low-diameter spanning trees P_{stree}^h is defined by the property which is true if L constitutes the spanning tree with diameter less than or equal to h. In this section, a simple modification of the proof in Section 4 gives a $\Omega(\min\{\sqrt{n}/\log n, h\})$-round (randomized) lower bound for P_{stree}^h.

Assume that h is a power of 2. Then, we consider a graph $\Gamma^h(N)$ for $h \le N$, which is almost same as $\Gamma(N)$ but the only difference is that (1) the length of P_i for each $i \in [0, 4N-1]$ is h, and (2) $AT(N)$ is replaced by a complete binary tree with h leaves (referred as $T(N)$). Then, we have the following corollary:

Corollary 1. *For any algorithm solving the networked N-bit set-disjointness in $\Gamma^h(N)$ with high probability, its worst-case running time is $\min\{\Omega(N/\log n), h\}$ rounds.*

The proof of the corollary is done in the same way as Theorem 3: We also show the lemma below, which is an analogy of Lemma 1:

Lemma 2. *If $P_{stree}^{(8h+8)}$ is solvable within r rounds with high probability, then the networked N-bit set-disjointness problem in $\Gamma(N)$ is solvable within $r+1$ rounds with high probability.*

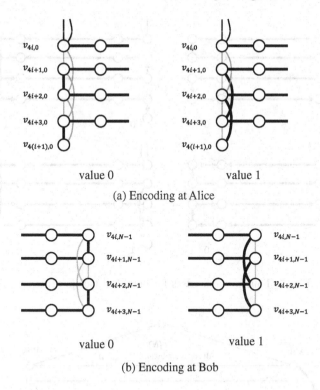

value 0   value 1

(a) Encoding at Alice

value 0   value 1

(b) Encoding at Bob

Fig. 4. The encoding of each bit at Alice and Bob

Proof. Except for the construction of $L_{(\mathbf{a},\mathbf{b})}$, the proof almost follows the one for Lemma 1. The difference in the construction of $L_{(\mathbf{a},\mathbf{b})}$ is the three points mentioned below:

- Exclude edge $(u'_0, v_{(0,0)})$.
- All edges in $T(N)$ are contained in $L_{(\mathbf{a},\mathbf{b})}$.
- Instead of the step 3 in the construction of $L_{(\mathbf{a},\mathbf{b})}$, we encode the value $\mathbf{a}[i]$ as follows: If $\mathbf{a}[0] = 0$ ($i \in [0, N-1]$), we add edges $(v_{(4i+1,0)}, v_{(4i+2,0)})$ and $(v_{(4i+3,0)}, u'_0)$. Otherwise, we add edges $(v_{(4i+1,0)}, v_{(4i+3,0)})$ and $(v_{(4i+2,0)}, u'_0)$. Note that the encoding of $\mathbf{b}$ follows the original one.

The construction is illustrated in Figure 6 Letting $\Gamma_i^h(N) = (V_i, E_i)$ be the subgraph induced by u'_0 and the nodes in P_{4i}, P_{4i+1}, P_{4i+2}, and P_{4i+3}, $L_{(\mathbf{a},\mathbf{b})} \cap E_i$ constitutes a cycle-free spanning subgraph of $\Gamma_i^h(N)$ with diameter $4h + 4$. Since $\Gamma_0^h(N), \Gamma_1^h(N), \cdots, \Gamma_{N-1}^h(N)$ are all connected only at the node u'_0, $L_{\mathbf{a},\mathbf{b}}$ constitutes a spanning tree of diameter $8h + 8$ if and only if $(\mathbf{a}, \mathbf{b})$ is disjoint (note that for the part of $T(N)$ the graph induced by $L_{(\mathbf{a},\mathbf{b})}$ is always a tree with height $\log h$). $\qquad \square$

By the corollary and the lemma above, we lead the following theorem:

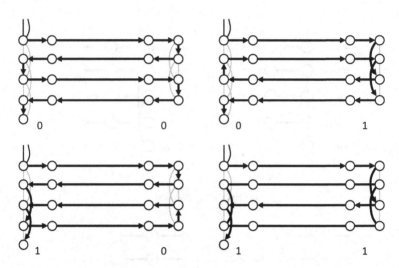

Fig. 5. Four possible cases for one-bit set-disjointness

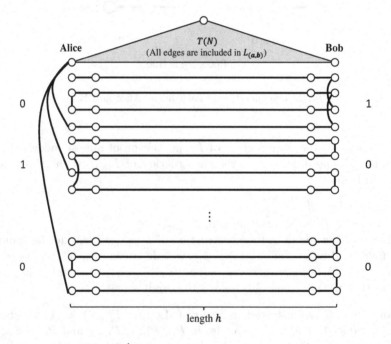

Fig. 6. The construction of $\Gamma^h(N)$ and $L_{(\mathbf{a},\mathbf{b})}$ for low-diameter spanning tree verification

Theorem 5. *For any randomized algorithm solving P^h_{stree} for any $h \leq n$, its worst-case running time is $\Omega(\min\{\sqrt{n}/\log n, h\})$ rounds.*

6 Concluding Remarks

In this paper, we have shown a new elementary proof of $\Omega(\sqrt{n}/\log n + D)$-round randomized lower bound for the spanning-tree verification problem. We have also shown $\Omega(\min\{\sqrt{n}/\log n, h\})$-round lower bound for the verification problem of *low-diameter* spanning trees, that is, verifying the property of "spanning tree with diameter less than h" for any given h. This result implies that the naive BFS-based algorithm is optimal for verifying the spanning trees of diameter $o(\sqrt{n}/\log n)$.

Acknowledgement. The author thank to the anonymous reviewer for pointing out the work by Elkin et al. [3].

References

1. Das Sarma, A., Holzer, S., Kor, L., Korman, A., Nanongkai, D., Pandurangan, G., Peleg, D., Wattenhofer, R.: Distributed verification and hardness of distributed approximation. In: Proc. of the 43rd Annual ACM Symposium on Theory of Computing, pp. 363–372 (2011)
2. Elkin, M.: An unconditional lower bound on the hardness of approximation of distributed minimum spanning tree problem. In: Proc. the 30th ACM Symposium on Theory of Computing (STOC), pp. 331–340 (2004)
3. Elkin, M., Klauck, H., Nanongkai, D., Pandurangan, G.: Quantum distributed network computing: Lower bounds and techniques. In: Proc. of the 2014 ACM Symposium on Principles of Distributed Computing, PODC (2014)
4. Frischknecht, S., Holzer, S., Wattenhofer, R.: Networks cannot compute their diameter in sublinear time. In: Proc. of the 23rd Annual ACM-SIAM Symposium on Discrete Algorithms (SODA), pp. 1150–1162 (2012)
5. Garay, J.A., Kutten, S., Peleg, D.: A sublinear time distributed algorithm for minimum-weight spanning trees. SIAM Journal on Computing 27(1), 302–316 (1998)
6. Ghaffari, M., Kuhn, F.: Distributed minimum cut approximation. In: Afek, Y. (ed.) DISC 2013. LNCS, vol. 8205, pp. 1–15. Springer, Heidelberg (2013)
7. Holzer, S., Wattenhofer, R.: Optimal distributed all pairs shortest paths and applications. In: Proc. of the 2012 ACM Symposium on Principles of Distributed Computing (PODC), pp. 355–364 (2012)
8. Kalyanasundaram, B., Schnitger, G.: The probabilistic communication complexity of set intersection. SIAM Journal on Discrete Mathematics 5(4), 545–557 (1992)
9. Kor, L., Korman, A., Peleg, D.: Tight bounds for distributed minimum-weight spanning tree verification. Theory of Computing Systems 53(2), 318–340 (2013)
10. Lenzen, C., Patt-Shamir, B.: Fast routing table construction using small messages: Extended abstract. In: Proc. of the 45th Annual ACM Symposium on Theory of Computing (STOC), pp. 381–390 (2013)
11. Lenzen, C., Peleg, D.: Efficient distributed source detection with limited bandwidth. In: Proc. of the 2013 ACM Symposium on Principles of Distributed Computing (PODC), pp. 375–382 (2013)
12. Lotker, Z., Patt-Shamir, B., Peleg, D.: Distributed mst for constant diameter graphs. Distributed Computing 18(6), 453–460 (2006)

13. Nanongkai, D.: Distributed approximation algorithms for weighted shortest paths. In: Proc. of the 46th ACM Symposium on Theory of Computing (STOC) (2014)
14. Nanongkai, D., Das Sarma, A., Pandurangan, G.: A tight unconditional lower bound on distributed random walk computation. In: Proc. of the 30th Annual ACM SIGACT-SIGOPS Symposium on Principles of Distributed Computing (PODC), pp. 257–266 (2011)
15. Peleg, D.: Distributed Computing: A Locality-sensitive Approach. Society for Industrial and Applied Mathematics (2000)
16. Peleg, D., Roditty, L., Tal, E.: Distributed algorithms for network diameter and girth. In: Czumaj, A., Mehlhorn, K., Pitts, A., Wattenhofer, R. (eds.) ICALP 2012, Part II. LNCS, vol. 7392, pp. 660–672. Springer, Heidelberg (2012)
17. Peleg, D., Rubinovich, V.: A near-tight lower bound on the time complexity of distributed minimum-weight spanning tree construction. SIAM Journal on Computing 30(5), 1427–1442 (2000)
18. Razborov, A.A.: On the distributional complexity of disjointness. Theoretical Computer Science 106(2), 385–390 (1992)
19. Yao, A.C.-C.: Some complexity questions related to distributive computing (preliminary report). In: Proc. of the 11th Annual ACM Symposium on Theory of Computing (STOC), pp. 209–213 (1979)

Lessons from the Congested Clique Applied to MapReduce*

James W. Hegeman and Sriram V. Pemmaraju

Department of Computer Science
The University of Iowa
Iowa City, Iowa 52242-1419, USA
{james-hegeman,sriram-pemmaraju}@uiowa.edu

Abstract. The main results of this paper are (I) a simulation algorithm which, under quite general constraints, transforms algorithms running on the Congested Clique into algorithms running in the MapReduce model, and (II) a distributed $O(\Delta)$-coloring algorithm running on the Congested Clique which has an expected running time of $O(1)$ rounds, if $\Delta \geq \Theta(\log^4 n)$; and $O(\log \log \log n)$ rounds otherwise. Applying the simulation theorem to the Congested Clique $O(\Delta)$-coloring algorithm yields an $O(1)$-round $O(\Delta)$-coloring algorithm in the MapReduce model.

Our simulation algorithm illustrates a natural correspondence between per-node bandwidth in the Congested Clique model and memory per machine in the MapReduce model. In the Congested Clique (and more generally, any network in the $\mathcal{CONGEST}$ model), the major impediment to constructing fast algorithms is the $O(\log n)$ restriction on message sizes. Similarly, in the MapReduce model, the combined restrictions on memory per machine and total system memory have a dominant effect on algorithm design. In showing a fairly general simulation algorithm, we highlight the similarities and differences between these models.

1 Introduction

The $\mathcal{CONGEST}$ model of distributed computation is a synchronous, message-passing model in which the amount of information that a node can transmit along an incident edge in one round is restricted to $O(\log n)$ bits [15]. As the name suggests, the $\mathcal{CONGEST}$ model focuses on *congestion* as an obstacle to distributed computation. Recently, a fair amount of research activity has focused on the design of distributed algorithms in the $\mathcal{CONGEST}$ model assuming that the underlying communication network is a *clique* [2,5,12,14]. Working with such a *Congested Clique* model completely removes from the picture obstacles that might be due to nodes having to acquire information from distant nodes (since any two nodes are neighbors), thus allowing us to focus on the problem of congestion alone. Making this setting intriguing is also the fact that no non-trivial lower bounds for computation on a Congested Clique have been proved. In fact, in a recent paper, Lenzen [12] showed how to do load-balancing deterministically so as to route up to n^2 messages (each of size $O(\log n)$) in $O(1)$ rounds

---

* This work is supported in part by National Science Foundation grant CCF 1318166.

M. Halldórsson (Ed.): SIROCCO 2014, LNCS 8576, pp. 149–164, 2014.
© Springer International Publishing Switzerland 2014

in the Congested Clique setting, provided each node is the source of at most n messages and the sink for at most n messages. Thus a large volume of information can be moved around the network very quickly and any lower-bound approach in the Congested Clique setting will have to work around Lenzen's routing-protocol result. While Lotker et al. [13] mention overlay networks as a possible practical application of distributed computation on a Congested Clique, as of now, research on this model is largely driven by a theoretical interest in exploring the limits imposed by congestion.

MapReduce [4] is a tremendously popular parallel-programming framework that has become the tool of choice for large-scale data analytics at many companies such as Amazon, Facebook, Google, Yahoo!, etc., as well as at many universities. While the actual time-efficiency of a particular MapReduce-like implementation will depend on many low-level technical details, Karloff et al. [9] have attempted to formalize key constraints of this framework to propose a *MapReduce model* and an associated MapReduce complexity class $(\mathcal{MRC})$. Informally speaking, a problem belongs to $\mathcal{MRC}$ if it can be solved in the MapReduce framework using: (i) a number of machines that is substantially sublinear in the input size, i.e., $O(n^{1-\epsilon})$ for constant $\epsilon > 0$, (ii) memory per machine that is substantially sublinear in the input size, (iii) $O(\text{poly}(\log n))$ number of map-shuffle-reduce rounds, and (iv) polynomial-time local computation at each machine in each round. Specifically, a problem is said to be in $\mathcal{MRC}^i$ if it can be solved in $O(\log^i n)$ map-shuffle-reduce rounds, while maintaining the other constraints mentioned above. Karloff et al. [9] show that *minimum spanning tree* (MST) is in $\mathcal{MRC}^0$ (i.e., MST requires $O(1)$ map-shuffle-reduce rounds) on non-sparse instances. Following up on this, Lattanzi et al. [11] show that other problems such as *maximal matching* (with which the distributed computing community is very familiar) are also in $\mathcal{MRC}^0$ (again, for non-sparse instances). We give a more-detailed description of the MapReduce model in Section 1.1.

The volume of communication that occurs in a Shuffle step can be quite substantial and provides a strong incentive to design algorithms in the MapReduce framework that use very few map-shuffle-reduce steps. As motivation for their approach (which they call *filtering*) to designing MapReduce algorithms, Lattanzi et al. [11] mention that past attempts to "shoehorn message-passing style algorithms into the framework" have led to inefficient algorithms. While this may be true for distributed message-passing algorithms in general, we show in this paper that algorithms designed in the Congested Clique model provide many lessons on how to design algorithms in the MapReduce model. We illustrate this by first designing an expected-$O(1)$-round algorithm for computing a $O(\Delta)$-coloring for a given n-node graph with maximum degree $\Delta \geq \log^4 n$ in the Congested Clique model. We then simulate this algorithm in the MapReduce model and obtain a corresponding algorithm that uses a constant number of map-shuffle-reduce rounds to compute an $O(\Delta)$-coloring of the given graph. While both of these results are new, what we wish to emphasize in this paper is the *simulation* of Congested Clique algorithms in the MapReduce model. Our simulation can also be used to obtain efficient MapReduce-model algorithms for other problems such as 2-*ruling sets* [2] for which an expected-$O(\log \log n)$-round algorithm on a Congested Clique was recently developed.

1.1 Models

The Congested Clique Model. The Congested Clique is a variation on the more general $\mathcal{CONGEST}$ model. The underlying communication network is a size-n clique, i.e., every pair of nodes can directly communicate with each other. Computation proceeds in synchronous rounds and in each round a node (i) receives all messages sent to it in the previous round; (ii) performs unlimited local computation; and then (iii) sends a, possibly distinct, message of size $O(\log n)$ to each other node in the network. We assume that nodes have distinct IDs that can each be represented in $O(\log n)$ bits. We call this the *Congested Clique* model.

Our focus in this paper is graph problems and we assume that the input is a graph G that is a spanning subgraph of the communication network. Initially, each node in the network knows who its neighbors are in G. Thus knowledge of G is distributed among the nodes of the network, with each node having a particular local view of G. Note that G can be quite dense (e.g., have $\Omega(n^2)$ edges) and therefore any reasonably fast algorithm for the problem will have to be "truly" distributed in the sense that it cannot simply rely on shipping off the problem description to a single node for local computation.

The MapReduce Model. Our description of the MapReduce model borrows heavily from the work of Karloff et al. [9] and Lattanzi et al. [11]. Introduced by Karloff et al. [9], the MapReduce model is an abstraction of the popular MapReduce framework [4] implemented at Google and also in the popular Hadoop open-source project by Apache.

The basic unit of information in the MapReduce model is a $(key, value)$-pair. At a high level, computation in this model can be viewed as the application of a sequence of functions, each taking as input a collection of $(key, value)$-pairs and producing as output a new collection of $(key, value)$-pairs. MapReduce computation proceeds in rounds, with each round composed of a map phase, followed by a shuffle phase, followed by a reduce phase. In the map phase, $(key, value)$ pairs are processed individually and the output of this phases is a collection of $(key, value)$-pairs. In the shuffle phase, these $(key, value)$-pairs are "routed" so that all $(key, value)$-pairs with the same *key* end up together. In the last phase, namely the reduce phase, each key and all associated values are processed together. We next describe each of the three phases in more detail.

- The computation in the Map phase of round i is performed by a collection of *mappers*, one per $(key, value)$ pair. In other words, each mapper takes a $(key, value)$ pair and outputs a collection of $(key, value)$ pairs. Since each mapper works on an individual $(key, value)$ pair and the computation is entirely "stateless" (i.e., not dependent on any stored information from previous computation), the mappers can be arbitrarily distributed among machines. In the MapReduce model, keys and values are restricted to the word size of the system, which is $\Theta(\log n)$. Because of this restriction, a mapper takes as input only a constant number of words.
- In the *Shuffle* phase of round i, which runs concurrently with the Map phase (as possible), key-value pairs emitted by the mappers are moved from the machine that produced them to the machine which will run the reducer for which they are

destined; i.e., a key-value pair (k, v) emitted by a mapper is physically moved to the machine which will run the reducer responsible for key k in round i. The Shuffle phase is implemented entirely by the underlying MapReduce framework and we generally ignore the Shuffle phase and treat data movement from one machine to another as a part of the Map phase.

- In the *Reduce* phase of round i, reducers operate on the collected key-value pairs sent to them; a reducer is a function taking as input a pair $(k, \{v_{k,j}\}_j)$, where the first element is a key k and the second is a multiset of values $\{v_{k,j}\}_j$ which comprises all of the values contained in key-value pairs emitted by mappers during round i and having key k. Reducers emit a multiset of key-value pairs $\{(k, v_{k,l})\}_l$, where the key k in each pair is the same as the key k of the input.

For our purposes, the concepts of a machine and a reducer are interchangeable, because reducers are allowed to be "as large" as a single machine on which they compute.

The MapReduce model of Karloff et al. [9] tries to make explicit three key resource constraints on the MapReduce system. Suppose that the problem input has size n (note that this is *not* referring to the input size of a particular reducer or mapper). We assume, as do Karloff et al. [9] and Lattanzi et al. [11], that memory is measured in $O(\log n)$-bit-sized words.

1. Key-sizes and value-sizes are restricted to a $\Theta(1)$ multiple of the word size of the system. Because of this restriction, a mapper takes as input only a constant number of words.
2. Both mappers and reducers are restricted to using space consisting of $O(n^{1-\epsilon})$ words of memory, and time which is polynomial in n.
3. The number of machines, or equivalently, the number of reducers, is restricted to $O(n^{1-\epsilon})$.

Given these constraints, the goal is to design MapReduce algorithms that run in very few – preferably constant – number of rounds. For further details on the justifications for these constraints, see [9].

Since our focus is graph algorithms, we can restate the above constraints more specifically in terms of graph size. Suppose that an n-node graph $G = (V, E)$ is the input. Following Lattanzi et al. [11], we assume that each machine in the MapReduce system has memory $\eta = n^{1+\epsilon}$ for $\epsilon \geq 0$. Since $n^{1+\epsilon}$ needs to be "substantially" sublinear in the input size, we assume that the number of edges m of G is $\Omega(n^{1+c})$ for $c > \epsilon$. Thus the MapReduce results in this paper are for non-sparse graphs.

1.2 Contributions

The main contribution of this paper is to show that fast algorithms in the Congested Clique model can be translated via a simulation theorem into fast algorithms in the MapReduce framework. As a case study, we design a fast graph-coloring algorithm running in the Congested Clique model and then apply the simulation theorem to this algorithm and obtain a fast MapReduce algorithm. Specifically, given an n-node graph G with maximum degree $\Delta \geq \log^4 n$, we show how to compute an $O(\Delta)$-coloring of G in expected $O(1)$ rounds in the Congested Clique model. We also present an algorithm for small Δ; for $\Delta < \log^4 n$ we present an algorithm that computes a $\Delta + 1$ coloring in $O(\log\log\log n)$ rounds with high probability on a Congested Clique. The implication of this result to the MapReduce model (via the simulation theorem) is that for any n-node graph with $\Omega(n^{1+c})$ edges, for constant $c > 0$, there is a MapReduce algorithm that runs in $O(1)$ map-shuffle-reduce rounds using $n^{1+\epsilon}$ memory per machine, for $0 \leq \epsilon < c$ and $n^{c-\epsilon}$ machines. Note that the even when using n memory per machine and n^c machines the algorithm still takes $O(1)$ rounds. This is in contrast to examples in Lattanzi et al. [11] such as maximal matching which require $O(\log n)$ rounds if the memory per machine is n.

The coloring algorithms in both models are new and faster than any known in the respective models, as far as we know. However, the bigger point of this paper is the connection between models that are studied in somewhat different communities.

1.3 Related Work

The earliest interesting algorithm in the Congested Clique model is an $O(\log\log n)$-round deterministic algorithm to compute a minimum spanning tree, due to Lotker et al. [13]. Gehweiler et al. [7] presented a random $O(1)$-round algorithm in the Congested Clique model that produced a constant-factor approximation algorithm for the *uniform* metric facility location problem. Berns et al. [2,3] considered the more-general non-uniform metric facility location in the Congested Clique model and presented a constant-factor approximation running in expected $O(\log\log n)$ rounds. Berns et al. reduce the metric facility location problem to the problem of computing a 2-ruling set of a spanning subgraph of the underlying communication network and show how to solve this in $O(\log\log n)$ rounds in expectation. In 2013, Lenzen presented a routing protocol to solve a problem called an *Information Distribution Task* [12]. The setup for this problem is that each node $i \in V$ is given a set of $n' \leq n$ messages, each of size $O(\log n)$, $\{m_i^1, m_i^2, \ldots, m_i^{n'}\}$, with destinations $d(m_i^j) \in V, j \in \{1, 2, \ldots, n'\}$. Messages are globally lexicographically ordered by their source i, destination $d(m_i^j)$, and j. Each node is also the destination of at most n messages. Lenzen's routing protocol solves the Information Distribution Task in $O(1)$ rounds.

Our main sources of reference on the MapReduce model and for graph algorithms in this model are the work of Karloff et al. [9] and Lattanzi et al. [11] respectively. Besides these, the work of Ene et al. [6] on algorithms for clustering in MapReduce model and the work of Kumar et al. [10] on greedy algorithms in the MapReduce model are relevant.

2   Coloring on the Congested Clique

In this section we present an algorithm, running in the Congested Clique model, that takes an n-node graph G with maximum degree Δ and computes an $O(\Delta)$-coloring in expected $O(\log\log\log n)$ rounds. In fact, for high-degree graphs, i.e., when $\Delta \geq \log^4 n$, our algorithm computes an $O(\Delta)$-coloring in $O(1)$ rounds. This algorithm, which we call Algorithm HIGHDEGCOL, is the main contribution of this section. For graphs with maximum degree less than $\log^4 n$ we appeal to an already-known coloring algorithm that computes a $(\Delta + 1)$ coloring in $O(\log \Delta)$ rounds and then modify its implementation so that it runs in $O(\log \log \log n)$ rounds on a Congested Clique.

We first give an overview of Algorithm HIGHDEGCOL. The reader is advised to follow the pseudocode given in Algorithm 1 as they read the following. The algorithm repeatedly performs a simple random trial until a favorable event occurs. Each trial is independent of previous trials. The key step of Algorithm HIGHDEGCOL is that each node picks a *color group* k from the set $\{1, 2, \ldots, \lceil \Delta / \log n \rceil\}$ independently and uniformly at random (Step 4). We show (in Lemma 1) that the expected number of edges in the graph G_k induced by nodes in color group k is at most $O(\frac{n \log^2 n}{\Delta})$. Of course, some of the color groups may induce far more edges and so we define a *good color group* as one that has at most n edges. The measure of whether the random trial has succeeded is the number of good color groups. If most of the color groups are good, i.e., if at most $2 \log n$ color groups are not good then the random trial has succeeded and we break out of the loop. We then transmit each graph induced by a good color group to a distinct node in constant rounds using Lenzen's routing scheme [12] (Step 11). Note that this is possible because every good color group induces a graph that requires $O(n)$ words of information to completely describe. Every node that receives a graph induced by a good color group locally computes a proper coloring of the graph using one more color than the maximum degree of the graph it receives (Step 12). Furthermore, every such coloring in an iteration employs a distinct palette of colors. Since there are very few color groups that are not good, we are able to show that the residual graph induced by nodes not in good color groups has $O(n)$ edges. As a result, the residual graph can be communicated in its entirety to a single node for local processing. This completes the coloring of all nodes in the graph.

We now analyze Algorithm HIGHDEGCOL and show that (i) it terminates in expected-$O(1)$ rounds and (ii) it uses $O(\Delta)$ colors. Subsequently, we discuss an $O(\log \log \log n)$ algorithm to deal with the small Δ case.

Lemma 1. *For each k, the expected number of edges in G_k is $\frac{n \log^2 n}{2\Delta}$.*

Proof: Consider edge $\{u, v\}$ in G. The probability that both u and v choose color group k is at most $\frac{\log n}{\Delta} \cdot \frac{\log n}{\Delta} = \frac{\log^2 n}{\Delta^2}$. Since G has at most $\frac{1}{2}\Delta \cdot n$ edges, the expected number of edges in G_k is at most $\frac{n \log^2 n}{2\Delta}$. $\square$

Lemma 2. *The expected number of color-group graphs G_k having more than n edges is at most $\log n$.*

Proof: By Lemma 1 and Markov's inequality, the probability that color group k has more than n edges is at most $\frac{n \log^2 n}{2\Delta \cdot n} = \frac{\log^2 n}{2\Delta}$. Since there are $\lceil \Delta / \log n \rceil$ groups, the expected number of G_k having more than n edges is bounded above by $2 \frac{\Delta}{\log n} \cdot \frac{\log^2 n}{2\Delta} = \log n$. $\square$

Lemma 3. *With high probability, every color group has $\frac{5n \log n}{\Delta}$ nodes.*

Proof: The number of color groups is $\lceil \Delta / \log n \rceil$. Thus, for any k, the expected number of nodes in G_k, denoted $|V(G_k)|$, is at most $n \cdot \frac{\log n}{\Delta}$. An application of a Chernoff bound

---

Algorithm 1. HIGHDEGCOL

Input: An n-node graph $G = (V, E)$, of maximum degree Δ
Output: A proper node-coloring of G using $O(\Delta)$ colors

1. Each node u in G computes and broadcasts its degree to every other node v in G.
2. **If** $\Delta \leq \log^4 n$ **then** use Algorithm LOWDEGCOL instead.
3. **while** *true* **do**
4.   Each node u chooses a *color group* k from the set $\{1, 2, \ldots, \lceil \Delta/\log n \rceil\}$ independently and uniformly at random.
5.   Let G_k be the subgraph of G induced by nodes of color group k.
6.   Each node u sends its choice of color group to all neighbors in G.
7.   Each node u computes its degree within its own color-group graph G_{k_u} and sends its color group and degree within color group to node 1.
8.   Node 1, knowing the partition of G into color groups and also knowing the degree of every node u ($u \in G_k$) within the induced subgraph G_k, can compute the number of edges in G_k for each k. Thus node 1 can determine which color-group graphs G_k are *good*, i.e., have at most n edges.
9.   If at most $2 \log n$ color-group graphs are not good, node 1 broadcasts a "break" message to all nodes causing them to **break** out of loop;
    endwhile
10. Node 1 informs every node u in a good group of the fact that u's color group is good
11. Using Lenzen's routing protocol, distribute all information about all good color-group graphs G_k to distinct nodes of G.
12. For each good G_k, the recipient of G_k computes a coloring of G_k using $\Delta(G_k) + 1$ colors. The color palette used for each G_k is distinct.
13. The residual graph $\overline{G}$ of uncolored nodes has size $O(n)$ with high probability, and can thus be transmitted to a single node (for local proper coloring) in $O(1)$ rounds.
14. Each node that locally colors a subgraph informs each node in the subgraph the color it has been assigned.

---

then gives, for each k,

$$\mathbf{P}\left(|V(G_k)| > 5n \cdot \frac{\log n}{\Delta}\right) \leq 2^{-5n \cdot \frac{\log n}{\Delta}} < 2^{-5\log n} = \frac{1}{n^5}$$

Taking the union over all k completes the proof. □

Lemma 4. *With high probability, no node u in G has more than $5 \log n$ neighbors in any color group.*

Proof: Node u has maximum degree Δ, so for any k, the expected number of neighbors of u which choose color group k is bounded above by $\log n$. Therefore, applying a Chernoff bound gives

$$\mathbf{P}\left(|N(u) \cap G_k| > 5 \log n\right) \leq 2^{-5 \log n} = \frac{1}{n^5}$$

Taking the union over all k and u shows that, with probability at least $1 - \frac{1}{n^3}$, the assertion of the lemma holds. □

Lemma 5. *The residual graph $\overline{G}$, induced by groups that are good, has $O(n)$ edges, with high probability.*

Proof: The residual graph $\overline{G}$ is a graph induced by at most $2\log n$ color groups, since the algorithm is designed to terminate only when it has performed a trial resulting in at most $2\log n$ groups that are not good. With high probability, no node u in $\overline{G}$ has more than $5\log n$ neighbors in any of the (at most) $2\log n$ color groups that make up $\overline{G}$, so therefore with high probability no node u has degree greater than $10\log^2 n$ in $\overline{G}$. Since $\overline{G}$ has at most $(2\log n)\cdot\frac{5n\log n}{\Delta}$ nodes with high probability, it follows that the number of edges in $\overline{G}$ is at most

$$(2\log n)\cdot\frac{5n\log n}{\Delta}\cdot 10\log^2 n = \frac{100n\log^4 n}{\Delta}$$

which is $O(n)$ when $\Delta \geq \log^4 n$.   □

Lemma 6. *Algorithm* HIGHDEGCOL *runs in a constant number of rounds, in expectation.*

Proof: By Lemma 2 and Markov's inequality, the expected number of color-group partitioning attempts required before the number of "bad" color groups (i.e., color groups whose induced graphs G_k contain more than n edges) is less than or equal to $2\log n$ is two. It is easy to verify that each iteration of the **while**-true loop requires $O(1)$ rounds of communication.

When $\Delta \geq \log^4 n$, the residual graph $\overline{G}$ is of size $O(n)$ with high probability, and can thus be communicated in its entirety to a single node in $O(1)$ rounds. That single node can then color $\overline{G}$ deterministically using $\Delta+1$ colors and then inform every node of $\overline{G}$ of its determined color in one further round.   □

Lemma 7. *Algorithm* HIGHDEGCOL *uses $O(\Delta)$ colors.*

Proof: A palette of size $O(\log n)$ colors suffices for each good color group because we showed in Lemma 4 that with high probability the maximum degree in any color group is $5\log n$. Since there are a total of $\lceil \Delta/\log n\rceil$ color groups and we use a distinct palette of size $O(\log n)$ for each good color group, we use a total of $O(\Delta)$ colors for the good color groups. The residual graph induced by not-good color groups is colored in the last step and it requires an additional $O(\Delta)$ colors.   □

Coloring low-degree graphs. Now we describe an algorithm that we call LOWDEGCOL that, given an n-node graph G with maximum degree Δ, computes a proper $(\Delta+1)$-coloring with high probability in $O(\log\log\log n)$ rounds in the Congested Clique model. The first phase of the algorithm is the simple, natural, randomized coloring algorithm first analyzed by Johannson [8] and more recently by Barenboim et al. [1]. Each node u starts with a color palette $C_u = \{1, 2, \ldots, \Delta+1\}$. In each iteration, each as-yet uncolored node u makes a tentative color choice $c(u) \in C_u$ by picking a color from C_u independently and uniformly at random. If no node in u's neighborhood picks color $c(u)$ then u colors itself $c(u)$ and $c(u)$ is deleted from the palettes of all neighbors of u. Otherwise, u remains uncolored and participates in the next iteration of the algorithm.

Barenboim et al. [1] show (as part of the proof of Theorem 5.1) that if the above algorithm is run for $O(\log \Delta)$ iterations, then with high probability the uncolored nodes induce connected components of size $O(\text{poly}(\log n))$. Since we are considering a situation in which $\Delta < \log^4 n$, this translates to using $O(\log \log n)$ iterations to reach a state with small connected components. Once all connected components become polylogarithmic in size, then things become quite easy and in the second phase of our algorithm, each connected component is shipped off to a single node to be colored locally. Since each connected component has $O(\text{poly}(\log n))$ edges, this phase can be completed in $O(1)$ rounds.

The first phase in the above described algorithm takes $O(\log \log n)$ rounds with high probability. But, notice that this algorithm uses only the edges of G – the graph being colored – for communication. By utilizing the entire bandwidth of the underlying clique communication network, it is possible to speed up this algorithm significantly. The trick to doing this is to gather, at each node u, all information needed by node u to execute the algorithm locally. Since we are interested in running $O(\log \Delta)$ iterations of the algorithm, the information needed by each node u is *all* the color choices made in $O(\log \Delta)$ iterations by all nodes within $O(\log \Delta)$ hops in G from u. There are at most $\Delta^{O(\log \Delta)}$ nodes with $O(\log \Delta)$ hops from u. Also, each node picks a color from $\{1, 2, \ldots, \Delta + 1\}$ and thus the $O(\log \Delta)$ color choices can be encoded in $O(\log^2 \Delta)$ bits. Since $\Delta < \log^4 n$, $\Delta^{O(\log \Delta)} = o(n)$ and $O(\log^2 \Delta) = O(\log n)$. Thus each node u needs to gather a single word of information each from $o(n)$ nodes. This could be done in $O(1)$ rounds using Lenzen's routing protocol [12] except for the fact that the nodes do not know hop distances in G and therefore do not know which nodes to exchange information with. However, we can leverage the fact that $\Delta < \log^4 n$ and use a "ball growing" algorithm running in $O(\log \log \log n)$ rounds to gather relevant information at each node u. More specifically, suppose that at some point in the algorithm, each node u knows all relevant information about all nodes within t hops from it. Then, by using Lenzen's routing protocol, each node can learn all relevant information about all nodes within $2t$ hops from it in $O(1)$ additional rounds. Since $\Delta < \log^4 n$ and $t = O(\log \log n)$ the pre-conditions for applying Lenzen's routing protocol are satisfied. Also, since $t = O(\log \log n)$, the algorithm terminates in $O(\log \log \log n)$ rounds. This discussion leads to the following theorem.

Lemma 8. *Given an n-node graph G with maximum degree $\Delta \leq \log^4 n$, Algorithm* LOWDEGCOL *computes a proper $(\Delta + 1)$-coloring in $O(\log \log \log n)$ rounds in the Congested Clique model.*

Combining Lemmas 6 and 7 along with Lemma 8 gives the following theorem.

Theorem 1. *Given an n-vertex input graph $G = (V, E)$ with maximum degree $\Delta \geq \log^4 n$, Algorithm* HIGHDEGCOL *computes an $O(\Delta)$-coloring in $O(1)$ rounds (in expectation) in the Congested Clique model. For arbitrary Δ, an $O(\Delta)$-coloring can be computed in $O(\log \log \log n)$ rounds in expectation in the Congested Clique model.*

3   MapReduce Algorithms from Congested Clique Algorithms

In this section, we prove a *simulation* theorem establishing that Congested Clique algorithms (with fairly weak restrictions) can be efficiently implemented in the MapReduce model. The simulation ensures that a Congested Clique algorithm running in T rounds can be implemented in $O(T)$ rounds (more precisely, $3 \cdot T + O(1)$ rounds) in the MapReduce model, if certain communication and "memory" conditions are met. The technical details of this simulation are conceptually straightforward, but the details are a bit intricate.

We will now precisely define restrictions that we need to place on Congested Clique algorithms in order for the simulation theorem to go through. We assume that each node in the Congested Clique possesses a word-addressable memory whose words are indexed by the natural numbers. For an algorithm $\mathcal{A}_{CC}$ running in the Congested Clique, let $I_u^{(j)} \subset \mathbb{N}$ be the set of memory addresses *used* by node u during the local computation in round j (not including the sending and receipt of messages).

After local computation in each round, each node in the Congested Clique may send (or not send) a distinct message of size $O(\log n)$ to each other node in the network. In defining notation, we make a special distinction for the case where a node u sends in the *same* message to every other node v in a particular round; i.e., node u sends a *broadcast* message. The reason for this distinction is that broadcasts can be handled more efficiently on the receiving end in the MapReduce framework than can distinct messages sent by u. Let $m_{u,v}^{(j)}$ denote a message sent by node u to node v in round j and let $D_u^{(j)} \subseteq V$ be the set of destinations of messages sent by node u in round j. Let $M_u^{(j)} = \{m_{u,v}^{(j)} : v \in D_u^{(j)} \subset V\}$ be the set of messages *sent* by node u in round j of algorithm $\mathcal{A}_{CC}$, except let $M_u^{(j)} = \emptyset$ if u has chosen to broadcast a message $b_u^{(j)}$ in round j. Similarly, let $\overline{M}_u^{(j)} = \{m_{v,u}^{(j)} : u \in D_v^{(j)}$ and v is not broadcasting in round $j\}$ be the set of messages *received* by node u in round j, except that we exclude messages $b_v^{(j)}$ from nodes v that have chosen to broadcast in round j. We say that $\mathcal{A}_{CC}$, running on an n-node Congested Clique, is (K, N)-*lightweight* if

(i)   for each round j (in the Congested Clique), $\sum_{u \in V}(|\overline{M}_u^{(j)}| + |I_u^{(j)}|) = O(K)$;

(ii) there exists a constant C such that for each round j and for each node u, $I_u^{(j)} \subseteq \{1, 2, \ldots, \lceil C \cdot N \rceil\}$; and

(iii) each node u performs only polynomial-time local computation in each round.

In plain language: no node uses more than $O(N)$ memory for local computation during a round; the total amount of memory that all nodes use and the total volume of messages nodes receive in any round is bounded by $O(K)$. Regarding condition (iii), traditional models of distributed computation such as the $\mathcal{CONGEST}$ and $\mathcal{LOCAL}$ models allow nodes to perform arbitrary local computation (e.g., taking exponential time), but since the MapReduce model requires mappers and reducers to run in polynomial time, we need this extra restriction.

Theorem 2. *Let ϵ, c satisfy $0 \le \epsilon \le c$, and let $G = (V, E)$ be a graph on n vertices having $O(n^{1+c})$ edges. If $\mathcal{A}_{CC}$ is a $(n^{1+c}, n^{1+\epsilon})$-lightweight Congested Clique-model algorithm running on input G in T rounds, then $\mathcal{A}_{CC}$ can be implemented in the*

MapReduce model with $n_r = n^{c-\epsilon}$ machines and $m_r = \Theta(n^{1+\epsilon})$ (words of) memory per machine such that the implementation runs in $O(T)$ Map-Shuffle-Reduce rounds on G.

Proof: The simulation that will prove the above theorem contains two stages: the *Initialization* stage and the *Simulation* stage. In the Initialization stage, the input to the MapReduce system is transformed from the assumed format (an unordered list of edges and vertices of G) into a format in which each piece of information, be it an edge, node, or something else, that is associated with a node of G is gathered at a single machine. After this gathering of associated information has been completed, the MapReduce system can emulate the execution of the Congested Clique algorithm.

Initialization. stage. Input (in this case, the graph G) in the MapReduce model is assumed to be presented as an unordered sequence of tuples of the form $(\varnothing, u)$, where u is a vertex of G, or $(\varnothing, (u, v))$, where (u, v) is an edge of G. The goal of the Initialization stage is to partition the input G among the n_r reducers such that each reducer r receives a subset $P_r \subseteq V$ and all edges E_r incident on nodes in P_r such that $|P_r| + |E_r|$ is bounded above by $O(n^{1+\epsilon})$. This stage can be seen as consisting of two tasks: (i) every reducer r learns the degree $\deg_G(u)$ of every node u in G and (ii) every reducer computes a partition (the same one) given by the partition function $F_0 : V \longrightarrow \{1, 2, \ldots, n_r\}$, defined by

$$F_0(x) = \begin{cases} 1, & \text{if } x = 1 \\ F_0(x-1), & \text{if } \sum_{v \in L(x)} \deg_G(v) \le n^{1+\epsilon}, \\ F_0(x-1) + 1, & \text{otherwise} \end{cases}$$

Here $L(x) = \{j < x : F_0(j) = F_0(x-1)\}$. All nodes in the same group in the partition are mapped to the same value by F_0. Since the degree of each node is bounded above by n, it is easy to see that for any $r \in \{1, 2, \ldots, n_r\}$, $F_0^{-1}(r)$ is a subset of nodes of G such that $|F_0^{-1}(r)| + \sum_{u \in F_0^{-1}(r)} \deg_G(u)$ is $O(n^{1+\epsilon})$. Each of the two tasks mentioned above can be implemented in a (small) constant number of MapReduce rounds as follows.

In Map Phase 1, each node and each edge in the input is mapped to a reducer chosen uniformly at random. With high probability each reducer receives $O(n^{1+\epsilon})$ nodes and edges. Let E_r be the edges received by reducer r. Each reducer r computes a "partial degree" $d_{u,r}$ of each node u with respect to E_r, i.e., the number of edges in E_r incident on u. In Map Phase 2, each $d_{r,u}$ is mapped to reducer $u \bmod n_r$. (Here u refers to a numeric ID and therefore $u \bmod n_r$ reduces this ID to the space $\{0, 1, \ldots, n_r - 1\}$.) Since there are $n_r = n^{c-\epsilon}$ reducers, there are a total of $O(n^{1+c-\epsilon})$ $d_{r,u}$ values and thus each reducer is the destination of $O(n)$ such values. Each reducer r then aggregates the partial degrees that it receives into $\deg_G(u)$ values; note that all partial degrees associated with a node u are sent to the same reducer. In Map Phase 3, each node degree is mapped to each reducer. Then each reducer r, knowing all node degrees, computes the partition function F_0 defined above. In Map Phase 4, which is the final phase of the Initialization stage, nodes and edges are mapped to reducers according to the partition function F_0.

Simulation. stage. At a high level, a Reduce phase serves as the "local computation" phase of the Congested Clique simulation, whereas a Map phase (together with the subsequent shuffle phase) serves as the "communication" phase of the simulation. However, there is, in general, a constant-factor slow-down because it may be that the sending and receiving of messages in $\mathcal{A}_{CC}$ could cause the subset of nodes assigned to a reducer to aggregate more than $O(n^{1+\epsilon})$ memory, necessitating a re-partitioning of the nodes among the reducers so as not to violate the memory-per-machine constraint.

Recall that $I_u^{(i)}$ denotes the set of memory addresses used by a node u in round i of $\mathcal{A}_{CC}$. Let $h_{u,j}^{(i)}$ be the value of word $j \in I_u^{(i)}$ in the memory of node u after node u has completed local computation in round i of $\mathcal{A}_{CC}$, but before messages have been sent and received in this round. For $i > 0$, define a tuple set

$$\mathcal{H}_u^{(i)} = \{(F_i(u), (u, i, h_{u,j}^{(i)})) : j \in I_u^{(i)}\}$$

where $F_i(\cdot)$ is the partition function used in round i. Like F_0, defined in the Initialization stage, F_i partitions G into n_r groups, one per reducer, so that reducer memory constraints are not violated in round i. The collection of tuples $\mathcal{H}_u^{(i-1)}$ is a representation, in the MapReduce key-value format, of the information necessary to simulate the computations of node u in round i of the Congested Clique algorithm $\mathcal{A}_{CC}$. The use of $F_i(u)$ as the key in each of the tuples in $\mathcal{H}_u^{(i)}$ ensures that all information needed to simulate a local computation at u in $\mathcal{A}_{CC}$ goes to the same reducer. Additionally, note that the inclusion of the identifier of u with the values allows the words from u's memory to be reassembled and distinguished from information associated with other nodes $v \in F_i^{-1}(u)$. We assume that $\mathcal{H}_u^{(0)}$ is the information in tuple format that node u has initially about graph G. In other words, $\mathcal{H}_u^{(0)} = \{(F_0(u), u)\} \cup \{(F_0(u), (u, v)) : v \text{ is a neighbor of } u\}$.

Once an initial partition function $F_0(\cdot)$ has been computed and the initial collections $\mathcal{H}_u^{(0)}$ have been assembled the main goals of our simulation algorithm are to (i) provide a mechanism for transforming $\mathcal{H}_u^{(i-1)}$ into $\mathcal{H}_u^{(i)}$ during the reduce phase of a MapReduce round; and (ii) provide a means of transmitting messages to reducers of a subsequent round (corresponding to messages transmitted in the Congested Clique at the end of each round). Since we assume messages to be sent and received after local computation has occurred during a Congested Clique round, $\mathcal{M}_u^{(i)}$ can be determined from $\mathcal{H}_u^{(i)}$; in turn, $\mathcal{H}_u^{(i)}$ is a function of $\mathcal{H}_u^{(i-1)}$ and $\overline{\mathcal{M}}_u^{(i-1)}$.

We describe the details of the simulation of a single round (round i) of a Congested Clique algorithm $\mathcal{A}_{CC}$ below. Let $j = 3i - 1$. Round i of $\mathcal{A}_{CC}$ is simulated by three MapReduce rounds (a total of six Map or Reduce phases) – Reduce $j - 1$, Map j, Reduce j, Map $j + 1$, Reduce $j + 1$, and Map $j + 2$. We assume inductively that as input to Reduce phase $j - 1$ below, each reducer receives, in addition to data tuples, $O(n)$ metadata tuples containing a description of a partition function $F_{i-1}(\cdot)$ such that for each r, $\sum_{u \in P_r}(|\mathcal{H}_u^{(i-1)}| + |\overline{\mathcal{M}}_u^{(i-1)}|) = O(n^{1+\epsilon})$, where $P_r = F_{i-1}^{-1}(r)$.

- **Reduce phase $j - 1$:** In Reduce phase $j - 1$, a reducer r receives input consisting of $\mathcal{H}_u^{(i-1)}$ together with $\overline{\mathcal{M}}_u^{(i-1)}$ for each $u \in P_r$; for each such u, reducer r performs the following steps:

(i) Reducer r simulates the local computation of Round i of $\mathcal{A}_{CC}$ at u.

(ii) Reducer r computes $\mathcal{H}_u^{(i)}$ from $\mathcal{H}_u^{(i-1)}$ and $\overline{\mathcal{M}}_u^{(i-1)}$, *but does not yet output* any tuples of $\mathcal{H}_u^{(i)}$; rather, reducer r outputs only a tuple (r, u, s_u) containing the size of the information $s_u = |\mathcal{H}_u^{(i)}|$.

(iii) Reducer r computes $\mathcal{M}_u^{(i)}$ from $\mathcal{H}_u^{(i)}$, but again, *does not output* any elements of $\mathcal{M}_u^{(i)}$. Reducer r then computes, for each $v \in V$, the aggregate count $c_{r,v}$ of messages emanating from nodes in P_r and destined for v, and outputs the tuple $(r, v, c_{r,v})$.

(iv) Reducer r outputs the exact same tuples it received as input, $\mathcal{H}_u^{(i-1)}$ and $\overline{\mathcal{M}}_u^{(i-1)}$.

– **Map phase** j: Before message tuples can be generated and aggregated (as a collection $\overline{\mathcal{M}}_u^{(i)}$ at reducer $F(u)$) a rebalancing of the nodes to reducers must be performed to ensure that the reducer-memory constraint is not violated. In Map phase j, a mapper forwards tuples from either a $\mathcal{H}_u^{(i-1)}$ or a $\overline{\mathcal{M}}_u^{(i-1)}$ through unchanged. However, for each tuple of the form $(r, u, c_{r,u})$, a mapper outputs the tuple $(u \bmod n_r, u, c_{r,u})$. In addition, for each tuple of the form (r, u, s_u), a mapper outputs n_r tuples (r', u, s_u) – one for each reducer r' – so that every reducer can know the future size of $\mathcal{H}_u^{(i)}$.

– **Reduce phase** j: In Reduce phase j, a reducer r receives as input nearly the exact same input (and output) of reducer r in the previous MapReduce round – the union of $\mathcal{H}_u^{(i-1)}$ and $\overline{\mathcal{M}}_u^{(i-1)}$ for each $u \in P_r$ – except that instead of receiving tuples of the form $(r, u, c_{r,u})$ for each $u \in V$, reducer r receives *all* partial message counts for the subset of vertices u for which $u \bmod n_r = r$; as well, each reducer receives n tuples of the form (r, u, s_u) describing the amount of memory required by node u in round i of $\mathcal{A}_{CC}$. Reducer r aggregates tuples of the form $(u \bmod n_r, u, c_{r,u})$ and outputs $(r, u, |\overline{M}_u^{(i)}|)$, since $|\overline{M}_u^{(i)}|$ is precisely the sum of the partial message counts $c_{r,u}$. (Notice that a reducer r receives $O(n)$ such tuples.) Reducer r forwards all other tuples through unchanged to the next MapReduce round.

– **Map phase** $j + 1$: In Map phase $j + 1$, a mapper continues to forward all tuples through unchanged to Reduce phase $j + 1$, except that for each tuple of the form $(r, u, |\overline{M}_u^{(i)}|)$, a mapper outputs n_r tuples $(r', u, |\overline{M}_u^{(i)}|)$ – one for each reducer r'. In this way, each reducer in Reducer phase $j + 1$ can come to know all n message counts for each node $u \in V$.

– **Reduce phase** $j + 1$: In Reduce phase $j + 1$, each reducer receives all n message counts (for each node $u \in V$) in addition to the sizes s_u of the state needed by each node u in round i of $\mathcal{A}_{CC}$. Each reducer thus has enough information to determine the next partition function $F_i : V \longrightarrow \{1, \ldots, n_r\}$, defined by

$$F_i(x) = \begin{cases} 1, & \text{if } x = 1 \\ F_i(x-1), & \text{if } \sum_{v \in L(x)} (s_v + |\overline{M}_v^{(i)}|) \leq n^{1+\epsilon}, \\ F_i(x-1) + 1, & \text{otherwise} \end{cases}$$

Here $L(x) = \{v \mid v < x \text{ and } F_i(v) = F_i(x-1)\}$. After determination of the new partition function F_i, reducers are now able to successfully output the "packaged memory" $\mathcal{H}_u^{(i)}$ of round i of $\mathcal{A}_{CC}$, as well as the new messages $m_{u,v}^{(i)}$ sent in round i, because the new partition function F_i is specifically designed to correctly load-balance these tuple sets across the reducers while satisfying the memory constraint. Therefore:

(i) Reducer r now simulates the local computation at each $u \in P_r$ and thus outputs the set $\mathcal{H}_u^{(i)}$ (which can be computed from $\mathcal{H}_u^{(i-1)}$ and $\overline{\mathcal{M}}_u^{(i-1)}$). It is important to recall here that because mappers operate on key-value pairs one at a time in the MapReduce model, there is no restriction on the size of the output from any reducer r in any MapReduce round (other than that it be polynomial). [9] Therefore, a reducer r may output (and thus free-up its memory) each tuple set $\mathcal{H}_u^{(i)}$ as it is created (as reducer r processes the nodes in P_r one at a time), and so there is no concern about reducer r attempting to maintain in memory all sets $\mathcal{H}_u^{(i)}$ for $u \in P_r$ at once. Note that $\mathcal{H}_u^{(i)}$, as generated by a reducer r, should contain tuples of the form $(r, F_i(u), u, h_{u,l}^{(i)})$ so that mappers in MapReduce round $j+2$ can correctly deliver $\mathcal{H}_u^{(i)}$ to reducer $F_i(u)$. Recall that $h_{u,l}^{(i)}$ denotes the contents of the word with address l in node u's memory at the end of local computation in round i.

(ii) As a reducer r processes, and simulates the computation at, each node $u \in P_r$ one at a time, generating $\mathcal{H}_u^{(i)}$, reducer r also uses $\mathcal{H}_u^{(i)}$ to generate the messages $M_u^{(i)}$ to be sent by node u in round i of $\mathcal{A}_{CC}$. Reducer r encapsulates $M_u^{(i)}$ in the tuple set $\mathcal{M}_u^{(i)}$ and outputs it alongside $\mathcal{H}_u^{(i)}$ before moving on to the next node in P_r. As with $\mathcal{H}_u^{(i)}$, tuples in $\mathcal{M}_u^{(i)}$ should initially be generated by a reducer r in the form $(r, F_i(v), u, v, m_{u,v}^{(i)})$ so that mappers in MapReduce round $j+2$ can correctly deliver the set $\overline{\mathcal{M}}_v^{(i)}$ to reducer $F_i(v)$.

(iii) Lastly regarding the simulation procedure, whenever a node $u \in P_r$ being simulated broadcasts a message $b_u^{(i)}$, reducer r outputs the tuple $(r, u, b_u^{(i)})$.

(iv) After simulation of each node $u \in P_r$ is complete, reducer r also outputs a description of the new partition function F_i.

- **Map $j+2$:** In Map phase $j+2$, a mapper simply transforms the key in a data tuple as appropriate: for each tuple $(r, F_i(u), u, h_{u,l}^{(i)})$, a mapper simply emits the tuple $(F_i(u), u, h_{u,l}^{(i)})$; for each tuple $(r, F_i(v), u, v, m_{u,v}^{(i)})$, a mapper simply emits the tuple $(F_i(v), u, v, m_{u,v}^{(i)})$. The exception to this is that tuples $(r, u, b_u^{(i)})$ containing broadcast messages are expanded: for each, a mapper emits n_r tuples $(r', u, b_u^{(i)})$ – one for each reducer r' – so that every reducer in Reducer phase $j+2$ receives a single copy of each message broadcast during round i of $\mathcal{A}_{CC}$.

- Tuples carrying metadata describing the (new) partition function F_i are forwarded unchanged, because there already exists one copy of each such metadata tuple for each reducer, and there need be only one such copy per reducer as well. After Map phase $j+2$, tuples from the sets $\mathcal{H}_u^{(i)}$ and $\overline{\mathcal{M}}_u^{(i)}$ have been emitted with keys $F_i(u)$, and for each broadcast message $b_u^{(i)}$, one tuple containing a copy of $b_u^{(i)}$ has been emitted for each reducer as well; thus, in Reduce phase $j+2$, simulation of round $i+1$ of algorithm $\mathcal{A}_{CC}$ can begin.

It remains to comment on the memory-per-machine constraint which must be satisfied during each MapReduce round. Observe that, inductively, for each r, the sum $\sum_{u \in P_r}(|\mathcal{H}_u^{(i-1)}| + |\overline{\mathcal{M}}_u^{(i-1)}|) = O(n^{1+\epsilon})$. These data tuples are forwarded unchanged until Reduce phase $j + 1$, in which the new partition function $F_i(\cdot)$ for the next round of simulation is computed, and then collectively $\mathcal{H}_u^{(i-1)}$ and $\overline{\mathcal{M}}_u^{(i-1)}$ are transformed into $\mathcal{H}_u^{(i)}$ and $\mathcal{M}_u^{(i)}$. By construction of the partition functions F_{i-1} and F_i, and by the assumption that $\mathcal{A}_{CC}$ is a $(n^{1+c}, n^{1+\epsilon})$-*lightweight* algorithm, it follows that these data tuples are never present on any reducer a number that exceeds $\Theta(n^{1+\epsilon})$. Secondly, it should be mentioned that because broadcast messages are not duplicated at any reducer r, no reducer will ever receive more than $n = O(n^{1+\epsilon})$ tuples containing broadcast messages. Thirdly, tuples containing state or message counts are never present in a number exceeding n at any reducer, and *partial* message counts are explicitly load-balanced so that only $O(n)$ such information is passed to a single reducer as well. Finally, metadata tuples describing a partition function never exceed $\Theta(n)$ on any reducer because the domain of each partition function has size n.   □

4   Coloring in the MapReduce Framework

Using the simulation theorem of Section 3, we can simulate Algorithm HIGHDEGCOL in the MapReduce model and thereby achieve an $O(\Delta)$-coloring MapReduce algorithm running in expected-$O(1)$ rounds. As in Lattanzi et al. [11], we consider graphs with $\Omega(n^{1+c})$ edges, $c > 0$.

Theorem 3. *When the input graph G has $\Omega(n^{1+c})$ edges, and $0 \leq \epsilon < c$, there exists an $O(\Delta)$-coloring algorithm running in the MapReduce model with $\Theta(n^{c-\epsilon})$ machines and $\Theta(n^{1+\epsilon})$ memory per machine, and having an expected running time of $O(1)$ rounds.*

Proof: It is easy to examine the lines of code in Algorithm HIGHDEGCOL to ascertain that the total amount of non-broadcast communication in any round in bounded above by $O(n^{1+c})$. Specifically, the total non-broadcast communication corresponding to only two lines of code – Lines 6 and 11 – can be as high as $\Theta(n^{1+c})$. For all other lines of code, the volume of total non-broadcast communication is bounded by $O(n)$. Similarly, it is easy to examine the lines of code in Algorithm HIGHDEGCOL to verify that the total memory (in words) used by all nodes for their local computations in any one round is bounded above by $O(n^{1+c})$. Finally, it is also easy to verify that the maximum amount of memory used by a node in any round of computation is $O(n)$.

Thus, Algorithm HIGHDEGCOL is an (n^{1+c}, n)-lightweight algorithm on a Congested Clique and applying the Simulation Theorem (Theorem 1) to this algorithm yields the claimed result.   □

It is worth emphasizing that the result holds even when $\epsilon = 0$; in other words, even when the per machine memory is $O(n)$, the algorithm can compute an $O(\Delta)$-coloring in $O(1)$ rounds. This is in contrast with the results in Lattanzi et al. [11], where $O(1)$-round algorithms were obtained (e.g., for maximal matching) with $n^{1+\epsilon}$ per machine memory, only when $\epsilon > 0$. In their work, setting $\epsilon = 0$ (i.e., using $\Theta(n)$ memory per machine) resulted in $O(\log n)$ round algorithms.

References

1. Barenboim, L., Elkin, M., Pettie, S., Schneider, J.: The locality of distributed symmetry breaking. In: Proc. of IEEE FOCS (2012)
2. Berns, A., Hegeman, J., Pemmaraju, S.V.: Super-Fast Distributed Algorithms for Metric Facility Location. In: Czumaj, A., Mehlhorn, K., Pitts, A., Wattenhofer, R. (eds.) ICALP 2012, Part II. LNCS, vol. 7392, pp. 428–439. Springer, Heidelberg (2012)
3. Berns, A., Hegeman, J., Pemmaraju, S.V.: Super-Fast Distributed Algorithms for Metric Facility Location. CoRR, abs/1308.2473 (August 2013)
4. Dean, J., Ghemawat, S.: Mapreduce: Simplified data processing on large clusters. Commun. ACM 51(1), 107–113 (2008)
5. Dolev, D., Lenzen, C., Peled, S.: "Tri, tri again": Finding triangles and small subgraphs in a distributed setting. In: Aguilera, M.K. (ed.) DISC 2012. LNCS, vol. 7611, pp. 195–209. Springer, Heidelberg (2012)
6. Ene, A., Im, S., Moseley, B.: Fast clustering using mapreduce. In: Proceedings of the 17th ACM SIGKDD International Conference on Knowledge Discovery and Data Mining, KDD 2011, pp. 681–689. ACM, New York (2011)
7. Gehweiler, J., Lammersen, C., Sohler, C.: A Distributed O(1)-approximation Algorithm for the Uniform Facility Location Problem. In: Proceedings of the Eighteenth Annual ACM Symposium on Parallelism in Algorithms and Architectures, SPAA 2006, pp. 237–243. ACM, New York (2006)
8. Johansson, Ö.: Simple distributed $(\delta + 1)$-coloring of graphs. Inf. Process. Lett. 70(5), 229–232 (1999)
9. Karloff, H., Suri, S., Vassilvitskii, S.: A model of computation for mapreduce. In: Proceedings of the Twenty-first Annual ACM-SIAM Symposium on Discrete Algorithms, SODA 2010, pp. 938–948. Society for Industrial and Applied Mathematics, Philadelphia (2010)
10. Kumar, R., Moseley, B., Vassilvitskii, S., Vattani, A.: Fast greedy algorithms in mapreduce and streaming. In: Proceedings of the Twenty-fifth Annual ACM Symposium on Parallelism in Algorithms and Architectures, SPAA 2013, pp. 1–10. ACM, New York (2013)
11. Lattanzi, S., Moseley, B., Suri, S., Vassilvitskii, S.: Filtering: A method for solving graph problems in mapreduce. In: Proceedings of the Twenty-third Annual ACM Symposium on Parallelism in Algorithms and Architectures, SPAA 2011, pp. 85–94. ACM, New York (2011)
12. Lenzen, C.: Optimal Deterministic Routing and Sorting on the Congested Clique. In: Proceedings of the 2013 ACM Symposium on Principles of Distributed Computing, PODC 2013, pp. 42–50 (2013)
13. Lotker, Z., Patt-Shamir, B., Peleg, D.: Distributed MST for Constant Diameter Graphs. Distributed Computing 18(6), 453–460 (2006)
14. Patt-Shamir, B., Teplitsky, M.: The round complexity of distributed sorting: Extended abstract. In: Proceedings of the 30th Annual ACM SIGACT-SIGOPS Symposium on Principles of Distributed Computing, PODC 2011, pp. 249–256. ACM, New York (2011)
15. Peleg, D.: Distributed Computing: A Locality-Sensitive Approach, vol. 5. Society for Industrial Mathematics (2000)

Oblivious Rendezvous
in Cognitive Radio Networks

Zhaoquan Gu[1], Qiang-Sheng Hua[1],
Yuexuan Wang[2], and Francis Chi Moon Lau[2]

[1] Institute for Interdisciplinary Information Sciences,
Tsinghua University, Beijing, P.R. China
[2] Department of Computer Science, The University of Hong Kong,
Pokfulam, Hong Kong, P.R. China

Abstract. Rendezvous is a fundamental process in the operation of a Cognitive Radio Network (CRN), through which a secondary user can establish a link to communicate with its neighbors on the same frequency band (channel). The licensed spectrum is divided into N non-overlapping channels, and most previous works assume all users have the same label for the same channel. This implies some degree of centralized coordination which might be impractical in distributed systems such as a CRN. Thus we propose *Oblivious Rendezvous* where the users may have different labels for the same frequency band.

In this paper, we study the oblivious rendezvous problem for M users (ORP-M for short) in a multihop network with diameter D. We first focus on the rendezvous process between two users (ORP-2) and then extend the derived algorithms to ORP-M. Specifically, we give an $\Omega(N^2)$ lower bound for ORP-2, and propose two deterministic distributed algorithms solving ORP-2. The first one is the ID Hopping (IDH) algorithm which generates a fixed length sequence and guarantees rendezvous in $O(N \max\{N, M\})$ time slots; it meets the lower bound when $M = O(N)$. The second one is the Multi-Step Hopping (MSH) algorithm which guarantees rendezvous in $O(N^2 \log_N M)$ time slots by combing ID scaling and hopping with different steps; it meets the lower bound if M can be bounded by a polynomial function of N, which is true of large scale networks. The two algorithms are also applicable to non-oblivious rendezvous and the performance is comparable to the state-of-the-art results. Then we extend the algorithms to ORP-M with bounded rendezvous time by increasing the diameter D by a factor.

1 Introduction

1.1 Rendezvous and Oblivious Rendezvous

Cognitive Radio Network (CRN) is attracting more and more attention in both academia and industry, which was proposed to solve the spectrum scarcity problem [1]. A CRN consists of primary users (PUs) which own the licensed spectrum and secondary users (SUs) which can sense and access the portion of the licensed

M. Halldórsson (Ed.): SIROCCO 2014, LNCS 8576, pp. 165–179, 2014.
© Springer International Publishing Switzerland 2014

spectrum left unused by the PUs. Unless otherwise specified, 'user' in this paper refers to SU.

There have been many interesting works in the CRN community tackling such problems as neighbor discovery [10, 27], broadcasting [16, 25], data gathering [7], and routing [15]. All these works assume one fundamental process in the operation of a CRN, called *rendezvous*, which establishes a link on some frequency band (channel) needed for communication between two or more users. One can imagine that the licensed spectrum is divided into N non-overlapping channels; each user can sense a channel, and if it is not occupied by any PU, it is an *available* channel. For the convenience of our derivations, a CRN over time is time-slotted and each user can access an available channel in each time slot. Practical rendezvous processes consist of many detailed steps, such as beaconing and handshaking. In this paper, we focus on the step of multiple users meeting on the same available channel: we say that rendezvous between users is achieved if they can access the same available channel in the same time slot. We give distributed algorithms for rendezvous. *Time to Rendezvous* (*TTR*) is used to measure these rendezvous algorithms, which is the time for the users to (achieve) rendezvous on a common channel.

Previous works use either a central controller (such as a base station) or a Common Control Channel (CCC) [18, 22] to simplify the process. However, such centralization could lead to a bottleneck in practical situations when the number of users increases, is vulnerable to adversary attacks, and is not flexible. Therefore, many *blind rendezvous algorithms* have been proposed, where the word 'blind' refers to non-reliance on any central controller or CCC [8, 9, 12, 13, 20, 21, 23, 26]. They construct sequences based on the channels' labels (some also use the users' identifiers) and let users hop on the frequency bands according to the sequences. Obviously, all these blind algorithms assume that the users see the same labels for these licensed frequency bands (channels). These labels represent global knowledge that must be communicated, somehow, to all the participating users. This may imply that there must exist some centralized entity that maintains and disseminates the knowledge.

To do away with the assumption of existing blind rendezous solutions that there is a common set of labels shared by all users, we propose the *oblivious rendezvous problem* where different users may have different labels for the licensed channels. Technically, each user can only assign (local) labels to those sensed available channels and attempt rendezvous based on such local information. Correspondingly, we refer to those other schemes where the users share the same labels for the frequency bands as *non-oblivious rendezvous*.

The oblivious rendezvous problem poses several challenges. First of all, because each user may have different labels for the channels, traditional methods based on a common set of channel labels cannot be applied at all. Second, each user can join the network at any time slot, and thus the algorithm needs to guarantee the rendezvous asynchronously. Third, as the users do not have each other's information until they achieve rendezvous and establish a common link for communication, symmetric algorithms are preferred, which means that all

Table 1. MTTR Comparison for Two Users' Scenario

Algorithms	Non-Oblivious Rendezvous	Oblivious Rendezvous
Jump-Stay [21]	$3NP^2 + 3P = O(N^3)$	–
CRSEQ [23]	$P(3P - 1) = O(N^2)$	–
DRDS [12]	$3P^2 + 2P = O(N^2)$	–
Hop-and-Wait [9]	$O(N^2 \log M)$	–
MMC [26]	$ETTR = O(N^2)$	$ETTR = O(N^2)$
IDH (this paper)	$O(N \max\{N, M\})$	$O(N \max\{N, M\})$
MSH (this paper)	$O(N^2 \log_N M)$	$O(N^2 \log_N M)$

Remarks: 1) "–" means the method is not applicable to oblivious rendezvous; 2) ETTR means expected time to rendezvous (note: MMC cannot guarantee bounded time rendezvous); 3) P is the smallest prime number $P > N$, $P = O(N)$.

users should execute the same algorithm. Finally, for scenarios with many users in a large area, two users may not be connected directly, and so multihop communication needs to be considered. In this paper, we present algorithms that address all these issues.

1.2 Related Work

Non-oblivious rendezvous algorithms assume all users share the same labeling for the licensed channels. There are commonly three types of these algorithms: centralized algorithms, decentralized algorithms based on Common Control Channel (CCC), and blind rendezvous algorithms.

Centralized algorithms assume that a central controller or a CCC exists during the rendezvous process, which substantially simplifies the problem [18, 22]. For practical deployment, however, the central controller or the CCC could become a bottleneck and is vulnerable to adversary attacks. There are some decentralized algorithms based on establishing local CCCs through which each user can communicate with their neighbors [17, 19]. However, these algorithms incur too much overhead in establishing and maintaining local CCCs.

Blind rendezvous algorithms without CCC have been attractive to many researchers. Several state-of-the-art results are listed in Table 1; they construct a fixed length sequence for each user to hop through. Generated Orthogonal Sequence (GOS) [11] is a pioneering work which generates an $N(N + 1)$-length sequence based on random permutation of $\{1, 2, \ldots, N\}$. However, it assumes that all channels are available to the users. Quorum-based Channel Hopping (QCH) [4,5] is based on a quorum system for synchronous users. Asynchronous QCH [6] can work even when two users start in different time slots, but it is only applicable to two available channels.

Channel Rendezvous Sequence (CRSEQ) [23], Jump-Stay (JS) [21] and Disjoint Relaxed Difference Set (DRDS) [12] are three representative efficient blind rendezvous algorithms. CRSEQ picks the smallest prime $P > N$ and generates the sequence with P periods, each period containing $3P - 1$ numbers based on the triangle number (triangle number means $T_i = \frac{i(i+1)}{2}$, for any $i \in [1, N]$; see [23] for details) and the modular operation. Jump-Stay uses the same idea by picking the prime number P and it generates the sequence with P periods, where each period contains two *jump* frames and one *stay* frame and each frame is P in length. DRDS is a new method we proposed in [12], through constructing a disjoint relaxed difference set and transforming it into a CH sequence of length $3P^2$; two users can achieve rendezvous in $O(N^2)$ time slots.

All these works construct the same sequence for all users, which we call *global sequence*. Correspondingly, there are several works constructing different sequences for the users, which we call *local sequences* [13]. Hop-and-Wait (HW) [9] makes use of each user's ID to construct a sequence of length $3P^2 \log m$, where m is the size of the network. Local sequences based blind rendezvous algorithms have been presented in some recent works [8,13], which favor the scenario where each user's available channels are just a small fraction of all the available channels. However, their worst case rendezvous time could still be $O(N^2 \log \log N)$ and $O(N^2)$ respectively.

Oblivious rendezvous assumes that different users have different labels for the licensed channels, which obviates the need to establish, maintain and communicate a global set of labels. Nearly all previous algorithms cannot be applied to oblivious rendezvous. To our knowledge, Modified Modular Clock (MMC) [26] is the only one that may work and achieve oblivious rendezvous for two users. MMC firstly counts the number of available channels (n) and picks a prime number $n \leq P \leq 2n$ randomly. Then the user generates a sequence based on P. It is claimed that using MMC, two users can achieve rendezvous within $O(N^2)$ time slots with high probability. However, it cannot guarantee bounded rendezvous. As a step forward, this paper offers deterministic distributed algorithms for bounded oblivious rendezvous.

1.3   Our Contributions

In this paper, we initiate the study of *oblivious rendezvous* in Cognitive Radio Networks. In this problem, each user has a distinct identifier (ID) within the range $[1, M]$ where M is the number of number of secondary users. First, we derive an $\Omega(N^2)$ rendezvous lower bound for any two asynchronous users by introducing the Adversary Assignment Graph, where N is the number of all licensed channels. Then, two deterministic distributed algorithms for the oblivious rendezvous problem for 2 users (ORP-2) are proposed, which subsequently serve as the building block for the cases with more users in a multihop network. The first algorithm is called ID hopping (IDH) which generates a sequence of N frames and each frame consists of $2P$ elements (P is the smallest prime larger than both N and M). We show that each user can repeat accessing the channels by the sequence and rendezvous is guaranteed in $O(N \max\{N, M\})$ time slots. The other one, called Multi-Step Hopping (MSH), is more complicated as it aims at a shorter sequence; it scales the user's ID and then hops among the channels with different steps (scaled values). MSH guarantees rendezvous in $O(N^2 \log_N M)$ time slots, which is much better than IDH, especially when the network size is large. These upper bounds match the presented lower bound if $M = O(N)$ for the IDH algorithm and $M = N^c$ (c can be an arbitrary large constant) for the MSH algorithm. We then extend these algorithms to the multiuser multihop networks with bounded time to rendezvous. Finally, we compare our algorithms with the state-of-the-art rendezvous algorithms through extensive simulations (details are in the full version [14]) which also validate our theoretical analyses.

2   Model and Problem Definitions

2.1   System Model

We consider a multihop Cognitive Radio Network (CRN) with M users (SUs) who coexist with some PUs, and the network diameter is D [1]. Each user has a distinct identifer (ID) $I \in [1, M]$. Suppose the licensed spectrum owned by the PUs is divided into $N(N \geq 1)$ non-overlapping channels where each channel represents certain frequency band (e.g., $470 - 478$ MHz in the TV white space). Each user is equipped with cognitive radios to sense the spectrum for available channels, where a channel is *available* if it is not occupied by any nearby PUs.

Through spectrum sensing, each user can obtain a set of available channels (frequency bands), and all previous blind rendezvous algorithms assume the labels of all these channels are known to all the users. We have already pointed out in the above some possible disadvantages of imposing a common set of labels. We propose the oblivious rendezvous problem where each user labels the sensed channels locally and attempts rendezvous with such local information. More specifically, we rewrite the available channel set for user a as $C_a = \{c_a(1), c_a(2), \ldots, c_a(n_a)\}$ (similarly for user b, as $C_b = \{c_b(1), c_b(2), \ldots, c_b(n_b)\}$) where $n_a = |C_a|, n_b = |C_b|$. Channel $c_a(i) \in C_a$ or $c_b(i) \in C_b$ represents a certain frequency band (channel), where i is a local label in these two users, respectively, but note that $c_a(i)$ and $c_b(i)$ may or may not be the same frequency band (Fig. 1 is an example).

Time is divided into slots of equal length of $2t$, where t is the time duration for establishing a link for communication. According to IEEE 802.22 [24], $t = 10ms$ and thus each time slot has a duration of $20ms$. Then we can consider the system slot-aligned because an overlap of t for link establishment exists even if the start times of different users are not aligned.

In each time slot, the user can access an available channel and attempt rendezvous with its potential neighbors. We use *Time to Rendezvous* (TTR) to denote the number of time slots it takes for users to achieve rendezvous once all users have begun the process. Since all users are physically dispersed and the wake-up time of each user may be different, the rendezvous algorithm should be designed to be applicable to both synchronous and asynchronous users. In this paper, we use *Maximum Time to Rendezvous* (MTTR) as a measure for the worst possible situation for the algorithms and we say rendezvous can be guaranteed if MTTR is bounded.

---

[1] The minimum number of hops between any two users is no larger than D.

(a) $\delta = 1$ (b) $\delta = 2$

Fig. 1. An example of ORP-2 for different δ values

2.2 Problem Definition

We define the *Oblivious Rendezvous Problem (ORP)* as follows:

ORP-M: Given a multihop CRN with M users, denote the available channel set for user i as C_i and its ID as I_i. Let $G = \cap_i C_i$, and $G \neq \emptyset$. Design a strategy for the users such that they are guaranteed to hop onto the same channel in the same time slot, no matter when they begin their attempts.

In order to tackle the above problem, we first focus on designing deterministic distributed algorithms for two users' (out of the M users) rendezvous (ORP-2), and then extend these algorithms to the multiuser multihop scenario (ORP-M) (cf. Section 5).

ORP-2: Given available channel set C and ID set I, design an algorithm over time slots $t : f(t) \in [1, |C|]$ such that for any two users a and b with C_a, C_b, $C_a \cap C_b \neq \emptyset$, $I_a, I_b \in [1, M], I_a \neq I_b$, and $\forall \delta \geq 0$:

$$\exists T \text{ s.t. } c_a(f_a(T + \delta)) = c_b(f_b(T)) \in C_a \cap C_b.$$

where $f_a(T)$ (or $f_b(T)$) represents the the output when user a (or b) runs the algorithm.

The TTR value is T and user b starts the process δ time slots later than user a. The $MTTR$ value of algorithm f is $MTTR_f = max_{\forall \delta} T$. The goal is to find an algorithm f with bounded $MTTR$ and which guarantees rendezvous.

Remark 1. If user b starts the rendezvous process earlier than user a, set $\delta < 0$ in the description of ORP-2 and $TTR = T + \delta$.

Fig. 1 shows an example of ORP-2. Suppose user a has two available channels, $C_a = \{c_a(1), c_a(2)\}$ and user b has four, $C_b = \{c_b(1), c_b(2), c_b(3), c_b(4)\}$. However, only one common channel exists between them, which is $c_a(1) = c_b(4)$. Consider a simple algorithm: each user accesses the channels by repeating the sequence $\{1, 2, \ldots, n\}$ where n is the number of available channels. Thus user a repeats accessing the channels $\{c_a(1), c_a(2), c_a(1), c_a(2), \ldots\}$ until rendezvous, and similarly for user b. For the asynchronous scenario, supposing that user b starts the attempt $\delta = 1$ time slot later, rendezvous is achieved as depicted in Fig. 1(a) at time slot 4 since $c_a(1) = c_b(4)$. However, it is easy to see that the above simple algorithm cannot guarantee rendezvous for all scenarios such as when $\delta = 2$, as in Fig. 1(b). Our goal is to design deterministic distributed algorithms with bounded $MTTR$ value for all δ values.

3   Lower Bound for ORP-2

Theorem 1. *For any deterministic distributed algorithm solving ORP-2, there exist $C_a, C_b, C_a \cap C_b \neq \emptyset$ such that the MTTR value is $\Omega(N^2)$.*

Proof. For any deterministic distributed algorithm $\mathcal{F}$ on the basis of $C, I : f \mapsto [1, n]$ ($n = |C|$), suppose users a and b have different IDs $I_a \neq I_b$ and let $|C_a| = |C_b| = \lceil N/2 \rceil$, $|C_a \cap C_b| = 1$. Equivalently, denote the only common channel between the users as c^* and there exists $1 \leq i, j \leq \lceil N/2 \rceil$ such that $c_a(i) = c_b(j) = c^*$.

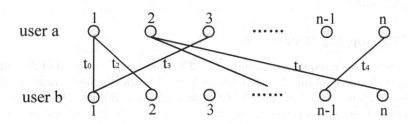

Fig. 2. Adversary Assignment Graph

We introduce the Adversary Assignment Graph (AAG), as in Fig. 2. There are two rows of nodes in the graph and the number of nodes in each row is $n = \lceil N/2 \rceil$. The upper row represents user a's local labels of the available channels with indices $\{1, 2, \ldots, n\}$ and the bottom row represents user b's labels. Let a_t, b_t be the outputs of the algorithm in time slot t, respectively, thus:

$$a_t = f(a_1, a_2, \ldots, a_{t-1}, n, I_a)$$
$$b_t = f(b_1, b_2, \ldots, b_{t-1}, n, I_b)$$

Without loss of generality, suppose user b begins δ slots later; accordingly, we connect node $a_{t+\delta}$ in the upper row with b_t in the other row with an edge having the label t (if the two nodes are already connected, then we just update the label on the edge). For example, $(1, 1)$ is connected in t_0 as depicted in Fig. 2 and $(2, n), (1, 2), (3, 1), (n, n-1)$ are also connected.

Supposing there exists an adversary who can assign licensed channels from the set $U = \{u_1, u_2, \ldots, u_N\}$ to C_a and C_b, rendezvous will not be achieved if the common channel c^* in the upper row is not connected to c^* in the lower row. Since the inputs to the algorithm $\mathcal{F}$ are fixed (for example, the inputs for user a are I_a and $|C_a|$), the lower bound of $MTTR$ is the smallest T such that (c^*, c^*) is connected in every adversary assignment.

Let δ_a be the smallest degree of the upper nodes. If $\delta_a < n$, the adversary can find a node i in the upper row and j in the lower row such that (i, j) is not connected, and then assigns c^* to them, which implies that rendezvous is not achieved. (Then it is easy to assign the other non-intersecting channels to other nodes.) We can verify that $\delta_a < n$ exists if $T < n^2$ and thus the lower bound of $MTTR$ is $n^2 = \Omega(N^2)$. Thus such C_a and C_b can be constructed by the adversary, which implies $MTTR = \Omega(N^2)$. □

4 Algorithms for ORP-2

In this section, we propose two deterministic distributed algorithms for ORP-2, which can meet the lower bound under certain conditions. The first one is based on the channel hopping method where the hopping step is based directly on the ID. The second method scales the user's ID and hops among the channels using different values.

4.1 ID Hopping Rendezvous

Alg. 1 generates a sequence of length $T = 2N\hat{P}$, which is composed of N frames and each frame contains $2\hat{P}$ elements, where $\hat{P}$ is the smallest prime number larger than both N and M. For the i-th frame ($0 \le i < N$), the $2\hat{P}$ elements are constructed as follows (Lines 5–6): set $i + 1$ to the 0-th element and $(i + j \cdot I)$ mod $\hat{P} + 1$ to the j-th element. This procedure can be thought of as picking numbers from a cycle with labels $\{0, 1, \cdots, \hat{P} - 1\}$, where the first one (the 0-th element) is $i+1$ and the second one is I steps later under the modular operation. We refer to this number as the *hopping step* and I is the hopping step in Alg. 1. Since only n available channels exist, elements in $[n + 1, \hat{P}]$ are mapped to $[1, n]$ to accelerate the process, as in Line 7.

Algorithm 1. ID Hopping Algorithm

1: Find the smallest prime $\hat{P}$ such that $\hat{P} > \max\{N, M\}$;
2: $T := 2N\hat{P}$, $t := 0$, $n = |C|$;
3: **while** Not rendezvous **do**
4:     $t' := t \bmod T$;
5:     $x := \lfloor \frac{t'}{2\hat{P}} \rfloor$, $y := t' \bmod 2\hat{P}$;
6:     $z = (x + yI) \bmod \hat{P} + 1$;
7:     $z' = (z - 1) \bmod n + 1$, access channel $c(z')$ in C;
8:     $t := t + 1$;
9: **end while**

For users a and b, the available channel sets are C_a, C_b and their IDs are I_a, I_b respectively. Denote the sequences generated in Alg. 1 (before mapping) as $S_a = \{a_0, a_1, \ldots, a_{T-1}\}$ and $S_b = \{b_0, b_1, \ldots, b_{T-1}\}$ where $T = 2N\hat{P}$. Without loss of generality, suppose user b is $\delta \ge 0$ time slots later than user a:

Lemma 1. *Consider sequences S_a, S_b: $\forall \delta \ge 0$ and $\forall i, j \in [1, \hat{P}]$; there exists $t < T$ such that:*

$$a_{(\delta+t) \bmod T} = i \text{ and } b_t = j.$$

Proof. The users repeat the generated sequence every T time slots, and thus we only need to consider $0 \le \delta < T$. Let $x_1 = \lfloor \frac{\delta}{2\hat{P}} \rfloor$, $y_1 = \delta \bmod 2\hat{P}$. Two situations are analyzed on the basis of y_1:

Case 1: $0 \le y_1 < \hat{P}$. Consider $t = x_2 \cdot 2\hat{P} + y_2$, $0 \le x_2 < N, 0 \le y_2 < \hat{P}$. Let $x_2 + y_2 I_b + 1 \equiv j \bmod \hat{P}$, and thus:

$$y_2 = (j - x_2 - 1)I_b^{-1} \bmod \hat{P}. \tag{1}$$

Here $I_b^{-1}(I_b I_b^{-1} \equiv 1 \bmod \hat{P})$ exists because I_b and $\hat{P}$ are co-primes. We enumerate x_2 from 0 to $N - 1$; y_2 can be computed from Eq. (1) and we denote the value as y_2^h when $x_2 = h$. Then these N values comprise the set $Y = \{y_2^0, y_2^1, \ldots, y_2^{N-1}\}$, and denote the set of corresponding time slots as $T_B = \{t_0, t_1, \ldots, t_{N-1}\}$ where $t_h = h \cdot 2\hat{P} + y_2^h$.

It is clear that $\forall t_h \in T_B$, $0 \le h < N$, $t_h < T$ and $b_{t_h} = j$. Let $T_A = \{t'_0, t'_1, \ldots, t'_{N-1}\}$ where $t'_h = (t_h + \delta) \bmod T$. Then we show that there exists $g \in [0, N)$ such that $a_{t'_g} = i$. Considering any two time slots $t'_g, t'_h \in T_A$ where user a accesses different channels:

$$a_{t'_g} = (x_1 + g) + (y_1 + y_2^g)I_a \bmod \hat{P} + 1$$
$$a_{t'_h} = (x_1 + h) + (y_1 + y_2^h)I_a \bmod \hat{P} + 1$$

Plugging in the expression of y_2^g, y_2^h as in Eq. (1), we can derive:

$$a_{t'_g} - a_{t'_h} \equiv (g - h)(I_a I_B^{-1} - 1) \ne 0 \bmod \hat{P}.$$

Here $I_a \ne I_b$, $I_a, I_b < \hat{P}$ implies $I_a I_b^{-1} \ne 1$. So $a_{t'_g} \ne a_{t'_h}$. As $|T_A| = |T_B| = N$, there are N different values for the N time slots in T_B, and thus there exists t'_g such that $a_{t'_g} = i$, which concludes the lemma.

Case 2: $\hat{P} \le y_1 < 2\hat{P}$. Consider $t = x_2 \cdot 2\hat{P} + y_2$ where $0 \le x_2 < N$ and $\hat{P} \le b_2 < 2\hat{P}$. Using the same technique as in Case 1, we can find $t < T$ such that $a_{(\delta+t) \bmod T} = i$ and $b_t = j$. Thus the lemma holds.   □

Theorem 2. *Alg. 1 guarantees rendezvous between two asynchronous users of ORP-2 in $MTTR = 2N\hat{P}$ time slots, where $\hat{P} \le 2\max\{N, M\}$.*

Proof. Since $C_a \cap C_b \ne \emptyset$, and supposing channel $c^* \in C_a \cap C_b$, there exists $i \in [1, n_a]$ and $j \in [1, n_b]$ such that $a_i = c^*$ and $b_j = c^*$, where $n_a = |C_a|, n_b = |C_b|$. Without loss of generality, and supposing user b is δ time slots later than user a, from Lemma 1, there exists $t < T$ such that they both access channel c^*, and thus rendezvous can be guaranteed in $T = 2N\hat{P}$ time slots no matter when they start the process.   □

Remark 2. P is shown to be $\hat{P} \le 2\max\{M, N\}^2$, and thus $MTTR = O(N\max(N, M))$. If $M = O(N)$ in Alg. 1, $MTTR = O(N^2)$, which meets the lower bound.

---
[2] *Bertrand-Chebyshev Theorem:* $\forall k > 1$, at least one prime p exists such that $k < p < 2k$.

4.2   Multi-Step Channel Hopping Rendezvous

Alg. 1 works well when $M = O(N)$. However, when the number of users increases, this algorithm becomes inefficient (for example, when $M = N^3$). The reason is that the user's ID is used as the hopping step and it enlarges the TTR when M is large. Therefore, we propose a new algorithm which is more efficient for large scale networks, by combining two techniques: ID scaling and hopping with different steps.

Algorithm 2. ID Scale Function

1: **Input:** I;
2: **Output:** $d = \{d(1), d(2), \ldots, d(l)\}$;
3: $l := \lfloor \log_N M \rfloor + 1$, $i := 1$, $cur(0) := I$;
4: **while** $i \leq l$ **do**
5:     $d(i) := cur(i-1) \bmod N + 1$;
6:     $cur(i) := \lfloor cur(i-1)/N \rfloor$
7:     $i := i + 1$;
8: **end while**

As shown in Alg. 2, the ID is scaled to $\lfloor \log_N M \rfloor + 1$ bits and each bit ranges from 1 to N^3. For example, for $N = 8, M = 100, I = 30$, the scaled values are $d = \{7, 4, 1\}$. The scale function plays a key role in the rendezvous algorithm design and the scaled values are used as the hopping steps in Alg. 3.

Algorithm 3. Multi-Step Channel Hopping Algorithm

1: Find the smallest prime P such that $P > N$;
2: $T := 2NP$, $t := 0$, $n = |C|$, $l := \lfloor \log_N M \rfloor + 1$;
3: Invoke Alg. 2 on the user's ID and get the output $d = \{d(1), d(2), \ldots, d(l)\}$;
4: **while** Not rendezvous **do**
5:     **if** $t < T$ **then**
6:         $z := \lfloor t/2P \rfloor + 1$;
7:     **else**
8:         $t' := (t - T) \bmod (2lT)$;
9:         $x := \lfloor t'/2T \rfloor + 1$, $y := t' \bmod 2T$;
10:         $y_1 := y \bmod (2P)$, $y_2 := (\lfloor y/(2P) \rfloor \bmod N + 1$;
11:         $z := (y_2 + y_1 \cdot d(x) - 1) \bmod P + 1$;
12:     **end if**
13:     $z' := (z - 1) \bmod n + 1$, access channel $c(z')$ in C;
14:     $t := t + 1$;
15: **end while**

Alg. 3 can be thought of as generating two types of sequences. The first one is a Scale Sequence (SS) which is composed of 0 and repetitions of l scaled values

---

[3] Here, 'bit' does not mean 0 or 1, but represents a value in $[1, N]$.

(since two users can start the rendezvous process asynchronously, bit 0 is added as a special flag to represent the start of the user):

$$SS = \{0, \underbrace{d(1), d(2)\ldots, d(l)}_{l}, \underbrace{d(1), d(2), \ldots, d(l)}_{l}, \ldots \ldots\}$$

The other one is a Channel Hopping Sequence which is composed of different frames based on SS, as shown in Fig. 3. There are $N+1$ different types of frames, $F(0), F(1), \ldots, F(N)$, and each type of frame is composed of N segments. For example, $F(i)$ has N segments and each segment contains $2P$ elements. The 0-th element of the j-th segment is j and the k-th element is $(j + ki - 1) \mod P + 1$ (the construction of each segment of $F(i)$ can be seen as accessing channel in $[1, P]$ by hopping i steps). For example, $F(0)$ and $F(1)$ are constructed as follows:

$$F(0) = \underbrace{1, 1, \ldots, 1}_{2P}, \underbrace{2, 2, \ldots, 2}_{2P}, \ldots, \underbrace{N, N, \ldots, N}_{2P}$$

$$F(1) = \underbrace{1, 2, \ldots, P}_{2P}, \underbrace{2, 3, \ldots, P, 1}_{2P}, \ldots, \underbrace{N, N+1, \ldots, N-1}_{2P}$$

As shown in Fig. 3, the first element 0 is special because it does not appear in other positions of SS and it corresponds to $F(0)$ once, while the other elements in SS correspond to each type of frames twice.

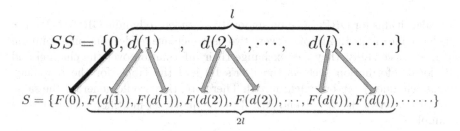

Fig. 3. Construction of Channel Hopping Sequence

Supposing users a and b run Alg. 3 with their local information (C_a, I_a) and (C_b, I_b) where $C_a \cap C_b \neq \emptyset, I_a \neq I_b$, let $n_a = |C_a|, n_b = |C_b|$, denote $d_a = \{d_a(1), d_a(2), \ldots, d_a(l)\}$, $d_b = \{d_b(1), d_b(2), \ldots, d_b(l)\}$ as the outputs of ID Scale function, denote SS_a, SS_b as the scale sequences (as constructed above), and denote $S_a = \{a_0, a_1, \ldots, a_t, \ldots\}$, $S_b = \{b_0, b_1, \ldots, b_t, \ldots\}$ as the Channel Hopping Sequences. Without loss of generality, suppose user b starts the process $\delta \geq 0$ time slots later than user a. we have the following Lemmas 2, 3 and 4. Due to the lack of space, the proofs are included only in the full version [14].

Lemma 2. *Consider SS_a, SS_b: $\forall \delta' \in Z$, there exists $i \geq 0, i + \delta' \geq 0$ such that:*

$$SS_a(i) \neq SS_b(i + \delta)$$

Lemma 3. *Consider* S_a, S_b; *for any pair* (i, j) *where* $1 \le i \le n_a$, $1 \le j \le n_b$, *if* $0 \le \delta < T$,

$$\exists t \le 2lT \quad s.t. \quad a_{(\delta+t)} = i \text{ and } b_t = j.$$

Lemma 4. *Consider* S_a, S_b, *for any pair* (i, j) *where* $1 \le i \le n_a$, $1 \le j \le n_b$, *if* $\delta \ge T$,

$$\exists t \le T \quad s.t. \quad a_{(\delta+t)} = i \text{ and } b_t = j.$$

Theorem 3. *Alg. 3 guarantees rendezvous between two asynchronous users of ORP-2 in* $MTTR = 4lNP = O(N^2 \log_N M)$ *time slots, where* $P \le 2N$.

Proof. As assumed, $G = C_a \cap C_b \neq \emptyset$, supposing $c^* \in G$ and there exists $1 \le i \le n_a$, $1 \le j \le n_b$ such that $c_a(i) = c^*, c_b(j) = c^*$. Without loss of generality, suppose user b starts the process δ time slots later. If $\delta < T$, from Lemma 3, rendezvous is guaranteed in $2lT$ time slots; if $\delta \ge T$, rendezvous is guaranteed in T time slots. Thus $MTTR \le 2lT = 4lNP = O(N^2 \log_N M)$. $\quad\square$

Generally speaking, if M is (bounded by) a polynomial function of the total number of licensed channels N, the length of scaled bits is a constant and two users can be guaranteed to rendezvous in $O(N^2)$ time slots, which meets the lower bound of ORP-2. Moreover, this result is also comparable to even state-of-the-art non-oblivious rendezvous algorithms as shown in Table 1.

5   Algorithm for ORP-M

The algorithms for ORP-2 can be smoothly extended to handle ORP-M. We use the basic idea in [9,12,21]: once every two users achieve rendezvous on a common channel successfully, they can exchange their information over the channel and the local information such as the user's ID and the labels for the frequency bands (channels) can be synchronized. Therefore, they would generate the same sequence afterwards. We extend Alg. 3 to the multiuser multihop scenario as an example.

Algorithm 4. Algorithm for Multiuser Multihop Scenario

1: **while** Not terminated **do**
2:     Run Alg. 3 with local information (I, C);
3:     **if** Rendezvous with user - (I', C') **then**
4:         $I := \min(I, I')$;
5:         $C := C \cap C'$;
6:         Synchronize labels for the channels as the user with smaller ID;
7:     **end if**
8: **end while**

In Alg. 4, the user runs Alg. 3 with local information (I, C). Once rendezvous is achieved with another user with (I', C'), they exchange their information and three operations are executed:

- Change I to be the smaller value between I, I';
- Change C to be the intersection of C and C';
- Synchronize the labels for the available channels with the user with smaller I value such that $\forall i \in [1, |C|]$, $c(i) = c'(i)$.

After these three steps, the local information of the two users are the same and they access the channels with the same sequence until rendezvous with others. Supposing that the network diameter of the CRN in ORP-M is D, the $MTTR$ value can be guaranteed as in Theorem 4 (pleas refer to [14] for the proof).

Theorem 4. *Alg. 4 guarantees that all users can achieve rendezvous in* $MTTR = 4lNPD = O(N^2 D \log_N M)$ *time slots, where* D *is the diameter of the CRN.*

6 Oblivious Rendezvous Applications

Oblivious rendezvous is not only practical in a Cognitive Radio Network (CRN), but also suitable for several other (theoretical) problems. For example, the telephone coordination problem [2]: there are n telephones in each of two rooms, where the telephones are connected pairwisely by some unknown rules. Each room has a player who can pick up one telephone and say 'hello' in each time slot until they hear each other. They do not have any common labels of the telephones by which they can coordinate, and the aim is to minimize the time slots required for the players to meet. This problem only considers two synchronous users and each has exactly n telephones. In our settings, once each user is assigned a distinct identifier, a deterministic algorithm for this problem can be designed even for asynchronous users and some of these telephones are broken. Another problem is rendezvous search on the graph [3], where different users are placed on the graph and they attempt to meet each other as quickly as possible. Our oblivious rendezvous problem is a little different as we can consider the users in the CRN being restricted to walk in a given clique (the set of available channels), and thus the time to rendezvous can be easily extended. For other more general rendezvous search problems, the method in this paper could be used as a basis for their study.

7 Conclusion

We introduce the oblivious rendezvous problem which is believed to be more practical in constructing Cognitive Radio Networks. In contrast to existing, non-oblivious rendezvous problem, the users in our setting have different labels for the licensed frequency bands (channels), and we derive rendezvous algorithms that is based on each user's local information.

For oblivious rendezvous, we first derive an $\Omega(N^2)$ rendezvous time lower bound. Then we propose two deterministic distributed algorithms: the ID Hopping (IDH) algorithm which can achieve rendezvous between two users in $O(N \max(M, N))$ time slots, where M is number of users in the network; and the Multi-Step channel Hopping (MSH) algorithm which guarantees oblivious rendezvous in $O(N^2 \log_N M)$ time slots. The IDH algorithm works efficiently when M is small, while the MSH algorithm performs much better for larger M, which implies large scale networks with many users. The upper bounds of two algorithms match the presented lower bound if $M = O(N)$ for the IDH algorithm and if $M = N^c$ (c is a constant) for the MSH algorithm. Third, we extend these two algorithms to multiuser multihop networks. We have conducted extensive simulations for both two-user rendezvous and multihop multiusers rendezvous using our algorithms (details in the full version [14]).

Although our algorithms are designed for oblivious rendezvous, the simulation results show that they are comparable to the state-of-the-art non-oblivious rendezvous algorithms and they even perform much better under some circumstances. For oblivious rendezvous, our two proposed algorithms also outperform the MMC algorithm, and the MSH algorithm performs the best as the number of rendezvous users increases.

N, M are the number of licensed channels and users in the network respectively; one future direction is to design fully distributed rendezvous algorithms without knowing these values. We also want to explore randomized distributed algorithms which can achieve bounded rendezvous time with high probability.

Acknowledgement. We thank the anonymous reviewers for their very helpful comments which helped improve the presentation of this paper. This work was supported in part by the National Basic Research Program of China Grant 2011CBA00300, 2011CBA00301, the National Natural Science Foundation of China Grant 61103186, 61033001, 61361136003, Hong Kong RGC-GRF grant 714311, and the Shu Shengman Special Research Fund.

References

1. Akyildiz, I., Lee, W., Vuran, M., Mohanty, S.: NeXt Generation/Dynamic Spectrum Access/Cognitive Radio Wireless Networks: A Survey. Computer Networks 50(13), 2127–2159 (2006)
2. Alpern, S., Pikounis, M.: The Telephone Coordination Game. Game Theory Appl. 5, 1–10 (2000)
3. Anderson, E.J., Weber, R.R.: The Rendezvous Problem on Discrete locations. Journal of Applied Probability 28, 839–851 (1990)
4. Bian, K., Park, J.-M., Chen, R.: A Quorum-Based Framework for Establishing Control Channels in Dynamic Spectrum Access Networks. In: Mobicom (2009)
5. Bian, K., Park, J.-M.: Asynchronous Channel Hopping for Establishing Rendezvous in Cognitive Radio Networks. In: IEEE INFOCOM (2011)
6. Bian, K., Park, J.-M.: Maximizing Rendezvous Diversity in Rendezvous Protocols for Decentralized Cognitive Radio Networks. IEEE Transactions on Mobile Computing 12(7), 1294–1307 (2013)
7. Cai, Z., Ji, S., He, J., Bourgeois, A.G.: Optimal Distributed Data Collection for Asynchronous Cognitive Radio Networks. In: ICDCS (2012)
8. Chen, S., Russell, A., Samanta, A., Sundaram, R.: Deterministic Blind Rendezvous in Cognitive Radio Networks. In: ICDCS (2014)

9. Chuang, I., Wu, H.-Y., Lee, K.-R., Kuo, Y.-H.: Alternate Hop-and-Wait Channel Rendezvous Method for Cognitive Radio Networks. In: INFOCOM (2013)
10. Dai, Y., Wu, J., Xin, C.: Virtual Backbone Construction for Cognitive Radio Networks without Common Control Channel. In: INFOCOM (2013)
11. DaSilva, L., Guerreiro, I.: Sequence-Based Rendezvous for Dynamic Spectrum Access. In: DySPAN (2008)
12. Gu, Z., Hua, Q.-S., Wang, Y., Lau, F.C.M.: Nearly Optimal Asynchronous Blind Rendezvous Algorithm for Cognitive Radio Networks. In: SECON (2013)
13. Gu, Z., Hua, Q.-S., Dai, W.: Local Sequence Based Rendezvous Algorithms for Cognitive Radio Networks. In: SECON (2014)
14. Gu, Z., Hua, Q.-S., Wang, Y., Lau, F.C.M.: Oblivious Rendezvous in Cognitive Radio Networks, http://i.cs.hku.hk/~qshua/sirocco2014full.pdf
15. Huang, X., Lu, D., Li, P., Fang, Y.: Coolest Path: Spectrum Mobility Aware Routing Metrics in Cognitive Ad Hoc Networks. In: ICDCS (2011)
16. Ji, S., Beyah, R., Cai, Z.: Minimum-Latency Broadcast Scheduling for Cognitive Radio Networks. In: SECON (2013)
17. Jia, J., Zhang, Q., Shen, X.: HC-MAC: A Hardware-Constrained Cognitive MAC for Efficient Spectrum Management. IEEE Journal on Selected Areas in Communications 26(1), 106–117 (2008)
18. Kondareddy, Y., Agrawal, P., Sivalingam, K.: Cognitive Radio Network Setup without a Common Control Channel. In: MILCOM (2008)
19. Lazos, L., Liu, S., Krunz, M.: Spectrum Opportunity-Based Control Channel Assignment in Cognitive Radio Networks. In: SECON (2009)
20. Lin, Z., Liu, H., Chu, X., Leung, Y.-W.: Enhanced Jump-Stay Rendezvous Algorithm for Cognitive Radio Networks. IEEE Communications Letters 17(9), 1742–1745 (2013)
21. Liu, H., Lin, Z., Chu, X., Leung, Y.-W.: Jump-Stay Rendezvous Algorithm for Cognitive Radio Networks. IEEE Transactions on Parallel and Distributed Systems 23(10), 1867–1881 (2012)
22. Perez-Romero, J., Salient, O., Agusti, R., Giupponi, L.: A Novel On-Demand Cognitive Pilot Channel enabling Dynamic Spectrum Allocation. In: DySPAN (2007)
23. Shin, J., Yang, D., Kim, C.: A Channel Rendezvous Scheme for Cognitive Radio Networks. IEEE Communications Letters 14(10), 954–956 (2010)
24. Stevenson, C.R., Chouinard, G., Lei, Z., Hu, W., Shellhammer, S.J., Caldwell, W.: IEEE 802.22: The First Cognitive Radio Wireless Regional Area Network Standard. IEEE Communications Magazine 47(1), 130–138 (2009)
25. Song, J., Xie, J., Wang, X.: A Novel unified Analytical Model for Broadcast Protocols in Multihop Cognitive Radio Ad Hoc Networks. IEEE Transaction on Mobile Computing (2013)
26. Theis, N.C., Thomas, R.W., DaSilva, L.A.: Rendezvous for Cognitive Radios. IEEE Transactions on Mobile Computing 10(2), 216–227 (2011)
27. Zhang, D., He, T., Ye, F., Ganti, R., Lei, H.: EQS: Neighbor Discovery and Rendezvous Maintenance with Extended Quorum System for Mobile Sensing Applications. In: ICDCS (2012)

Local Broadcasting with Arbitrary Transmission Power in the SINR Model

Fabian Fuchs and Dorothea Wagner

Karlsruhe Institute for Technology
Karlsruhe, Germany
{fabian.fuchs,dorothea.wagner}@kit.edu

Abstract. In the light of energy conservation and the expansion of existing networks, wireless networks face the challenge of nodes with heterogeneous transmission power. However, for more realistic models of wireless communication only few algorithmic results are known. In this paper we consider nodes with arbitrary, possibly variable, transmission power in the so-called physical or SINR model. Our first result is a bound on the probabilistic interference from all simultaneously transmitting nodes on receivers. This result implies that current local broadcasting algorithms can be generalized to the case of non-uniform transmission power with minor changes. The algorithms run in $\mathcal{O}(\Gamma^2 \Delta \log n)$ time slots if the maximal degree Δ is known, and $\mathcal{O}((\Delta + \log n)\Gamma^2 \log n)$ otherwise, where Γ is the ratio between the maximal and the minimal transmission range. The broad applicability of our result on bounding the interference is further highlighted, by generalizing a distributed coloring algorithm to this setting.

1 Introduction

One of the most fundamental problems in wireless ad hoc networks is to enable efficient communication between neighboring nodes. This problem recently received increasing attention among the distributed algorithm community, as more refined models of wireless communication became established in algorithms research. Among these models, the so-called physical or *signal-to-interference-and-noise* (SINR) model is most prominent and promising, due to its common use in the engineering literature. However, so far most algorithmic work in the SINR model is restricted to the case of uniform transmission power. In this case, *local broadcasting* [7,16,18,6] provides initial communication by enabling each node to transmit one message such that all intended receivers (i.e., neighbors) are able to decode the message.

In this work we consider the problem of local broadcasting in the SINR model under arbitrary transmission power assignment, i.e., each node has its individual, possibly variable, transmission power. We are the first to consider this setting from an algorithmic perspective. While some distributed node coloring algorithms do consider the transmission power to be variable [2,17], they still increase the transmission power synchronously and thus effectively operate on an uniform power network. The sole line of research that leverages non-uniform transmission

M. Halldórsson (Ed.): SIROCCO 2014, LNCS 8576, pp. 180–193, 2014.

power is on link scheduling and capacity maximization [8,10]. However, there, each node is usually considered to be either transmitter or receiver. If a node has multiple roles it might have to adapt its transmission power frequently. On the other hand, the effects of heterogeneous transmission power are considered in simulation-based studies for example in [5,14], while the case of unidirectional communication links, which are a result of heterogeneous transmission powers, is studied even more frequently [19,15].

We assume the harsh environment of an wireless ad hoc network just after deployment. In particular, we consider multi-hop networks, where the nodes do initially not have any information about whether other nodes are awake, have already started the algorithm or in which phase of the algorithm they are. The only knowledge they may have is an upper bound on the number of neighbors, and a rough bound on the total number of nodes in the network. Note that our model does not assume a *collision detection* mechanism. Additionally to this harsh model, we also considered some recent ideas regarding practical matters of algorithms for wireless networks by Kuhn *et. al.* [1]. They promoted the use of lower and upper bounds for important network parameters such as α, β and N (cf. Section 2). This is an important step towards practicability of the algorithms as upper and lower bounds to these values are well-represented in the literature, however, exact values vary depending on the network environment.

1.1   Contributions

We are the first to consider arbitrary transmission powers in the SINR model. Our main contribution provides an abstract method for bounding the interference in these networks. We prove that transmissions are feasible based on the sum of local transmission probabilities. This result is widely applicable, as verifying that the sum of local transmission probabilities is bounded as required is relatively simple. By applying this result to known results on local broadcasting we are able to generalize current local broadcasting algorithms to our setting. The algorithms run in $\mathcal{O}(\Gamma^2 \Delta \log n)$ time slots if the maximal degree Δ is known and $\mathcal{O}((\Delta + \log n)\Gamma^2 \log n)$ time slots otherwise, where Γ is the ratio between the maximal and the minimal transmission range. Thus, they match the runtime of the algorithms for uniform transmission power as $\Gamma = 1$ in this case. Additionally we discuss the case of variable transmission power in Section 4.2, which achieves similar bounds, but allows nodes to change the transmission power in each time slot instead of fixing it for each round of local broadcasting. The applicability of our result on bounding the interference is further emphasized by a brief description of how a well-known coloring algorithm can be generalized to arbitrary transmission power networks. Note that the algorithms are fully operational under asynchronous node wake-up and sleep.

1.2   Related Work

The study of local broadcasting, and interference in general, has only recently emerged. Especially in classical distributed message passing models such as

$\mathcal{LOCAL}$ or $\mathcal{CONGEST}$ [13], the transmission of a message to neighbors is guaranteed. However, this is not the case for wireless networks. Hence interference in general and local broadcasting in particular must be considered in the more realistic SINR model of interference. Goussevskaia $et.\ al.$ [6] were the first to present local broadcasting algorithms in the SINR model. Their first algorithm assumes an upper bound Δ on the number of neighbors to be known by the nodes and solves local broadcasting with high probability in $\mathcal{O}(\Delta \log n)$ time, while the second algorithm does not assume this knowledge and requires $\mathcal{O}(\Delta \log^3 n)$ time. The second algorithm has subsequently been improved by Yu $et.\ al.$ to run in $\mathcal{O}(\Delta \log^2 n)$ [18], and again to $\mathcal{O}(\Delta \log n + \log^2 n)$ [16]. This bound has been matched by Halldórsson and Mitra in [7] using a more robust algorithm, along with an algorithm that leverages carrier sensing to achieve a time complexity of $\mathcal{O}(\Delta + \log n)$. For related work regarding distributed node coloring, we refer to the full version [4].

2   Preliminaries

We consider a wireless network consisting of n nodes, that are placed arbitrarily on the Euclidean plane. We assume that all nodes in the network know their ID and an upper bound $\tilde{n}$ on n, with $\tilde{n} \leq n^c$ for some constant $c \geq 1$. As the upper bound influences our results only by a constant factor we usually write n even though only $\tilde{n}$ may be known by the nodes. Also, we assume that nodes know lower and upper bounds on the transmission power or the transmission ranges. This assumption is realistic, as lower bounds for reasonable minimal transmission ranges can be computed while upper bounds (for specified frequencies) are often regulated by public authorities.

In the geometric SINR model a transmission from node v to node w is successful iff the SINR condition holds:

$$\frac{\frac{P_v}{\text{dist}(v,w)^\alpha}}{\sum_{u \in \mathcal{I}} \frac{P_u}{\text{dist}(u,w)^\alpha} + N} \geq \beta \tag{1}$$

where P_v (P_u) denotes the transmission power of node v (u), α is the attenuation coefficient, which depends on the environment and characterizes how fast the signal fades. The SINR-threshold $\beta \geq 1$ is a hardware-defined constant, N is the environmental noise and $\mathcal{I}$ is the set of nodes transmitting simultaneously with v. As introduced in [1] and motivated by the hardness of determining exact network parameters we restrict our nodes knowledge to upper and lower bounds of the values α, β and N and denote them by e.g. $\underline{\alpha}$ and $\bar{\alpha}$ for the minimal and maximal values.

Based on the SINR constraints, we define the $maximum\ transmission\ range$ of a node v to be $\bar{R}_v = (\frac{P_v}{N\beta})^{1/\bar{\alpha}}$. Note that this is maximal under the restriction that this range can be reached regardless of the actual network parameters α, β, N. The global maximum transmission range in the network is denoted by $\bar{R}$, the minimum range by $\underline{R}$ and the ratio between $\bar{R}$ and $\underline{R}$ by $\Gamma = \frac{\bar{R}}{\underline{R}}$. Due to the SINR constraints, a node v cannot reach another node w which is located at the maximum transmission range of v, as soon v transmits simultaneously with any other node in the network. As having only one simultaneous transmission in the network is not desired, we use a parameter $\delta > 1$ to determine the distance up to which the nodes messages should be received. We call this distance the *broadcasting range* $R_v = (\frac{P_v}{\delta N \beta})^{1/\alpha}$ and the region within this range from v the broadcasting region B_v. We denote the maximum number of nodes within the transmission range $\bar{R}_v$ of any v as Δ. This is an upper bound on the number of nodes reachable from v, since the broadcasting range R_v is fully contained in the transmission range. Note that Δ is known by the nodes only if stated with the corresponding algorithms. We define the *proximity region* around v as the area closer than $3\bar{R}$ to v. Note that even though we use time slots in our analysis, we do not require a global clock or synchronized time slots in our algorithm. Decent local clocks are sufficient, while time slots are only required in the analysis.

Roadmap: In the following section we bound the probabilistic interference of nodes outside the proximity region based on the sum of transmission probabilities from within each transmission region. In Section 4 we apply this result to previous results on local broadcasting and thereby transfer current algorithms to the more general model. The applicability of our results is highlighted in Section 5, as we consider the problem of distributed node coloring and generalize a well-known algorithm from the case of uniform transmission powers. We conclude this paper in Section 6 with some final remarks.

3   Bounding the Interference

In contrast to other models for interference in wireless communication such as the protocol model, the SINR model captures the global aspect of interference and reflects that even interference from far-away nodes can add up to a level that prevents the reception of transmissions from relatively close nodes. To ensure that a given transmission can be decoded by all nodes within the broadcasting range, one usually proves that reception within a certain time interval is successful *with high probability* (w.h.p.—with probability at least $1 - \frac{1}{n^c}$ for a constant $c > 1$). Such a proof can be split in two parts

1. The probability that a node transmits within a proximity region around a sender is constant
2. Let $P_{2\text{high}}(v)$ be the event that the interference from all nodes outside of the proximity region of v on nodes in the broadcasting region of v is too high. Show that $P_{2\text{high}}(v)$ has constant probability.

We shall follow this scheme by considering the transmission of an arbitrary node and proving that both conditions hold with constant probability in each time slot, and hence a local broadcast is successful with high probability.

   In order to make the result general and applicable to many different settings, we make only one very general assumption. Namely we assume the sum of transmission probabilities from within a broadcasting region to be bounded by a constant. It is very common to require algorithms in the SINR model to ensure this, which allows us to apply the analysis from this section in the following Sections 4 and 5 to generalize algorithms designed for the uniform transmission power case to the more general case considered in this paper.

Definition 1. *Given a network of n nodes with at most Δ nodes in each transmission region. Let γ be the upper bound on the sum of transmission probabilities from within one transmission region.*

Let the upper bound on the sum of transmission probabilities from within each transmission region be

$$\gamma := \frac{(\delta - 1)}{120 \bar{\beta} \Gamma^2 \sum_{i=1}^{n} \frac{1}{i^{\bar{\alpha}-1}}}. \tag{2}$$

Note that this bound can be realized, for example by requiring nodes to transmit with probability γ/Δ. Another option is the so-called slow-start technique, cf. Section 4.1. The constant is of the stated form, mainly to bound the interference from all other nodes in the network in the proof of Theorem 4. It holds that $\gamma \leq 1$[1]. Let us now prove a bound on the probability that a close-by node transmits, which is also required for the main theorem of this section.

Lemma 2. *Given an arbitrary node v. The probability that no node in the proximity region of v transmits in a given time slot is at least 1/4.*

Proof. Let $3\bar{R}(v)$ denote the set of nodes that are closer to v than $3\bar{R}$ in this argument. This is the set of nodes in the proximity region of v. The probability that a node in $3\bar{R}(v)$ transmits in a single time slots is

$$P_{\text{none}}^{3\bar{R}(v)} \geq \prod_{u \in 3\bar{R}(v)} (1 - p_u) \geq \left(\frac{1}{4}\right)^{\sum_{u \in 3\bar{R}(v)} p_u} \geq \left(\frac{1}{4}\right)^{49\Gamma^2 \cdot \gamma} \geq \left(\frac{1}{4}\right),$$

where the second inequality holds due to a fact used in the proof of Lemma 4.2 in [9]. The third inequality due to a simple geometric argument about the number of independent nodes within distance $3\bar{R}$ of v and the bound on the sum of transmission probabilities from within each transmission region. The last inequality holds since $49\Gamma^2 \cdot \gamma < 1$.

Let us now consider nodes that are not in the proximity region of the transmitting node. In order to bound the interference originating from these nodes, we use rings around the transmitting node and bound the probabilistic interference from within each ring. Note that although our definition of the proximity region and rings differ, similar arguments are made, for example, in [7,6].

---

[1] This may not be true for a large δ. Thus for $\delta > 2$ we use $\gamma := \frac{1}{120 \bar{\beta} \Gamma^2 \sum_{i=1}^{n} \frac{1}{i^{\bar{\alpha}-1}}}$.

Definition 3. *For a node v, the ring C_i^v, $i \geq 0$, is defined as the set of nodes with distance at least $(i+1) \cdot \bar{R}$ and at most $(i+2) \cdot \bar{R}$. For a ring C_i^v, the extended ring C_{i+}^v is defined as the set of nodes with distance at least $i \cdot \bar{R}$ and at most $(i+3) \cdot \bar{R}$.*

Note that for a ring C_i^v, the extended ring C_{i+}^v is defined such that the nodes in the transmission region of an arbitrary node $w \in C_i^v$ are contained in C_{i+}^v. If it is clear to which node v the rings refer, we write C_i and C_{i+} for brevity.

Theorem 4. *Let the sum of transmission probabilities from each transmission region be upper bounded by γ. Given a node v, the probabilistic interference from nodes outside the proximity region of v is upper bounded by $(\delta - 1)N$.*

Proof. Let us first bound the interference from a single ring C_i. By a simple geometric argument it holds that the maximal number of independent nodes in the extended ring C_{i+} is at most $(6i + 9)\bar{R}^2/\underline{R}^2$. By combining this number with the sum of transmission probabilities from within each broadcasting region, we can bound the interference from the nodes in C_i. As each node in the ring C_i has distance greater than $i \cdot \bar{R}$ from any node in B_v, it follows that the probabilistic interference on any node $u \in B_v$ is at most

$$\Psi_{C_i} \leq \sum_{w \in C_{i+}} \frac{p_w P_w}{(i\bar{R})^{\bar{\alpha}}} \leq \frac{4(6i+9)\bar{R}^2 \gamma \bar{\beta} \bar{N}}{\underline{R}^2 i^{\bar{\alpha}}} \cdot \left(\frac{\bar{R}}{\underline{R}}\right)^{\bar{\alpha}} \leq \frac{60\gamma\bar{\beta}\bar{N}}{i^{\bar{\alpha}-1}} \cdot \left(\frac{\bar{R}}{\underline{R}}\right)^2.$$

The second inequality holds since at most $(6i + 9)\bar{R}^2/\underline{R}^2$ independent nodes are in C_{i+}, from each broadcasting region around such an independent node the sum of transmission probabilities is at most γ, and by $P_w \leq \bar{\beta}\bar{N}\bar{R}^{\bar{\alpha}}$. The last inequality simplifies the fraction and holds since $i > 1$. Summing over all rings it follows

$$\Psi_{w \notin 3\bar{R}(v)} \leq \sum_{i=2}^{\infty} \Psi_{C_i} \leq 60\gamma\bar{\beta}\bar{N}\Gamma^2 \sum_{i=1}^{n} \frac{1}{i^{\bar{\alpha}-1}} \leq \frac{(\delta - 1)\bar{N}}{2},$$

where the second inequality holds by inserting the bound on Ψ_{C_i} and the fact that there are at most n non-empty rings. The last inequality follows from the upper bound on γ, stated in Equation 2.

4   Local Broadcasting

In the previous section we have shown how to bound the probabilistic interference from nodes outside of the proximity region based on an upper bound on the sum of transmission probabilities from within each transmission region. Such bounds are known for many algorithms in the case of uniform transmission power, and hence we can plug our results into a large body of related work, and transfer results with minimal additional efforts to the case of arbitrary but fixed transmission power. In the following section we briefly state our results regarding local broadcasting along with proof sketches as required. In Section 4.2 we discuss our results regarding variable transmission power.

4.1   Arbitrary But Fixed Transmission Power

The current results on local broadcasting with the knowledge of Δ are based on transmitting with a fixed probability in the order of $1/\Delta$ for a sufficient number of time slots in $\mathcal{O}(\Delta \log n)$, while results that do not assume the maximal degree Δ to be known are usually based on a so-called slow-start mechanism.

With Knowledge of the Maximal Degree Δ. Let us first consider the case, in which each node knowns the maximal degree Δ. Using the result on local broadcasting by Goussevskaia, Moscibroda and Wattenhofer [6], it is easy to show that local broadcasting can be realized in $\mathcal{O}(\Gamma^2 \Delta \log n)$ time slots by simply adapting the transmission probability to our requirements.

Theorem 5. *Let the transmission probability of each node be* $p = \gamma/\Delta$, *and* $c > 1$ *an arbitrary constant. A node* v *that transmits with probability* p *for* $8c/p \log n = \mathcal{O}(\Gamma^2 \Delta \log n)$ *time slots successfully transmits to its neighbors whp.*

Proof. Since the transmission probability is chosen such that the sum of transmission probabilities from within each proximity range is at most γ, we can directly apply Theorem 4. Using the theorem, combined with the standard Markov inequality, the probability that the interference from nodes outside of the proximity region is too high (i.e., higher than $(\delta - 1)\bar{N}$) is less than $1/2$. Lemma 2 states that the probability that no node within the proximity range of a node transmits is greater than $\frac{1}{4}$. Combining both probabilities with the transmission probability of p implies that the probability of a successful broadcast is at least $p/8$ in each time slot. Thus transmitting for $8c/p \log n$ time slots results in a successful local broadcast with probability at least $1 - \frac{1}{n^c}$. A detailed proof can be found in the full version [4]. □

Without Knowledge of Δ. Let us now consider the case that the nodes are not given a bound on the maximum degree Δ. In contrast to the previous algorithm for local broadcasting, the "optimal" transmission probability is initially unknown.

In order to create local broadcasting algorithms for this model, a slow start mechanism can be used [7,16,18,6]. In such a mechanism each node starts with a very low transmission probability in the range of $\mathcal{O}(1/n)$ and doubles the probability until a certain number of transmissions are received, and the probability is reset to a smaller value. With such a mechanism, local broadcasting in the (uniform-powered) SINR model can be achieved in $\mathcal{O}(\Delta \log n + \log^2 n)$ [7,16]. Although different forms of the slow start mechanisms are used they reset the transmission probabilities such that the sum of transmission probabilities in each transmission region can be upper bounded by a constant.

Let us now consider the algorithm of Halldórsson and Mitra, described in [7]. We can adapt the algorithm so that local broadcasting provably works with high probability in the more general model considered in this paper. This can be done by modifying the maximal transmission probability to be $\gamma/16$ instead of $1/16$, which can be done by simply changing Line 7 of Algorithm 1 in [7] from $p_y \leftarrow \min\{\frac{1}{16}, 2p_y\}$ to $p_y \leftarrow \min\{\frac{\gamma}{16}, 2p_y\}$. This minimal adaptation allows us to bound the sum of transmission probabilities similar to how it is done in the original paper.

Lemma 6. *Let $\mathcal{N}$ be a network with arbitrary transmission power assignment, asynchronous node wake-up and let all nodes execute Algorithm 1 from [7] with maximal transmission probability be set to $\gamma/16$. Then the sum of transmission probabilities from within each proximity region is upper bounded by γ.*

By combining this result with Theorem 4, Lemma 2, and a similar argumentation as in the previous section, the transmission is successful at least once with high probability. The correctness of the algorithm follows with the original argumentation in [7]. Using the modified Algorithm 1 from [7], we get for the more general case of arbitrary transmission power assignment

Theorem 7. *There exists an algorithm for which the following holds whp: Each node v successfully performs a local broadcast within $\mathcal{O}((\Delta + \log n)\Gamma^2 \log n)$.*

Remark: Note that the local broadcasting algorithm by Yu *et. al.* [16] has the same runtime guarantees as the algorithm by Halldórsson and Mitra [7], but was proposed slightly earlier. However, their algorithm cannot be transfered to the case of arbitrary transmission power as it heavily relies on bidirectional communication to operate. Specifically, their algorithm computes an MIS, acquires information about dominated nodes and then assigns transmission intervals to the dominated nodes. Thus, it requires (at least) significant changes to generalize it to networks of arbitrary transmission power.

4.2   Variable Transmission Power

For local broadcasting, the transmission power is required to be fixed for at least one full round of local broadcasting. In this section, we consider a more general setting and allow the nodes to change the transmission power for each time slot. As it is not initially clear which nodes should be considered as intended receivers in such a setting, our result states the achieved broadcasting range, based on the number of times certain transmission power levels were exceeded within the considered time interval. Note that we assume Δ to be known to the nodes in this section. We shall now briefly discuss the notation required in this section. We consider the time slots in one interval $(1, \ldots, t)$. For multiple time intervals that are not continuous, a transmission power of 0 can be added to fill the gaps. Let $\{0 = \mathrm{P}_v^{[0]}, \mathrm{P}_v^{[1]}, \ldots, \mathrm{P}_v^{[k]}\}$ the set of transmission powers used by v (plus 0), such that $\mathrm{P}_v^{[j]} < \mathrm{P}_v^{[j+1]}$ for $j = 0, 1, \ldots, k - 1$. We denote the number of time slots, v used a transmission power of at least $\mathrm{P}_v^{[j]}$ by T_j. Let $\mathrm{R}_v^{[j]}$ be the broadcasting range corresponding to $\mathrm{P}_v^{[j]}$.

Theorem 8. *Let all the nodes in the network transmit with probability at most $p = \gamma/\Delta$ and a variable transmission power between $\underline{\mathrm{R}}$ and $\bar{\mathrm{R}}$. Let v be an arbitrary node that transmits with variable transmission powers during the interval*

$(1, \ldots, t)$. For j maximal such that $T_j > 8c/p \log n$, all nodes closer to v than $R_v^{[j]}$ received v's message whp for an arbitrary constant $c > 1$.

Proof. Let j be maximal such that $T_j > 8c/p \log n$. It holds that v transmits with probability p and transmission power at least $P_v^{[j]}$ in at least $8c/p \log n$ time slots. Let us consider such a time slot i. As the sum of transmission probabilities from within each proximity range is obviously bounded by at most γ, we can apply our method to bound the interference. It holds due to Theorem 4 and Lemma 2 that a message transmitted by v in time slot i is received by nodes closer to v than $R_v^{[j]}$ with probability at least $1/8$. Combined with the transmission probability p and considered over $8c/p \log n$ time slots, this results in a success probability of at least $1 - \frac{1}{n^c}$ with an argumentation similar to the that in the proof of Theorem 5 in the full version [4].

5   Distributed Node Coloring

We shall demonstrate the applicability of our results to existing algorithmic results in the uniform SINR model in this section. Therefore we consider a distributed node coloring algorithm[2], and briefly show how this algorithm can be transfered to the case of arbitrary transmission powers. Distributed node coloring is a fundamental problem in wireless networks, as a node coloring can be used to compute a schedule of transmissions by assigning each color to a different time slot. Thus, efficient transmissions based on a *time-division-multiple-access* (TDMA) schedule can be reduced to a node coloring. The algorithm we consider computes a node coloring that ensures that two nodes with the same color cannot communicate directly. This does not necessarily result in a transmission schedule that is feasible in the SINR model, however, one can use additional techniques like those described in [2] or [3] to transform such a node coloring to a local broadcasting schedule that is feasible in the SINR model. Let us now define some notation required for the coloring problem. For two nodes $v, u \in V$ we say that there is a *communication link* from v to u if u is in the broadcasting region of v. We say that there is a *unidirectional* communication link from v to u if there is a communication link from v to u, but not from u to v. In this case v *dominates* u. If both communication links are available we say that it is *bidirectional*. We call two nodes u and v *independent* if there is no communication link between u and v. Accordingly, a set is independent if each two nodes in the set are mutually independent. A *node coloring* is *valid* if each color forms an independent set.

Before stating the algorithms, we shall briefly characterize the communication graph implied by arbitrary transmission powers in the SINR model. Obviously, it is still based on a disk graph, but, not a unit disk graph as in the uniform case. Additionally, there are two main characteristics that are introduced by directed communication links and are relevant for graph-based algorithms in this setting. First, unidirectional communication links can form long directed paths. This is formalized in the following definition.

Definition 9. *Given a network N and the induced communication graph $G =$ (V, E). Let G' be the graph that remains after deleting all bidirectional edges from G. The* longest directed path *in the network is defined as the longest simple path in G'. We denote the length of the longest directed path in a network by ℓ.*

Second, these directed paths cannot form a directed circuit. This holds since in any circle in the communication graph, there must be a bidirectional communication link. Consider a directed path consisting of the nodes $(v_1, \ldots, v_\ell)$. It holds that the transmission range decreases monotonically, i.e., $\bar{R}_{v_i} \geq \bar{R}_{v_{i+1}}$ for $i = 1, \ldots, \ell - 1$. If a node v_i can be reached from v_j with $i \leq j$, there must be a bidirectional communication link as v_i reaches v_j as well due to $\bar{R}_{v_i} \geq \bar{R}_{v_j}$.

5.1   The MW-Coloring Algorithm

The algorithm we adapted to the case of arbitrary transmission power is based on the coloring algorithm by Moscibroda and Wattenhofer [12,11], originally designed for unstructured radio networks. Derbel and Talbi [2] modified the algorithm to fit the case of uniform transmission powers in the SINR model. The original MW-coloring algorithm starts with an initial leader election. Afterwards, each non-leader node queries a nearby leader for a block of colors. As neighbors may query different leaders, they may end up with the same color. Hence a final competition for colors within such a block ensures valid colors between neighbors. In order to generalize the algorithm for the case of non-uniform transmission power, we must modify both the transmission probabilities and the algorithm itself to handle issues introduced by directional communication links. Although the communication between the nodes can be easily adapted by slightly modifying the transmission probabilities, the algorithm itself depends on unidirectional communication in a nontrivial way. Due to space restrictions, we briefly introduce the algorithm and show that our results on bounding the interference and thus enabling successful communication can be applied easily. A more thorough description of the algorithm along with the analysis of the correctness can be found in the full version of this paper [4].

Algorithm Description. Let us now describe an execution of the algorithm at node v, which is also depicted in Figure 1. As the MW-coloring algorithm requires bidirectional communication, v starts with an initial neighborhood learning, which allows v to know whether it is dominated or not. Once all nodes that dominate v are colored, v enters a wait and listen phase, which is long enough so that v knows the current status of all other nodes that are awake and are able to communicate to v. Afterwards, if there is a leader w to which bidirectional communication is possible, v enters the request state, and requests a color from w. After w answers the request by assigning the first color j of a block of colors, v tries to verify the assigned color j. If this is not successful, v increases j by one and retries. Note that this can happen only a constant number of times. Once v is successful, it announces its success so that neighboring nodes know that v colors itself with color j. On the other hand, if there is no leader that can communicate bidirectionally with

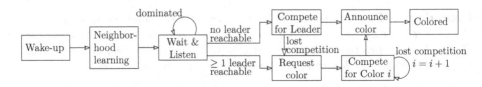

Fig. 1. State diagram of the MW-coloring algorithm

v, v tries to compete for the leader status. If this is not successful, v enters the request state and proceeds as above, as there must be a leader with bidirectional communication available now. If v is successful in becoming leader, it selects a free leader color and announces its choice so that all neighbors of v are informed. After the announcement phase, v is a leader and will only periodically transmit its color and serve color requests as they arrive.

Transmissions Are Successful. Although the transmission probabilities of leader and non-leader nodes are different in the algorithm, we shall show in this section, that we can easily adapt the transmission probabilities used in the algorithm such that communication is successful in the general case of arbitrary transmission powers. In order to apply the bound on the interference shown in Section 3, we need to prove that the sum of transmission probabilities from within each transmission region is at most γ.

As leader nodes need to serve up to Δ color requests, they are allowed to transmit with a probability that is a Δ factor higher than the transmission probability of non-leader nodes. In order to fit the requirements of communication with arbitrary transmission powers, let the transmission probability for non-leader nodes be $p_s = \gamma/(2\Delta)$, and the transmission probability for leader nodes be $p_l = \gamma/(18\Gamma^2)$. Due to space restrictions, we the proof of the following lemma. A proof is given in the full version [4].

Lemma 10. *Let v be an arbitrary leader node. Then there are at most $9\Gamma^2$ other leader nodes in the transmission range of v.*

The lemma follows from the fact that two nodes that can communicate bidirectionally will not both become leaders. Thus, disks of size $\underline{R}/2$ around each leader node in v's neighborhood would not intersect. Hence there can be at most $\frac{(\bar{R}+\underline{R}/2)^2}{(\underline{R}/2)^2} \leq 9\Gamma^2$ leader nodes in a maximal transmission range. We shall now prove the bound on the transmission probabilities.

Lemma 11. *Let leader nodes send with probability p_l and non-leader nodes with probability p_s, then the sum of transmission probabilities from within each transmission region is upper bounded by γ.*

Proof. Let us consider an arbitrary node v and sum over the transmission probabilities from within v's transmission region

$$\sum_{w \in B_v} p_w \leq 9\Gamma^2 p_l + \Delta p_s \leq \gamma$$

This holds as at most $9\Gamma^2$ leader nodes from each transmission region may transmit with probability p_l due to Lemma 10, while at most Δ other nodes in v's neighborhood transmit with probability at most p_s.

The corollary follows from the lemma along with the argumentation for Theorem 5. It shows that the limited number of leader nodes are able to communicate to their neighbors in $\mathcal{O}(\log n)$ time slots, while non-leader nodes require $\mathcal{O}(\Delta \log n)$ time slots. Overall it implies that all transmissions in the algorithm are successful w.h.p.

Corollary 12. *A message transmitted by an arbitrary node with probability p_l (p_s) for $8c/p_l \log n$ ($8c/p_s \log n$) time slots reaches its intended receivers w.h.p.*

This shows that communication is successful with high probability even in this more general case. Combined with the algorithmic changes and the refined analysis in the full version of this paper [4], the modified MW-coloring algorithm computes a coloring with $\mathcal{O}(\Gamma^2 \Delta)$ colors such that each color forms an independent set in $\mathcal{O}((\Delta + \ell)\Gamma^4 \Delta \log n)$ time slots. This highlights the applicability of our method to bound the interference in networks of nodes with arbitrary transmission powers.

6 Conclussion

In this paper we have proven a bound on the interference in wireless ad hoc networks with arbitrary transmission power assignments. We believe that this generic result will be of use in many algorithms designed for such networks. We have shown that local broadcasting can be transfered to the general case of arbitrary transmission powers with minor efforts due to this result. Additionally, we considered variable transmission power, which allows each node to change its transmission power in each time slot. To highlight the applicability of our results on communication in networks with arbitrary transmission power, we presented a distributed node coloring algorithm that is fully adapted to characteristics of directed communication networks such as unidirectional communication links. For future directions, we would like to investigate, whether the dependence on the neighborhood learning algorithm is required and whether the dependence on Γ could be decreased.

Acknowledgments. This work was supported by the German Research Foundation (DFG) within the Research Training Group GRK 1194 "Self-organizing Sensor-Actuator Networks".

References

1. Daum, S., Gilbert, S., Kuhn, F., Newport, C.: Broadcast in the ad hoc SINR model. In: Afek, Y. (ed.) DISC 2013. LNCS, vol. 8205, pp. 358–372. Springer, Heidelberg (2013),
http://dx.doi.org/10.1007/978-3-642-41527-2_25

2. Derbel, B., Talbi, E.G.: Distributed Node Coloring in the SINR Model. In: Proc. 30th Internat. Conf. on Distributed Computing Systems (ICDCS 2010), pp. 708–717. IEEE Computer Society (2010)

3. Fuchs, F., Wagner, D.: On Local Broadcasting Schedules and CONGEST Algorithms in the SINR Model. In: Flocchini, P., Gao, J., Kranakis, E., Meyer auf der Heide, F. (eds.) ALGOSENSORS 2013. LNCS, vol. 8243, pp. 170–184. Springer, Heidelberg (2014)

4. Fuchs, F., Wagner, D.: Arbitrary transmission power in the SINR model: Local broadcasting, coloring and mis. CoRR abs/1402.4994 (2014)

5. Fujii, T., Takahashi, T., Bandai, T., Udagawa, T., Sasase, T.: An efficient MAC protocol in wireless ad-hoc networks with heterogeneous power nodes. In: 5th Internat. Symp. Wireless Personal Multimedia Communications (WPMC 2002), vol. 2, pp. 776–780. IEEE (2002)

6. Goussevskaia, O., Moscibroda, T., Wattenhofer, R.: Local Broadcasting in the Physical Interference Model. In: Proc. 5th ACM Internat. Workshop on Foundations of Mobile Computing (DialM-POMC 2008), pp. 35–44. ACM Press (2008)

7. Halldórsson, M.M., Mitra, P.: Towards Tight Bounds for Local Broadcasting. In: Proc. 8th ACM Internat. Workshop on Foundations of Mobile Computing (FOMC 2012). ACM Press (July 2012)

8. Halldórsson, M.M., Mitra, P.: Wireless connectivity and capacity. In: Proc. 23rd Ann. ACM-SIAM Symp. Discrete Algorithms (SODA 2012), pp. 516–526. SIAM (2012), http://dl.acm.org/citation.cfm?id=2095116.2095160

9. Jurdziński, T., Stachowiak, G.: Probabilistic algorithms for the wakeup problem in single-hop radio networks. In: Bose, P., Morin, P. (eds.) ISAAC 2002. LNCS, vol. 2518, pp. 535–549. Springer, Heidelberg (2002)

10. Kesselheim, T., Vöcking, B.: Distributed contention resolution in wireless networks. In: Lynch, N.A., Shvartsman, A.A. (eds.) DISC 2010. LNCS, vol. 6343, pp. 163–178. Springer, Heidelberg (2010), http://dl.acm.org/citation.cfm?id=1888781.1888803

11. Moscibroda, T., Wattenhofer, M.: Coloring Unstructured Radio Networks. J. Distr. Comp. 21(4), 271–284 (2008)

12. Moscibroda, T., Wattenhofer, R.: Coloring unstructured radio networks. In: Proc 17th ACM Symp. on Parallelism in Algorithms and Architectures (SPAA 2005), pp. 39–48. ACM (2005)

13. Peleg, D.: Distributed Computing: A Locality-Sensitive Approach. Society for Industrial Mathematics (2000)

14. Poojary, N., Krishnamurthy, S.V., Dao, S.: Medium access control in a network of ad hoc mobile nodes with heterogeneous power capabilities. In: IEEE Internat. Conf. on Communications (ICC 2001), vol. 3, pp. 872–877. IEEE (2001)

15. Wang, G., Turgut, D., Bölöni, L., Ji, Y., Marinescu, D.C.: A MAC layer protocol for wireless networks with asymmetric links. Ad Hoc Networks 6(3), 424–440 (2008)

16. Yu, D., Hua, Q.S., Wang, Y., Lau, F.C.M.: An O(log n) Distributed Approximation Algorithm for Local Broadcasting in Unstructured Wireless Networks. In: Proc. 8th Internat. Conf. on Distributed Computing in Sensor Systems (DCOSS 2012), pp. 132–139. IEEE Computer Society (2012)

17. Yu, D., Wang, Y., Hua, Q.-S., Lau, F.C.M.: Distributed $(\Delta + 1)$-Coloring in the Physical Model. In: Erlebach, T., Nikoletseas, S., Orponen, P. (eds.) ALGOSEN-SORS 2011. LNCS, vol. 7111, pp. 145–160. Springer, Heidelberg (2012)
18. Yu, D., Wang, Y., Hua, Q.S., Lau, F.C.M.: Distributed Local Broadcasting Algorithms in the Physical Interference Model. In: Proc. 7th Internat. Conf. on Distributed Computing in Sensor Systems (DCOSS 2011), pp. 1–8. IEEE Computer Society (2011)
19. Zuhairi, M., Zafar, H., Harle, D.: On-demand routing with unidirectional link using path loss estimation technique. In: Proc. Wireless Telecommunications Symposium (WTS 2012), pp. 1–7 (2012)

Continuous Aggregation
in Dynamic Ad-Hoc Networks*

Sebastian Abshoff and Friedhelm Meyer auf der Heide

Heinz Nixdorf Institute & Computer Science Department,
University of Paderborn, Fürstenallee 11, 33102 Paderborn, Germany
{abshoff,fmadh}@hni.upb.de

Abstract. We study a scenario in which n nodes of a mobile ad-hoc network continuously collect data. Their task is to repeatedly update aggregated information about the data, e.g., the maximum, the sum, or the full information about all data received by all nodes at a given time step. This aggregated information has to be disseminated to all nodes.

We propose two performance measures for distributed algorithms for these tasks: The *delay* is the maximum time needed until the aggregated information about the data measured at some time is output at all nodes. We assume that a node can broadcast information proportional to a constant number of data items per round. A too large communication volume needed for producing an output can lead to the effect that the delay grows unboundedly over time. Thus, we have to cope with the restriction that outputs are computed not for all but only for a fraction of rounds. We refer to this fraction as the *output rate* of the algorithm.

Our main technical contributions are trade-offs between delay and output rate for aggregation problems under the assumption of T-stable dynamics in the mobile ad-hoc network: The network is always connected and is stable for time intervals of length $T \geq c \cdot \mathrm{MIS}(n)$ where $\mathrm{MIS}(n)$ is the time needed to compute a maximal independent set. For the maximum function, we are able to show that we can achieve an output rate of $\Omega(T/(n \cdot \mathrm{MIS}(n)))$ with delay $\mathcal{O}(n \cdot \mathrm{MIS}(n))$. For the sum, we show that it is possible to achieve an output rate of $\Omega(T^{5/2}/(n^2 \cdot \mathrm{MIS}(n)^3))$ with delay $\mathcal{O}(n^2 \cdot \mathrm{MIS}(n)^2/T^{3/2})$ if $T = \mathcal{O}(n^{2/3} \cdot \mathrm{MIS}(n)^{2/3})$, and if $T = \Omega(n^{2/3} \cdot \mathrm{MIS}(n)^{2/3})$, we can achieve an output rate of $\Omega(T/(n \cdot \mathrm{MIS}(n)^2))$ with delay $\mathcal{O}(n \cdot \mathrm{MIS}(n))$.

Keywords: Dynamic Networks, Aggregation, Token Dissemination.

1 Introduction

There are various devices that communicate wirelessly with each other and observe their environment. For example, many smartphones are able to communicate with close-by smartphones via technologies such as Bluetooth, WiFi, or

---

* This work was partially supported by the German Research Foundation (DFG) within the Priority Program "Algorithms for Big Data" (SPP 1736), by the EU within FET project MULTIPLEX under contract no. 317532, and the International Graduate School "Dynamic Intelligent Systems".

M. Halldórsson (Ed.): SIROCCO 2014, LNCS 8576, pp. 194–209, 2014.

Near Field Communication. In addition, these smartphones are equipped with more and more sensors nowadays, e.g. accelerometers, magnetometers, gyroscopic, light, temperature, pressure, and humidity sensors to name only a few. In this paper, we consider a scenario where these devices have to form an ad-hoc network to process the huge amount of data collected by their sensors. The links in such a network are unstable and they change over time, and thus, the network is dynamic – especially, if the nodes are mobile. We are interested in providing all nodes of the network with aggregated information about their sensor data.

We model the ad-hoc network as a T-stable dynamic network, which is controlled by an adaptive adversary (as introduced by Haeupler and Karger [7]). This adversary is able to change all edges of the network every T rounds, but is restricted to give a connected network. The set of nodes is fixed and nodes have unique IDs. A message of $\mathcal{O}(\log n)$ bits sent by some node in some round r is delivered to all its neighbors in the graph of the following round $r + 1$.

In this adversarial model, we study two aggregation problems: the extremum problem (e.g., the maximum) and the summation problem (e.g., the sum). Here, the nodes are given inputs (e.g., integers) and they have to compute a function of all inputs of the network. While both problems can be solved with existing algorithms for dissemination problems, which allow for full reconstruction of each input, we show that they can be solved faster with algorithms that aggregate information within the process. For this purpose, we exploit certain properties of the binary operations used to define the problems. For speeding up the summation, we make use of the commutativity and associativity, and we exploit the additional idempotence of the extremum which makes the problem simpler.

Our main focus lies on continuous versions of these problems where nodes receive a new input in each round and have to compute a function of all inputs for a single round. Here, we refer to the *delay* of an algorithm as the maximum number of rounds between the round when inputs arrive at all nodes and the round the last node outputs the result of the function of these inputs. We are interested in algorithms that have a high *output rate*, i.e., algorithms that output as many results for different rounds at all nodes as possible. One way to continuously produce outputs is to start the execution of the non-continuous algorithm, output one result and restart the execution of the algorithm again to produce the next result. However, we show how to use a pipelining technique to increase this trivial output rate but only slightly increase the delay.

1.1 Our Contribution

In static networks, all three (non-continuous) problems can be solved in a linear number of rounds (cf. Section 4). The continuous variants of the extremum and summation problem can be solved with constant output rate, and the continuous dissemination problem with rate $\Omega\left(\frac{1}{n}\right)$ while all delays remain linear. Note that these results are asymptotically tight in static networks.

For 1-stable dynamic networks, we show that the (non-continuous) extremum problem still can be solved in a linear number of rounds (cf. Section 5). To solve the other problems, we assume $T \geq c \cdot \text{MIS}(n)$ where c is a sufficiently large constant and $\text{MIS}(n)$ is the number of rounds required to compute a maximal independent set in a graph with n nodes (note that there are some restrictions under which this maximal independent set must be computed, cf. Section 5.1). Compared to the (non-continuous) dissemination problem, we are able to solve the (non-continuous) summation problem $\frac{T}{\text{MIS}(n)}$ times faster if $T = \mathcal{O}(\sqrt{n \cdot \text{MIS}(n)})$, and if $T = \Omega(\sqrt{n \cdot \text{MIS}(n)})$, this problem can be solved in a linear number of rounds. For the continuous extremum and summation problem, we prove non-trivial output rates, i.e., output rates that are higher than these obtained by executing the non-continuous algorithms over and over again. For the continuous extremum problem, we thereby increase the delay only slightly. If $T = \mathcal{O}(n^{2/3} \text{MIS}(n)^{2/3})$, we can achieve the same delay and a slightly smaller output rate for the continuous summation problem compared to the extremum problem. Besides these deterministic results (cf. Table 1), we show in the corresponding sections how randomization helps to improve these results.

Table 1. Overview about the deterministic results shown in this paper

(a) Static Networks.

	Extremum	Summation	Dissemination
non-continuous			
Running Time:	$\mathcal{O}(n)$	$\mathcal{O}(n)$	$\mathcal{O}(n)$
continuous			
Delay:	$\mathcal{O}(n)$	$\mathcal{O}(n)$	$\mathcal{O}(n)$
Output Rate:	$\Omega(1)$	$\Omega(1)$	$\Omega\left(\frac{1}{n}\right)$

(b) T-Stable Dynamic Networks with $T \geq c \cdot \text{MIS}(n)$.

	Extremum	Summation	Dissemination
non-continuous			
Running Time:			
if $T = \mathcal{O}(\sqrt{n \cdot \text{MIS}(n)})$	$\mathcal{O}(n)$	$\mathcal{O}\left(\frac{n^2 \cdot \text{MIS}(n)}{T^2}\right)$	$\mathcal{O}\left(\frac{n^2}{T}\right)$
if $T = \Omega(\sqrt{n \cdot \text{MIS}(n)})$		$\mathcal{O}(n)$	
continuous			
Delay:			
if $T = \mathcal{O}(n^{2/3} \text{MIS}(n)^{2/3})$	$\mathcal{O}(n \cdot \text{MIS}(n))$	$\mathcal{O}\left(\frac{n^2 \cdot \text{MIS}(n)^2}{T^{3/2}}\right)$	$\mathcal{O}\left(\frac{n^2}{T}\right)$
if $T = \Omega(n^{2/3} \text{MIS}(n)^{2/3})$		$\mathcal{O}(n \cdot \text{MIS}(n))$	
Output Rate:			
if $T = \mathcal{O}(n^{2/3} \text{MIS}(n)^{2/3})$	$\Omega\left(\frac{T}{n \cdot \text{MIS}(n)}\right)$	$\Omega\left(\frac{T^{5/2}}{n^2 \cdot \text{MIS}(n)^3}\right)$	$\Omega\left(\frac{T}{n^2}\right)$
if $T = \Omega(n^{2/3} \text{MIS}(n)^{2/3})$		$\Omega\left(\frac{T}{n \cdot \text{MIS}(n)^2}\right)$	

2 Models and Problems

We adapt the dynamic network model from Haeupler and Karger [7]: A dynamic network is a dynamic graph G_r, which consists of a fixed set of n nodes. Each node has a unique ID that can be encoded with $\mathcal{O}(\log n)$ bits. Time proceeds in discrete, synchronous rounds. An adaptive adversary chooses a set of undirected edges E_r defining the graph $G_r = (V, E_r)$ for round r. This adversary is only restricted to choose a connected graph in each round. In each round r, each node can send a message of $\Theta(\log n)$ bits, which is delivered to all neighbors in the graph G_{r+1} in the following round $r+1$. The computational power of each node is unbounded. A dynamic network is called T-stable if the adversary is restricted to change the network only once every T rounds. We assume throughout the paper that $T \leq n$. Furthermore, we assume that the nodes know both values T and n. Note that T-stability is a stronger assumption than T-interval connectivity that has been proposed by Kuhn et al. [12] because T-interval connectivity only requires a stable spanning subgraph for T rounds.

We study the following three (non-continuous) aggregation problems.

Problem 1 (Extremum). Let $(S, +)$ be a commutative and idempotent semi-group and let the elements of S be representable with $\mathcal{O}(\log n)$ bits. Each node i in the network receives as input an element $x_i \in S$. Let f be defined by $f(x_1, x_2, \ldots, x_n) := \sum_{i=1}^{n} x_i$. All nodes of the network have to output $f(x_1, x_2, \ldots, x_n)$.

Problem 2 (Summation). Let $(S, +)$ be a commutative semigroup and let the elements of S be representable with $\mathcal{O}(\log n)$ bits. Each node i in the network receives as input an element $x_i \in S$. Let f be defined by $f(x_1, x_2, \ldots, x_n) := \sum_{i=1}^{n} x_i$. All nodes of the network have to output $f(x_1, x_2, \ldots, x_n)$.

Problem 3 (Dissemination). Let S be a structure representable with $\mathcal{O}(\log n)$ bits. Each node i in the network receives as input an element $x_i \in S$ called token. All nodes of the network have to output $x_1, x_2, \ldots, x_n$.

In addition to that, we solve continuous versions of these problems where f has not to be computed only once but several times for different inputs.

Problems 4/5/6 (Continuous Extremum/Summation/Dissemination). Define f and $(S, +)$ as in the corresponding extremum/summation/dissemination problem. In each round r, each node i in the network receives as input an element $x_{i,r} \in S$. For a subset of rounds $R \subseteq \mathbb{N}$ (defined by the algorithm) and each $r' \in R$, all nodes have to output $f(x_{1,r'}, x_{2,r'}, \ldots, x_{n,r'})$.

For example, consider S to be a subset of $\mathbb{N}$ of size polynomial in n. Then, computing the sum is a summation problem whereas computing the maximum or the minimum of all inputs is an extremum problem. The dissemination problem is also known as the all-to-all token dissemination problem [12].

Note that solving these problems for one round, in general, requires more than just one round. Although it is possible to produce the result for each round

$r \in \mathbb{N}$, it could take longer and longer: Let T be the number of rounds required to produce one output. Then, it is possible to output the result of round r in round $r \cdot T$ by running the algorithm for each round one by one. Since we intend to run our algorithms for a long time, this is not a feasible approach and we do not want the number of rounds to produce one output to depend on the round when the computation is started.

Instead, we would like our algorithms to drop some rounds and produce outputs without this dependence. Intuitively, a good algorithm that continuously gives results produces as many results as possible and requires few rounds per output. This is captured by the following two definitions.

Definition 1 (Output Rate). *The* output rate *of an algorithm is defined as*

$$\lim_{r \to \infty} \frac{\#\text{results up to round } r}{r}.$$

Definition 2 (Delay). *The* delay *of an algorithm is defined as the maximum number of rounds between the round when inputs arrive and the round the function of these inputs is output by all nodes.*

3   Related Work

For static networks, one knows that many problems such as computing the maximum, sum, parity, or majority can be solved in linear time in a graph by first computing a spanning tree (see, e.g., Awerbuch [3]). More specifically, if D is the diameter of the graph, all these functions can be computed in $\mathcal{O}(D)$ rounds. Beyond that, more complicated problems have been studied, e.g., selection problems [11] or the problem of computing the mode (most frequent element) [10].

The dynamic network model was introduced by Kuhn et al. [12]. In contrast to the model we use, they assumed that the number of nodes in the network is not known beforehand. In their setting, they studied two problems, the token dissemination and the counting problem. In the token dissemination problem, each node receives as input a token that has to be disseminated to all nodes such that in the end all nodes know all tokens. In the counting problem, the nodes have to determine the exact number of nodes in the network. They solved both problems with so-called token forwarding algorithms that are only allowed to store and forward tokens. Especially, these algorithms are not allowed to annotate or combine tokens or to send any other message except the empty message. For T-interval connected dynamic networks, they gave a deterministic $\mathcal{O}(n(n+k)/T)$ algorithm. This algorithm can also be used to solve the counting problem in $\mathcal{O}(n^2/T)$ rounds. If n is known, k tokens can be disseminated in $\mathcal{O}(nk/T + n)$ rounds. On the negative side, they showed that a subclass of knowledge-based token-forwarding algorithms needs $\Omega(nk/T)$ rounds for solving the token dissemination problem. In addition to that, they showed that even a centralized deterministic token-forwarding algorithm needs $\Omega(n \log k)$ rounds.

In a subsequent paper, the lower bound by Kuhn et al. was improved by Dutta et al. [6]. They showed that any randomized (even centralized) token-forwarding algorithm requires $\Omega(nk/\log n + n)$ rounds. Furthermore, they gave an algorithm that can solve the k-token dissemination problem in $\mathcal{O}((n + k)\log n \log k)$ rounds w.h.p. in the presence of a weakly-adaptive adversary. For the offline case, they developed two randomized and centralized algorithms where one gives an $\mathcal{O}(n, \min\{k \log n\})$ schedule w.h.p. and the other gives an $\mathcal{O}((n + k)\log n^2)$ schedule w.h.p. if each node is allowed to send one token along each edge in each round. Haeupler and Kuhn [8] proved lower bounds when each node is allowed to send $b \leq k$ tokens per round or when the nodes have to collect a δ-fraction of the tokens only. Their results are applicable for T-interval connected dynamic networks and dynamic networks that are c-vertex connected in each round.

Abshoff et al. [1] adapted the model by Kuhn et al. and restricted the adversary in a geometric setting. Here, each node has a position in the Euclidean plane and is moved by the adversary with maximum velocity $v_{\max}$. The nodes are able to reach all nodes within distance $R > 1$ and the adversary must keep the unit disk graph w.r.t. radius 1 connected. The k-token dissemination problem can be solved in $\mathcal{O}(n \cdot k \cdot \min\{v_{\max}, R\} \cdot R^{-2})$ rounds. In a different paper, Abshoff et al. [2] established a relation between counting and token dissemination by showing that a special token dissemination problem is at most as hard as counting the number of nodes in a directed variant of dynamic networks.

Haeupler and Karger [7] applied network coding techniques to the domain of dynamic networks. Their algorithm solves the token dissemination problem in $\mathcal{O}(n^2/\log n)$ rounds w.h.p. when both the message size and token size have length $\Theta(\log n)$ bits. With T-stability, they achieve a T^2 speedup. In the deterministic case, they can solve k-token dissemination in $\mathcal{O}\left(\frac{1}{\sqrt{\log nT}} \cdot n \cdot \min\{k, \frac{n}{T}\} + n\right) \cdot 2^{\mathcal{O}(\sqrt{\log n})}$ rounds. In the randomized case, they can solve k-token dissemination in $\mathcal{O}\left(\min\{\frac{nk}{T^2} + T^2 n \log^2 n, \frac{nk \log n}{T^2} + Tn \log^2 n, \frac{n^2 \log n}{T^2} + n \log n\}\right)$ rounds.

Cornejo et al. [5] studied a different aggregation problem where tokens have to be gathered at a minimum number of nodes. On the one hand, they proved that there is no algorithm with a good competitive ratio compared to an optimal offline algorithm. On the other hand, under the assumption that every node interacts with at least a p-fraction of the nodes, they give an algorithm that aggregates the tokens to a logarithmic number of nodes with high probability.

Mosk-Aoyama and Shah [15] showed how so-called separable functions can be approximated with a gossiping algorithm based on exponential random variables. Their techniques can also be applied in dynamic networks as Kuhn et al. [12] showed for approximate counting.

A main building block of this paper is the construction of maximal independent sets (MIS). The distributed algorithm by Luby [14] computes an MIS in expected $\mathcal{O}(\log n)$ rounds. It can also be shown that the number of rounds is $\mathcal{O}(\log n)$ w.h.p. [9,4]. The best known distributed deterministic algorithm by Panconesi and Srinivasan [16] computes an MIS in $2^{\mathcal{O}(\sqrt{\log n})}$ rounds. In growth-bounded graphs, Schneider and Wattenhofer [17] showed how to deterministically create an MIS in $\mathcal{O}(\log^* n)$ communication rounds. This is asymptotically optimal as Linial [13] gave a corresponding $\Omega(\log^* n)$ lower bound.

4   Static Networks

For the sake of a simpler presentation of the algorithms for T-stable dynamic networks, we shortly discuss how the problems can be solved in static networks.

The extremum problem can be solved in $n-1$ rounds: Each node i initially broadcasts its input x_i. In every other round up to round $n-1$, each node i takes its incoming messages $m_1, \ldots, m_l$ and broadcasts $\sum_{j=1}^{l} m_j + x_i$. In round $n-1$, $\sum_{j=1}^{l} m_j + x_i = f(x_1, \ldots, x_n)$ since all inputs must be contained in this sum and multiplicities cancel out due to the idempotence of the semigroup.

To solve the summation problem, we can first find the node with the smallest ID in $n-1$ rounds (this is an extremum problem). Then, in further $n-1$ rounds, we can build up a shortest path tree rooted at the node with the smallest ID. Along this tree, starting from the leaves up to the root, we can sum up the inputs and finally broadcast the result back from the root to all elements. We thereby guarantee that each summand is only considered once.

Finally, to solve the dissemination problem, we can build up a tree as for the summation problem. Then, each node in the tree sends a token it has not yet sent upwards to the root of the tree in each round. After that the tokens are sent one after the other from the root to the leaves.

Proposition 1. *In static networks, the extremum, the summation, and the dissemination problem can be solved in $\mathcal{O}(n)$ rounds.*

We could have used the algorithm that solves the dissemination problem to solve the extremum and the summation problem. However, we chose the aforementioned algorithms since they are similar to those we will use to solve these problems in their continuous versions in dynamic networks.

To continuously solve both the extremum and the summation problem, we build up a tree as before and apply a pipelining technique. The leaves of the tree start sending their inputs from the first round upwards. In the next round, the leaves start sending their inputs from the second round upwards and so on until round n. The nodes with distance l to the leaves within the tree in round r sum up the incoming messages from the level below and add their input from round $r-l$ if $r-l > 0$. Then, after n rounds, the results for round 1 to n are sent one after the other from the root to the leaves. This gives n outputs in $\mathcal{O}(n)$ rounds. Since the best possible output rate is 1 and the delay cannot be better than the diameter of the network, we get the following result.

Proposition 2. *In static networks, the continuous extremum problem and the continuous summation problem can be solved with delay $\mathcal{O}(n)$ and output rate $\Omega(1)$. The delay and the output rate are asymptotically optimal.*

For the dissemination problem, we cannot achieve better delays and output rates than these we get if we just solve the non-continuous version over and over again. The delay is bounded by the diameter of the network. For the output rate, consider $|S| = n$, a line of n nodes and the edge e with $\lceil \frac{n}{2} \rceil$ nodes on the left an $\lfloor \frac{n}{2} \rfloor$ nodes on the right. If the output rate was $\omega(\frac{1}{n})$, then $\omega(\frac{r}{n})$ outputs must have been computed up to round r. We know that at most $\mathcal{O}(r \cdot \log n)$ bits could have passed e from the left to the right. These bits separate up to $n^{\mathcal{O}(r)}$ instances. However, there are $\binom{|S|}{n/2}^{\omega(r/n)} = n^{\omega(r)}$ possibilities to choose the tokens on the left side. Hence, at least one output must be wrong.

Proposition 3. *In static networks, the continuous dissemination problem can be solved with delay $\mathcal{O}(n)$ and output rate $\Omega\left(\frac{1}{n}\right)$. The delay and the output rate are asymptotically optimal.*

5 T-stable Dynamic Networks

We now show how to solve the problems in T-stable dynamic networks. First, we introduce a graph patching technique.

5.1   Graph Patching in T-stable Dynamic Networks

In this section, we show how a T-stable dynamic network can be partitioned into patches such that aggregation is possible. This partitioning will help speeding up the summation problem, the continuous extremum, and the continuous summation problem. The following patching idea is adapted from Haeupler and Karger [7].

Definition 3 (D-Patch, D-Patching). *A D-patch of a graph $G = (V, E)$ is a rooted tree in G that spans at least $\frac{D}{2}$ nodes and has depth at most $\frac{D}{2}$. A D-patching of a graph is a set of D-patches such that the sets of the nodes of all D-patches give a disjoint partition of V.*

Such a D-patching of G can be distributedly computed by

1. finding a set of nodes in G that form a maximal independent set (MIS) in G^D, i.e., the Dth power[1] of G,
2. computing breadth-first trees rooted in each node of the MIS, where each non-MIS node is assigned to its closest MIS node.

Existing distributed MIS algorithms can be adapted for this approach. Let $\text{MIS}(n)$ be the number of rounds necessary to compute an MIS in a graph with n nodes. If an MIS algorithm running in G^D is simulated in G, one needs to take care that one edge in G^D corresponds to a path of length up to D in G. Therefore, an MIS algorithm is slowed down by a factor of D. In addition to that, it is also important to consider the congestion in the nodes of G caused by paths that overlap during simulation. If an MIS algorithm can be modified appropriately, a D-patching can be computed in $\mathcal{O}(\text{MIS}(n) \cdot D)$ rounds.

Proposition 4. *[7,14,9,4] A graph G can be partitioned into D-patches of size $\Omega(D)$ in $\mathcal{O}(\log(n) \cdot D)$ rounds w.h.p. with Luby's randomized MIS algorithm.*

---

[1] $G^D = (V, E^D)$ with $E^D = \{\{u, v\} | \exists \text{path between } u, v \in V \text{ of length} \leq D \text{ in } G\}$.

Proposition 5. *[7,16] A graph G can be partitioned into D-patches of size $\Omega(D)$ in $\mathcal{O}(2^{\mathcal{O}(\sqrt{\log n})} \cdot D)$ rounds with Panconesi and Srinivasan's deterministic MIS algorithm.*

We would like to add that a patching can be computed faster in growth-bounded graphs.

Proposition 6. *A growth-bounded graph G can be partitioned into D-patches of size $\Omega(D)$ in $\mathcal{O}(\log^*(n) \cdot D)$ rounds with Schneider and Wattenhofer's deterministic MIS algorithm.*

Proof. The algorithm by Schneider and Wattenhofer [17] can be modified such that it can be executed in our setting: In each competition and whenever the states are updated, each competitor is interested in the competitor u in its neighborhood that has the minimum result r_u^{j-1} among all its neighbors. In addition to that, the nodes only need to know whether there exists a competitor, a ruler, or a dominator in their neighborhood. Therefore, each node only needs to flood the minimum result of a competitor, whether there exists a ruler, and whether there exists a dominator for D rounds.   □

5.2 Non-Continuous Extremum

Despite the presence of an adaptive adversary, the extremum problem can be solved without the need for a graph patching. This is a tight result since even in a static network the extremum problem cannot be solved faster.

Theorem 1. *In 1-stable dynamic networks, the extremum problem can be solved in $\mathcal{O}(n)$ rounds.*

Proof. The algorithm that solves the extremum problem in dynamic networks is the same as the algorithm used for static networks. Each node i initially broadcasts its input x_i. In every other round up to round $n - 1$, each node i takes all its incoming messages $m_1, \ldots, m_l$ and broadcasts $\sum_{j=1}^{l} m_j + x_i$. In round $n - 1$, the sum $\sum_{j=1}^{l} m_j + x_i$ is equal to $f(x_1, \ldots, x_n)$ because it contains all inputs (each node causally influenced each other node after $n - 1$ rounds [12]) and multiplicities cancel out due to the idempotence of the semigroup.   □

5.3 Non-Continuous Summation

Theorem 2. *In T-stable dynamic networks with $T \geq c \cdot \mathrm{MIS}(n)$ for a sufficiently large constant c, the summation problem can be solved in*

- $\mathcal{O}\left(\frac{n^2 \cdot \mathrm{MIS}(n)}{T^2}\right)$ *rounds if $T = \mathcal{O}(\sqrt{n \cdot \mathrm{MIS}(n)})$ and*
- $\mathcal{O}(n)$ *rounds if $T = \Omega(\sqrt{n \cdot \mathrm{MIS}(n)})$.*

Proof. Consider the following algorithm for which we choose $D = \Theta\left(\frac{T}{\mathrm{MIS}(n)}\right)$.

1. Compute a D-patching.
2. In each patch, compute the sum of all inputs of the nodes in the patch.
3. Disseminate all partial sums of the patches to all nodes and sum them up.

If c is large enough and D is chosen properly, then we can do the first and the second step in at most T rounds. Since each patch has size at least $\frac{D}{2}$ nodes, we have at most $\frac{2n}{D} = \mathcal{O}\left(\frac{n \cdot \mathrm{MIS}(n)}{T}\right)$ partial sums left. To disseminate them in the third step, we can use the token dissemination algorithm by Kuhn et al. [12] for T-interval connected dynamic networks. Thus, we solve the summation problem in $\mathcal{O}\left(\frac{n^2 \cdot \mathrm{MIS}(n)}{T^2} + n\right)$ rounds. $\qquad\square$

Corollary 1. *In T-stable dynamic networks with $T \geq 2^{c \cdot \sqrt{\log n}}$ for a sufficiently large constant c, the summation problem can be solved in*

- $\mathcal{O}\left(\frac{n^2}{T^2} \cdot 2^{c \cdot \sqrt{\log n}}\right)$ *rounds if $T = \mathcal{O}\left(\sqrt{n} \cdot \sqrt{2}^{c \cdot \sqrt{\log n}}\right)$ and*
- $\mathcal{O}(n)$ *rounds if $T = \Omega\left(\sqrt{n} \cdot \sqrt{2}^{c \cdot \sqrt{\log n}}\right)$.*

Randomization allows us to speed up this computation if we use Luby's algorithm to compute the patching and Haeupler and Karger's randomized network coding algorithm for dissemination.

Theorem 3. *Let L be the number of rounds Luby's algorithm needs to compute a maximal independent set with high probability. Then, in T-stable dynamic networks with $T \geq L$, the summation problem can be solved within the number of rounds as listed in Table 2a.*

Proof. Let $D = \frac{\frac{1}{2}T}{L+1}$. Then, we need at most $D \cdot L \leq \frac{1}{2}T$ rounds to compute a D-patching and have further $D \leq \frac{1}{2}T$ rounds to sum up all values in each patch. Now, we can use the randomized network coding algorithm by Haeupler and Karger [7] for dissemination. It needs

$$\mathcal{O}\left(\min\{\frac{nk}{T^2} + T^2 n \log^2 n, \frac{nk \log n}{T^2} + Tn \log^2 n, \frac{n^2 \log n}{T^2} + n \log n\}\right)$$

rounds to disseminate k tokens with high probability. For different ranges of T, we need the following number of rounds with high probability.

1. $\mathcal{O}\left(\frac{n^2 \log n}{T^3}\right)$ if $T = \mathcal{O}(n^{1/5} \log^{-1/5} n)$
2. $\mathcal{O}(T^2 n \log^2 n)$ if $\Omega(n^{1/5} \log^{-1/5} n) = T = \mathcal{O}(n^{1/5})$
3. $\mathcal{O}\left(\frac{n^2 \log^2 n}{T^3}\right)$ if $\Omega(n^{1/5}) = T = \mathcal{O}(n^{1/4})$
4. $\mathcal{O}(Tn \log^2 n)$ if $\Omega(n^{1/4}) = T = \mathcal{O}(n^{1/3} \log^{-1/3} n)$
5. $\mathcal{O}\left(\frac{n^2 \log n}{T^2}\right)$ if $\Omega(n^{1/3} \log^{-1/3} n) = T = \mathcal{O}(n^{1/2})$
6. $\mathcal{O}(n \log n)$ if $\Omega(n^{1/2}) = T$

Note that the number of rounds in the second and fourth range increase with T. However, a T-stable dynamic network is also $\frac{T}{l}$-stable for any $l > 1$. Therefore, we can replace T by the lower bound of the range.

1. $\mathcal{O}\left(\frac{n^2 \log n}{T^3}\right)$ if $T = \mathcal{O}(n^{1/5} \log^{-1/5} n)$
2. $\mathcal{O}(n^{7/5} \log^{8/5} n)$ if $\Omega(n^{1/5} \log^{-1/5} n) = T = \mathcal{O}(n^{1/5} \log n^{2/15})$
3. $\mathcal{O}\left(\frac{n^2 \log^2 n}{T^3}\right)$ if $\Omega(n^{1/5} \log n^{2/15}) = T = \mathcal{O}(n^{1/4})$
4. $\mathcal{O}(n^{5/4} \log^2 n)$ if $\Omega(n^{1/4}) = T = \mathcal{O}(n^{3/8} \log^{-1/2} n)$
5. $\mathcal{O}\left(\frac{n^2 \log n}{T^2}\right)$ if $\Omega(n^{3/8} \log^{-1/2} n) = T = \mathcal{O}(n^{1/2})$
6. $\mathcal{O}(n \log n)$ if $\Omega(n^{1/2}) = T$

This gives the results for the non-continuous summation in Table 2a.   □

Table 2. Summation in T-Stable Dynamic Networks with $T \geq L$

(a) (Non-Continuous) Summation.

Running Time	Range for T
$\mathcal{O}\left(\frac{n^2 \log n}{T^3}\right)$ w.h.p.	if $L \leq T = \mathcal{O}(n^{1/5} \log^{-1/5} n)$
$\mathcal{O}(n^{7/5} \log^{8/5} n)$ w.h.p.	if $\Omega(n^{1/5} \log^{-1/5} n) = T = \mathcal{O}(n^{1/5} \log n^{2/15})$
$\mathcal{O}\left(\frac{n^2 \log^2 n}{T^3}\right)$ w.h.p.	if $\Omega(n^{1/5} \log n^{2/15}) = T = \mathcal{O}(n^{1/4})$
$\mathcal{O}(n^{5/4} \log^2 n)$ w.h.p.	if $\Omega(n^{1/4}) = T = \mathcal{O}(n^{3/8} \log^{-1/2} n)$
$\mathcal{O}\left(\frac{n^2 \log n}{T^2}\right)$ w.h.p.	if $\Omega(n^{3/8} \log^{-1/2} n) = T = \mathcal{O}(n^{1/2})$
$\mathcal{O}(n \log n)$ w.h.p.	if $\Omega(n^{1/2}) = T \leq n$

(b) Continuous Summation.

Delay	Output Rate	Range for T
$\mathcal{O}\left(\frac{n^2}{T^2}\right)$ w.h.p.	$\Omega\left(\frac{T^3}{n^2}\right)$ w.h.p.	if $L \leq T = \mathcal{O}(n^{1/4} \log^{-1/2} n)$
$\mathcal{O}(n^{3/2} \log n)$ w.h.p.	$\Omega\left(\frac{T}{n^{3/2} \log n}\right)$ w.h.p.	if $\Omega(n^{1/4} \log^{-1/2} n) = T = \mathcal{O}(n^{1/4})$
$\mathcal{O}\left(\frac{n^2 \log n}{T^2}\right)$ w.h.p.	$\Omega\left(\frac{T^3}{n^2 \log n}\right)$ w.h.p.	if $\Omega(n^{1/4}) = T = \mathcal{O}(n^{1/2})$
$\mathcal{O}(n \log n)$ w.h.p.	$\Omega\left(\frac{T}{n \log n}\right)$ w.h.p.	if $\Omega(n^{1/2}) = T \leq n$

5.4   Continuous Extremum

Theorem 4. *In T-stable dynamic networks with $T \geq c \cdot \mathrm{MIS}(n)$ for a sufficiently large constant c, the continuous extremum problem can be solved with delay $\mathcal{O}(n \cdot \mathrm{MIS}(n))$ and output rate $\Omega\left(\frac{T}{n \cdot \mathrm{MIS}(n)^2}\right)$.*

Proof. Consider the following algorithm for which we choose $D = \Theta\left(\frac{T}{\mathrm{MIS}(n)}\right)$.

1. Each node $i \in V$ initializes $y_{i,r,0}$ with $x_{i,r}$ for $r = 1, \ldots, D$.
2. For $j = 1, \ldots, \frac{2n}{D}$ phases of T rounds do:
   (a) Compute a D-patching.
   (b) Each node i in each patch P, computes $y_{i,r,j}$ as the sum of $y_{i',r,j-1}$ for all nodes i' from P and all adjacent patches of P for $r = 1, \ldots, D$.
3. Each node $i \in V$ returns $y_{i,r,\frac{2n}{D}}$ for $r = 1, \ldots, D$.

If c is large enough and D is chosen properly, we can do a) and b) in a stable phase of T rounds. Consider any input $x_{i,r}$. We say a patch P knows $x_{i,r}$ iff $x_{i,r}$ is contained in any $y_{i',r,j}$ for $i' \in P$. If there is a patch P that does not know $x_{i,r}$ at the beginning a phase, then there is a patch P^* that does not know $x_{i,r}$ at the beginning of the phase but knows $x_{i,r}$ at the end of the phase. Thus, at least $\frac{D}{2}$ nodes learn about $x_{i,r}$ in each phase until all nodes know $x_{i,r}$. We can conclude that after $\frac{2n}{D}$ phases all inputs $x_{i,r}$ are contained in all $y_{i',r,\frac{2n}{D}}$. Therefore, after $\frac{2n}{D} \cdot T = \mathcal{O}(n \cdot \mathrm{MIS}(n))$ rounds, we have generated D outputs which gives the claimed delay and the output rate. $\qquad \square$

Corollary 2. *In T-stable dynamic networks with $T \geq 2^{c \cdot \sqrt{\log n}}$ for a sufficiently large constant c, the continuous extremum problem can be solved with delay $\mathcal{O}(n \cdot 2^{c \cdot \sqrt{\log n}})$ and output rate $\Omega\left(\frac{T}{n \cdot 2^{c \cdot \sqrt{\log n}}}\right)$.*

Again, randomization allows us to speed up this computation.

Theorem 5. *Let L be the number of rounds Luby's algorithm needs to compute a maximal independent set with high probability. Then, in T-stable dynamic networks with $T \geq L$, the continuous extremum problem can be solved with high probability with output rate $\Omega\left(\frac{T}{n \log n}\right)$ and delay $\mathcal{O}(n \log n)$.*

Proof. Let $D = \frac{\frac{1}{9}T}{L+1}$. Then, we need at most $D \cdot L \leq \frac{1}{2}T$ rounds to compute a D-patching and have further $9D \leq \frac{1}{2}T$ rounds to do the computations in the patch as we do in the proof of Theorem 4. If we repeat this $\frac{n}{D}$ times, then, w.h.p., we still have at least $\frac{n}{D}$ valid D-patchings. Therefore, w.h.p., after $\frac{n}{D} \cdot T = \mathcal{O}(n \log n)$ rounds, we can generate D outputs which gives the claimed delay and output rate. $\qquad \square$

5.5   Continuous Summation

Theorem 6. *In T-stable dynamic networks with $T \geq c \cdot \mathrm{MIS}(n)$ for a sufficiently large constant c, the continuous summation problem can be solved with delay*

- $\mathcal{O}\left(\frac{n^2 \cdot \mathrm{MIS}(n)^2}{T^{3/2}}\right)$ *if* $T = \mathcal{O}(n^{2/3} \cdot \mathrm{MIS}(n)^{2/3})$ *and*
- $\mathcal{O}(n \cdot \mathrm{MIS}(n))$ *if* $T = \Omega(n^{2/3} \cdot \mathrm{MIS}(n)^{2/3})$

and output rate

- $\Omega\left(\frac{T^{5/2}}{n^2 \cdot \mathrm{MIS}(n)^3}\right)$ *if* $T = \mathcal{O}(n^{2/3} \cdot \mathrm{MIS}(n)^{2/3})$ *and*
- $\Omega\left(\frac{T}{n \cdot \mathrm{MIS}(n)^2}\right)$ *if* $T = \Omega(n^{2/3} \cdot \mathrm{MIS}(n)^{2/3})$.

Proof. Consider the following algorithm for which we choose $D = \Theta\left(\frac{T}{\mathrm{MIS}(n)}\right)$.

1. Compute a D-patching.
2. In each patch, compute $\frac{D}{2}$ sums of all inputs of the nodes in the patch of $\frac{D}{2}$ rounds.
3. Disseminate all partial sums of the patches to all nodes and sum them up.

If c is large enough and D is chosen properly, then we can do the first and the second step in at most T rounds. Since each patch has size at least $\frac{D}{2}$, we have at most n partial sums left. Now, we use the network coding algorithm by Haeupler and Karger [7]. This algorithm is able to disseminate $k \leq n$ tokens in $\mathcal{O}\left(\left(\frac{n \cdot \mathrm{MIS}(n)}{\sqrt{T}} \cdot \min\{k \cdot \sqrt{\log n}, \frac{n}{T}\} + n\right) \cdot \mathrm{MIS}(n)\right)$ rounds. Thus, we can disseminate all up to n partial sums in $\mathcal{O}\left(\left(\frac{n^2 \cdot \mathrm{MIS}(n)}{T^{3/2}} + n\right) \cdot \mathrm{MIS}(n)\right)$ rounds. If $T = \mathcal{O}(n^{2/3} \cdot \mathrm{MIS}(n)^{2/3})$, we thereby generate $\frac{D}{2}$ outputs in $\mathcal{O}\left(T + \frac{n^2 \cdot \mathrm{MIS}(n)^2}{T^{3/2}}\right)$ rounds and achieve an output rate of $\Omega\left(\frac{T^{5/2}}{n^2 \cdot \mathrm{MIS}(n)^3}\right)$. If $T = \Omega(n^{2/3} \cdot \mathrm{MIS}(n)^{2/3})$, we are able to generate $\frac{D}{2}$ outputs in $\mathcal{O}(n \cdot \mathrm{MIS}(n))$ rounds and achieve an output rate of $\Omega\left(\frac{T}{n \cdot \mathrm{MIS}(n)^2}\right)$.   □

Corollary 3. *In T-stable dynamic networks with $T \geq 2^{c \cdot \sqrt{\log n}}$ for a sufficiently large constant c, the continuous summation problem can be solved with delay*

- $\mathcal{O}\left(\frac{n^2 \cdot 2^{2c \cdot \sqrt{\log n}}}{T^{3/2}}\right)$ *if* $T = \mathcal{O}\left(n^{2/3} \cdot 2^{c \cdot \frac{2}{3} \cdot \sqrt{\log n}}\right)$ *and*
- $\mathcal{O}\left(n \cdot 2^{c \cdot \sqrt{\log n}}\right)$ *if* $T = \Omega\left(n^{2/3} \cdot 2^{c \cdot \frac{2}{3} \cdot \sqrt{\log n}}\right)$

and output rate

- $\Omega\left(\frac{T^{5/2}}{n^2 \cdot 2^{3c \cdot \sqrt{\log n}}}\right)$ *if* $T = \mathcal{O}\left(n^{2/3} \cdot 2^{c \cdot \frac{2}{3} \cdot \sqrt{\log n}}\right)$ *and*
- $\Omega\left(\frac{T}{2^{c \cdot \sqrt{\log n}}}\right)$ *if* $T = \Omega\left(n^{2/3} \cdot 2^{c \cdot \frac{2}{3} \cdot \sqrt{\log n}}\right)$.

Again, we can use Luby's algorithm to compute the patching and Haeupler and Karger's randomized network coding algorithm for dissemination.

Theorem 7. *Let L be the number of rounds Luby's algorithm needs to compute a maximal independent set with high probability. Then, in T-stable dynamic networks with $T \geq L$, the continuous summation problem can be solved with the output rates and delays as listed in Table 2b.*

Proof. Let $D = \frac{\frac{1}{2}T}{L+1}$. Then, we need at most $D \cdot L \leq \frac{1}{2}T$ rounds to compute a D-patching and have further $2D \leq \frac{1}{2}T$ rounds to do the computations in the patch as we do in the proof of Theorem 6. Now, we can use the randomized network coding algorithm by Haeupler and Karger [7] for dissemination. It needs

$$\mathcal{O}\left(\min\{\frac{nk}{T^2} + T^2 n \log^2 n, \frac{nk \log n}{T^2} + Tn \log^2 n, \frac{n^2 \log n}{T^2} + n \log n\}\right)$$

rounds to disseminate k tokens with high probability. For different ranges of T and $k = n$, we need the following number of rounds with high probability.

1. $\mathcal{O}\left(\frac{n^2}{T^2}\right)$ if $T = \mathcal{O}(n^{1/4} \log^{-1/2} n)$
2. $\mathcal{O}(T^2 n \log^2 n)$ if $\Omega(n^{1/4} \log^{-1/2} n) = T = \mathcal{O}(n^{1/4} \log^{-1/4} n)$
3. $\mathcal{O}\left(\frac{n^2 \log n}{T^2}\right)$ if $\Omega(n^{1/4} \log^{-1/4} n) = T = \mathcal{O}(n^{1/2})$
4. $\mathcal{O}(n \log n)$ if $\Omega(n^{1/2}) = T$

Note that the number of rounds in the second range increases with T. However, a T-stable dynamic network is also $\frac{T}{l}$-stable for any $l > 1$. Therefore, we can replace T by the lower bound of the range.

1. $\mathcal{O}\left(\frac{n^2}{T^2}\right)$ if $T = \mathcal{O}(n^{1/4} \log^{-1/2} n)$
2. $\mathcal{O}(n^{3/2} \log n)$ if $\Omega(n^{1/4} \log^{-1/2} n) = T = \mathcal{O}(n^{1/4})$
3. $\mathcal{O}\left(\frac{n^2 \log n}{T^2}\right)$ if $\Omega(n^{1/4}) = T = \mathcal{O}(n^{1/2})$
4. $\mathcal{O}(n \log n)$ if $\Omega(n^{1/2}) = T$

This gives the results for the continuous summation in Table 2b.   □

6 Geometric Dynamic Networks

In the geometric dynamic network model by Abshoff et al. [1], nodes have positions in the Euclidean plane and the adversary is allowed to move the nodes with maximum velocity $v_{\max}$. Furthermore, the adversary must keep the unit disk graph w.r.t. radius 1 connected in each round and the nodes are able to reach all nodes within communication range $R > 1$. This special class of dynamic networks is $\left\lfloor \frac{R-1}{2 \cdot v_{\max}} \right\rfloor + 1$-interval connected because a node within distance 1 can increase its distance by at most $2v_{\max}$. If in addition to that $R \geq 2$, then the communication graph contains a spanning $\Theta(R)$-connected subgraph that is stable for $\Theta(R \cdot v_{\max}^{-1})$ rounds. If nodes know their positions (e.g., by using GPS) or if they at least have the ability to sense the distances to their neighbors, then they are able to determine the stable subgraphs and the algorithms presented in this paper can be applied. For the MIS computation, we can use the algorithm by Schneider and Wattenhofer [17] since the stable subgraphs are growth-bounded. This yields improved results for geometric dynamic networks with $\mathrm{MIS}(n) = \mathcal{O}(\log^* n)$ and $T = \Theta(R \cdot v_{\max}^{-1})$.

7   Conclusion and Future Prospects

We showed that both extremum and summation problems can be solved faster than dissemination problems in T-stable dynamic networks by exploiting properties such as commutativity, associativity, and idempotence. Especially, the idempotence seems to make the extremum problem a lot simpler. Future work could focus on new problems that have different properties and allow for aggregation. It would also be interesting to see if similar techniques could be applied to other dynamic models such as T-interval stable dynamic networks where only a connected subgraph must be stable for T rounds. Furthermore, we would like to investigate lower bounds for these problems. In case of the summation problem, this could lead to a non-trivial lower bound for the counting problem (if n is not known beforehand) since the counting problem can be reduced to a summation problem where each node starts with a 1 as input.

References

1. Abshoff, S., Benter, M., Cord-Landwehr, A., Malatyali, M., Meyer auf der Heide, F.: Token dissemination in geometric dynamic networks. In: Flocchini, P., Gao, J., Kranakis, E., Meyer auf der Heide, F. (eds.) ALGOSENSORS 2013. LNCS, vol. 8243, pp. 22–34. Springer, Heidelberg (2014)
2. Abshoff, S., Benter, M., Malatyali, M., Meyer auf der Heide, F.: On two-party communication through dynamic networks. In: Baldoni, R., Nisse, N., van Steen, M. (eds.) OPODIS 2013. LNCS, vol. 8304, pp. 11–22. Springer, Heidelberg (2013)
3. Awerbuch, B.: Optimal distributed algorithms for minimum weight spanning tree, counting, leader election and related problems (detailed summary). In: Aho, A.V. (ed.) STOC, pp. 230–240. ACM (1987)
4. Chaudhuri, S., Dubhashi, D.P.: Probabilistic recurrence relations revisited. Theor. Comput. Sci. 181(1), 45–56 (1997)
5. Cornejo, A., Gilbert, S., Newport, C.C.: Aggregation in dynamic networks. In: Kowalski, D., Panconesi, A. (eds.) PODC, pp. 195–204. ACM (2012)
6. Dutta, C., Pandurangan, G., Rajaraman, R., Sun, Z., Viola, E.: On the complexity of information spreading in dynamic networks. In: Khanna, S. (ed.) SODA, pp. 717–736. SIAM (2013)
7. Haeupler, B., Karger, D.R.: Faster information dissemination in dynamic networks via network coding. In: Gavoille, C., Fraigniaud, P. (eds.) PODC, pp. 381–390. ACM (2011)

8. Haeupler, B., Kuhn, F.: Lower bounds on information dissemination in dynamic networks. In: Aguilera, M.K. (ed.) DISC 2012. LNCS, vol. 7611, pp. 166–180. Springer, Heidelberg (2012)

9. Karp, R.M.: Probabilistic recurrence relations. J. ACM 41(6), 1136–1150 (1994)

10. Kuhn, F., Locher, T., Schmid, S.: Distributed computation of the mode. In: Bazzi, R.A., Patt-Shamir, B. (eds.) PODC, pp. 15–24. ACM (2008)

11. Kuhn, F., Locher, T., Wattenhofer, R.: Tight bounds for distributed selection. In: Gibbons, P.B., Scheideler, C. (eds.) SPAA, pp. 145–153. ACM (2007)

12. Kuhn, F., Lynch, N.A., Oshman, R.: Distributed computation in dynamic networks. In: Schulman, L.J. (ed.) STOC, pp. 513–522. ACM (2010)

13. Linial, N.: Locality in distributed graph algorithms. SIAM J. Comput. 21(1), 193–201 (1992)

14. Luby, M.: A simple parallel algorithm for the maximal independent set problem. SIAM J. Comput. 15(4), 1036–1053 (1986)

15. Mosk-Aoyama, D., Shah, D.: Computing separable functions via gossip. In: Ruppert, E., Malkhi, D. (eds.) PODC, pp. 113–122. ACM (2006)

16. Panconesi, A., Srinivasan, A.: Improved distributed algorithms for coloring and network decomposition problems. In: Kosaraju, S.R., Fellows, M., Wigderson, A., Ellis, J.A. (eds.) STOC, pp. 581–592. ACM (1992)

17. Schneider, J., Wattenhofer, R.: A log-star distributed maximal independent set algorithm for growth-bounded graphs. In: Bazzi, R.A., Patt-Shamir, B. (eds.) PODC, pp. 35–44. ACM (2008)

Network Creation Games
with Traceroute-Based Strategies*

Davide Bilò[1], Luciano Gualà[2], Stefano Leucci[3], and Guido Proietti[3,4]

[1] Dipartimento di Scienze Umanistiche e Sociali, Università di Sassari, Italy
[2] Dipartimento di Ingegneria dell'Impresa, Università di Roma "Tor Vergata", Italy
[3] Dipartimento di Ingegneria e Scienze dell'Informazione e Matematica,
Università degli Studi dell'Aquila, Italy
[4] Istituto di Analisi dei Sistemi ed Informatica, CNR, Rome, Italy
`davide.bilo@uniss.it, guala@mat.uniroma2.it,`
`{stefano.leucci,guido.proietti}@univaq.it`

Abstract. Network creation games model the autonomous formation of
an interconnected system of selfish users. In particular, when the network
will serve as a digital communication infrastructure, each user is identified by a node of the network, and contributes to the build-up process
by strategically balancing between her *building cost* (i.e., the number of
links she personally activates in the network) and her *usage cost* (i.e.,
some function of the distance in the sought network to the other players). When the corresponding game is analyzed, the generally adopted
assumption is that players have a *common and complete* information
about the evolving network topology, which is quite unrealistic though,
due to the massive size this may have in practice. In this paper, we thus
relax this assumption, by instead letting the players have only a *partial*
knowledge of the network. To this respect, we make use of three popular
traceroute-based knowledge models used in *network discovering* (i.e., the
activity of reconstructing the topology of an unknown network through
queries at its nodes), namely: (i) *distance vector*, (ii) *shortest-path tree
view*, and (iii) *layered view*. For all these models, we provide exhaustive
answers to the canonical algorithmic game theoretic questions: convergence, computational complexity for a player of selecting a best response,
and tight bounds to the price of anarchy, all of them computed w.r.t. a
suitable (and unifying) equilibrium concept.

Keywords: Network Creation Games, Local-Knowledge Equilibrium,
Convergence Dynamics, Price of Anarchy.

1 Introduction

The spontaneous construction of large-scale *communication networks*, such as ad-
hoc wireless networks or the Internet, involves the interaction of many competing

---

* This work was partially supported by the Research Grant PRIN 2010 "ARS Tech-
noMedia", funded by the Italian Ministry of Education, University, and Research.

M. Halldórsson (Ed.): SIROCCO 2014, LNCS 8576, pp. 210–223, 2014.

and selfish entities. Thus, in this elusive strategic setting, it naturally arises the problem of understanding the process of an *ex-novo* creation of a network. More formally, we are given a set of n nodes of a graph, each occupied by a player which is willing to connect to all the other players. This can be realized by each player either by connecting directly to another player through the costly and unilateral activation of a corresponding link (i.e., an edge of the graph), or by using also links which were activated by other players, along a shortest path in the graph. In this latter case, however, it cannot be disregarded the fact that the relative communication performances may decay, as lengthy paths may induce large delays. Thus, the challenging analysis of the player's trade-off between her *building cost* (i.e., the number of network links she personally activates) and her *usage cost* (i.e., some function of the distance in the network from the other players) results in the corresponding study of a *communication network creation game*, a.k.a. *network connection game* (simply NCG in the following).

The standard model for NCGs. The universally accepted model for a NCG is that developed by Fabrikant *et al.* [9]. There, the activation of each link has a fixed real cost $\alpha > 0$, and the usage cost for each player is given by the *sum* of distances to all the players. More formally, this game, which we call SumNCG, is as follows: We are given a set of n players, say V, where the strategy space of player $u \in V$ is the power set $2^{V \setminus \{u\}}$. Given a combination of strategies $\sigma = (\sigma_u)_{u \in V}$, let $G(\sigma)$ denote the underlying undirected graph whose node set is V, and whose edge set is $E(\sigma) = \{(u, v) : u \in V \land v \in \sigma_u\}$. If no confusion arises, we will use G to denote $G(\sigma)$. Then, the *cost* incurred by player u in σ is

$$C_u(\sigma) = \alpha \cdot |\sigma_u| + \sum_{v \in V} d_{G(\sigma)}(u, v) \qquad (1)$$

where $d_{G(\sigma)}(u, v)$ is the distance between u and v in $G(\sigma)$. Correspondingly, the *social cost* is given by the sum of all players' costs. When a player takes an action (i.e., activates a subset of incident edges), she aims to keep this cost as low as possible. Thus, a *Nash Equilibrium*[1] (NE) for the game is a strategy profile $\bar{\sigma}$ such that for every player u and every strategy σ_u, we have that $C_u(\bar{\sigma}) \leq C_u(\bar{\sigma}_{-u}, \sigma_u)$.[2] If we characterize the space of NE in terms of the *Price of Anarchy* (PoA), i.e., the ratio between the social cost of the costlier NE to the optimal (centralized) social cost, then it has been shown this is constant for all values of α except for $n^{1-\varepsilon} \leq \alpha \leq 65\,n$, for any $\varepsilon \geq 1/\log n$ (see [14,15]). Moreover, very recently, in [10] it was proven that for all constant non-integral $\alpha \geq 2$, the PoA is bounded by $1 + o(1)$.

A first natural variant of SumNCG was introduced in [7], where the authors redefined the player cost function as follows

$$C_u(\sigma) = \alpha \cdot |\sigma_u| + \max\{d_{G(\sigma)}(u, v) : v \in V\}. \qquad (2)$$

---

[1] In this paper, we only focus on *pure* strategies Nash equilibria.
[2] The strategy vector (σ_{-u}, σ'_u) is defined from σ by replacing component σ_u with σ'_u.

This variant, named MaxNCG, received further attention in [15], where the authors improved the PoA of the game on the whole range of values of α, obtaining in this case that the PoA is constant for all values of α except for $129 > \alpha = \omega(1/\sqrt{n})$.

Other models for NCGs. Several variants of these two basic models have been defined, each one of them aiming to better characterize some specific aspect of the network creation process. They range from limiting the modification a player can do on her current strategy (see [1,13,16]), to budgeting either the number of edges a player can activate or her eccentricity (see [12,8,6]), and finally to constraining to a host graph the set of activable links (see [4]). Generally speaking, in all the above models the obtained results on the PoA are asymptotically worse than those we get in the two basic models, and we refer the reader to the cited papers for the actual bounds.

Observe that all these models, except for those given in [13,16], share with the basic model a severe restriction, namely the NP-hardness for a player to select a best-response strategy. Besides that, they also all assume that players have a *common and complete* information about the ongoing network. While this is feasible for small-size instances of the game, this becomes unrealistic for large-size networks. This is rather problematic, given the growing size of the inputs in the practice. Moreover, quite paradoxically, the full-knowledge assumption is not simplifying at all: it makes it computationally unfeasible for a player to select a best-response strategy, as said before, or even to check whether she is actually in a NE! To address these delicate issues, very recently in [5] it has been then introduced a new model which limits a player's full knowledge of the network structure up to a given radius k from herself, without even knowing the size n of the network (in distributed computing terminology, the system is *uniform*). In such a setting, the authors provided a comprehensive set of upper and lower bounds to the *Price of Anarchy* (PoA) for the entire range of values of k, as computed w.r.t. to a new suitable equilibrium concept. More precisely, the authors observed that a player has a partial (defective) view of the network, and thus before taking a step, she has to evaluate whether such a choice is convenient in *every* realizable network which is compatible with her current view. Then, let σ_u be the strategy played by player u, and define $\Sigma|_{\sigma_u}$ to be the set of strategy profiles $\sigma = (\sigma_{-u}, \sigma_u)$ of the players such that the network $G(\sigma)$ is realizable according to player u's view. Let

$$\Delta(\sigma_u, \sigma'_u) = \max_{\sigma \in \Sigma|_{\sigma_u}} \{C_u(\sigma_{-u}, \sigma'_u) - C_u(\sigma)\} \tag{3}$$

denote the worst possible cost difference u would have in switching from σ_u to σ'_u. Then, the *Local Knowledge Equilibrium* (LKE) is defined as a strategy profile $\bar{\sigma}$ such that for every player u and every strategy σ_u, we have $\Delta(\bar{\sigma}_u, \sigma_u) \geq 0$. Notice that our equilibrium concept is actually weaker than the classical NE concept as we have that every NE is also a LKE.

Our new local-view models for NCGs. In this paper, we move ahead along the direction of studying new NCG models with limited players' information. To this aim, we consider the most qualified *local-knowledge* models adopted in the field of *network discovery* (see [2,3]), i.e., the problem of fully identifying a large unknown network (i.e., all its edges and all its non-edges) through a minimum number of queries at its nodes (where the response to a query is exactly the information that the node can return according to the selected local-knowledge model). More precisely, we consider the following *traceroute-based* view models, which all find a motivation in the practice of probing the map of a network by tracing the route of packets:[3]

($\mathcal{M}_1$) *Distance vector*: besides her incident edges, each player knows only the distances in $G(\sigma)$ to all the other players (notice this is the minimal knowledge that a player must have in order to know her current cost);

($\mathcal{M}_2$) *Shortest-Path Tree (SPT) view*: each player knows the set of edges belonging to a given SPT of $G(\sigma)$ rooted at herself;

($\mathcal{M}_3$) *Layered view*: each player knows the set of edges belonging to *all* the SPTs of $G(\sigma)$ rooted at herself.

Notice in all these models a player has a partial (defective) view of the network, and therefore the LKE fits perfectly as solution concept. However, differently from the model provided in [5], we now have that a player is fully aware of her actual cost, which in all the three models is completely defined by the set of distances to all the other nodes (in other words, $C_u(\sigma)$ is constant for all $\sigma \in \Sigma|_{\sigma_u}$). For all the models we consider the iterated version of the game and we study the convergence of improving and best-response dynamics. In doing this, we assume that the players other than being *myopic* are also *oblivious*, namely at each time they only argue about the current view, without taking care of previous views. Moreover, we study the players' tension between the degree of knowledge and the computational feasibility of selecting a best-response strategy. Finally, we characterize the space of equilibria with respect to the social optimum through the study of upper and lower bounds to the PoA. Our results are summarized in Table 1.

The paper is organized as follows: in Section 2 we focus on convergence issues, while in Section 3 we analyze the computational complexity of finding a best-response. Finally, in Section 4 we study the PoA. All the sections are structured in subsections, according to the various view models.

2   Convergence

2.1   Model $\mathcal{M}_1$

Observe that in $\mathcal{M}_1$, besides her adjacent edges, a player u can only infer the existence of all the edges (x, y) such that $d_G(u, y) = d_G(u, x) + 1$ and x is the unique vertex having distance $d_G(u, x)$ from u. Thus, if u has two or more

---

[3] According to the spirit of the game, we assume that in all the models, the players initially sit on a connected network.

Table 1. Summary of our results (and open problems). In the first column, convergence (and divergence) are reported w.r.t. either improving or best-response dynamics (IRD and BRD, resp.). In the second column, we report the time complexity of selecting a best-response strategy.

	Convergence	Best-response complexity	PoA
$\mathcal{M}_1$	SUM: Yes ($\forall$ *improving* response dynamics (IRD)) MAX: Yes ($\forall$ IRD)	SUM: Open MAX: Polynomial	SUM: $\Theta(\min\{1+\alpha, n\})$ MAX: $\Theta(n)$ if $\alpha = \Omega(1)$ $\Theta(1+\alpha n)$ if $\alpha = O(1)$
$\mathcal{M}_2$	SUM: No ($\exists$ *best response* dynamics (BRD) cycle) MAX: No ($\exists$ BRD cycle)	SUM: Polynomial MAX: Polynomial	SUM: $\Theta(\min\{1+\alpha, n\})$ MAX: $\Theta(n)$ if $\alpha = \Omega(1)$ $\Theta(1+\alpha n)$ if $\alpha = O(1)$
$\mathcal{M}_3$	SUM: No ($\exists$ BRD cycle) MAX: No ($\exists$ BRD cycle)	SUM: NP-hard MAX: NP-hard	SUM: $\Theta(\min\{1+\alpha, n\})$ MAX: $\Theta(n)$ if $\alpha = \Omega(1)$ $\Theta(1+\alpha n)$ if $\frac{1}{n-1} \leq \alpha = O(1)$

adjacent vertices, *swapping* (i.e., replacing an owned edge with another one) any of her edges is not a plausible action since this might disconnect the network. This implies the following:

Lemma 1. *In both* SUMNCG *and* MAXNCG *in* $\mathcal{M}_1$, *the only players who can swap edges have degree 1 (and own the incident edge).*

Theorem 1. *Any improving response dynamics for* SUMNCG *in* $\mathcal{M}_1$ *converges to an equilibrium.*

Proof. Notice that, in any configuration of strategies, no edge can be removed.[4] As the maximum number of edges is bounded we only need to show that the dynamic restricted to swap-only moves converges.

Let G be the actual network and u be a player who modifies her strategy from σ_u to σ'_u (where $|\sigma_u| = |\sigma'_u|$) thus improving her cost (in the worst-case). Call G' the resulting network and notice that the usage cost of u in G' must be less than her usage cost in G. We now define $\Phi(G) = \sum_{x \in V} \sum_{v \in V} d_G(x, v)$, and show that whenever an improving move is made the value of Φ decreases. As the domain of Φ is finite this suffices to prove the claim.

By Lemma 1, u must have degree 1 in G, hence:

$$\Phi(G) - \Phi(G') = \sum_{x \in V} \sum_{v \in V} d_G(x, v) - \sum_{x \in V} \sum_{v \in V} d_{G'}(x, v)$$

$$= \sum_{x \in V \setminus \{u\}} \sum_{v \in V \setminus \{u\}} d_G(x, v) + 2 \sum_{v \in V} d_G(u, v) - \sum_{x \in V \setminus \{u\}} \sum_{v \in V \setminus \{u\}} d_{G'}(x, v) - 2 \sum_{v \in V} d_{G'}(u, v)$$

$$= 2 \left(\sum_{v \in V} d_G(u, v) - \sum_{v \in V} d_{G'}(u, v) \right) > 0.$$

---

[4] Except edges owned by both endpoints, which could arise in the starting network.

where we used the fact that $d_G(x,v) = d_{G'}(x,v)$ for every $x,v \in V \setminus \{u\}$, as u has degree 1 in both G and G'. □

Concerning MAXNCG, the following can be proven:

Theorem 2. *Any improving response dynamics for* MAXNCG *in* $\mathcal{M}_1$ *converges to an equilibrium.*

Proof. As for SUMNCG, we only need to show that the dynamic restricted to swap-only moves converges. We use an approach similar to that in [11]. Given a profile of strategies, consider the associated graph G and the n-dimensional vector ε_G of the *eccentricities* of the vertices in G. Now sort the elements of the vector in decreasing order. We claim that a function that maps each profile of strategies to such a vector can only decrease after an improving response (comparisons are done in lexicographic order).

By Lemma 1, only players having degree 1 can move. Let u be a player who modifies her strategy in such a way that the corresponding graph changes from G to G'. As u has not removed any edge besides the swap, $\varepsilon_G(u) > \varepsilon_{G'}(u)$ must hold.

Let $\varepsilon_H(x)$ denote the eccentricity of a vertex $x \in V(H)$ in the graph H. To prove the claim we show that, when u changes her strategy, if another player x increases her eccentricity as a consequence, then $\varepsilon_{G'}(x) < \varepsilon_{G'}(u)$. Indeed, suppose $\varepsilon_G(x) < \varepsilon_{G'}(x)$; as u was a vertex having degree 1 in G, the same holds in G', and we must have $\varepsilon_{G'}(x) = d_{G'}(x,u) \leq \varepsilon_{G'}(u)$. This ends the proof. □

2.2 Models $\mathcal{M}_2$ and $\mathcal{M}_3$

In this section we show that the best-response dynamics might not converge in $\mathcal{M}_2$ and $\mathcal{M}_3$. In particular, we will exhibit cycles of best responses in both models for both SUMNCG and MAXNCG. Notice that in some cases, a best-response strategy of a player will contain a subset of factitious moves that are only instrumental to generate the sought cycle. We conjecture that these moves could be completely avoided by means of more complicated examples. Remember that all the graphs are undirected, that is, an edge can be used in both directions, regardless of the player who bought it. In our figures, however, some edges might be drawn as directed with the arrow exiting from the buyer. The vertex that is changing her strategy is highlighted in dark grey while the edges of her current view are shown in bold.

Theorem 3. *Both* SUMNCG *and* MAXNCG *in* $\mathcal{M}_2$ *admit best-response cycles.*

Proof. Regarding MAXNCG, consider the sequence shown in Figure 1 for $\alpha \geq 6$. From graph (a) to (d) each player changes her current strategy with an improving best response. The latter graph is isomorphic to the first one: vertex a in graph (d) is playing the role of vertex b in graph (a) and vice-versa. Therefore, by repeating a very similar best-response dynamics starting from the graph (d) the players will reach again the configuration (a), hence the cycle.

Regarding SUMNCG, Figure 2 shows a cycle of best responses for $\alpha \geq 6$. □

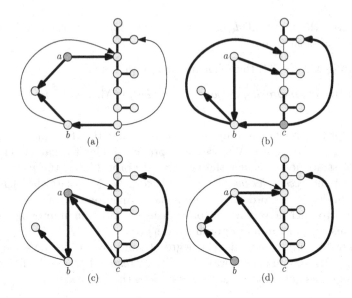

Fig. 1. Cycle of best responses for MAXNCG in $\mathcal{M}_2$ for $\alpha \geq 6$

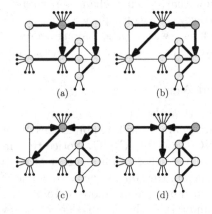

Fig. 2. Cycle of best responses for SUMNCG in $\mathcal{M}_2$ for $\alpha \geq 6$

We also have:

Theorem 4. *Both* SUMNCG *and* MAXNCG *in* $\mathcal{M}_3$ *admit best-response cycles.*

Proof. Regarding SUMNCG, a best-response cycle is shown in Figure 3 for $\alpha = 15$. Notice that the graph in (a) is isomorphic to the graph in (d) where vertex c plays the role of vertex a and vice-versa.

Regarding MAXNCG, Figure 4 shows a cycle of best responses for $\alpha = 2 - \epsilon$ for a small value of $\epsilon > 0$. $\qquad\square$

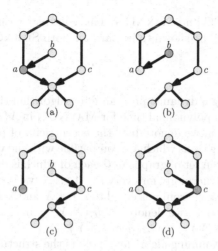

Fig. 3. Cycle of best responses for SumNCG in $\mathcal{M}_3$ for $\alpha = 15$

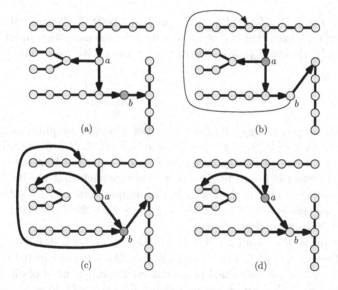

Fig. 4. Cycle of best responses for MaxNCG in $\mathcal{M}_3$ for $\alpha = 2 - \epsilon$

3   Complexity of Computing a Best Response

In this section we consider the complexity of computing a best-response strategy. Concerning $\mathcal{M}_1$ we prove that a best response can be computed in polynomial time only for MaxNCG, while for SumNCG, for which we conjecture that the same holds, the problem remains open. Regarding $\mathcal{M}_2$, we show that situation is more favorable, in that a best response strategy can be computed in polynomial time for both SumNCG and MaxNCG. This contrasts with the results for $\mathcal{M}_3$ where the problem will be shown to be hard for both SumNCG and MaxNCG.

3.1   Model $\mathcal{M}_1$

We start by describing a dynamic programming algorithm that computes a best response of a player in polynomial time for MaxNCG in $\mathcal{M}_1$. For the rest of the proof, we fix a player u, we denote by ℓ the eccentricity of player u in $G(\sigma)$, we denote by $\mathcal{G}$ the set of graphs which are compatible with the view of player u in σ, and we denote by $\mathcal{G}'$ the set of graphs in $\mathcal{G}$ each of which is deprived of the edges incident to u in σ. Furthermore, for every $i = 0, \ldots, \ell$, we denote by L_i the set of all vertices whose distance from u is equal to i. Let X and Y be two subsets of V. For a fixed graph G on V, we denote by $d_G(X, Y) = \max_{y \in Y} \min_{x \in X} d_G(x, y)$, and we denote by $\delta(X, Y) = \max_{G \in \mathcal{G}'} d_G(X, Y)$.

The dynamic programming algorithm exploits the structural properties of the worst possible graph and best-response strategies as highlighted in the following lemmas.

Lemma 2. *Let $L_i' \subsetneq L_i$ and $y \in L_j$, with $i, j > 0$. If there exists an index $0 < t \leq \min\{i, j\}$ such that $|L_t| = 1$, then let h be the maximum index such that $0 < h \leq \min\{i, j\}$ and $|L_h| = 1$; otherwise, let $h = \perp$. We have that*

$$\delta(L_i', y) = \begin{cases} j + i - 2h & \text{if } h \neq \perp; \\ +\infty & \text{otherwise.} \end{cases}$$

Proof. First, we prove the claim for $h \neq \perp$. Let z be the unique vertex in L_h. It is easy to see that $\delta(L_i', y) \leq j + i - 2h$ as every graph $G \in \mathcal{G}'$ contains a path of length $i - h$ from z to every vertex in L_i, and a path of length $j - h$ from z to y. To complete the proof, let $G \in \mathcal{G}$ be any tree such that the least common ancestor of every vertex in L_i' and y is the unique vertex in L_h (observe that such a tree always exists). Clearly, $d_G(x, y) = j + i - 2h$ for every $x \in L_i'$, and thus $\delta(x, y) \geq d_G(x, y) = i + i - 2h$.

Now, we prove the claim for $h = \perp$. Let $G \in \mathcal{G}'$ be any disconnected graph such that the vertices of L_i are all contained in the same connected component of G, say G', while y is contained in a connected component of G different from G' (observe that such a graph always exists). Clearly, $\delta(L_i', y) = \infty$. $\square$

Lemma 3. *Let $y \in L_j$, with $j > 0$. If there exists an index $0 < t \leq j$ such that $|L_t| = 1$, then let h be the maximum index such that $0 < h \leq j$ and $|L_h| = 1$; otherwise, let $h = \perp$. For every $i = 1, \ldots, \ell$, we have that*

$$\delta(L_i, y) = \begin{cases} j - i & \text{if } i \leq j; \\ j + i - 2h & \text{if } h \neq \perp \text{ and } i > j; \\ +\infty & \text{otherwise.} \end{cases}$$

Proof. We already proved the case $h = \perp$ and $i > j$ in the proof of Lemma 2. If $i \leq j$, then $d_G(L_i, y) = j - i$ for every graph $G' \in \mathcal{G}$. If $i > j$, then let $G \in \mathcal{G}$ be any tree such that the least common ancestor of x and y is the unique vertex in L_h. Clearly for every $x \in L_i$, $d_G(x,y) = j + i - 2h$, and thus $\delta(x,y) = j + i - 2h$. $\square$

Let $S = \{v \in L_1 : u \in \sigma_v\}$ be the set of players who have bought an edge towards u in σ. We have the following:

Lemma 4. *There exists a best-response strategy σ_u^* such that, for every $i = 2,\ldots,\ell$, either $L_i \cap \sigma_u^* = \emptyset$ or $L_i \subseteq \sigma_u^*$. Moreover either $|L_1| = 1$ or $L_1 \subseteq (\sigma_u^* \cup S)$ (or both).*

Proof. First of all notice that if $|L_1| \geq 2$ and $L_1 \not\subseteq (\sigma_u^* \cup S)$ then there exists a network $G \in \mathcal{G}$ such that, when u changes her strategy from σ to σ^*, the resulting networks is disconnected. This contradicts the fact that σ^* is a best response. Now, let f be a function such that $f(\sigma_u') = |\{i \in \{1,\ldots,\ell\} \text{ s.t. } \sigma_u' \cap L_i \neq \emptyset \text{ and } L_i \not\subseteq \sigma_u'\}|$, for any strategy σ_u'. Let σ_u' be any strategy such that $f(\sigma_u') > 0$. We prove the claim by showing that there exists a strategy σ_u'' such that $\Delta(\sigma_u, \sigma_u') \geq \Delta(\sigma_u, \sigma_u'')$ and $f(\sigma_u'') < f(\sigma_u')$. Let i be an index such that neither $L_i \cap \sigma_u' = \emptyset$ nor $L_i \subseteq \sigma_u'$ (such an index always exists as $f(\sigma_u') > 0$). Let $L_i' := L_i \cap \sigma_u'$ and let $\sigma_u'' := \sigma_u' \setminus L_i'$. We modify σ_u'' as follows. Let j be the maximum index such that $0 < j < i$ and $|L_j| = 1$, if any. If such a j exists, then add L_j to σ_u''. Clearly, $f(\sigma_u'') < f(\sigma_u')$ as well as $|\sigma_u''| \leq |\sigma_u'|$. Furthermore, from Lemma 2 and Lemma 3, we have that $\delta(L_j, V) \leq \delta(L_i', V)$. Therefore, $\delta(\sigma_u'', V) \leq \delta(\sigma_u', V)$ and thus, for every $G \in \mathcal{G}$:

$$C_u((\sigma_{-u}, \sigma_u''), G) = \alpha \cdot |\sigma_u''| + \delta(\sigma_u'', V) \leq \alpha \cdot |\sigma_u'| + \delta(\sigma_u', V) = C_u((\sigma_{-u}, \sigma_u'), G),$$

i.e., $\Delta(\sigma_u, \sigma_u') \geq \Delta(\sigma_u, \sigma_u'')$. $\square$

We are now ready to prove the following:

Theorem 5. *The best response of a player in $\mathcal{M}_1$ can be computed in polynomial time for MaxNCG.*

Proof. The following dynamic programming algorithm computes a best-response strategy of the form described in Lemma 4. For every $i = 0,\ldots,\ell$ and every $\eta = 0,\ldots,n-2$, we define $A[i,\eta]$ as the size of a minimum-size set of vertices X such that $L_i \subseteq X$ and $\delta(X, \bigcup_{h=1,\ldots,i} L_h) \leq \eta - 2$. Intuitively, $A[i,\eta]$ represents the cost of a cheapest strategy that allows the player u to have all vertices in $\bigcup_{h=1,\ldots,i} L_h$ at a distance of at most $\eta + 1$ from it by buying only the $|X|$ edges towards all vertices in X. However, $A[i,\eta]$ restricts the player u to buy edges only towards sets X such that $L_i \subseteq X \subseteq \bigcup_{h=1,\ldots,i} L_h$.

With a little abuse of notation, we assume that $L_0 = \emptyset$. For each pair of values i and η, we denote by $g(i,\eta)$ the minimum index such that $0 \leq g(i,\eta) \leq i$ and $\delta(L_{g(i,\eta)} \cup L_i, \bigcup_{h=g(i,\eta),\ldots,i} L_h) \leq \eta$. The cost of a best-response strategy is equal to

$$1 + \min_{\eta=0,\ldots,n-2} \{\eta + \alpha \cdot \min_{\max\{1,g(\ell,\eta)\} \leq j \leq \ell} A[j,\eta]\}.$$

Clearly, for every $\eta = 0, \ldots, n - 1$, $A[0, \eta] = 0$ while $A[1, \eta] = |\sigma_u \cap L_1|$, as the edges connecting u with the vertices in $L_1 \setminus \sigma_u$ have already been bought by other players. Furthermore, it is easy to see that, for every $i = 1, \ldots, \ell$ and every $\eta = 0, \ldots, n - 2$, $A[i, \eta]$ can be computed efficiently as follows

$$A[i, \eta] = |L_i| + \min_{g(i, \eta) \leq j < i} A[j, \eta].$$

Once the cost of an optimal strategy has been computed, it is easy to construct the strategy itself by proceeding backwards.   □

3.2 Model $\mathcal{M}_2$

We prove the following:

Theorem 6. *The best response of a player in $\mathcal{M}_2$ can be computed in polynomial time for both* SumNCG *and* MaxNCG.

Proof. Let u be a player, let σ_u be her current strategy, and let $T(u)$ be the SPT (rooted at u) currently seen by u. We will describe a polynomial-time dynamic programming algorithm that computes a best response for u. For MaxNCG and SumNCG, this algorithm will need to solve (a variant of) the *k-center* and the *k-median* problem on trees, respectively. Recall that in the k-center (resp., k-median) problem, we are given a graph H and we want to select a set $S \subseteq V(H)$ of exactly k vertices such that the *maximum* (resp., *average*) distance from each node in $V(H)$ to a closest node in S is minimized.

If we remove u from $T(u)$, the tree will split into a forest F containing a number h of trees. We will refer to those trees by $T_1, \ldots, T_h$, and to the corresponding roots by $r_1, \ldots, r_h$. W.l.o.g., let us assume that $T_1, \ldots, T_z$ are exactly the trees such that u owns the edge $(u, r_i), i = 1, \ldots, z$, while edges $(u, r_i), i = z+1, \ldots, h$, are owned by r_i.

We define $A[i, k]$ with $i \leq z$ and $0 \leq k < n$ to be the measure of an optimal solution to the k-center (resp., k-median) problem on T_i. In a similar manner, we define $A[i, k]$ with $z < i \leq h$ and $0 \leq k < n$ to be the measure of an optimal solution to the r_i-*constrained* k-center (resp., k-median) problem on T_i, i.e., that in which we constrain the solution to include the vertex r_i. In both cases, we consider the measure to be $+\infty$ when there is no feasible solution (i.e., $k \geq |V(T_i)|$). It is well known that all these problems can be solved in polynomial time on trees.

We now consider all the ways of selecting a subset of j vertices as centers (resp., medians) for solving the k-center (resp., k-median) problem on the first i trees, and let $B[i, j]$ be the minimum cost that we obtain. Since $B[1, j] = A[1, j]$, it follows that, for $i > 1$, we can efficiently compute $B[i, j]$ as follows: for the k-center problem we have

$$B[i, j] = \min_{1 \leq t \leq j-i+1} \max \{A[i, t], B[i - 1, j - t]\},$$

while for the k-median problem we have

$$B[i, j] = \min_{1 \leq t \leq j-i+1} \{A[i, t] + B[i - 1, j - t]\}.$$

In order to determine the best response for u, we notice that the values $1 + B[h,j]$ (resp., $n - 1 + B[h,j]$), for every possible $j \geq h - z$, are exactly the best usage costs that u can attain if she wants to buy exactly $j - (h - z)$ edges (indeed, $h - z$ edges are adjacent to $r_i, i = z + 1, \ldots, h$, and they are already owned by r_i). Therefore, a best response for u can be found (in polynomial time) by computing the following

$$\arg \min_{h-z \leq j < n} \{B[h,j] + \alpha(j - h + z)\},$$

and by proceeding backwards on the values of $B[i,j]$. Finally, the edges to be activated by u can be found by looking at the optimal solution of the corresponding k-center (resp., k-median) problem. □

3.3 Model $\mathcal{M}_3$

As far as model $\mathcal{M}_3$ is concerned, the problem of computing a best response is NP-hard for both SumNCG and MaxNCG, as shown by the following results:

Theorem 7. *Computing a best response for a player in $\mathcal{M}_3$ for* SumNCG *is NP-hard.*

Sketch of proof. The proof is a simple modification of the reduction from the *Minimum Dominating Set problem* (MDSP) shown in [9]. Consider a bipartite graph G and an additional player u that is buying all the edges towards the vertices on one side of the bipartition. Notice that the view of u contains all the edges of G. As shown in [9], a best response of u for $1 < \alpha < 2$ is that of buying the edges towards a minimum dominating set of G. The claim follows as the MDSP remains NP-hard even on bipartite graphs. □

Theorem 8. *Computing a best response for a player in $\mathcal{M}_3$ for* MaxNCG *is NP-hard.*

Sketch of proof. The proof is similar to the one shown for Theorem 7 by starting from the reduction given in [15]. □

4   Price of Anarchy

As far as the PoA is concerned, models $\mathcal{M}_1$, $\mathcal{M}_2$ and $\mathcal{M}_3$ are equivalent, as the following two results show.

Theorem 9. *The PoA for* SumNCG *in $\mathcal{M}_1$, $\mathcal{M}_2$, and $\mathcal{M}_3$ is $\Theta(\min\{1+\alpha, n\})$ for every α.*

Proof. First of all, observe that for every actual network and for every player u, the set of networks that are compatible with the view of u can only shrink when we move from $\mathcal{M}_1$ to $\mathcal{M}_2$ to $\mathcal{M}_3$. This implies that the PoA in $\mathcal{M}_1$ can only be greater than the PoA in $\mathcal{M}_2$ which, in turn, can only be greater than the PoA in $\mathcal{M}_3$, hence it suffices to prove an upper bound on $\mathcal{M}_1$ and a lower bound on $\mathcal{M}_3$.

Concerning the lower bound, notice that the complete graph is an equilibrium in $\mathcal{M}_3$ as the view of each player is a star (and no player can remove any edge). The social cost of the complete graph is at least $\Omega(\alpha n^2 + n^2)$, while the social cost of a star is $O(\alpha n + n^2)$. Thus, the PoA is $\Omega(\min\{n, 1 + \alpha\})$.

Concerning the upper bound, we first prove an upper bound of $O(1 + \sqrt{\alpha})$ on the diameter of an equilibrium graph. Indeed, consider a player u with eccentricity $D \geq 4$, and let v be a vertex such that $d(u, v) = D$. The player u knows that by buying (u, v) then she can decrease the distances of at least $\frac{D}{4}$ vertices on the path between u and v by at least $\frac{3}{4}D - \frac{1}{4}D - 1 = \frac{1}{2}D - 1$, hence she would save $\Omega(D^2)$. Notice that this reasoning does not require for u to know the actual path towards v. As u has not bought the edge (u, v) we conclude that $\alpha = \Omega(D^2)$, so $D = O(1 + \sqrt{\alpha})$.

We can now bound the PoA as follows:

$$O\left(\frac{\alpha n^2 + n^2(1 + \sqrt{\alpha})}{\alpha n + n^2}\right) = O(\min\{n, 1 + \alpha\}).$$

$\square$

Theorem 10. *The PoA for* MaxNCG *in* $\mathcal{M}_1$, $\mathcal{M}_2$, *and* $\mathcal{M}_3$ *is* $\Theta(n)$ *for* $\alpha = \Omega(1)$, *and* $\Theta(1 + \alpha n)$ *if* $\alpha = O(1)$.

Proof. As shown in the proof of Theorem 9 it suffices to prove an upper bound on $\mathcal{M}_1$ and a lower bound on $\mathcal{M}_3$.

As far as the lower bound is concerned, we consider again the complete graph. We already argued that it is an equilibrium and the same arguments also hold for MaxNCG; moreover, its social cost is $\Omega(\alpha n^2 + n)$, while the cost of a star is $O(\alpha n + n)$. We thus obtain PoA $= \Omega\left(\frac{\alpha n^2 + n}{\alpha n + n}\right)$, which is $\Omega(n)$ when $\alpha = \Omega(1)$, and $\Omega(1 + \alpha n)$ otherwise.

Concerning the upper bound, we first prove an upper bound of $O(1 + \alpha n)$ on the diameter of an equilibrium graph. Indeed, consider a player u with eccentricity $D \geq 4$, and let η be the number of vertices at distance roughly $D/2$. If u buys an edge towards each of these vertices, then her eccentricity decreases by at least $\Omega(D)$. Since we are in an equilibrium, we have that $\alpha n \geq \alpha \eta = \Omega(D)$. We then have PoA $= \frac{O(\alpha n^2 + n + n^2)}{\Omega(\alpha n + n)}$, which gives the bounds of the claim. $\square$

Acknowledgements. The authors wish to thank the anonymous referees for their insightful and useful comments.

References

1. Alon, N., Demaine, E.D., Hajiaghayi, M., Leighton, T.: Basic network creation games. In: Proc. of the 22nd ACM Symp. on Parallelism in Algorithms and Architectures (SPAA 2010), pp. 106–113. ACM Press (2010)

2. Bampas, E., Bilò, D., Drovandi, G., Gualà, L., Klasing, R., Proietti, G.: Network verification via routing table queries. In: Kosowski, A., Yamashita, M. (eds.) SIROCCO 2011. LNCS, vol. 6796, pp. 270–281. Springer, Heidelberg (2011)
3. Beerliova, Z., Eberhard, F., Erlebach, T., Hall, A., Hoffman, M., Mihalák, M., Ram, S.: Network discovery and verification. IEEE Journal on Selected Areas in Communications 24(12), 2168–2181 (2006)
4. Bilò, D., Gualà, L., Leucci, S., Proietti, G.: The max-distance network creation game on general host graphs. In: Goldberg, P.W. (ed.) WINE 2012. LNCS, vol. 7695, pp. 392–405. Springer, Heidelberg (2012)
5. Bilò, D., Gualà, L., Leucci, S., Proietti, G.: Locality-based network creation games. In: Proc. of the 26th ACM Symp. on Parallelism in Algorithms and Architectures (SPAA 2014), pp. 277–286. ACM Press (2014)
6. Bilò, D., Gualà, L., Proietti, G.: Bounded-distance network creation games. In: Goldberg, P.W. (ed.) WINE 2012. LNCS, vol. 7695, pp. 72–85. Springer, Heidelberg (2012)
7. Demaine, E.D., Hajiaghayi, M., Mahini, H., Zadimoghaddam, M.: The price of anarchy in network creation games. In: Proc. of the 36th Annual ACM Symp. on Principles of Distributed Computing (PODC 2007), pp. 292–298. ACM Press (2007)
8. Ehsani, S., Fazli, M., Mehrabian, A., Sadeghabad, S.S., Saghafian, M., Shokatfadaee, S., Safari, M.: On a bounded budget network creation game. In: Proc. of the 23rd ACM Symp. on Parallelism in Algorithms and Architectures (SPAA 2011), pp. 207–214. ACM Press (2011)
9. Fabrikant, A., Luthra, A., Maneva, E., Papadimitriou, C.H., Shenker, S.: On a network creation game. In: Proc. of the 22nd Symp. on Principles of Distributed Computing (PODC 2003), pp. 347–351. ACM Press (2003)
10. Graham, R., Hamilton, L., Levavi, A., Loh, P.-S.: Anarchy is free in network creation. In: Bonato, A., Mitzenmacher, M., Prałat, P. (eds.) WAW 2013. LNCS, vol. 8305, pp. 220–231. Springer, Heidelberg (2013)
11. Kawald, B., Lenzner, P.: On dynamics in selfish network creation. In: Proc. of the 25th ACM Symp. on Parallelism in Algorithms and Architectures (SPAA 2013), pp. 83–92. ACM Press (2013)
12. Laoutaris, N., Poplawski, L.J., Rajaraman, R., Sundaram, R., Teng, S.-H.: Bounded budget connection (BBC) games or how to make friends and influence people, on a budget. In: Proc. of the 27th ACM Symp. on Principles of Distributed Computing (PODC 2008), pp. 165–174. ACM Press (2008)
13. Lenzner, P.: Greedy selfish network creation. In: Goldberg, P.W. (ed.) WINE 2012. LNCS, vol. 7695, pp. 142–155. Springer, Heidelberg (2012)
14. Mamageishvili, A., Mihalák, M., Müller, D.: Tree Nash equilibria in the network creation game. In: Bonato, A., Mitzenmacher, M., Prałat, P. (eds.) WAW 2013. LNCS, vol. 8305, pp. 118–129. Springer, Heidelberg (2013)
15. Mihalák, M., Schlegel, J.C.: The price of anarchy in network creation games is (mostly) constant. In: Kontogiannis, S., Koutsoupias, E., Spirakis, P.G. (eds.) SAGT 2010. LNCS, vol. 6386, pp. 276–287. Springer, Heidelberg (2010)
16. Mihalák, M., Schlegel, J.C.: Asymmetric swap-equilibrium: A unifying equilibrium concept for network creation games. In: Rovan, B., Sassone, V., Widmayer, P. (eds.) MFCS 2012. LNCS, vol. 7464, pp. 693–704. Springer, Heidelberg (2012)

Patrolling by Robots Equipped with Visibility[*]

Jurek Czyzowicz[1], Evangelos Kranakis[2], Dominik Pajak[3], and Najmeh Taleb[2]

[1] Département d'informatique, Université du Québec en Outaouais, Gatineau, Canada
[2] School of Computer Science, Carleton University, Ottawa, Canada
[3] LaBRI, Inria Bordeaux Sud-Ouest, France

Abstract. We study the problem of mobile robots with distinct visibility ranges patrolling a curve. Assume a set of k mobile robots (patrolmen) $a_1, a_2, \cdots, a_k$ walking along a unit-length curve in any of the two directions, not exceeding their maximal speeds. Every robot a_i has a range of visibility r_i, representing the distance from its current position at which the robot can see in each direction along the curve. The goal of the patrolling problem is to find the perpetual movement of the robots minimizing the maximal time when a point of the curve remains unseen by any robot.

We give the optimal patrolling algorithms for the case of close curve environment (known as the boundary patrolling problem in the robotics literature) and open curve (fence patrolling), when all robots have the same maximal speed. We briefly discuss the case of distinct speeds, showing that the boundary patrolling problem for robots with distinct visibility ranges is essentially different than the case of point visibility robots. We also give the optimal algorithm for fence patrolling by two robots with distinct speeds and visibility ranges.

For the case when the environment in which the robots operate is a general graph, we show that the patrolling problem for robots with distinct visibility ranges is NP-hard, while it is known that the same problem for point-visibility robots has been known to have a polynomial-time solution.

Keywords: Cycle, Graph, Mobile Robots, Patrolling, Segment, Speed, Visibility.

1 Introduction

A set of k mobile robots $a_1, a_2, \cdots a_k$, each one able to observe some neighborhood of its current position, has to protect (*patrol*) a given region. For this purpose the robots move perpetually around the region in order to see each point of their environment as often as possible. In this paper we study robots, moving with speeds not exceeding a certain maximal velocity, inside a uni-dimensional region represented by a unit segment or a unit-length cycle. The objective of this paper is to design algorithms producing the movements of the robots which minimize the time interval when some points of the environment remain unseen by all robots, taken over all points of the given domain being patrolled.

[*] This work was partially supported by NSERC grants. D. Pajak was supported by LaBRI project "mobilité junior" and LIRCO.

M. Halldórsson (Ed.): SIROCCO 2014, LNCS 8576, pp. 224–234, 2014.
© Springer International Publishing Switzerland 2014

1.1 Preliminaries and Notation

Each robot a_i is equipped with visibility allowing it to see its environment within its visibility radius r_i in both directions from its current position. The visibility ranges of all robots may be different. During the movement of the set of robots, at each time t a point p of the environment is called *protected* if it is seen by at least one robot. In other words, if robot a_i protects point p at time t the distance of the robot a_i from p at time t must be at most equal to r_i. Given a perpetual movement of all robots produced by some patrolling algorithm, by the *idle time* of point p we mean the smallest value $I(p)$, such that in every time interval $[t, t + I(p)]$ point p is protected by some robot. By the idle time of such an algorithm we mean the maximal value of $I(p)$ taken over all points of the environment. Throughout the paper we will assume that $\sum_{i=1}^{k} 2r_i < 1$, since otherwise agents could constantly observe the environment without need to move.

1.2 Related Work

Patrolling has been intensely studied in the last decade by the robotics community (cf. [3,17,18]). It is defined as the act of monitoring consisting in traveling around an environment in order to protect or supervise it. Patrolling is useful, e.g., to identify humans or objects of interest that need to be rescued from a disaster. Network administrators use patrolling by mobile robots to determine web pages which must be indexed by search engines or to detect network deficiencies.

Earlier work on patrolling was mainly experimental and studying heuristic methods (cf. [14,12] though [6] brings up a theoretical analysis of the methods of patrolling. The two basic strategies to patrolling discussed in [6] are *cyclic strategy* and the *partition strategy*. In a cyclic approach, a cycle inside the environment is identified and the robots walk around this cycle in the same direction. In the partition approach the environment is divided into subregions (which may be sometimes overlapping) that are assigned to different robots. [6] was first to introduce the notion of *idleness*, which has been most often used to measure the performance of patrolling. Several other issues related to the patrolling problem were also studied, e.g. coordination and cooperation of multi-agent teams ([2,15,16]), dynamically changing environments or robot teams ([18,19]) dealing with adversarial environments ([1,4]) and many others.

Recently several interesting algorithmic issues related to patrolling were investigated. The ant-like mobile agents were used by [20] to realize an interesting distributed strategy attaining patrolling by agents traversing an Eulerian cycle of an input graph, while [10] used ant-like agents to partition the graph to patrol among them.

The optimality of the fundamental partition strategy (in the case of fence patrolling) and the cyclic strategy (for boundary patrolling) has been proven for robot teams having the same maximal speed and for small sets of robots with distinct maximal speeds (cf. [8,13]). However, for distinct-speed sets of robots, both these strategies have been proven sub-optimal for boundary patrolling by at least three robots, [8], and fence patrolling by at least six robots, [13]. [5] presented examples with several other non-standard strategies for fence and boundary patrolling.

Same-speed robots were used in [7] for fence and boundary patrolling where some *neutral* regions may be left unprotected and in [9], where an optimal patrolling algorithm for graphs is proposed.

[13] also considered *weighted* patrolmen, i.e. such that any point p could be left unvisited for a time equal to the weight of the robot which was last to visit p. To the best of our knowledge, patrolmen equipped with visibility have not studied in the scientific literature before.

1.3   Outline and Results of the Paper

In Section 2 we show that a version of the cyclic strategy is optimal for boundary patrolling. In Section 3 we prove that the partition strategy achieves the optimal idle time for fence patrolling. To this end we show that any patrolling algorithm using robots equipped with visibility may be converted to a strategy for robots without visibility achieving at least the same idle time on a fence shortened by twice the sum of all robots' visibility radii. This could suggest that the patrolling problem for robots with visibility is equivalent to patrolling with zero-visibility robots considered elsewhere. However the hardness of the problem for general graphs shown in Section 4, in view of the polynomial solution from [9] contradicts this supposition. In Sections 2.2 and 3.2 we discuss the cases of two robots with distinct speeds and visibility radii. In Section 3.2 we show that it is possible to extend the proof of optimality of the partition strategy (cf. [8,13]) on the case of two visibility-equipped, distinct-speed robots. However the example from Section 2.2 shows that for the circle patrolling with two distinct-speed robots the cyclic strategy from [8] is no longer optimal when the robots are equipped with visibility. This is another evidence that patrolling with visibility-equipped robots presents new challenges, even for the case of robots with the same maximal speed. All missing proofs will appear in the full version of the paper.

2   Circle Patrolling

In this section we investigate patrolling of a circle. First we give the optimal patrolling for the case of any number of robots with identical speeds. In the second subsection we give the optimal algorithm for the case of two robots with distinct speeds. The case of three or more robots remains open.

2.1   Equal Speeds

We start by considering the case of agents with equal speeds. We will assume that the maximum speed of each agent is equal to 1. Recall that 1 is also the length of the environment i.e., we consider unit circles and unit intervals.

Proposition 1. *Algorithm A_1 achieves the idle time $T = \frac{1-\sum_{i=1}^{k} 2r_i}{k}$.*

Theorem 1. *Any patrolling algorithm for k robots with speeds $v_1,\ldots,v_k$ and visibilities $r_1,\ldots r_k$, patrolling unit circle achieves idle time at least $I_{opt} \geq \frac{1-\sum_{i=1}^{k} 2r_i}{\sum_{i=1}^{k} v_i}$.*

Proof. Consider any algorithm A and its idle time I_A. Take any moment of time t. Regardless of the positions of the robots, the total length of the subset of the circle being

---

Algorithm A_1 [for k robots with the same speed and different visibility to patrol a circle]

1. If $\sum_{i=1}^{k} 2r_i < 1$, place the robots $a_1, a_2 \ldots a_k$ such that the distance between robots i and $i+1$ is equal to $\frac{1 - \sum_{i=1}^{k} 2r_i}{k} + r_i + r_{i+1}$ around the circle in the counterclockwise direction and distance between robots k and 1 is $\frac{1 - \sum_{i=1}^{k} 2r_i}{k} + r_k + r_1$ in the counterclockwise direction.
2. For each $i = 1, \ldots, k$ robot a_i moves perpetually counterclockwise around the circle at maximum possible speed 1.

---

within the radius of visibility of some robot is at most $\sum_{i=1}^{k} 2r_i$. Thus the total length of points not being observed at time t is $1 - \sum_{i=1}^{k} 2r_i$. Denote this set of not observed points at time t by U. Take the interval of time

$$J = \left[t, t + \frac{1 - \sum_{i=1}^{k} 2r_i}{\sum_{i=1}^{k} v_i} - \varepsilon \right],$$

for any $\varepsilon > 0$. The interval has length $\frac{1 - \sum_{i=1}^{k} 2r_i}{\sum_{i=1}^{k} v_i} - \varepsilon$ thus the set of all points from set U patrolled by the robot i within interval J has length at most $v_i \frac{1 - \sum_{i=1}^{k} 2r_i}{\sum_{i=1}^{k} v_i} - v_i \varepsilon$. Thus the set of all points from set U patrolled by all robots within interval J has total length at most

$$\sum_{i=1}^{k} v_i \frac{1 - \sum_{i=1}^{k} 2r_i}{\sum_{i=1}^{k} v_i} - \sum_{i=1}^{k} v_i \varepsilon = 1 - \sum_{i=1}^{k} 2r_i - \sum_{i=1}^{k} v_i \varepsilon = |U| - \sum_{i=1}^{k} v_i \varepsilon < |U|.$$

Since within interval J robots are unable to patrol all points from set U thus the idle time of algorithm A is bounded from below by the length of the interval J

$$I_A \geq |J| = \frac{1 - \sum_{i=1}^{k} 2r_i}{\sum_{i=1}^{k} v_i} - \varepsilon.$$

Therefore the claim of the theorem is obtained by passing to the limit $\varepsilon \to 0$. ∎

Corollary 1. *Algorithm A_1 achieves an optimal idle time in the case of robots with equal speeds and possibly different visibilities.*

2.2 Different Speeds

In the case of equal speeds our results are the same as for the problem without visibility on a circle of length $1 - \sum_{i=1}^{k} 2r_i$. It turns out that this is not the case any more for the case of different speeds. Consider the optimal algorithm for two robots without visibilities with speeds $v_1 > v_2$. It is either an algorithm where both robots are at antipodal positions and move with the slower speed v_2 or it is an algorithm where the faster robot goes around the circle with his maximum speed v_1 and the movement of the slower robot irrelevant. It was proven in [8] that the idle time of such algorithm is optimal. But in our problem in the case of different visibilities in some cases neither of these algorithms is optimal. In particular, when one robot is very fast with small visibility radius, while the other robot is slow but it has a large visibility radius, the partition strategy, when both robots zigzag, protecting two interior-disjoint segments of the circle, such strategy may give a better idle time bound that those obtained by the two algorithms mentioned above. It is easy to verify that this is the case for $v_1 = 5$, $r_1 = 1/12$, $v_2 = 1$ and $r_2 = 1/3$.

3   Segment Patrolling

In this section we investigate patrolling of a segment. First we give the optimal patrolling for the case of any number of robots with identical speeds. In the second subsection we give the optimal algorithm for the case of two robots with distinct speeds.

3.1   Equal Speeds

Proposition 2. If $\sum_{i=1}^{k} 2r_i < 1$ then patrolling algorithm A_2 achieves idle time $T = 2\frac{1-\sum_{i=1}^{k} 2r_i}{k}$.

Algorithm A_2 [for k robots with the same speed and different visibility to patrol a segment]

1. If $\sum_{i=1}^{k} 2r_i < 1$, partition the unit segment into k segments, such that the length of the i-th segment s_i equals $\frac{1-\sum_{i=1}^{k} 2r_i}{k} + 2r_i$.
2. For each $i = 1, ..., k$ place robot a_i at the center of the segment s_i.
3. For each $i = 1, ..., k$ robot a_i moves perpetually at maximal speed and changes its direction when being at distance r_i from an endpoint of s_i.

We want to prove optimality of the algorithm A_2. For any patrolling algorithm A^{vis} with visibility, operating on a segment of length 1 we will construct an algorithm $A^{\overline{vis}}$ with no visibility, working on a segment of length $1 - \sum_{i=1}^{k} 2r_i$. The construction will ensure that the idle time of A^{vis} is bigger than or equal to the idle time of $A^{\overline{vis}}$. Then, since it is straightforward to show the optimal algorithm for robots with the same speed and no visibility, we will obtain a desired lower bound.

Take any algorithm A^{vis} for k robots with different visibilities $r_1, r_2, ..., r_k$ and the same speed 1 working on the line segment of length 1. Assume that $\sum_{i=1}^{k} 2r_i < 1$. Denote by $A_i^{vis}(t)$ the position of robot i in time t. Define functions $L_i(t) = A_i^{vis}(t) - r_i$ and $R_i(t) = A_i^{vis}(t) + r_i$ for any robot i and time t. Function $L_i(t)$ denotes the leftmost point visible to robot i in time t, except the case when robot i is too close to the left endpoint 0. Similarly, function $R_i(t)$ denotes the rightmost point visible to robot i in time t, except the case when robot i is too close to the right endpoint 1.

Definition of mobile intervals. A mobile interval is a pair of functions $\langle L(t), R(t) \rangle$ denoting its left and right endpoints respectively. The first step of our construction is the definition of a set of mobile intervals on the line. The goal of the construction is to obtain a dynamic process such that at any moment of time the union of all mobile intervals covers all points that are visible to all robots. We will require that mobile intervals move with speed not exceeding 1 and total length of the intervals is always equal to $\sum_{i=1}^{k} 2r_i$. If at some time t for all $i \in \{1, 2, \ldots, k\}$, the length of the set of points visible to a robot i is equal to $2r_i$, (no robot is too close to the endpoint) and all sets of visible points are disjoint (the visibility regions of two robots are always interior-disjoint), then mobile intervals will simply have k intervals with lengths equal to areas of visibilities. However, for such time moments t when the areas of visibilities of the robots are overlapping then the mobile intervals at time t will cover some points that are not being observed by any robot. Now we will present an algorithm that determines the positions of mobile intervals depending on the positions of robots in the algorithm A^{vis} at any time moment t.

Procedure 1. Construction of mobile intervals

1: $J \leftarrow \{\langle L_i(t), R_i(t), 1 \rangle, i \in 1, 2, \ldots k\}$ ▷ initially intervals are equal to areas of visibilities
2: **while** there exists $\langle L, R, h \rangle \in J$ such that $L < 0$ **do**   ▷ if left endpoint is beyond the point 0
3:      $J \leftarrow J \setminus \{\langle L, R, h \rangle\} \cup \{\langle 0, R - L, h \rangle\}$ ▷ move the interval to the right
4: **end while**
5: **while** there exists $\langle L, R, h \rangle \in J$ such that $R > 1$ **do**   ▷ if right endpoint is beyond the point 1
6:      $J \leftarrow J \setminus \{\langle L, R, h \rangle\} \cup \{\langle 1 - (R - L), 1, h \rangle\}$ ▷ move the interval to the left
7: **end while**
8: **while** there exist $\langle L, R, h \rangle, \langle L', R', h' \rangle \in J$ such that $L < L' \leq R$ **do**   ▷ overlapping intervals
9:      $j \leftarrow \langle L, R + R' - L', h + h' \rangle$ ▷ merge the pair into one long interval
10:     **if** $R + R' - L' > 1$ **then**   ▷ if the new interval would go beyond the endpoint
11:         $j \leftarrow \langle 1 - (R - L) - (R' - L'), 1, h + h' \rangle$ ▷ move it to the left
12:     **end if**
13:     $J \leftarrow (J \setminus \{\langle L, R, h \rangle, \langle L', R', h' \rangle\}) \cup \{j\}$ ▷ replace the pair with newly constructed interval
14: **end while**
15: $J(t) \leftarrow \{j_1, j_2, \ldots, j_l\} \leftarrow$ sort set J according to left endpoints of the intervals
16: **return** $J(t)$

If t is the real variable denoting time then $J(t)$ is a dynamic process in which the mobile intervals are moving on the interval $[0, 1]$. In the following lemma we prove some properties of this process.

Lemma 1. *At any time moment t, the sequence of intervals $J(t)$ returned by the Procedure 1 satisfy*

(1) *the intervals from $J(t)$ cover all points visible to the robots in the algorithm A^{vis} at time t,*
(2) *the intervals from $J(t)$ are interior-disjoint and have total length equal to $\sum_{i=1}^{k} 2r_i$,*
(3) *the velocity of any interval $j \in J(t)$ is either equal to velocity of some robot a or it is equal to 0.*

Thus any mobile interval either moves with the same speed and direction as some robot, or merges with other interval or splits into multiple intervals.

Based on the positions of the intervals we can define positions of the robots in algorithm A^{vis} at any step t.

Definition of the algorithm A^{vis} . Take any time moment t. Based on the positions of robots in the algorithm A^{vis} at this moment we construct the mobile intervals $J(t)$. Consider the output of the Procedure 1, namely the set $J(t) = \{j_1, j_2, \ldots, j_l\}$. Each j_i is a tuple $j_i = \langle L_i, R_i, h_i \rangle$, where L_i is the left endpoint, R_i is the right endpoint and h_i is the number of robots whose areas of visibilities are being covered by interval j_i. Recall that $J(t)$ is sorted thus $L_1 < R_1 < L_2 < R_2 < \cdots < L_l < R_l$. To obtain positions of the robots in A^{vis} we intuitively cut the terrain that is covered by the mobile intervals (see Figure 1). The amount of terrain that is left is $1 - \sum_{i=1}^{k} 2r_i$. Define points $p_1 = L_1$,

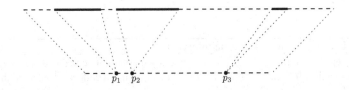

Fig. 1. Positions of robots in the algorithm A^{vis}

$p_i = L_i - \sum_{j=1}^{i-1}(R_j - L_j)$ for $i = 2, 3, \ldots, l$ as the total length of all points to the left of point L_i not covered by any mobile interval. Points p_i will be positions of the robots in the algorithm A^{vis} . Number of robots located at the point p_i will be equal to h_i. Let $s_i = \sum_{j=1}^{i} h_j$ for $i = 1, 2, \ldots, l$. Positions of robots in algorithm A^{vis} are defined as follows.

$$A_a^{vis}(t) = p_1 \quad \text{for } a = 1, 2, \ldots, s_1$$
$$A_a^{vis}(t) = p_2 \quad \text{for } a = s_1 + 1, s_1 + 2, \ldots, s_2$$

$$\vdots$$

$$A_a^{vis}(t) = p_l \quad \text{for } a = s_{l-1} + 1, s_{l-1} + 2, \ldots, s_l$$

This completes the construction of algorithm A^{vis} . We want to prove that algorithm A^{vis} has idle time not bigger than A^{vis}. First we need to show that robots in the algorithm A^{vis} move with speeds not exceeding 1.

Lemma 2. *For any time moment t and robot a_i*

(1) $A_i^{vis}(t) \in [0, 1 - \sum_{i=1}^{k} 2r_i]$,
(2) *speed of robot a_i is at most 1,*
(3) *the trajectory followed by robot a_i is continuous.*

Lemma 2 shows that algorithm A^{vis} is a correct patrolling algorithm for segment of length $1 - \sum_{i=1}^{k} 2r_i$ and robots with the same speed 1. Now we need to show that the transformation does not increase the idle time.

Lemma 3. *The idle time of algorithm A^{vis} is not larger that the idle time of algorithm A^{vis}.*

Theorem 2. *The optimal traversal algorithm for k robots with the same speed and different visibility, patrolling unit segment $S = [0, 1]$ achieves idle time $I_{opt} = 2\frac{1 - \sum_{i=1}^{k} 2r_i}{k}$.*

3.2 Different Speeds

Proposition 3. *Traversal algorithm A_3 achieves idle time $T = 2\frac{1 - 2(r_1 + r_2)}{v_1 + v_2}$.*

Algorithm A_3 [for 2 robots with different speed and visibility to patrol a segment]

1. Partition the unit segment into 2 segments, such that the length of the i-th segment s_i equals $\frac{v_i(1 - 2(r_1 + r_2))}{v_1 + v_2} + 2r_i$.
2. For each $i = 1, 2$ place robot a_i at the center of the segment s_i.
3. For each $i = 1, 2$ robot a_i moves perpetually at maximal speed and changes its direction when being at distance r_i from an endpoint of s_i.

We prove below that the algorithm A_3 is optimal.

Theorem 3. *The optimal traversal algorithm for two robots with different speed and visibility, patrolling unit segment $S = [0, 1]$ achieves idle time $I_{opt} = 2\frac{1 - 2(r_1 + r_2)}{v_1 + v_2}$.*

Proof. We suppose, by contradiction, that there exists an algorithm A with an idle time of $I_A = T - \varepsilon$ for some $\varepsilon > 0$. Without loss of generality, we may assume that $v_1 \le v_2$. Observe that a_1 must see one of the endpoints $(0, 1)$. By symmetry suppose that a_1 sees endpoint 0 at some time t_1. Let $L_1 = r_1$, $R_1 = r_1 + \frac{1 - 2(r_1 + r_2)}{v_1 + v_2}v_1$, $L_2 = 1 - r_2 - \frac{1 - 2(r_1 + r_2)}{v_1 + v_2}v_2$, $R_2 = 1 - r_2$, and $B = R_1 + r_1 = L_2 - r_2$. At time t_1, a_1 is within $[0, L_1]$. Considering the speed of a_1,

$$v_1 \frac{T - \varepsilon}{2} + r_1 < |B - 0|$$

and a_1 cannot see B within time $[t_1 - \frac{T-\varepsilon}{2}, t_1 + \frac{T-\varepsilon}{2}]$. So, a_2 has to see B at some time t_2. We will show that neither a_2 nor a_1 can see 1 within $[t_2 - \frac{T-\varepsilon}{2}, t_2 + \frac{T-\varepsilon}{2}]$. Considering the speed of a_2, $v_2 \frac{T-\varepsilon}{2} + r_2 < |B - 1|$ and a_2 cannot see 1 within the time interval $[t_2 - \frac{T-\varepsilon}{2}, t_2 + \frac{T-\varepsilon}{2}]$. Now, we show that a_1 cannot see 1 neither. Since $|t_1 - t_2| < \frac{T}{2}$, the rightmost point p at which a_1 can be at time t_2 is $p < L_1 + \frac{T}{2}v_1$. We show that from the moment when a_1 sees point p, it is not possible for a_1 to see 1 within time $\frac{T-\varepsilon}{2}$, this would prove that in interval $[t_2 - \frac{T-\varepsilon}{2}, t_2 + \frac{T-\varepsilon}{2}]$, a_1 cannot see 1. The rightmost point q

that a_1 can visit within time $t_2 + \frac{T-\varepsilon}{2}$ is $q < p + \frac{T-\varepsilon}{2}v_1$. Hence, the rightmost point that a_1 can see within time $t_2 + \frac{T-\varepsilon}{2}$ is

$$q + r_1 < L_1 + \frac{T}{2}v_1 + \frac{T-\varepsilon}{2}v_1 = L_1 + Tv_1 - \frac{\varepsilon}{2}v_1,$$

which proves the theorem. ∎

4 Hardness Results

Let us recall the definition of the PARTITION problem.

Instance: Finite set A and size $s(a) \in \mathbb{Z}^+$ for each $a \in A$.
Question: Is there a subset $A' \subset A$ such that $\sum_{a \in A'} s(a) = \sum_{a \in A \setminus A'} s(a)$?

The PARTITION problem remains NP-complete even if we require that $|A'| = |A|/2$ [11]. We will refer to the PARTITION problem with this additional condition as RESTRICTED PARTITION problem.

Theorem 4. *For some graphs the problem of deciding, for any set of robots $a_1, a_2, \ldots, a_k$ with equal speeds and different visibilities, whether there exists a patrolling algorithm with idle time 0 is NP-hard.*

When the idle time is strictly positive the construction of the example showing NP-hardness is more involved.

Theorem 5. *For any fixed I and for some graphs the problem of deciding, for any set of robots $a_1, a_2, \ldots, a_k$ with equal speeds and different visibilities, whether there exists a patrolling algorithm with idle time at most I is NP-hard.*

Proof. Fix any I. Assume that there exists a polynomial algorithm deciding for any set of robots whether it is possible to patrol a graph obtaining idle time at most I. We want to show that it would imply existence of a polynomial algorithm for the RESTRICTED PARTITION problem.

Let a multiset of integers $S = \{x_1, x_2, \ldots, x_{2k}\}$ be an instance from the RESTRICTED PARTITION problem. We construct an instance of patrolling problem consisting of $2k + 1$ robots in the following way. Let the radius of visibility r_i of i-th robot be $r_i = I/4 \sum_{j=1}^{2k} x_i$ for $i = 1, 2, \ldots, 2k$ and let the radius of visibility of $(2k+1)$-st robot (call it a^*) be $r_{2k+1} = I$. We take the following graph H (see Figure 2). The length of interval AB is $I/2$, the length of each interval AA_i is I for $i = 1, 2, \ldots k+1$, the length of each interval BB_i is I for $i = 1, 2, \ldots k+1$ and the radius of each circle is $k/2 + I/4$. We ask if such collection of robots can patrol graph H with idle time at most I.

We will argue that the answer can be yes if and only if the RESTRICTED PARTITION problem has a solution. First observe that the robot a^* has to walk perpetually between nodes A and B. Note that a robot has to visit node A at least once in every interval of time of length I. If it does not visit A in some interval of length I then in this interval some of the nodes $A_1, A_2, \ldots A_{k+1}$ will not be patrolled by any robot. Since the distance between two nodes A_i, A_j ($i \neq j$) is $2I$ thus any robot different from a^*

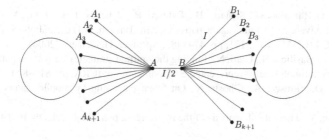

Fig. 2. Graph H

can patrol at most one node among $A_1, A_2, \ldots A_{k+1}$ in time I. And since we have $k+1$ nodes A_i then a^* has to visit node A once every I time steps. Since the same applies to B, robot a^* has to perpetually walk between A and B. To patrol circles we have to use the remaining robots. First note that, if during some time interval of size I $k-1$ or less robots will be on one of the circles then the idle time must be larger than I. It is because $k-1$ robots can patrol in time I intervals of total length at most $(k-1)I + I/2$ (because $I/2$ is an upper bound on the total length of the visibilities) which is less than the length of the circle. Therefore, idle time I will be achieved if and only if the two sums of diameters of visibilities of groups assigned to both circles are the same. But this corresponds exactly to the solution of the RESTRICTED PARTITION problem. ∎

5   Conclusion

In this paper we investigated the problem of mobile robots with visibility patrolling a curve. We gave optimal patrolling algorithms for the case of boundary patrolling and fence patrolling when all robots have the same maximal speed and discussed briefly the case of distinct speeds, thus showing that the boundary patrolling problem for robots with distinct visibility ranges is entirely different than the case of point visibility robots. We also give the optimal algorithm for fence patrolling by two robots with distinct speeds and visibility ranges. If the underlying domain in which the robots operate is a general graph, then the patrolling problem for robots with distinct visibility ranges is shown to be NP-hard; this contrasts sharply with point-visibility robots which has been known to have a polynomial-time solution [9].

There are several open problems but the most interesting class of problems seems to be related to the analysis of approximation as well as online and offline patrolling algorithms for robots with distinct visibilities and/or speeds.

References

1. Agmon, N.: On events in multi-robot patrol in adversarial environments. In: AAMAS, pp. 591–598 (2010)
2. Agmon, N., Fok, C.-L., Emaliah, Y., Stone, P., Julien, C., Vishwanath, S.: On coordination in practical multi-robot patrol. In: ICRA, pp. 650–656 (2012)

3. Almeida, A., Ramalho, G., Santana, H., Tedesco, P., Menezes, T., Corruble, V., Chevaleyre, Y.: Recent advances on multi-agent patrolling. In: Bazzan, A.L.C., Labidi, S. (eds.) SBIA 2004. LNCS (LNAI), vol. 3171, pp. 474–483. Springer, Heidelberg (2004)
4. Amigoni, F., Basilico, N., Gatti, N.: Finding the optimal strategies for robotic patrolling with adversaries in topologically-represented environments. In: ICRA, pp. 819–824 (2009)
5. Chen, K., Dumitrescu, A., Ghosh, A.: On fence patrolling by mobile agents. In: CCCG (2013)
6. Chevaleyre, Y.: Theoretical analysis of the multi-agent patrolling problem. In: IAT, pp. 302–308 (2004)
7. Collins, A., Czyzowicz, J., Gasieniec, L., Kosowski, A., Kranakis, E., Krizanc, D., Martin, R., Ponce, O.M.: Optimal patrolling of fragmented boundaries. In: SPAA, pp. 241–250 (2013)
8. Czyzowicz, J., Gąsieniec, L., Kosowski, A., Kranakis, E.: Boundary patrolling by mobile agents with distinct maximal speeds. In: Demetrescu, C., Halldórsson, M.M. (eds.) ESA 2011. LNCS, vol. 6942, pp. 701–712. Springer, Heidelberg (2011)
9. Czyzowicz, J., Gasieniec, L., Kosowski, A., Kranakis, E., Pajak, D.: Optimal patrolling by mobile agents in arbitrary continuous graphs (in preparation, 2014)
10. Elor, Y., Bruckstein, A.M.: Autonomous multi-agent cycle based patrolling. In: Dorigo, M., et al. (eds.) ANTS 2010. LNCS, vol. 6234, pp. 119–130. Springer, Heidelberg (2010)
11. Garey, M.R., Johnson, D.S.: Computers and Intractability; A Guide to the Theory of NP-Completeness. W. H. Freeman & Co., New York (1990)
12. Hazon, N., Kaminka, G.A.: On redundancy, efficiency, and robustness in coverage for multiple robots. Robotics and Autonomous Systems 56(12), 1102–1114 (2008)
13. Kawamura, A., Kobayashi, Y.: Fence patrolling by mobile agents with distinct speeds. In: Chao, K.-M., Hsu, T.-s., Lee, D.-T. (eds.) ISAAC 2012. LNCS, vol. 7676, pp. 598–608. Springer, Heidelberg (2012)
14. Machado, A., Ramalho, G., Zucker, J.-D., Drogoul, A.: Multi-agent patrolling: An empirical analysis of alternative architectures. In: Sichman, J.S., Bousquet, F., Davidsson, P. (eds.) MABS 2002. LNCS (LNAI), vol. 2581, pp. 155–170. Springer, Heidelberg (2003)
15. Pasqualetti, F., Durham, J.W., Bullo, F.: Cooperative patrolling via weighted tours: Performance analysis and distributed algorithms. IEEE Transactions on Robotics 28(5), 1181–1188 (2012)
16. Pasqualetti, F., Franchi, A., Bullo, F.: On cooperative patrolling: Optimal trajectories, complexity analysis, and approximation algorithms. IEEE Transactions on Robotics 28(3), 592–606 (2012)
17. Portugal, D., Rocha, R.: A survey on multi-robot patrolling algorithms. In: Camarinha-Matos, L.M. (ed.) DoCEIS 2011. IFIP AICT, vol. 349, pp. 139–146. Springer, Heidelberg (2011)
18. Portugal, D., Rocha, R.P.: Multi-robot patrolling algorithms: examining performance and scalability. Advanced Robotics 27(5), 325–336 (2013)
19. Smith, S.L., Schwager, M., Rus, D.: Persistent robotic tasks: Monitoring and sweeping in changing environments. IEEE Transactions on Robotics 28(2), 410–426 (2012)
20. Yanovski, V., Wagner, I.A., Bruckstein, A.M.: A distributed ant algorithm for efficiently patrolling a network. Algorithmica 37(3), 165–186 (2003)

Distributed Barrier Coverage with Relocatable Sensors*

Mohsen Eftekhari[1], Paola Flocchini[2], Lata Narayanan[1],
Jaroslav Opatrny[1], and Nicola Santoro[3]

[1] Dept. of Comp. Science and Soft. Eng., Concordia University, Montréal, Canada
m_eftek@encs.concordia.ca, {lata,opatrny}@cs.concordia.ca
[2] School of El. Eng. and Computer Science, University of Ottawa, Ottawa, Canada
flocchin@site.uottawa.ca
[3] School of Computer Science, Carleton University, Ottawa, Canada
santoro@scs.carleton.ca

Abstract. A wireless sensor can detect the presence of an intruder in its sensing range, and is said to cover the portion of a given barrier that intersects with its sensing range. Barrier coverage is achieved by a set of sensors if every point on the barrier is covered by some sensor in the set. Assuming n identical, anonymous, and relocatable sensors are placed initially at arbitrary positions on a line segment barrier, we are interested in the following question: under what circumstances can they independently make decisions and movements in order to reach final positions whereby they collectively cover the barrier? We assume each sensor repeatedly executes Look-Compute-Move cycles: it looks to find the positions of sensors in its visibility range, it computes its next position, and then moves to the calculated position. We consider only oblivious or memoryless sensors with sensing range r and visibility range $2r$ and assume that sensors can move at most distance r along the barrier in a move. Under these assumptions, it was shown recently that if the sensors are fully synchronized, then there exists an algorithm for barrier coverage even if sensors are unoriented, that is, they do not distinguish between left and right [7]. In this paper, we prove that orientation is critical to being able to solve the problem if we relax the assumption of tight synchronization. We show that if sensors are unoriented, then barrier coverage is unsolvable even in the semi-synchronous setting. In contrast, if sensors agree on a global orientation, then we give an algorithm for barrier coverage, even in the completely asynchronous setting. Finally, we extend the result of [4] and show that convergence to barrier coverage by unoriented sensors in the semi-synchronous model is possible with bounded visibility range $2r + \rho$ (for arbitrarily small $\rho > 0$) and bounded mobility range r.

Keywords: sensor networks, barrier coverage, distributed algorithms.

1 Introduction

1.1 The Problem

A wireless sensor network consists of several sensors, each equipped with a sensing module. Among the many applications of sensor networks (e.g., [15]), the establishment

---

* Research supported by NSERC, Canada.

M. Halldórsson (Ed.): SIROCCO 2014, LNCS 8576, pp. 235–249, 2014.

of *barrier coverage* has an important place, and it has been studied intensively in the literature; it guarantees that any intruder attempting to cross the perimeter of a protected region (e.g., crossing an international border) is detected by one or more of the sensors (e.g., see [1, 2, 5, 6, 11, 16, 18]). By protecting the access to the region, barrier coverage provides a less expensive alternative to a complete coverage of the region (e.g., [18]). A barrier can be modelled as a line segment of length $L \in Z$ covering the interval $[0, L]$ on the x-axis; sensors are deployed along the barrier. Intruders may traverse the line segment at any point; an intruder is detected only if it is within the sensing range r of at least one sensor. The barrier is *covered* if no intruder can cross the line segment without being detected. Clearly, at least $\bar{n} = \lceil \frac{L}{2r} \rceil$ sensors are needed, where r is the sensing range.

Barrier coverage, in the case of *static sensors*, can be achieved by careful (i.e., non ad hoc) deterministic deployment of $\bar{n}$ sensors, but this could be unfeasible in some situations. Alternatively, a large number $N \gg \bar{n}$ of sensors can be randomly deployed, but barrier coverage can only be probabilistically guaranteed [11–13]. Finally, in ad hoc deployment of sensors, the sensors are initially located at *arbitrary* positions on the line. In sensor networks composed of *relocatable sensors*, every sensor has a movement module that enables the sensor to move along the barrier. Hence, although initially they are located at arbitrary positions on the line without providing barrier coverage, they may move to new points on the line so that the entire barrier is covered (e.g., [3, 5–7, 17]). In this paper we study the problem of barrier coverage with relocatable sensors.

The *centralized* version of the problem has been studied and solutions proposed, focusing on minimizing some cost measures (e.g., traveled distances) [3, 5, 6, 14]. In these centralized solutions, the algorithm knows the initial positions of all sensors, and uses this information to determine the final positions that the sensors should occupy; notice that $\bar{n}$ sensors suffice for a centrally directed relocation of sensors. However, in the context of sensor networks deployed in an ad hoc manner, typically there is no central control or authority, and no global knowledge of the locations of the sensors is available. Indeed, the sensors might not even know the total number of sensors deployed, or the length of the barrier. Thus every sensor must make decisions on whether and where to move, based only on local information in an autonomous and decentralized way.

In order to develop a solution protocol for a *distributed* setting, it is first of all necessary to model such a setting. Following the approach used in the research on autonomous mobile robots (e.g., [10]), sensors are modelled as mobile computational entities. The entities are anonymous and identical, have no centralized coordination, have a sensing range as well as a visibility[1] range: their decisions are made solely based on their observations of their surroundings. Each entity alternates activity with inactivity. When becoming active, it executes a Look-Compute-Move operational cycle and then becomes inactive. In a cycle, an entity determines the positions of the other entities in its visibility range (Look); then it computes its own next position (Compute); and finally it moves to this new position (Move). In the cases of sensor networks, the visibility range v is limited [9]; we assume $v = 2r$, which is the minimum visibility radius necessary for sensors to determine local gaps in coverage. The movements of the sensors are said

---

[1] Combined with mobility, it provides stigmergic communication between sensors within range.

to be *bounded* if there is a maximum distance they can move in each cycle, and *rigid* if they are not interrupted (e.g., by an adversary).

Depending on the assumptions on the activation schedule and the duration of the cycles, three main settings are identified. In the *fully-synchronous* setting (FSYNC), all sensors are activated simultaneously, and each cycle is instantaneous. The *semi-synchronous* setting (SSYNC) is like the fully synchronous one except that each activation might involve only a subset of the sensors; activations are fair: each sensor will be activated infinitely often. In the *asynchronous* setting (ASYNC), no assumption is made on timing of activation, other than fairness, nor on the duration of each computation and movement, other than it is finite.

The first *distributed* algorithmic investigation of the barrier coverage problem has been recently presented for the discrete line [7], solving the problem in the fully synchronous setting, FSYNC. Interestingly, it is shown that the sensors can be totally *oblivious*, that is, at the beginning of a cycle, a sensor does not (need to) have any recollection of previous operations and computations. Furthermore, the sensors are completely unoriented; they have no concept of left and right. Finally the algorithm terminates for any $n \geq \bar{n}$, hence even with the minimal number used by centralized solutions.

Notice that when $L/2r$ is an integer and $n = \bar{n}$, the barrier coverage problem is equivalent to the *uniform deployment* problem (studied for lines and circles, see [4, 8, 9]) on a line segment, which requires the oblivious sensors to move to equidistant positions between the borders of the segment. This problem has been studied on a line [4] assuming that a sensor can always see the sensors that are closest to it, regardless of their distance, and it always reaches its destination, regardless of its distance; in other words, both visibility and movements are a priori unbounded. Under these assumptions, an SSYNC distributed protocol that *converges* with rigid movements to uniform covering (and thus to barrier coverage) was given in [4]. However, equidistant positions are not required for barrier coverage when $n > L/2r$.

1.2 Main Contributions

In this paper we first of all investigate under what conditions $\bar{n}$ oblivious sensors can actually achieve barrier coverage in the complex semi-synchronous and asynchronous settings, without requiring unbounded visibility or mobility range.

We prove that a crucial factor for solvability of the barrier coverage problem is whether the network is *oriented* or *unoriented*. In an oriented network, each sensor has a notion of "left-right", and this notion is globally consistent; in a unoriented network, sensors have no "left-right" direction.

In particular, we prove that the problem is *unsolvable* by $\bar{n}$ oblivious sensors in SSYNC (and thus ASYNC) if the network is unoriented. The result holds even if all movements are rigid. On the other hand, we prove that, if the network is oriented, the problem is solvable even in ASYNC and even if movements are not rigid (i.e., they can be interrupted by an adversary). The proof is constructive: we present an ASYNC protocol that allows any $n \geq \bar{n}$ oblivious sensors to achieve barrier coverage within finite time and terminate, even if movements are non-rigid. In other words, we show that, with orientation, it is possible to achieve barrier coverage in a totally local and decentralized way, asynchronously, obliviously, and with movements interruptible by an adversary;

furthermore, this is achievable with the same number of sensors of the optimal totally centralized solution with global knowledge of all parameters.

We also show that allowing a slightly larger visibility range (e.g., $v = 2r + \rho$ for an arbitrary small ρ), $\bar{n}$ unoriented and oblivious sensors can converge with rigid movements to barrier coverage in SSYNC, extending the result of [4] to fixed limited visibility and bounded movements.

2   Model and Notation

We model the barrier with a line segment of length $L \in Z$ covering the interval $[0,L]$ on the x-axis. A sensor network consists of a set of n sensors $\{s_1, s_2, \ldots, s_n\}$ located on the segment.

A sensor is modelled as a computational entity capable of moving along the segment; it is equipped with a sensing module and a visibility module. A sensor can sense an intruder if and only if it lies within the sensor's sensing range; it can see another sensor if and only if it it lies within the sensor's visibility range. In this paper, we assume that all sensors have the same sensing range r and the same visibility range v.

Sensors are autonomous, anonymous and identical (i.e., without central authority, distinct markers or identifiers); they all execute the same algorithm. Sensors are said to be *oriented* if and only if all sensors agree on a global left and right; they are called *unoriented* if they do not have a sense of left and right.

Let s_i^t denote sensor s_i at time t located at x_i^t. We assume that for every sensor $r \leq x_i^0 \leq L - r$, and that for $i \neq j$, we have $x_i^0 \neq x_j^0$. For convenience, we assume that $x_1^0 < x_2^0 \cdots < x_n^0$. We emphasize that while these names and positions of sensors facilitate our proofs, they are not known to any of the sensors. In addition, we assume there are two special sensors s_0 and s_{n+1} that are immobile, and are always located at $-r$ and $L + r$; while these special sensors do not require any sensing capabilities or visibility, the other sensors in the network cannot distinguish these special sensors from any other sensors; the entire set of sensors is denoted by $S = \{s_0, s_1, \ldots, s_n, s_{n+1}\}$.

The sensors can be *active* or *inactive*. When *active*, a sensor performs a *Look-Compute-Move* cycle of operations: the sensor first observes the portion of the segment within its visibility range obtaining a snapshot of the positions of the sensors in its range at that time (*Look*); using the snapshot as an input, the sensor then executes the algorithm to determine a destination point (*Compute*); finally, the sensor moves towards the computed destination, if different from the current location (*Move*). After that, it becomes *inactive* and stays idle until the next activation. Sensors are *oblivious*: when a sensor becomes active, it does not remember any information from previous cycles.

A move is said to be *non-rigid* if it may be stopped by an adversary before the sensor reaches its destination; the only constraint on the adversary is that, if interrupted before reaching its destination, a robot moves at least a minimum distance $\delta > 0$ (otherwise, no destination can ever be reached). If no such an adversary exists, the moves are said to be *rigid*.

A sensor can detect the presence of an intruder in its sensing range r, and is said to *cover* the portion of the segment within its sensing range; therefore the *coverage length* of a sensor is $2r$. *Barrier coverage* is achieved if every point on the segment is covered by some sensor. An *overlap* is a maximal interval on $[0, L]$ such that every point in the interval is within the sensing range of more than one sensor. A *coverage gap* is a maximal interval of the segment where no point is within the sensing range of any sensor. We say that ε-*approximate barrier coverage* is achieved if the length of any coverage gap is $\leq \varepsilon$.

The goal of an algorithm for barrier coverage is to move sensors to final positions so that the entire barrier is covered. Observe that if $2rn > L$, then the final positions are not necessarily equidistant. We say an algorithm $\mathcal{A}$ for barrier coverage *terminates* on input S at time t if and only if when running $\mathcal{A}$ on S, no sensor in S moves at any time $t' \geq t$. We say that algorithm $\mathcal{A}$ *solves* the barrier coverage problem if there is a time t at which the algorithm terminates on any input S and barrier coverage is achieved. We say an algorithm $\mathcal{A}$ *converges* to barrier coverage on input S if and only if for any $\varepsilon > 0$ there is a time t such that at any time $t' \geq t$ the size of any coverage gap is at most ε. We say that algorithm $\mathcal{A}$ solves the ε-*approximate* barrier coverage problem for $\varepsilon > 0$ if and only if it converges on any input S.

Unless specified otherwise we assume $v = 2r$, which is the minimum visibility radius necessary for sensors to determine local gaps in coverage. More precisely, sensor s_i^t is able to see all other sensors located in $[x_i^t - 2r, x_i^t + 2r]$. For convenience, we say s_i^t sees s_j^t on its right if and only if $0 < x_j^t - x_i^t \leq 2r$ and s_i^t sees s_k^t on its left if and only if $0 < x_i^t - x_k^t \leq 2r$. Observe that a sensor is able to detect when its sensing area overlaps with another sensor's sensing area.

Note that in our figures, each sensor is represented by a rectangle which shows the interval that the sensor covers on the line barrier. Also for convenience, two sensors whose coverage lengths overlap are placed at different levels in the illustration; however in our assumptions, all sensors have circular sensing area and are initially placed on the barrier and can only move on the barrier.

3   Impossibility without Orientation

In this section we consider the case where sensors are unoriented. We show that there is no algorithm for barrier coverage in the SSYNC model with $\bar{n}$ sensors.

We give an adversary argument, by creating input arrangements and activation schedules that force any algorithm in the SSYNC model to either not terminate, or terminate without coverage. All movements will be assumed to be rigid; a sensor can always reach the destination it has computed. We focus on three types of sensors (see Figure 1): (a) sensors that have an overlap on one side, and a gap on the other side, (b) sensors that are attached to the next sensor on one side and a gap on the other side and (c) sensors that have an overlap on one side and are attached to the next sensor on the other side. Any algorithm for barrier coverage must specify rules for movement in each of these situations. Note that with $2r$ visibility range, sensors can only determine whether there exists a gap with a neighboring sensor but cannot determine anything about the length of such a gap. Thus, the magnitude of the movement of a sensor can only be a function of an overlap, if any, with a neighboring sensor, and cannot be a function of the length of an adjacent gap. We show that there exist arrangements and activation schedules for the sensors that defeat all possible combinations of these rules.

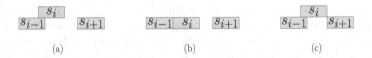

(a)   (b)   (c)

Fig. 1. The three types of sensors under consideration

First we study the behavior of a sensor s_i with $1 \leq i \leq n$ that has an overlap of e with the sensor on its left, and has a gap on its right, as in Figure 1(a). We show that such a sensor must move right; if the gap is at least as big as the overlap, the sensor must eventually move so as to exactly remove the overlap, and if the gap is smaller than the overlap, the sensor must move at least enough distance to remove the gap.

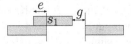

Fig. 2. Arrangement for proof of Lemma 1; $n = 1$

Lemma 1. *Consider an algorithm $\mathcal{A}$ for barrier coverage in SSYNC model and a sensor s_i^t with $dist(s_{i-1}^t, s_i^t) = 2r - e$ and $dist(s_i^t, s_{i+1}^t) = 2r + g$, with $e, g > 0$. If s_{i-1} and s_{i+1} are deactivated and only s_i is activated, there exists a time step $t' > t$ such that:*

(a) $x_i^{t'} = x_i^t + e$ if $g \geq e$ and
(b) $x_i^t + g \leq x_i^{t'} \leq x_i^t + e$ if $g < e$.

Proof. First we observe that the sensor s_i must eventually move at least distance $min(g, e)$ to the right. If not, the algorithm $\mathcal{A}$ does not terminate with barrier coverage on the arrangement shown in Figure 2, since s_1 is the only sensor that can move in the arrangement. Next we show that $x_i^{t'} \leq x_i^t + e$ for some $t' > t$. For the sake of contradiction, assume that there is a value of overlap e, such that according to $\mathcal{A}$, sensor s_i moves more than e; that is s_i moves $e + a$ to the right, with $a > 0$. Then we can construct an activation schedule such that $\mathcal{A}$ never terminates on the input shown in Figure 3. Choose $n = \lceil a/e \rceil$. A single sensor is activated in each step. Starting with configuration C_1, the sensors s_n to s_1 are activated in consecutive steps, yielding configurations $C_2, C_3, \ldots C_{n+1}$ in turn, and then the sequence of activations is reversed. It is easy to verify that at the end of the activation schedule, the initial arrangement C_1 is obtained again. The schedule can be repeated *ad infinitum*, forcing non-termination of the algorithm. It follows that in the case when $g \geq e$, whatever the overlap e with s_{i-1}, we can force s_i to move exactly e to the right.

Next we consider the behavior of a sensor s_i that is attached to its neighbor on its left, and has a gap on its right as in Figure 1(b). We activate s_i and keep s_{i-1} and s_{i+1} deactivated. If s_i moves left, it creates an overlap with s_{i-1} and by Lemma 1(a), it will eventually move to the right to remove that overlap, and return to the same position. Alternatively, s_i may not move at all, or may move to the right. If it moves to the right,

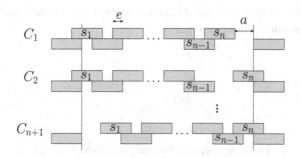

Fig. 3. Arrangement for proof of Lemma 1; $n = \lceil a/e \rceil$

since it does not know the distance of the gap with s_{i+1} and has no overlap with s_{i-1}, it can only move a fixed constant distance, say b. The lemma below is a consequence of the preceding discussion.

Lemma 2. *Let $\mathcal{A}$ be an algorithm for barrier coverage and s_i be a sensor with* $dist(s_{i-1}^t, s_i^t) = 2r$ *and* $dist(s_i^t, s_{i+1}^t) > 2r$. *If s_{i-1} and s_{i+1} are both kept deactivated and s_i is activated, there exists a time $t' > t$ such that $x_i^{t'} = x_i^t + h$ with $h \geq 0$.*

Finally, we consider the behavior of a sensor s_i that has an overlap e with s_{i-1} and is attached to sensor s_{i+1}, as shown in Figure 1(c). As before, we activate only s_i and keep both s_{i-1} and s_{i+1} deactivated. If s_i moves left, it creates a gap with s_{i+1}. By Lemma 1(b), s_i must eventually move right, either returning to its initial position, or moving further right. If it moves right by more than the value of the overlap, then it creates a gap to its left, and once again by Lemma 1(b) , it must move back left until the gap is removed. If for all values of the overlap, s_i makes a move to the right that does not eliminate the overlap, then we show below that the algorithm cannot achieve barrier coverage, leading to the conclusion that there must exist some value of overlap such that such a sensor will either not move, or move to exactly eliminate the overlap.

Lemma 3. *Consider an algorithm $\mathcal{A}$ for barrier coverage. There exists an overlap c with $0 < c < 2r$ such that for any sensor s_i with $dist(s_{i-1}^t, s_i^t) = 2r - c$ and $dist(s_i^t, s_{i+1}^t) = 2r$, if s_i is the only one of $\{s_i, s_{i-1}, s_{i+1}\}$ to be activated, there exists a time step $t' > t$ such that either $x_i^{t'} = x_i^t + c$ (s_i moves right to exactly eliminate the overlap) or $x_i^{t'} = x_i^t$ (s_i returns to the same position).*

Proof. Assume the contrary. By the discussion preceding the lemma, we can conclude that for any overlap e, there exists a time step t' such that $x_i^{t'} = x_i^t + d$ with $0 < d < e$. Consider the arrangement of sensors shown in Figure 4. We first activate s_1 until it moves distance d to the right. By assumption, there remains an overlap of $e - d$ between s_0 and s_1, and now there is an overlap of d between s_1 and s_2. We now keep s_1 deactivated, and activate s_2. Lemma 1 implies that sensor s_2 eventually moves exactly d to the right and eliminates the overlap completely. Observe that at this point, the arrangement repeats with only a different value of overlap. The new value of the overlap between s_0 and s_1 is strictly greater than zero and and the distance between s_1 and s_2

is exactly $2r$. This activation schedule can be repeated *ad infinitum*, and algorithm $\mathcal{A}$ never terminates with barrier coverage.

Fig. 4. Arrangement for proof of Lemma 3; $n = 2$

We proceed to prove our main result:

Theorem 1. *Let $s_1, s_2, \ldots, s_n$ be n sensors with sensing range r initially placed at arbitrary positions on a line segment. If the sensors are unoriented and have visibility radius $2r$, there is no algorithm for barrier coverage in the SSYNC model.*

Proof. Consider the arrangement of sensors shown in Figure 5 with c chosen as in Lemma 3. If the value of h as specified in Lemma 2 is zero, then choose $b = c$, otherwise, choose $b = h$, and fix $n = 1 + 2\lceil b/c \rceil$. We create an activation schedule with three phases with a different set of sensors being activated in each phase, such that the sensors return to arrangement C_1 at the end of each phase. At each phase we only activate a subset of sensors and all other sensors are kept deactivated. We first activate only the sensor s_1. By Lemma 2, there is a future time step when either s_1 is in the same position (if $h = 0$), or it moves distance b to the left to yield arrangement C_2. In the second case, since sensors are unoriented, it will subsequently return to arrangement C_1. In the second phase, we activate only the sensors $\{s_3, s_5, \ldots, s_n\}$. By Lemma 3, there is a future time when either these sensors return to arrangement C_1, or they have moved right by a distance c to reach arrangement C_3. In the second case, they will eventually return to arrangement C_1. In the third phase, we activate only the set of sensors $\{s_2, s_4, \ldots, s_{n-1}\}$. Using the same logic, they will return to arrangement C_1, possibly via arrangement C_4. Observe that all sensors have been activated at least once during the schedule. By repeating the above schedule *ad infinitum*, we can force sensors to repeatedly return to the arrangement C_1, thus completing the proof.

Since an adversary in the ASYNC model has at least the power it has in the SSYNC model, obviously the impossibility result also holds for the ASYNC model.

4   Possibility with Orientation

In this section, we present and analyze an algorithm, ORIENTED SENSORS for barrier coverage by any $n \geq \bar{n}$ oblivious oriented sensors in the ASYNC model; that is, all sensors agree on left and right, but are completely asynchronous.

We proceed to prove the correctness of algorithm ORIENTED SENSORS. A *collision* occurs if two distinct sensors move to exactly the same position. Since sensors are identical and anonymous, from the time a collision of two sensors happens, they cannot be

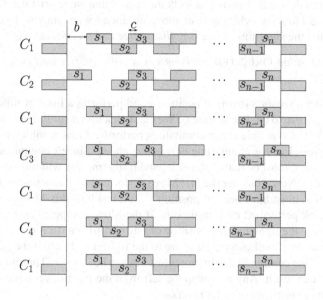

Fig. 5. Arrangement for proof of Theorem 1; $n = 1 + 2\lceil b/c \rceil$

Algorithm 1. ORIENTED SENSORS

Algorithm for sensor $s_i \in S$

$\varepsilon \leq r$ is a fixed positive (arbitrarily small) constant

if s_{i-1} is not visible to s_i (there is a gap to its left) **then**

    s_i moves distance r to the left.

else

    $a := 2r - dist(s_{i-1}, s_i)$ (amount of overlap with previous sensor's range)

    if $dist(s_i, s_{i+1}) \geq 2r$ (no overlap from right) and $a > 0$ **then**

        s_i moves distance $min(r - \varepsilon, a)$ to its right.

    else

        do nothing

    end if

end if

---

distinguished and will behave exactly the same if they have the same activation schedule. Therefore a collision is fatal for a barrier coverage algorithm, and must be avoided by the algorithm designer. This is precisely the reason that we restrict the distance of a move to the right to $r - \varepsilon$, while sensors move distance r when moving to the left. We show below that the algorithm above is *collision-free* and *order-preserving*.

Lemma 4. *Algorithm* ORIENTED SENSORS *is a collision-free and order-preserving protocol.*

Proof. Consider a sensor s that is at position x and performs a Look at time t_1 and the corresponding Move to the *left* at time t_2. We claim that no sensor s' that is at a position $x' < x$ (to the left of s) at time t_1 can compute or perform a Move resulting in a collision or an order reversal with s at any time between t_1 and t_2. Since s computes a Move to the left at time t_1, it must be that $x' < x - 2r$, and furthermore, s will move to a position $\geq x - r$ at time t_2. Now, consider the *last* Move performed by s' at a time $\leq t_1$. Observe that the sensor s' must have been at position x' as a result of this Move. Consider the subsequent Look performed by s' at time t_3. If the Move computed as a result of this Look is a move to the right, the next position of s' is $\leq x' + r - \varepsilon < x - r$. Any subsequent Look performed by s' will compute a move to the right if and only if the position of s' is $\leq x - 2r$ and the computed destination must always be $< x - r$. Thus no collision or order reversal can result. Any moves to the left from the positions reachable by s' can clearly not cause collisions or order reversals.

Next we show that no sensor s' that is at a position $x' > x$ (to the right of s) at time t_1 can compute or perform a Move resulting in a collision or order reversal with s at any time between t_1 and t_2. Clearly, if $x' > x + r$, any move to the left can only bring it to a position $> x$. Suppose $x < x' \leq x + r$. If s' performs a Look after time t_1, then it can see s in its visibility range and therefore would not perform a Move to the left. So s' must have performed a Look at a time $t_3 < t_1$. For s' to have computed a move to the left, the position of s at time t_3 must have been $< x' - 2r$. As argued above, s cannot have subsequently arrived at position x at time t_1.

A similar argument shows that for a sensor s that is at position x and performs a Look at time t_1 and the corresponding Move to the *right* at time t_2, neither a sensor on its left nor a sensor on its right can compute a move resulting in a collision or order reversal with s.

Next we show that there is a time after which no sensors will move left, and after this time, the sensors provide contiguous coverage of some part of the barrier including the sensor s_0.

Lemma 5. *For every sensor $s_i \in S - \{s_{n+1}\}$ there is a time t_i such that s_i never moves left at any time after t_i. Furthermore, there is no coverage gap between s_0 and s_i at any time after t_i.*

Proof. We prove the claim inductively. Clearly it is true for s_0. Suppose there is a time t_i such that s_i never moves left at any time after t_i, and there is no gap between s_0 and s_i at any time after t_i. Consider any Look of s_{i+1} after time t_i. If there is a gap between s_i and s_{i+1}, then s_{i+1} will move at least δ towards s_i. Let t_{i+1} be the time of the first Look of s_{i+1} after time t_i when there is no gap between s_i and s_{i+1}. If there is an overlap with

s_i, then s_{i+1} will move right, but observe that this Move can never create a gap between s_i and s_{i+1} since s_i does not move left by the inductive assumption, and s_{i+1} moves right by at most the amount of the overlap. It follows that after time t_{i+1}, the sensor s_{i+1} will never move left, and furthermore, there is no gap in coverage between s_0 and s_{i+1}.

After time t_n, then, none of the sensors moves left, and furthermore there is no coverage gap between s_0 and s_n. The next two lemmas show that after this time, a sensor moves right under some circumstances, but can only move a finite number of times.

Lemma 6. *Assume s_i and s_{i+1} have an overlap of e at some time after t_n. Then for any j with $i+1 \leq j \leq n$, if the sensors s_{i+1} to s_j are in attached position, and there is no overlap between s_j and s_{j+1}, then sensor s_j will eventually move at least $min(\delta, e)$ to the right.*

Proof. Let $t > t_n$ be a time when s_{i+1} performs a Look and s_i and s_{i+1} have an overlap of e. Clearly s_{i+1} will move right in the corresponding Move, creating an overlap between s_{i+1} and s_{i+2}. Inductively it can be seen that when s_{j-1} moves to the right, it creates an overlap with s_j, causing s_j to move at least $min(\delta, e)$ to the right.

Lemma 7. *Every sensor makes a finite number of moves to the right after time t_n.*

Proof. We give an inductive proof. Clearly this is true for sensor s_0. Suppose sensor s_i has an overlap of e with sensor s_{i-1} at time t_n. Observe that s_{i-1} cannot move until and unless this overlap is removed. Since every time s_i moves to the right, it reduces this overlap by at least $min(e, \delta)$, it is clear that s_i can make at most $\lceil e/\delta \rceil$ moves to the right. If these moves remove the overlap, then s_i may move again only if s_{i-1} subsequently moves to the right and creates an overlap with s_i. Assuming inductively that s_{i-1} makes a finite number of moves to the right, we conclude that sensor s_i moves to the right a finite number of times.

The above lemmas lead to the following theorem:

Theorem 2. *Let $s_1, s_2, \ldots, s_n$ be $n \geq \bar{n}$ sensors with sensing range r initially placed at arbitrary positions on a line segment. If the sensors have the same orientation and visibility radius of $2r$, Algorithm ORIENTED SENSORS always terminates with the barrier fully covered in the ASYNC model.*

Proof. Lemma 5 assures that after time t_n, no sensor moves left, and there is no coverage gap between sensors s_0 and s_n. It follows from Lemma 7 that there is a time, say $t' > t_n$, after which no sensor will move right. However, if there is a gap between s_n and s_{n+1} at time t', since there are enough sensors to cover the barrier, there must be an overlap between two sensors s_i and s_{i+1} for some $0 \leq i < n$. But Lemma 6 implies that the sensor s_n must eventually move to the right, a contradiction. It follows that after time t', there is no gap between s_n and s_{n+1} and therefore no gap between any sensors in S, that is, Algorithm ORIENTED SENSORS terminates with barrier coverage.

5 On Visibility and Convergence

We have seen that, without orientation, barrier coverage with $\bar{n}$ sensors is impossible even in SSYNC (Theorem 1). Observe that the impossibility proof holds when the visibility range is precisely $2r$. So the question naturally arises of what happens in SSYNC if the visibility range is larger.

It is known that in SSYNC, it is possible for $\bar{n}$ sensors to *converge* with rigid movements to equidistant positions *if* a sensor can always see the sensors that are closest to it, regardless of their distance (thus without a priori restrictions on the visibility range) and it can move to destination regardless of its distance (thus without a priori restrictions on the mobility range) [4]. In our setting these conditions do not hold. In this section, we show how that result can be extended to our setting. In fact, we prove that $\bar{n}$ oblivious sensors can converge with rigid movements to barrier coverage in SSYNC if $v = 2r + \rho$, where ρ is an arbitrarily small positive constant; furthermore they can do so with rigid movements of length at most r.

Consider Algorithm CONVERGENT COVERAGE shown below; it operates by first removing all visibility gaps within finite time, and then behaving as the algorithm of [4].

Algorithm 2. CONVERGENT COVERAGE

Algorithm for sensor $s_i \in S$
if only one sensor $s_j \in \{s_{i+1}, s_{i-1}\}$ is visible to s_i and $d = dist(s_i, s_j) < 2r$ **then**
     s_i moves distance $\frac{2r-d}{2} + \frac{\rho}{2}$ away from s_j.
else
     if both s_{i+1}, s_{i-1} are visible. **then**
         if $d_1 = dist(s_{i-1}, s_i) < d_2 = dist(s_{i+1}, s_i)$
         (resp. $d_1 = dist(s_{i+1}, s_i) < d_2 = dist(s_{i-1}, s_i)$) **then**
             s_i moves $\frac{d_2 - d_1}{2}$ toward s_{i+1} (resp. toward s_{i-1}).
         end if
     end if
end if

Lemma 8. *If $s_j \in \{s_{i+1}, s_{i-1}\}$ is in the visibility range of s_i at time t, for any time $t' > t$, s_j is still in the visibility range of s_i.*

Proof. According to the algorithm, a movement is performed by s_i in a cycle only in two situations:
Case 1: Only one sensor s_j is visible to s_i and $d = dist(s_i, s_j) < 2r$. The worst case is when also s_j is activated in this cycle and it sees only s_i. In this case, both sensors move at most $\frac{2r-d}{2} + \frac{\rho}{2}$ away from each other. After the movement we have that $dist(s_i, s_j)$ has become: $dist(s_i, s_j) = 2(\frac{2r-d}{2} + \frac{\rho}{2}) + d = 2r + \rho$. So, sensors s_i and s_j are still within visibility.
Case 2: Both s_{i+1} and s_{i-1} are visible to s_i. Let, without loss of generality, $d_1 = dist(s_j, s_i) < d_2 = dist(s_k, s_i)$ where $s_i, s_k \in \{s_{i-1}, s_{i+1}\}$. The worst case is when s_j is

also activated and it does not see the other neighbouring sensor. In this case s_i moves at most $\frac{d_2-d_1}{2}$ toward s_k, and s_j moves at most $\frac{2r-d}{2} + \frac{\rho}{2}$ away from s_i. After the movement, we have that $dist(s_i,s_j)$ has increased as follows:

$$dist(s_i,s_j) = \frac{d_2 - d_1}{2} + d_1 + \frac{2r - d}{2} + \frac{\rho}{2} = \frac{d_2}{2} + r + \rho < 2r + \rho$$

So, sensors s_i and s_j are still within visibility.

Lemma 9. *Within finite time there will be no visibility gaps.*

Proof. By Lemma 8, visibility is never lost once gained. Consider the visibility gaps. After each activation of a sensor next to a visibility gap, the size of that visibility gap is reduced by at least $\frac{\rho}{2}$. As a consequence, within a finite number of activations, all visibility gaps will be eliminated and all sensors will be within visibility to their neighbours.

Lemma 10. *If at time t there are no visibility gaps, within finite time all coverage gaps will be of size at most ε, for any $\varepsilon > 0$.*

Proof. If there are no visibility gaps at time t then, by Lemma 8, for all $t' \geq t$ there will be no visibility gaps. Hence at all times $t' \geq t$ each sensor s_i when active sees its two neighbours s_{i-1} and s_{i+1}; furthermore, since the distance between two neighbours is at most $2r$, the computed destination of a robot is at at most at distance r. Notice that at this point the algorithm behaves exactly as the protocol of [4]. Since the conditions for its correct behaviour, visibility of neighbours and reaching destination are met, the lemma follows.

By Lemmas 9 and 10, and by the definition of approximate barrier coverage, the claimed result immediately follows:

Theorem 3. *Let $s_1, s_2, \ldots, s_n$ be n sensors with sensing range r initially placed at arbitrary positions on a line segment. If the sensors have no orientation and visibility radius $2r + \rho$, there is an algorithm for ε-approximate barrier coverage in SSYNC with rigid movements of length at most r.*

6   Conclusions

The results of this paper provide a first insight into the nature of the complexity and computability of distributed barrier coverage problems. Not surprisingly, it poses many new research questions. Here are some of them.

We have shown that barrier coverage is unsolvable in SSYNC with $\bar{n}$ unoriented sensors, but solvable in ASYNC with oriented sensors. Oriented sensors have a globally consistent sense of "left-right" while unoriented sensors have no sense of "left-right". Hence the first immediate question is whether something weaker than global consistency would suffice. More precisely, if each sensor has a local orientation (i.e. a private sense of "left-right") but there is no global consistency, is barrier coverage possible, at least in SSYNC? Is it impossible, at least in ASYNC?

Even in the presence of local orientation, solutions that work for unoriented sensors are desirable because they can tolerate the class of faults called *dynamic compasses*: a sensor is provided with a private sense of "left-right", but this might change at each cycle (e.g., [19]). The open problem is to determine conditions which would make coverage possible under such conditions, at least in SSYNC. In particular, observing that the impossibility is established for $\bar{n}$ unoriented sensors, a relevant open question is what happens if $n > \bar{n}$ sensors are available ? Would barrier coverage become possible in SSYNC ?

For SSYNC we have shown that ε-approximate coverage is possible if $v > 2r$: is it possible to achieve the same result with $v = 2r$? In the case of unoriented sensors, no positive result exists in ASYNC. Is a higher visibility range sufficient for ε-approximate coverage in ASYNC ?

References

1. Balister, P., Bollobas, B., Sarkar, A., Kumar, S.: Reliable density estimates for coverage and connectivity in thin strips of finite length. In: Proceedings of MobiCom 2007, pp. 75–86 (2007)
2. Bhattacharya, B., Burmester, M., Hu, Y., Kranakis, E., Shi, Q., Wiese, A.: Optimal movement of mobile sensors for barrier coverage of a planar region. Theoretical Computer Science 410(52), 5515–5528 (2009)
3. Chen, D.Z., Gu, Y., Li, J., Wang, H.: Algorithms on minimizing the maximum sensor movement for barrier coverage of a linear domain. In: Fomin, F.V., Kaski, P. (eds.) SWAT 2012. LNCS, vol. 7357, pp. 177–188. Springer, Heidelberg (2012)
4. Cohen, R., Peleg, D.: Local spreading algorithms for autonomous robot systems. Theoretical Computer Science 399, 71–82 (2008)
5. Czyzowicz, J., Kranakis, E., Krizanc, D., Lambadaris, I., Narayanan, L., Opatrny, J., Stacho, L., Urrutia, J., Yazdani, M.: On minimizing the maximum sensor movement for barrier coverage of a line segment. In: Ruiz, P.M., Garcia-Luna-Aceves, J.J. (eds.) ADHOC-NOW 2009. LNCS, vol. 5793, pp. 194–212. Springer, Heidelberg (2009)
6. Czyzowicz, J., Kranakis, E., Krizanc, D., Lambadaris, I., Narayanan, L., Opatrny, J., Stacho, L., Urrutia, J., Yazdani, M.: On minimizing the sum of sensor movements for barrier coverage of a line segment. In: Nikolaidis, I., Wu, K. (eds.) ADHOC-NOW 2010. LNCS, vol. 6288, pp. 29–42. Springer, Heidelberg (2010)
7. Eftekhari, M., Kranakis, E., Krizanc, D., Narayanan, L., Opatrny, J., Shende, S.: Distributed algorithms for barrier coverage using relocatable sensors. In: Proceedings of PODC 2013, pp. 383–392 (2013)
8. Flocchini, P., Prencipe, G., Santoro, N.: Self-deployment of mobile sensors on a ring. Theoretical Computer Science 402(1), 67–80 (2008)
9. Flocchini, P., Prencipe, G., Santoro, N.: Computing by Mobile Robotic Sensors. In: ch. 21 [15] (2011)
10. Flocchini, P., Prencipe, G., Santoro, N.: Distributed Computing by Oblivious Mobile Robots. Morgan & Claypool (2012)
11. Kumar, S., Lai, T.H., Arora, A.: Barrier coverage with wireless sensors. In: Proceedings of MobiCom 2005, pp. 284–298 (2005)
12. Li, L., Zhang, B., Shen, X., Zheng, J., Yao, Z.: A study on the weak barrier coverage problem in wireless sensor networks. Computer Networks 55, 711–721 (2011)
13. Liu, B., Dousse, O., Wang, J., Saipulla, A.: Strong barrier coverage of wireless sensor networks. In: Proceedings of MobiHoc 2008, pp. 411–419 (2008)

14. Mehrandish, M., Narayanan, L., Opatrny, J.: Minimizing the number of sensors moved on line barriers. In: Proceedings of WCNC, pp. 1464–1469 (2011)
15. Nikoletseas, S., Rolim, J. (eds.): Theoretical Aspects of Distributed Computing in Sensor Networks. Springer (2011)
16. Saipulla, A., Westphal, C., Liu, B., Wang, J.: Barrier coverage of line-based deployed wireless sensor networks. In: Proceedings of IEEE INFOCOM 2009, pp. 127–135 (2009)
17. Shen, C., Cheng, W., Liao, X., Peng, S.: Barrier coverage with mobile sensors. In: Proceedings of I-SPAN 2008, pp. 99–104 (2008)
18. Wang, B.: Coverage Control in Sensor Networks. Springer (2010)
19. Yamamoto, K., Izumi, T., Katayama, Y., Inuzuka, N., Wada, K.: The optimal tolerance of uniform observation error for mobile robot convergence. Theoretical Computer Science 444, 77–86 (2012)

Exploration of Constantly Connected Dynamic Graphs Based on Cactuses*

David Ilcinkas, Ralf Klasing, and Ahmed Mouhamadou Wade

LaBRI, CNRS and Bordeaux University
{ilcinkas,klasing,wade}@labri.fr

Abstract. We study the problem of exploration by a mobile entity (agent) of a class of dynamic networks, namely constantly connected dynamic graphs. This problem has already been studied in the case where the agent knows the dynamics of the graph and the underlying graph is a ring of n vertices [5]. In this paper, we consider the same problem and we suppose that the underlying graph is a cactus graph (a connected graph in which any two simple cycles have at most one vertex in common). We propose an algorithm that allows the agent to explore these dynamic graphs in at most $2^{O(\sqrt{\log n})}n$ time units. We show that the lower bound of the algorithm is $2^{\Omega(\sqrt{\log n})}n$ time units.

Keywords: Exploration, Dynamic graphs, Mobile agent, Connectivity over time.

1 Introduction

Exploration of a graph by a mobile agent (physical or software) is the task that the mobile agent, starting at a vertex of the graph, visits all vertices at least once. In practice, many concrete systems can be modeled by graphs. This is what makes the use of graphs very versatile. For example, graphs can be used to model pipeline systems, underground tunnels, roads networks, etc. In this case, the exploration is performed by a mobile robot. Graphs can also be used to model more abstract environments such as computer networks. In this case, the mobile entities used to explore these environments are software agents, that is to say a program running in the environment.

This fundamental problem in distributed computing by mobile agents has been extensively studied since the seminal paper by Claude Shannon [12]. However, the majority of the work concerns static graphs, while new generations of interconnected environments tend to be extremely dynamic. To take into account the dynamism of these extreme environments, for a decade, researchers have begun to model these dynamic environments with dynamic graphs. Several models have been developed. The interested reader may find in [2] a comprehensive overview of the different models and studies of dynamic graphs (see also [7]).

* Partially supported by the ANR project DISPLEXITY (ANR-11-BS02-014). This study has been carried out in the frame of "the Investments for the future" Programme IdEx Bordeaux – CPU (ANR-10-IDEX-03-02).

M. Halldórsson (Ed.): SIROCCO 2014, LNCS 8576, pp. 250–262, 2014.

One of the first models developed, and also one of the most classic, is the model of evolving graphs [4]. For simplicity, given a static graph G, called underlying graph, an evolving graph $\mathcal{G}$ based on G is a (possibly infinite) sequence of (spanning but not necessarily connected) subgraphs of G (see Section 2 for the precise definitions). This model is particularly suited for modeling *synchronous* dynamic networks.

In this paper, we study the problem of exploration of dynamic graphs considering the model of constantly connected evolving graphs. An evolving graph $\mathcal{G}$ is called *constantly connected* if each graph $\mathcal{G}_i$ which composes it is connected. This class of graphs was used in [10] to study the problem of information dissemination. In 2010, Kuhn, Lynch and Oshman [6] generalize this class of dynamic graphs by introducing the notion of T-interval-connectivity. Roughly speaking, given an integer $T \geq 1$, a dynamic graph is T-interval-connected if for any window of T time units, there is a connected spanning subgraph that is stable throughout the period. (The notion of constant connectivity is equivalent to the notion of 1-interval-connectivity.) This new concept, which captures the connection stability over time, allows to derive interesting results: the T-interval-connectivity allows a savings of a factor about $\Theta(T)$ on the number of messages necessary and sufficient to achieve a complete exchange of information between all vertices [3,6].

It turns out that the problem of exploration is much more complex in dynamic graphs than in static graphs. Indeed, let us consider for example the scenario where the dynamic graph is known. The worst-case exploration time of n-node static graphs is clearly in $\Theta(n)$ (worst case $2n-3$). On the other hand, the worst-case exploration time of n-node (1-interval-connected) dynamic graphs remains largely unknown. No lower bound better than the static bound is known, while the best known upper bound is quadratic, and follows directly from the fact that the temporal diameter of these graphs is bounded by n.

The problem of exploration of constantly connected dynamic graphs has already been studied in the case where the underlying graph of the dynamic graph is a ring of n vertices [5]. That article shows that if the agent knows the dynamics of the graph, $2n - 3$ units of time are necessary and sufficient to solve the problem. The goal of this paper is to extend these results to larger families of underlying graphs. Unfortunately, the problem turns out to be much more difficult than it seems. We will see that proving that any dynamic graph based on a tree of cycles (a cactus) can be explored in time $O(n)$ is already a challenging problem. The difficulty of the exploration problem in general dynamic graphs is further underlined by the fact that the exploration problem for static graphs is the well-known GRAPH TSP problem (see e.g. [8,9,11]), which is already APX HARD in general graphs.

Our results. At a first instance, we will give two exploration methods that are efficient for exploring a very large set of constantly connected dynamic graphs based on a cactus, when the agent knows the dynamics of the graph. We will then combine these two exploration methods. We show that the combination of the two methods yields an algorithm that explores all constantly connected dynamic graphs based on a cactus of n vertices in $2^{O(\sqrt{\log n})}n$ time units, and we derive a lower bound of $2^{\Omega(\sqrt{\log n})}n$ time units for the algorithm.

2   Preliminaries

This section provides precise definitions of the concepts and models discussed informally earlier. We also give some previous results from the literature on the problem studied.

Definition 1 (Dynamic graph). *A* dynamic graph *is a pair* $\mathcal{G} = (V, \mathcal{E})$, *where* V *is a static set of* n *vertices, and* $\mathcal{E}$ *is a function which maps to every integer* $i \geq 0$ *a set* $\mathcal{E}(i)$ *of undirected edges on* V.

Definition 2 (Underlying graph). *Given a dynamic graph* $\mathcal{G} = (V, \mathcal{E})$, *the static graph* $G = (V, \bigcup_{i=0}^{\infty} \mathcal{E}(i))$ *is called the* underlying graph *of* $\mathcal{G}$. *Conversely, the dynamic graph* $\mathcal{G}$ *is said to be* based *on the static graph* G.

In this paper, we consider dynamic graphs based on a cactus of size n. We also assume that the agent knows the dynamics of the graph, that is to say, the times of appearance and disappearance of the edges of the dynamic graph.

Definition 3 (Constant connectivity). *A dynamic graph is called* constantly connected *if for any integer* i, *the static graph* $G_i = (V, \mathcal{E}(i))$ *is connected.*

Definition 4 (Cactus). *A cactus is a graph* $G = (V, E)$ *in which two connected cycles have at most one vertex in common (see Figure 1).*

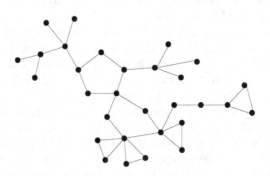

Fig. 1. Example of a cactus

A mobile entity, called *agent*, operates on these dynamic graphs. The agent can traverse at most one edge per time unit. It may also stay at the current node (typically to wait for an incident edge to appear). We say that an agent *explores* the dynamic graph if and only if it visits all the nodes.

Theorem 1. *[5] For every integer $n \geq 3$ and for every constantly connected dynamic graph based on a ring with n vertices, there exists an agent (algorithm),* EXPLORE-RING, *capable of exploring this dynamic graph in at most $2n - 3$ time units, when the agent knows the dynamics of the graph.*

Theorem 2. *[6] For every constantly connected dynamic graph on n vertices, at most $n - 1$ time units are sufficient for an agent to go from any vertex to any other vertex in the graph, when the agent knows the dynamics of the graph.*

Corollary 1. *For every constantly connected dynamic graph on n vertices, there exists an agent (algorithm) capable of exploring this dynamic graph in $O(n^2)$ time units, when the agent knows the dynamics of the graph.*

To give a simpler analysis of our algorithms, we consider the tree representation of a cactus given in [1].

For any given cactus, the set of all vertices V is partitioned into three subsets of vertices. Call *C-vertices* the vertices of degree 2 that belong to one and only one cycle, *G-vertices* the vertices that do not belong to any cycle, and *H-vertices* the other vertices (which belong to at least one cycle and have a degree ≥ 3) which we also call *attachment vertices*.

A *subtree* is a connected set consisting of H-vertices and G-vertices. A subtree is called *maximal* if the sets of H-vertices and G-vertices that it consists of cannot be extended. A *graft* is a maximal subtree that does not contain two H-vertices belonging to the same cycle. Finally, a *block* is a graft or a cycle.

It is not difficult to see that a cactus is formed by a set of blocks attached via H-vertices (see Figure 2.(a)).

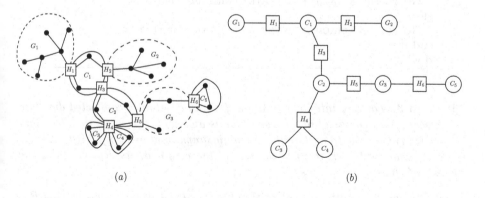

(a)   (b)

Fig. 2. Tree representation of a cactus

If we add an edge between the blocks and the H-vertices, we obtain the tree $T_G = (V_G, E_G)$ such that each element of V_G is a block or an H-vertex. Figure 2.(b) gives the tree representation of the cactus shown in Figure 1. We say that a cactus is *rooted* if the tree that represents it is rooted.

Given that constantly connected dynamic graphs based on trees (or grafts) are static, in this paper we consider cactuses that only consist of cycles and H-vertices. In the following, we will assume that the cactus is rooted at the block where the agent starts exploration. If the agent starts on an H-vertex, one of the blocks attached to the H-vertex will be the starting block.

In this paper, we use the classical formalism of static trees. We will talk about degree, child, parent, height or depth of a block.

3   Chain Method

In this section, we give a simple algorithm inspired by DFS to explore constantly connected dynamic graphs based on a cactus of n vertices. The principle of the algorithm is very simple. If the agent enters a ring it has not visited yet, it visits it using the algorithm EXPLORE-RING for exploring dynamic graphs based on the ring (see Theorem 1), then passes to the point of attachment of its closest unexplored child and explores it recursively. If all its children have already been explored and there is a ring not yet explored, then it goes to its parent.

Algorithm 1. CHAIN-METHOD()

---
1. **while** not all vertices have been visited **do**
2.   **if** the current ring is not yet explored **then**
3.   EXPLORE-RING (**current ring**)
4.   **end if**
5.   **if** there is a child not yet explored **then**
6.   GO-TO-THE-ATTACHMENT-VERTEX (with this child)
7.   **else**
8.   GO-TO-THE-ATTACHMENT-VERTEX (with the parent)
9.   **end if**
10. **end while**

Theorem 3. *For any integer $n \geq 3$, and for any constantly connected dynamic graph based on a cactus of n vertices, there is an agent, executing the algorithm* CHAIN-METHOD, *able to explore this dynamic graph in at most $\sum_{i=1}^{k}((d_i+2)n_i - (d_i + 3))$ time units, where n_i is the size of the ring i, d_i its degree, and k the number of rings of the cactus.*

Proof. An agent executing the algorithm CHAIN-METHOD pays on each ring R_{n_i} of the cactus at most $2n_i - 3$ units of time to explore it (see Theorem 1). To switch to the point of attachment of a child or the parent (if it has one), $n_i - 1$ time units are sufficient (see Theorem 2). As the degree of a block is equal to the number of incident edges, then on each ring R_{n_i} of the cactus, the agent pays at most $(d_i + 2)n_i - (d_i + 3)$ units of time. The cactus is composed of k rings, hence the agent pays at most $\sum_{i=1}^{k}((d_i + 2)n_i - (d_i + 3))$ units of time to explore the dynamic graph. □

Note that if the degree of each ring is constant, then the time to explore the dynamic graph using the CHAIN-METHOD is in $O(n)$, where n is the size of the cactus. Figure 3 presents a cactus of size n in which exploration using the CHAIN-METHOD takes time $\Omega(n^2)$. Indeed, any algorithm exploring this graph has to explore the $\Omega(n)$ attached cycles of length 3. However, when the CHAIN-METHOD is used, the adversary may choose the dynamicity of the graph such that changing from one attached cycle to another takes time $\Omega(n)$, hence the overall exploration time is $\Omega(n^2)$.

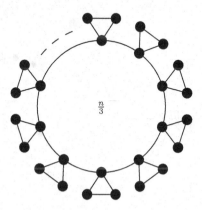

Fig. 3. Difficult graph for the CHAIN-METHOD

4   Star Method

Because the exploration method that we gave earlier is not effective for exploring constantly connected dynamic graphs based on cactuses with rings of large degree, this section provides an exploration technique to overcome this.

The algorithm we give here uses a similar technique as the exploration algorithm for dynamic graphs based on the ring. Assume that the agent starts exploring from some vertex of some constantly connected dynamic graph $\mathcal{G}$ based on a cactus C of n vertices. From the starting point, the agent explores the starting ring. The major difference with the exploration algorithm for dynamic graphs based on the ring is that when an agent arrives at a vertex where an unexplored subtree is attached, it explores the subtree recursively and then it returns to the point of attachment and continues its exploration. However, when returning to the point of attachment, the problem is that the agent cannot continue the exploration according to the basic exploration algorithm on the starting ring, as the dynamicity has changed on the ring.

In order to cope with this dynamicity problem, we need to refine the approach appropriately. We take into account the time needed to recursively explore the sub-cactuses by introducing the following transformation of $\mathcal{G}$ into another dynamic graph $\mathcal{G}'$, based on a ring $R_{n'}$ of larger size n'. The dynamic graph $\mathcal{G}'$ is constructed as follows. We retain the starting ring of C and the dynamics of the graph $\mathcal{G}$ based on this part. We replace every H-vertex of C with two C-vertices by adding a static path of length equal to twice the recursive cost of exploring the subtree attached to the H-vertex. Thus, we obtain a constantly connected dynamic graph based on a ring of size n' (see Figure 4). The dynamic graph $\mathcal{G}'$ is constantly connected because we retained the dynamics of the subgraph of $\mathcal{G}$ based on the starting ring of C, which respects the constant connectivity.

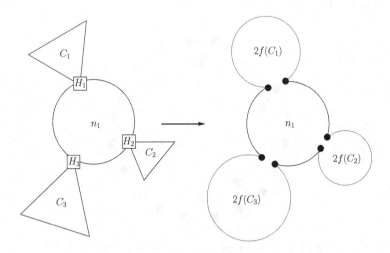

Fig. 4. Correspondence between the dynamic graph based on C and the dynamic graph based on $R_{n'}$

Theorem 4. *For any integer $n \geq 3$ and for any constantly connected dynamic graph based on a cactus C, there is an agent (algorithm) capable to explore this dynamic graph in at most $\sum_{i=1}^{k} 2^{p_i}(3n_i - 3)$ time units, where p_i is the depth of the ring i in the rooted tree, n_i is the size of the ring i in the rooted tree, k the number of rings of the cactus, and $n = \sum_{i=1}^{k} n_i - k + 1$ the number of vertices of the cactus.*

Proof. For some $n \geq 3$, let C be a cactus with n vertices and let $\mathcal{G}$ be a constantly connected dynamic graph based on C. Let us first determine the size of the dynamic graph $\mathcal{G}'$ based on $R_{n'}$ which is obtained from $\mathcal{G}$ by the above construction.

Suppose that C is rooted at the starting block. By construction, the size n' of $\mathcal{G}'$ is the sum of the size of the root ring plus the sum of twice the costs of the recursive exploration of the sub-cactuses that are attached, using the STAR-METHOD.

Denote by $f(C)$ the cost of exploring any constantly connected dynamic graph based on the cactus C using the STAR-METHOD. If C is reduced to a ring of size n, then $f(C) = 3n - 4$, because to explore a ring of size n and return to the starting vertex, an agent executing the algorithm EXPLORE-RING needs at most $3n - 4$

time units. Otherwise let n_1 be the size of the root ring, and let $C_1, C_2, \ldots, C_\ell$ be the sub-cactuses attached to the root, then we have

$$f(C) = 3(n_1 - 1) + 2 \sum_{i=1}^{\ell} f(C_i). \tag{1}$$

In order to obtain the recursive cost (1), we use the following algorithm for exploring a dynamic ring. For a constantly connected dynamic graph based on a ring R_N, one virtually deploys one agent on each vertex of the ring R_N, using $N - 1$ time units. The virtual agents then move in clockwise direction along the ring whenever they can. As there are N agents and in each round, only one agent can be held up by the adversary, after $N - 1$ rounds there is one (virtual) agent that has never been held up, hence this agent explores the ring in $N - 1$ additional time units. This agent is chosen as the actual exploration algorithm.

We consider a slightly modified version of this algorithm to explore the transformed dynamic graph $\mathcal{G}'$. Instead of allocating $n' - 1$ time units for the deployment phase, we assume that $n_1 - 1$ time units are sufficient. Now let Agent B be the virtual agent that is never held up in $\mathcal{G}'$. We define the Agent A following the STAR-METHOD as follows.

First Agent A uses $n_1 - 1$ time units to reach the starting node v of Agent B. If v is not a node of the starting ring, then Agent A goes to the attachment node in C corresponding to the static subpath containing v.

Now, whenever the (virtual) Agent B stays on a subpath P corresponding to some sub-cactus C_i for at least $f(C_i)$ consecutive time units, Agent A uses this time to recursively explore the sub-cactus C_i. If, after completing this exploration, Agent B is still lying on P, then Agent A simply waits on the attachment node. Whenever Agent B lies on the part corresponding to the starting ring (that is outside of the added subpaths), Agent A behaves exactly as Agent B. This part of the exploration of $\mathcal{G}$ takes at most $(n_1 - 1) + 2 \sum_{i=1}^{\ell} f(C_i)$ time units.

After that, Agent A returns to its starting position. This takes at most $n_1 - 1$ time units.

Solving recurrence (1), we obtain the bound announced in the theorem.   □

If the height of the rooted tree of the cactus is constant, then the time to explore the dynamic graph using the STAR-METHOD is $O(n)$ time units, where n is the size of the cactus. Figure 5 presents a cactus of size n in which exploration using the STAR-METHOD takes time $2^{\Omega(n)} n$. Indeed, when using the STAR-METHOD, from the starting point, the agent explores the starting cycle. When it reaches the rightmost vertex of the starting cycle, it explores the sub-cactus attached to the right recursively. However, the time allocated by the STAR-METHOD to do so corresponds to twice the exploration time of the sub-cactus. Hence, recursively, each additional cycle of length 4 will introduce an additional factor of 2 in the cost. As the number of cycles of length 4 is $\Omega(n)$ and the cycle of length $n/2$ to the right needs exploration time $\Omega(n)$, the overall exploration time is $2^{\Omega(n)} n$.

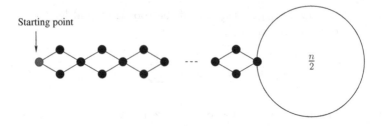

Fig. 5. Difficult graph for the STAR-METHOD

5   Mixed Method

Note that if the agent is on a block that has a subtree attached to it, then the extra cost of exploring the block plus the subtree is equal to the block size minus one if the agent uses the CHAIN-METHOD, and it is equal to the cost of exploring the subtree if the agent uses the STAR-METHOD. Because none of the two methods presented above alone allows to have a bound of $O(n)$ without further assumptions, in this section we introduce a combination of both methods, that is to say, on some blocks the agent will use the STAR-METHOD to explore, and on the remaining blocks it will use the CHAIN-METHOD. The use of the two methods is as follows. If the agent is on a block that has no child, then it uses the ring exploration algorithm. Otherwise, on a block and a given subtree, in order to choose its method of exploration, the agent will compare the cost of exploring the subtree with the block size. If the block size is greater than the cost of exploring the subtree, then the agent uses the STAR-METHOD to explore the block and the subtree, otherwise it uses the CHAIN-METHOD to explore them. In the following, we call this exploration algorithm MIXED-METHOD.

5.1   Upper Bound for the Algorithm MIXED-METHOD

In this section, we give an upper bound on the complexity of the algorithm MIXED-METHOD.

Theorem 5. *An agent executing the algorithm* MIXED-METHOD *needs at most* $2 \cdot 2^{2\sqrt{\log n}} \cdot n$ *time units to explore any constantly connected dynamic graph based on a cactus of n vertices.*

Proof. Fix an arbitrary constantly connected dynamic graph based on a cactus C of n vertices. In order to study the exploration used by the MIXED-METHOD, we will discuss another algorithm, denoted EXPLORE-CACTUS, which is less efficient but easier to analyze. The upper bound obtained for this less efficient algorithm will also give us a valid upper bound for the MIXED-METHOD. Given a parent ring R_{n_1} in the cactus, let $C_1, \ldots, C_\ell$ be its sub-cactus children. The MIXED-METHOD chooses for each child the best of the STAR-METHOD and the CHAIN-METHOD in terms of the time for exploring the sub-cactus and the size of the parent. The algorithm EXPLORE-CACTUS itself chooses the method to be used according to the criteria below. Assume without loss of generality that the sub-cactuses $C_1, \ldots, C_\ell$ are ranked in descending order of their number of vertices. The algorithm EXPLORE-CACTUS chooses the CHAIN-METHOD for the sub-cactuses $C_1, \ldots, C_{c-1}$, and the STAR-METHOD for the sub-cactuses $C_c, \ldots, C_\ell$, where $c = 2^{\sqrt{\log n}}$. According to the ordering of the sub-cactuses, the number of vertices of each sub-cactus $C_c, \ldots, C_\ell$ cannot exceed a fraction $1/c$ of the total number of vertices of the cactus rooted at the parent R_{n_1}. Therefore, a ring cannot have more than $\log_c n$ ancestors (potentially including itself) for which the STAR-METHOD was chosen. In summary, the total time used by the algorithm EXPLORE-CACTUS on the dynamic graph based on C is at most $2^{\log_c n}(c - 1 + 3)n \leq 2 \cdot 2^{2\sqrt{\log n}}n$ by definition of c. This concludes the proof of the theorem. $\square$

5.2   Lower Bound for the Algorithm MIXED-METHOD

It turns out that the algorithm MIXED-METHOD does not explore all constantly connected dynamic graphs based on a cactus of size n in $O(n)$ time units. We have the following theorem to prove it.

Theorem 6. *There is a constantly connected dynamic graph based on a cactus of n vertices such that the exploration of the dynamic graph by an agent executing the algorithm* MIXED-METHOD *takes at least $1/2 \cdot 2^{\sqrt{\log n}} \cdot n$ time units.*

Proof. Let h be an arbitrary even integer. Let $d = 2^{h+1}$. Consider a cactus based on a rooted complete d-ary tree of height h, that is to say all internal vertices have exactly d children and all of whose leaves are at distance h from the root (i.e. at depth h). For p between 0 and h, let $f_h(p) = d(2d + 3)^{h-p} - \sum_{i=0}^{h-p-1}(2d + 3)^i$. Any internal vertex of depth p is a ring of size $f_h(p + 1) + 1$. The leaves are cycles of size $\frac{d+4}{3}$ (which is an integer by definition of h). For any cycle, the points in common with the parent cycle and with each of the d child cycles, if they exist, are all different (see Figure 6). Let $t_h(p)$ be the time that algorithm MIXED-METHOD uses on a sub-cactus rooted at a cycle of depth $p \leq h$. We now prove that for any $p \leq h$, we have $t_h(p) = f_h(p)$. The proof is by induction on $(h - p)$. By Theorem 1, for $p = h$, we have $t_h(h) = 3\frac{d+4}{3} - 4 = d = f_h(h)$. Fix p such that $1 \leq p \leq h$ and suppose by induction hypothesis that $t_h(p) = f_h(p)$. At a cycle of depth $p - 1$, for each of its children, the two methods are equivalent. Hence, the time used by the algorithm MIXED-METHOD will be $t_h(p - 1) = 2\left(f_h(p) + 1 + d \cdot t_h(p)\right) - 3 + f_h(p) + 1 - 1$. After simplification, and using the induction hypothesis, we obtain $t_h(p - 1) = (2d + 3)f_h(p) - 1$, which is equal to $f_h(p - 1)$. This concludes the proof by induction. Hence, the total exploration time of the cactus by the algorithm MIXED-METHOD is $f_h(0)$. We now compute a lower bound on $f_h(0)$. We have

$$f_h(0) = d(2d+3)^h - \sum_{i=0}^{h-1}(2d+3)^i$$

$$= d(2d+3)^h - \frac{(2d+3)^h - 1}{2d+2}$$

$$\geq \frac{d-1}{2d+2} \cdot (2d+3)^h$$

$$\geq 2d^2 \cdot (2d+3)^{h-1}$$

$$\geq 2^h \cdot d^{h+1}$$

$$\geq \frac{d}{2} \cdot d^{h+1}$$

We now calculate the total number n of vertices of the cactus. According to the definition of the cactus, we have $n = \sum_{p=0}^{h-1}\left(d^p \cdot f_h(p+1)\right) + d^h \cdot \frac{d+4}{3} + 1$.

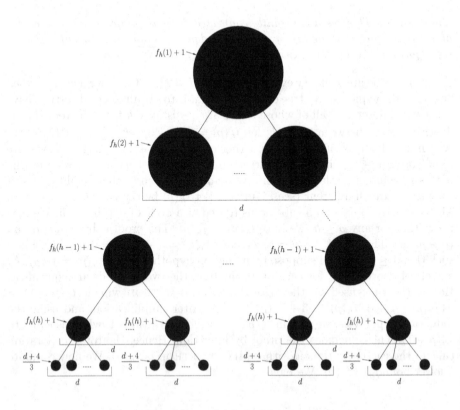

Fig. 6. Lower bound for the MIXED-METHOD

Therefore,

$$n \le \sum_{p=0}^{h-1} \left(d^p \cdot d(2d+3)^{h-p-1} \right) + d^{h+1}/3$$

$$\le d(2d+3)^{h-1} \sum_{p=0}^{h-1} (d/(2d+3))^p + d^{h+1}/3$$

$$\le 2d(2d+3)^{h-1} + d^{h+1}/3$$

$$\le (2d)^h (1 + 3/(2d))^{h-1} + d^{h+1}/3$$

$$\le d^{h+1}/2 \cdot 4/3 + d^{h+1}/3$$

$$\le d^{h+1}.$$

From this, we deduce that $d \ge 2^{\sqrt{\log n}}$. Combining all these bounds, we obtain $f_h(0) \ge 1/2 \cdot 2^{\sqrt{\log n}} \cdot n$, which concludes the proof.   □

6   Conclusion

In this paper, we studied the time complexity for exploring constantly connected dynamic graphs based on cactuses, under the assumption that the agent knows the dynamics of the graph. We gave an exploration algorithm for dynamic graphs that we called MIXED-METHOD, and we have shown that for exploring the whole class of constantly connected dynamic graphs based on cactuses of n vertices, with this algorithm, $2^{\Theta(\sqrt{\log n})} \cdot n$ units of time are necessary and sufficient. This study opens several perspectives.

In the short term, it would be interesting to find a new method in order to obtain a better upper bound on the exploration time of dynamic graphs based on cactuses. At a second stage, an interesting question to investigate would be if T-interval-connectivity (for $T > 1$) allows to save a significant factor in the exploration time of the cactuses. A natural further objective is to extend the family of underlying graphs. Note that the families of underlying graphs considered so far (ring and cactuses) have the property that at most one edge can be absent at a given time in every bi-connected component. Studying families of underlying graphs that do not possess this property seems to be a challenging problem.

A more general objective is to establish whether there is an agent which knows the dynamics of the graph and which is able to explore all T-interval-connected dynamic graphs where the underlying graph has m edges in time $O(m)$, or even $o(m)$. A further perspective is to consider the exploration problem of T-interval-connected dynamic graphs using more than one agent, assuming standard models of communication between the agents. The objective would be to study whether dynamic graph exploration can be performed more efficiently by using more than one agent. Finally, the computational complexity of the exploration problem for dynamic graphs is largely unknown. As noted in the Introduction, the exploration problem for static graphs is already APX HARD in general graphs, hence the exploration problem for dynamic graphs is at least APX HARD in general graphs. However, it is not known whether this non-approximability result for dynamic graphs is tight, and whether efficient approximation algorithms for the exploration problem in dynamic graphs can be derived.

References

1. Burkard, R., Krarup, J.: A Linear Algorithm for the Pos/Neg-Weighted 1-Median Problem on a Cactus. Computing 60(3), 193–216 (1998)
2. Casteigts, A., Flocchini, P., Quattrociocchi, W., Santoro, N.: Time-varying graphs and dynamic networks. International Journal of Parallel, Emergent and Distributed Systems 27(5) (2012)
3. Dutta, C., Pandurangan, G., Rajaraman, R., Sun, Z.: Information spreading in dynamic networks. CoRR, abs/1112.0384 (2011)
4. Ferreira, A.: Building a Reference Combinatorial Model for Dynamic Networks: Initial Results in Evolving Graphs. INRIA, RR-5041 (2003)
5. Ilcinkas, D., Wade, A.M.: Exploration of the T-Interval-Connected Dynamic Graphs: The Case of the Ring. In: Moscibroda, T., Rescigno, A.A. (eds.) SIROCCO 2013. LNCS, vol. 8179, pp. 13–23. Springer, Heidelberg (2013)
6. Kuhn, F., Lynch, N.A., Oshman, R.: Distributed computation in dynamic networks. In: 42nd ACM Symposium on Theory of Computing (STOC), pp. 513–522 (2010)
7. Kuhn, F., Oshman, R.: Dynamic networks: models and algorithms. ACM SIGACT News 42(1), 82–96 (2011)
8. Mömke, T., Svensson, O.: Approximating Graphic TSP by Matchings. In: 52nd IEEE Symposium on Foundations of Computer Science (FOCS), pp. 560–569 (2011)
9. Mucha, M.: 13/9-approximation for Graphic TSP. In: 29th Int. Symposium on Theoretical Aspects of Computer Science (STACS), pp. 30–41 (2012)
10. O'Dell, R., Wattenhofer, R.: Information dissemination in highly dynamic graphs. In: DIALM-POMC, pp. 104–110 (2005)
11. Sebö, A., Vygen, J.: Shorter Tours by Nicer Ears: 7/5-approximation for graphic TSP, 3/2 for the path version, and 4/3 for two-edge-connected subgraphs. Combinatorica (to appear)
12. Shannon, C.E.: Presentation of a maze-solving machine. In: 8th Conf. of the Josiah Macy Jr. Found (Cybernetics), pp. 173–180 (1951)

How Many Ants Does It Take to Find the Food?

Yuval Emek[1], Tobias Langner[2], David Stolz[2],
Jara Uitto[2], and Roger Wattenhofer[2]

[1] Technion, Israel
[2] ETH Zürich, Switzerland

Abstract. Consider the *Ants Nearby Treasure Search (ANTS)* problem, where n mobile agents, initially placed at the origin of an infinite grid, collaboratively search for an adversarially hidden treasure. The agents are controlled by deterministic/randomized finite or pushdown automata and are able to communicate with each other through constant-size messages. We show that the minimum number of agents required to solve the ANTS problem crucially depends on the computational capabilities of the agents as well as the timing parameters of the execution environment. We give lower and upper bounds for different scenarios.

1 Introduction

Recent research on understanding the behavior of insect colonies from a distributed computing perspective has mainly focused on questions like "How long does it take a large collection of ants to locate a food source?" [1, 2] or "How do the computational capabilities of a single ant within this collection affect the time until the food source is found?" [3–5].

In this paper, we take a computability point of view and, instead of focusing on *large* numbers of agents and on the time required to find a food source, analyze the *minimum* number of agents that is required to locate a food source within (expected) finite time. More precisely, we show that the minimally required number of agents crucially depends on the model assumptions, i.e., whether each agent is controlled by a finite automaton (FA) or a pushdown automaton (PDA), whether it has access to random bits or not, and whether the environment is synchronous or asynchronous.[1] For most combinations of the aforementioned characteristics, we establish lower and upper bounds on the number of agents required to locate the food. Our bounds are tight in most cases. We essentially present two different families of algorithms – rectangle/spiral and geometric searches – which are inspired by results of Emek et al. [1]. The main contributions of this paper, however, are the lower bounds for two deterministic FA- and one deterministic PDA-agent presented in sections 4.1 and 5.1, respectively. Table 1 at the end of the paper gives a complete picture of our findings.

---

[1] Notice the striking resemblance to the problem of finding the number of people needed to change a light bulb: For people, the answer usually depends on nationality and profession while for ants, it depends on timing and computational power.

M. Halldórsson (Ed.): SIROCCO 2014, LNCS 8576, pp. 263–278, 2014.

As border cases of our findings, we point out that in an asynchronous setting four agents are sufficient to solve the problem when their computational capabilities are most restricted, i.e., they are controlled by deterministic FAs. If we allow access to random bits and grant the agents slightly more computational power – a PDA – already one single agent can solve the problem. Note that neither of these results require the full computational power of a Turing machine.

We do not claim that our considerations are particularly relevant from a biological perspective – an ant hive generally consists of significantly more than four ants. However, our results show that powerful computational capabilities can be traded for primitive means of communication while still being able to solve complex problems – even for small number of agents.

Related Work. Our work is inspired by Feinerman et al. who proposed a problem called *ants nearby treasure search (ANTS)*, where n ants, or *agents*, are searching the plane [2, 3]. The agents are controlled by Turing machines and are not allowed to communicate with each other after leaving the origin. Assuming a knowledge of a constant approximation of n, the agents are able to locate the treasure in time $\mathcal{O}(D + D^2/n)$ where D is the distance to the treasure. Furthermore, Feinerman et al. observe a matching lower bound and prove that this lower bound cannot be matched without some knowledge of n.

There are two fundamental differences between the model studied by Feinerman et al. and our models. First, our agents are operated by finite automata or pushdown automata. The stronger computational model provided by Turing machines enables individual agents to accomplish tasks way beyond our capabilities, such as performing spiral searches and remembering the execution history. In a recent related work, Lenzen et al. study the effects that bounding the memory of the agents and the range of available probabilities have on the runtime [5]. Second, our agents are allowed to communicate outside the origin, yet only through constant-size messages – a model which was also studied by Emek et al. [1].

The general concept of graph exploration is widely studied in computer science. Typically, given a graph, the task is to visit all nodes by walking along the edges [6–10]. It is well-known that random walks allow a single agent to visit all nodes of a finite undirected graph in expected polynomial time [11]. Note that there are infinite graphs, such as a grid, where the expected time for a random walk to reach any designated node is infinite. Our problem can also be seen as a variant of the game of cops and robbers, where the robber remains dormant [12].

The classic example of a treasure finding problem is the *cow-path* problem. The task in the cow-path problem is to find a treasure on a line as quickly as possible. This task can be solved with a constant competitive ratio with a deterministic algorithm. The optimal algorithm for the 2-dimensional version is a simple spiral search [13]. The problem has also been studied in a multi-agent setting by López-Ortiz and Sweet [14].

Also finite automata searching a graph have been studied earlier [4]. Other work considering distributed computing by finite automata includes for example *population protocols* [15, 16]. Recently, a new general model of computation in graphs was introduced, where the nodes are controlled by finite automata instead of Turing machines [17]. The main connection to our work is that we use an equivalent communication model.

Model. We consider a variant of [2]'s ANTS problem, where a set of mobile *agents* search the infinite grid for an adversarially hidden treasure. Our model is an adapted version of the model used in a paper by Emek et al. [1]. Each agent is controlled either by a finite automaton or by a pushdown automaton, both either deterministic or randomized, with a common sense of direction and can communicate only with agents sharing the same grid cell.

More formally, consider n mobile agents that explore $\mathbb{Z}^2$. In the beginning of the execution, all agents are positioned in the same grid cell referred to as the *origin* (say, the cell with coordinates $(0,0) \in \mathbb{Z}^2$). In contrast to prior work, we do *not* assume that the agents can distinguish between the origin and the other cells.[2] We denote the cells with either x or y-coordinate being 0 as *north/east/south/west-axis*, depending on their location.

The *distance* $\mathrm{dist}(c, c')$ between two grid cells $c = (x, y)$ and $c' = (x', y')$ in $\mathbb{Z}^2$ is defined with respect to the ℓ_1 norm (a.k.a. Manhattan distance), that is, $|x - x'| + |y - y'|$. Two cells are called *neighbors* if the distance between them is 1. In each step of the execution, agent a positioned in cell $(x, y) \in \mathbb{Z}^2$ can either move to one of the four neighboring cells $(x, y+1), (x, y-1), (x+1, y), (x-1, y)$, or stay put in cell (x, y). The former four *position transitions* are denoted by the corresponding cardinal directions N, E, S, W, whereas the latter (stationary) position transition is denoted by P (standing for "stay put"). We point out that the agents have a common sense of orientation, i.e., the cardinal directions are aligned with the corresponding grid axes for every agent in every cell.

In an *asynchronous environment*, each agent's execution progresses in discrete (asynchronous) steps indexed by the non-negative integers and we denote the time at which agent a completes step $i > 0$ by $t_a(i) > 0$. Following common practice, we assume that the time stamps $t_a(i)$ are determined by the policy ψ of an adversary that knows the protocol but is oblivious to its random bits, whereas the agents do not have any sense of time. A *synchronous environment* corresponds to the special case where $t_a(i) = i$ for all agents a and all $i > 0$.

The communication and computational capabilities of the agents are limited. Specifically, in our model, an agent a positioned in cell $c \in \mathbb{Z}^2$ can communicate with all other agents positioned in cell c at the same time. This communication is limited though: agent a merely senses for each state q of its (finite or pushdown) automaton, whether there exists at least one agent $a' \neq a$ in cell c whose current state is q. Notice that this communication scheme is a special case of the one-two-many communication scheme introduced in [17] with bounding parameter $b = 1$.

Since we only consider instances with a constant number of agents, we allow each agent to run a different individual protocol. This is modeled by assigning

---

[2] The motivation behind this is that, in contrast to previous work, we consider constant numbers of agents. While models with large numbers can spare one agent to mark the origin without affecting their upper bounds, our upper bounds actually increase (by one) if such behavior is required. Consequently, we consider the weaker variant.

to each agent an individual initial state in the respective automaton (note that this is only relevant in the deterministic case as otherwise coin flips can be used to separate agents). The protocol is controlled by either a finite automaton or a pushdown automaton. We shall first explain the semantics of the former and then explain the additional capabilities of the latter.

FA-protocol. When an agent employs an *FA-protocol*, it has a constant memory and thus, in general, cannot store coordinates in $\mathbb{Z}^2$. Formally, the agent's protocol is captured by the 3-tuple $\Pi = \langle Q, s_0^a, \delta \rangle$, where Q is the finite set of *states*, $s_0^a \in Q$ is the *initial state* of agent a, and $\delta : Q \times 2^Q \to 2^{Q \times \{N,S,E,W,P\}}$ is the *transition function*. To allow the agents to perform different tasks also in the absence of randomization, each agent a has a unique start state s_0^a in which it resides at time 0. Suppose that at time $t_a(i)$, agent a is in state $q \in Q$ and positioned in cell $c \in \mathbb{Z}^2$. Then, the state $q' \in Q$ of agent a at time $t_a(i+1)$ and its corresponding movement $\tau \in \{N, S, E, W, P\}$ are dictated based on the transition function δ by picking the tuple (q', τ) uniformly at random from $\delta(q, Q_a)$, where $Q_a \subseteq Q$ contains state $p \in Q$ if and only if there exists some (at least one) agent $a' \neq a$ such that a' is in state p and positioned in cell c at time $t_a(i)$. A FA-protocol is *deterministic* if each step is deterministic, i.e., $|\delta(q, Q_a)| \leq 1$ for all $q \in Q$ and $Q_a \subseteq Q$. For simplicity, we assume that while Q_a (input to δ) is determined based on the status of cell c at time $t_a(i)$, the actual application of the transition function δ occurs instantaneously at the end of the step, i.e., agent a is considered to be in state q and positioned in cell c throughout the time interval $[t_a(i), t_a(i+1))$.

PDA-protocol. When an agent employs a *PDA-protocol*, it is controlled by a pushdown automaton with an infinite stack. The communication and movement model remains the same. The only addition is that in each step, an agent reads and removes the top-most symbol from the stack ("pop") – if the stack is empty, the agent reads the special symbol ε and the stack remains unchanged – and then adds a finite amount of symbols to the top of the stack ("push"). The symbol read from the stack serves as additional input to the agent. Formally, the agents' protocol is captured by the 4-tuple $\Pi = \langle Q, s_0^a, \Gamma, \delta \rangle$, where Q is the finite set of states, $s_0^a \in Q$ is the *initial state* of agent a, Γ is the finite *stack alphabet*, and $\delta : Q \times 2^Q \times \Gamma \cup \{\varepsilon\} \to 2^{Q \times \Gamma^* \times \{N,E,S,W,P\}}$ is the *transition function*. Suppose that at time $t_a(i)$, agent a is in state $q \in Q$, positioned in cell $c \in \mathbb{Z}^2$, and the top-most symbol on the stack is $\gamma \in \Gamma \cup \{\varepsilon\}$. Then, the state $q' \in Q$ of agent a at time $t_a(i+1)$, the word $\alpha \in \Gamma^*$ to be written to the stack, and the corresponding movement $\tau \in \{N, E, S, W, P\}$ are dictated based on the transition function δ by picking the tuple (q', α, τ) uniformly at random from $\delta(q, \gamma, Q_a)$, where $Q_a \subseteq Q$ is defined as in an FA-protocol.

Problem setting. We consider two different variants of the problem, where the goal in both is to locate an adversarially hidden *treasure*, i.e., to bring at least one agent to the cell in which the treasure is positioned while the distance of the treasure from the origin is denoted by D. In async-ANTS, the problem is to find the treasure in an arbitrary asynchronous environment while in the sync-ANTS problem the agents operate in a synchronous environment. A FA/PDA-protocol

$\mathcal{P}$ is *effective* if it allows the agents to locate the treasure in finite time if $\mathcal{P}$ is deterministic, or if the agents locate the treasure in expected finite time if $\mathcal{P}$ is randomized.

Preliminaries. For our deliberations we require a sequence of definitions. Let $\mathcal{A}$ be the set of agents. We denote by $E_a^{\mathcal{P}}(t)$ the cells that an agent a employing protocol $\mathcal{P}$ has visited until time t and furthermore $E^{\mathcal{P}}(t) = \bigcup_{a \in \mathcal{A}} E_a^{\mathcal{P}}(t)$. In the context of the sync-ANTS problem, we take the liberty to write $E_a^{\mathcal{P}}(i)$ for a (then global) step i as shorthand for $E_a^{\mathcal{P}}(t_a(i))$ and analogous for $E^{\mathcal{P}}(i)$. We omit $\mathcal{P}$ in the previous expressions if the considered protocol is clear from the context.

2 Four Agents

The goal of this section is to solve the async-ANTS problem without using randomization. We provide a simple protocol for four FA-agents that uses three of the four agents as landmarks for the fourth agent. The fourth agent discovers the whole grid in a spiraling fashion with increasing distance to the origin.

We begin by giving an informal description of the protocol. The landmark agents, referred to as Guides, position themselves in a triangle around the origin and after getting a signal from the searching agent, called the Explorer, move step by step further away from the origin. The Explorer moves to the Guides one by one signaling them to expand the triangle. This way the Explorer is able to guarantee that it can always reach one Guide after meeting another by simply walking a (possibly diagonal) straight line, even after the Guides are within a super-constant distance from each other and the origin.

All three Guides have specific roles and therefore we give them task-specific names: NorthGuide, WestGuide and EastGuide. The agents execute the following protocol, which is illustrated in Figure 1. The protocol is initialized by the NorthGuide moving once north, the WestGuide moving once west and the East-Guide moving once east. After the Explorer notices that the origin is empty, it moves once north.

NorthGuide. When the NorthGuide meets a WaitingExplorer it moves once north.

WestGuide. When the WestGuide meets a WaitingExplorer it moves once west and becomes a MovingWestGuide. The MovingWestGuide first moves once west and then once south and becomes a WestGuide again.

EastGuide. When the EastGuide meets a WaitingExplorer it moves once south and becomes a MovingEastGuide. The MovingEastGuide moves twice east and becomes again an EastGuide.

Explorer. The Explorer continuously performs *triangle searches* in increasing distances. It continuously moves into a given direction, starting with south-west (by alternatingly moving south and west). When the Explorer meets a WestGuide,

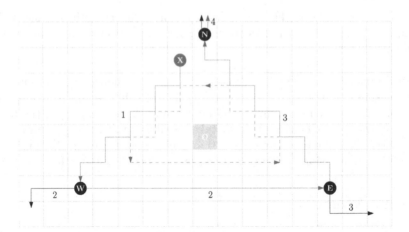

Fig. 1. Four agents are discovering the grid and currently are performing a triangle search in distance 3. The origin is denoted by a gray square, the Explorer (X) by a red circle and the NorthGuide (N), WestGuide (W) and EastGuide (E) by black circles labeled with the corresponding initial letters. The numbers indicate the order of movements, i.e., moves along the arrow labeled with i are performed only after the moves along the arrow labeled with $i-1$ are finished. The dashed red line indicates the path of the Explorer in distance 2.

it changes its moving direction to east and becomes a WaitingExplorer. When it meets an EastGuide, it changes the direction to north-west and becomes a WaitingExplorer. Finally, when the Explorer meets a NorthGuide, it changes its moving direction to south-west (alternates between west and south) and becomes a WaitingExplorer. Notice that the Explorer meets the NorthGuide in the starting position of the triangle search in the next distance. Whenever the Explorer meets a MovingWestGuide or a MovingEastGuide in cell c, it waits until c is empty before continuing to move.

WaitingExplorer. When the WaitingExplorer resides in a cell that does not contain an EastGuide, a NorthGuide, or a WestGuide, it becomes an Explorer and continues moving.

We index the triangle searches by their distances, i.e., if the Explorer meets the NorthGuide in cell $(0, i)$ and starts moving south-west, we index the corresponding triangle search by index i and denote it by TS_i. A triangle search in distance i starts when the Explorer leaves cell $(0, i)$ by moving west and ends when the Explorer meets a NorthGuide. Furthermore, we say that TS_i *works correctly*, if the Explorer meets the WestGuide only in cell $(-2i + 1, -i + 1)$, the EastGuide only in cell $(2i - 1, -i + 1)$ and the NorthGuide only in cell $(0, i + 1)$ during TS_i.

Lemma 1. *(Proof deferred to full version) Every triangle search works correctly.*

To show that the treasure eventually gets discovered, we need two more auxiliary observations. First, we show that every cell in distance d is discovered latest during TS_{d+1}. Second, we show that each triangle search finishes within finite time. We call the set of cells along which the Explorer moves during TS_i the path of rectangle search i.

Observation 2. *(Proof deferred to full version) Every cell c within distance d to the origin is discovered latest during TS_{d+1}.*

Observation 3. *(Proof deferred to full version) Every triangle search ends within finite time.*

We can now combine the results from this section. Let D be the distance to the treasure. By Observation 2, the treasure is found latest during TS_{D+1}. As the duration of each search is finite by Observation 3 and by Lemma 1 each triangle is eventually searched, we get the following theorem.

Theorem 1. *There exists an effective deterministic FA-protocol for async-ANTS for $n = 4$.*

3 Three Agents

3.1 Deterministic Protocol for sync-ANTS

In this section, we first show that we can get rid of one of the FA-agents by giving the agents a common notion of time. In other words, if we assume that the execution of the algorithm is synchronous, three agents suffice to discover the treasure. Our goal is to prove the following theorem.

Theorem 2. *There exists an effective deterministic FA-protocol for sync-ANTS for $n = 3$.*

The idea of the three-agent protocol is similar to the protocol from Section 2. Again, one of the agents, the Explorer, performs the actual searching and the two other agents work as Guides. The task of one of the Guides, called OriginGuide, is simply to stand still and mark the origin throughout the execution. The task of the other Guide is to tell the Explorer when it hits an axis. On the first round of the execution, the Explorer and the other Guide move one step north to cell $(0, 1)$ and then start the execution of the following protocol.

Explorer. The Explorer repeatedly performs *rectangle searches* in increasing distances. It starts the first rectangle search in distance 1 by diagonally moving south-west, i.e., alternating between moving west and south. When it meets a Guide, it alters its movement direction by 90° counter-clockwise. At the end of a complete rectangle (i.e., when meeting a Guide again at the starting point), it moves one step outwards starting a new rectangle search with a larger distance. During a rectangle search in distance d, the Explorer discovers all cells that have distance d to the origin.

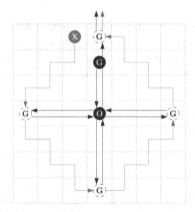
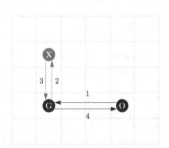

Fig. 2. Three agents can discover the entire grid under a synchronous environment. The dashed circles indicate the locations where the Explorer (X) meets the Guide (G). The OriginGuide (O) marks the origin.

Fig. 3. Three agents are performing a geometric search on the north-west quarter plane. Moves along the black arrows are executed by both the Explorer (X) and the Guide (G) while the OriginGuide (O) states at the origin. Moves along the red arrows are executed only by the Explorer.

Guide. The Guide starts by moving towards the OriginGuide that marks the origin. When it meets the OriginGuide, it alters its direction by 90° clockwise and moves outwards. When it meets the Explorer, it turns around and moves inwards towards the OriginGuide. The Guide also moves one step north with the Explorer when they meet in the end of searching a rectangle and starts walking towards the OriginGuide afterwards.

The execution of our protocol is illustrated in Figure 2. To prove Theorem 2, we only need to show that every time the Explorer enters a cell on an axis, it meets a Guide. To see why this is sufficient, consider any cell c on the plane with distance d to the origin. Then c is searched (latest) during rectangle search in distance d. Therefore, assuming that each rectangle search is performed correctly, the whole plane is eventually discovered.

It is fairly easy to see that the Explorer and the Guide never fail to meet. Consider round r when the Explorer and a Guide meet on an axis during rectangle search in distance d. Then the distance that both of them have to move until the next meeting point is $2d$. Since both agents move exactly once per round, the claim follows. Note that the assumption of a synchronous environment is crucial here.

3.2   Randomized Protocol for async-ANTS

We now show that if we are not restricted to deterministic state machines but allow randomization, we can find the treasure under an asynchronous environment with only 3 FA-agents. The fundamental idea behind our randomized protocol is that the agents use a fair coin to determine which cells to discover.

Again, we have two Guides and one Explorer and the task of one of the agents, the OriginGuide, is to simply stay in the origin. The Explorer performs the actual searching and starts by uniformly at random choosing either (north, east), (east, south), (south, west) or (west, north), i.e., it randomly chooses a quarter plane. Then, the Explorer performs a *geometric search* on that quarter plane.

Consider the case of choosing (east, south) as the quarter-plane (the search in the other quarter-planes works analogously). The Guide and the Explorer execute the following protocols.

Explorer. The Explorer starts by moving once east. Then on every step the Explorer tosses a fair coin and if it shows heads, it moves east. When the coin shows tail, the Explorer stops and becomes a WaitingExplorer until its cell is occupied by a WaitingGuide. When the WaitingGuide appears, the WaitingExplorer moves one cell south, becomes an Explorer, and continues tossing coins but now moves one cell south every time the coin shows head instead of east. When the coin shows tails, the Explorer turns back, i.e., starts moving north. After the Explorer reaches a cell with a WaitingGuide, it stops and moves west (until it reaches an OriginGuide) whenever its cell contains no WaitingGuide.

Guide. The Guide moves east on every step if its cell is not occupied by an Explorer. When it meets a WaitingExplorer, it turns into a WaitingGuide. When the WaitingGuide meets an Explorer, it becomes a Guide again and moves west whenever its cell is not occupied by an Explorer until it meets an OriginGuide.

After all the agents reach the origin, they restart the process. The protocol is illustrated in Figure 3. It is easy to see that each geometric search has a finite duration with probability 1 since the Explorer throws a finite number of heads in every search with probability 1. Assume that the number of heads is finite. Then the Explorer becomes a WaitingExplorer in finite time. After the Explorer becomes a WaitingExplorer, the Guide moves towards the cell of the WaitingExplorer in every step and therefore reaches it in finite time. Similarly, the Explorer returns to the WaitingGuide in finite time and they both reach the OriginGuide in finite time.

Theorem 3. *There exists an effective randomized FA-protocol for async-ANTS for $n = 3$.*

Proof. Assume that the treasure is located in cell $c = (x, y)$ in the north-east quarter plane with $D = x + y$. Let us index the geometric searches, i.e., the iterations of the algorithm, by the positive integers. Clearly, the protocol is defined so that if the treasure is found in search i, then search $j > i$ is not needed, however, for the sake of the analysis, we assume that the agents keep performing the searches indefinitely and bound the time until the treasure is found – let T be the random variable that captures this time. Given this view, we know that search i is independent of all searches other than i.

Let A_i be the event that the Explorer finds the treasure in search i. This happens if it chooses the right quarter plane, throws heads exactly $x - 1$ times before throwing tails once and then throws heads $y - 1$ times. Hence, $\Pr(A_i) = \frac{1}{4} \cdot 2^{-(x-1)} \cdot \frac{1}{2} \cdot 2^{-(y-1)} = 2^{-(D+1)}$. Let $B_i = \neg A_1 \wedge \cdots \neg A_{i-1} \wedge A_i$ be the event that the treasure is found in search i and not in any search $j < i$. We rely on the following equations that hold for every $i \geq 1$ and $1 \leq j < i$:

(1) $\Pr(A_i) = 2^{-(D+1)}$
(2) $\Pr(B_i) = (1 - 2^{-(D+1)})^{i-1} 2^{-(D+1)}$
(3) $\mathbb{E}[L_i \mid B_i] = \mathbb{E}[L_i \mid A_i] = \mathcal{O}(D)$
(4) $\mathbb{E}[L_j \mid B_i] = \mathbb{E}[L_j \mid \neg A_j] = \mathcal{O}(1)$

Therefore,

$$\mathbb{E}[T] = \sum_{i=1}^{\infty} \mathbb{E}[T \mid B_i] \cdot \Pr(B_i)$$

$$= \sum_{i=1}^{\infty} \left(\sum_{j=1}^{i-1} \mathbb{E}[L_j \mid B_i] + \mathbb{E}[L_i \mid B_i] \right) \cdot (1 - 2^{-(D+1)})^{i-1} 2^{-(D+1)}$$

$$= \sum_{i=1}^{\infty} (\mathcal{O}(i) + \mathcal{O}(D)) \cdot (1 - 2^{-(D+1)})^{i-1} 2^{-(D+1)}$$

$$= 2^{-(D+1)} \cdot \sum_{i=1}^{\infty} \mathcal{O}(i) \cdot (1 - 2^{-(D+1)})^{i-1}$$

$$+ \mathcal{O}(D) \cdot 2^{-(D+1)} \cdot \sum_{i=1}^{\infty} (1 - 2^{-(D+1)})^{i-1}$$

$$= 2^{-(D+1)} \cdot \mathcal{O}(2^{2D}) + \mathcal{O}(D) \cdot 2^{-(D+1)} \cdot 2^{D+1} = \mathcal{O}(2^D) \ . \qquad \square$$

4   Two Agents

Our goals in this section are to show, on the negative side, that two deterministic FA-agents *cannot* solve sync-ANTS, and, on the positive side, that one deterministic FA-agent together with one deterministic PDA-agent *can* solve sync-ANTS.

4.1   No Deterministic FA-Protocol

We start off with proving the first result. Before doing so, we define the notion of a *band* in $\mathbb{Z}^2$. A band is the discrete version of a fat line in Euclidean space, i.e., the set of cells that have at most a certain distance from a line.

Definition. *A band $B = (s, m, e)$, $s = (x_s, y_s) \in \mathbb{Z}^2$ with slope $m = (m_x, m_y) \in \mathbb{Z}^2$ of extent $e \in \mathbb{N}_{>0}$ consists of all cells c for which there exists a point $p = (s_x + \lambda m_x, s_y + \lambda m_y)$ for some $\lambda \in \mathbb{R}$ such that $\|c - p\|_1 \leq e$ where $\|x\|_1$ denotes the ℓ_1-norm of x.*

Observation 4. *Let $\mathcal{B}$ be a finite set of bands with finite extent. Then $\mathbb{Z}^2 \setminus \bigcup_{B \in \mathcal{B}} B \neq \emptyset$.*

Proof. Assume for the sake of a contradiction that the bands in $\mathcal{B}$ cover $\mathbb{Z}^2$ completely. Let e^* be the maximum extent of the bands in $\mathcal{B}$. Consider a square region S of $\mathbb{Z}^2$ with ℓ^2 cells for $\ell > 2|\mathcal{B}|e^*$ and a fixed band $B = (s, m, e) \in \mathcal{B}$. Assume wlog. that $|m_x| \leq |m_y|$. Observe that $|B \cap S| \leq \ell \cdot 2e^*$ since S vertically extends over ℓ cells and the horizontal width of $B \cap S$ is at most $2e^*$. Let $A = \bigcup_{B \in \mathcal{B}} B$ and we get $|A \cap S| \leq 2|\mathcal{B}|e^* \cdot \ell < \ell^2 = |S|$. Thus, the bands in $\mathcal{B}$ do not even cover the cells in S, a contradiction. □

We denote by $M(\mathcal{P}) = (t_i)_{i>0}$ the strictly increasing sequence of all points in time when two agents meet during the execution of protocol $\mathcal{P}$. An important ingredient for the proof is the following lemma, which holds for an arbitrary amount of agents.

Lemma 5. *If $\mathcal{P}$ is an effective deterministic FA-protocol for sync-ANTS, then $|M(\mathcal{P})| = \infty$.*

Proof. Assume for the sake of contradiction that $\mathcal{P}$ is an effective deterministic protocol with finite $|M(\mathcal{P})|$. Thus, there exists a largest point in time $t^* = \max(M(\mathcal{P}))$ when two agents meet and after which no two agents meet anymore and the number of cells explored until t^* is finite. Consider now agent a and let q be the state that has been entered by agent a twice after t^* at the earliest time. Let $(t_i)_{i>0}$ be the strictly increasing sequence of points in time after t^* when a enters state q and denote $I_i = [t_i, t_{i+1}]$. Observe that the behavior of a in each interval I_i is identical, hence a will keep on repeating the same transitions and movements as in I_1 forever. Observe further that a can only move a finite distance in each I_i as it has a finite length.

Consider the vector $v_i(a) = C_a(t_{i+1}) - C_a(t_i)$ describing the net-translation of a during I_i and observe that by the above argument $v_i(a) = v_1(a)$ for all $i > 0$. There are two cases: If $v_1(a) = 0$, then agent a explores only a constant amount of cells for $t \to \infty$. If $v_1(a) \neq 0$, then a exhibits a net-movement into the direction of $v_1(a)$ in each I_i and since it only explores a constant amount of cells in each I_i, agent a explores only cells in a band with finite width after t^*. By Observation 4, the agents cannot explore all cells in $\mathbb{Z}^2$ and the claim follows. □

Theorem 4. *There exists no effective deterministic FA-protocol for sync-ANTS for $n = 2$.*

Proof. Assume for the sake of contradiction that $\mathcal{P}$ is an effective deterministic protocol for two agents a_1 and a_2. By Lemma 5 we know that $|M(\mathcal{P})| = \infty$. Let Q_1 and Q_2 be the set of states of the two FAs controlling a_1 and a_2. We denote by $Q_1(t) \in Q_1$ and $Q_2(t) \in Q_2$ the state of agent a_1 and a_2 at time t and further $Q(t) = (Q_1(t), Q_2(t))$. Observe that since $|M(\mathcal{P})| = \infty$, there must be a pair of states $(q_1, q_2) \in Q_1 \times Q_2$ such that the sub-sequence $T = (\tau_i)_{i>0}$ of $M(\mathcal{P})$ that consists of all $\tau \in M(\mathcal{P})$ such that $Q(\tau) = (q_1, q_2)$, is infinite.

We denote the intervals $I_i = [\tau_i, \tau_{i+1}]$ and observe that a_1 and a_2 (individually) perform exactly the same state transitions and movements in each interval I_i (agent a_1 and a_2 might meet between τ_i and τ_{i+1} in different states, but their behavior is fully determined by their states at time τ_i). Thus, there is a fixed vector $v = C_{a_1}(\tau_{i+1}) - C_{a_1}(\tau_i)$ representing the translation of the meeting cell of a_1 and a_2 during some I_i and furthermore a fixed constant $\vartheta > 0$ such that $\tau_{i+1} - \tau_i = \vartheta$. Consequently, a_1 and a_2 can only explore cells in a band with finite width after τ_1. Since $E(\tau_1)$ is finite, Observation 4 yields a contradiction.   □

4.2   Deterministic FA/PDA-Protocol for sync-ANTS

The second result of this section establishes that while two agents controlled by a FA do not allow for an effective deterministic protocol for sync-ANTS, one FA-agent and one PDA-agent do so.

The protocol is essentially an adapted version of the protocol from Section 3.1. The Explorer behaves identically to Section 3.1 and performs rectangle searches with increasing distances to the origin. The second PDA-agent replaces the two Guides by walking along the axis in order to signal to the Explorer when the search in a quarter-plane is complete and it should therefore alter its movement direction. The trick here is that the Guide tracks its distance from the origin using the stack. More precisely, the Guide pushes a symbol onto the stack whenever it performs a movement outwards on one of the axes and pops one symbol from the stack whenever it moves towards the origin. Using this trick, the Guide can detect when it has arrived at the origin by verifying whether the stack is empty, i.e., the read symbol is ε. Then the algorithm works as follows:

At time $t = 0$, the Guide and the Explorer both move one cell north (and the Guide records this move on the stack). Whenever the two agents are located together on the north-axis in cell $(0, d)$, the Explorer starts a diagonal walk towards south-west while the Guide moves south towards the origin until it arrives there, which it can track using the stack. Upon arriving there, it moves west until it meets the Explorer. As the length of the two (different) paths from cell $(0, d)$ to cell $(-d, 0)$ is equal, both the Guide and the Explorer arrive in cell $(0, -d)$ at the same time. Now the Explorer changes its movement direction and the Guide moves back to the origin after which it moves south to meet the Explorer on the south axis in cell $(0, -d)$. They repeat this process to meet on the west axis in cell $(d, 0)$ and on the north axis in cell $(0, d)$. When the Explorer has completed the rectangle search of level d by arriving at cell $(0, d)$ again, it moves together with the Guide to cell $(0, d + 1)$ and the search of level $d + 1$ begins.

It is easy to see that the above algorithm guarantees that the Explorer meets the Guide every time it crosses an axis and that therefore any level d is explored in finite time.

Theorem 5. *There exists an effective deterministic protocol for sync-ANTS for $n = 2$ that uses one FA-protocol and one PDA-protocol.*

4.3 Deterministic PDA-Protocol for async-ANTS

Since two PDAs can simulate a Turing machine [18] by using both their stacks to represent the infinite band of the Turing machine, it is not too surprising that two PDAs allow for an effective deterministic protocol for async-ANTS. The two agents a and b employ the following protocol: Both agents walk "hand-in-hand", i.e., have a distance of at most 1 at all times, and perform a spiral search with increasing distances from the origin (cf. Section 3.1). At any time during the execution, they maintain the invariant that the sum of the number of symbols on both stacks equals their distance from the origin. They start from the cell $(0, 1)$ with the stack of agent a containing one symbol. When the two agents start a spiral search from cell $(0, i)$, agent a has i symbols on is stack. When a and b walk south-west, agent a removes a symbol from its stack every other step while agent b pushes one symbol to its stack every other step. When the stack of agent a is empty, agent b's stack contains i symbols and the agents have arrived at the cell $(-i, 0)$ on the west axis. Then they reverse their roles and move together to the south, east, and again north axis in the same fashion to finish the search in distance i. Thereafter, they move one cell north, push one additional symbol to the stack to account for the increased distance and start a new search in distance $i + 1$. It is easy to see that this protocol can be implemented to work in an asynchronous environment and guarantees that the two agents locate the treasure.

Theorem 6. *There exists an effective deterministic PDA-protocol for async-ANTS for $n = 2$.*

5 One Agent

In this section we show that neither a single randomized FA-agent nor a single deterministic PDA-agent can find the treasure in finite time while a randomized PDA-agent is able to do so.

Theorem 7. *(Proof deferred to full version) There exists no effective randomized FA-protocol for sync-ANTS for $n = 1$.*

5.1 No Deterministic PDA-Protocol

Consider a single agent controlled by a *deterministic* PDA-protocol. We denote by $S(i)$ the size of the stack, i.e., the number of symbols on the stack (directly) after step i and by $C(i) = (q, \gamma)$ the tuple of the state $q \in Q$ and the top-most stack symbol $\gamma \in \Gamma$ (directly) after step i. Let $\mathcal{C} = Q \times \Gamma$ be the set of all configurations and observe that $|\mathcal{C}|$ is constant. As the behavior of a PDA is fully determined by its state and the top-most stack symbol, the following observation is immediate.

Observation 6. *Let $0 < i_1 < i_2$ be two different steps with $C(i_1) = C(i_2)$ and let i_2 be the smallest such index. If $S(i) \geq S(i_1)$ for all $i_1 \leq i \leq i_2$, then $C(j) = C(j + k \cdot (i_2 - i_1))$ for all $i_1 \leq j \leq i_2$ and $k \in \mathbb{N}_0$.*

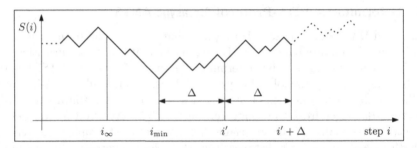

Fig. 4. The size $S(i)$ of the stack varies for the different steps. All configurations entered after step i_∞ are entered infinitely often. The stack exhibits its minimal size after i_∞ at step i_{min} while $C(i_{min})$ is entered again for the first time at time i'. Then the PDA will keep repeating its behavior after i_{min} with period $\Delta = i' - i_{min}$.

Note that the observation also implies that the agent executes the identical sequence of actions between step i_1 and i_2.

Observe that, since any protocol must be able to run for an arbitrary time, we can partition the set $\mathcal{C}$ into the configurations $\mathcal{C}_f$ containing all configurations that are entered finitely often and the configurations $\mathcal{C}_\infty$ that are entered infinitely often during the execution of a given protocol. Observe that there exists step i_∞ such that $C(i) \in \mathcal{C}_\infty$ for any step $i > i_\infty$. The following lemma essentially states that after a certain step $i_r > i_\infty$, the PDA will keep on repeating its behavior with a finite period Δ (see Figure 4 for an illustration).

Lemma 7. *There exists an index $i_r > i_\infty$ and a period $\Delta \in \mathbb{N}_0$ such that for all steps i with $i_r \le i < i_r + \Delta$ we have $C(i + k \cdot \Delta) = C(i)$ for all $k \in \mathbb{N}_0$.*

Proof. Let $s_{min} \in \mathbb{N}_0$ be the minimum stack size after i_∞ and let i_{min} be the smallest index $i > i_\infty$ for which $S(i) = s_{min}$. Let $i' > i_{min}$ be the smallest step such that $C(i') = C(i_{min})$. By definition of i_{min} there exists no index $i > i_{min}$ with $S(i) < S(i_{min})$. Thus, i_{min} and i' satisfy the preconditions of Observation 6 and the claim follows for $i_r = i_{min}$ and $\Delta = i' - i_{min}$.   □

As the PDA keeps on repeating its behavior after step i_r with constant period Δ, the agent can only explore cells in a band of finite width after i_r. As i_r is finite and thus $E(i_r)$ is also finite, Observation 4 implies the following theorem.

Theorem 8. *There exists no effective deterministic PDA-protocol that for sync-ANTS for $n = 1$.*

5.2   Randomized PDA-Protocol for async-ANTS

The randomized protocol is an adapted version of the randomized FA-protocol for three agents from Section 3.2. There, one agent repeatedly performs geometric searches to a random cell in a geometrically distributed distance. It uses the two other agents to find its way back to the origin in order to start the next iteration of the search. A single agent employing a randomized PDA-protocol can do the same by using the stack to record its distance to the origin and thereby, it can perform a geometric search and then return to the origin for the next iteration. More precisely, the agent performs a geometric search as in Section 3.2 but whenever moving north/east/south/west, it pushes N/E/S/W, respectively, to the stack. When one geometric search ends, the agent can re-track its steps by walking north/east/south/west when reading S/W/N/E, respectively, and ends up at the origin when the stack is empty. Then, it can start the next iteration. It is easy to see that the analysis from Section 3.2 applies identically.

Theorem 9. *There exists an effective randomized PDA-protocol for* async-ANTS *for* $n = 1$.

Table 1. The symbol × indicates that the given combination does not allow for an effective protocol while ✓ states that there does exist an effective protocol. Empty cells follow immediately from other entries while cells marked with ? represent open problems. The numbers in the superscript refer to the theorem establishing the respective result.

Problem	FA				PDA			
	sync		async		sync		async	
	det	rand	det	rand	det	rand	det	rand
One agent		×[7]		×[7]	×[8]	✓[9]	×[8]	✓[9]
Two agents	×[4]	?	×[4]	?	✓[5,6]		✓[6]	
Three agents	✓[2]	✓[3]	?	✓[3]				
Four agents			✓[1]					

Conclusion

The variety of results of this paper are summarized in Table 1. While our findings almost completely cover the landscape of problem configurations, Table 1 essentially shows two gaps, which, in our opinion, represent interesting open problems: Can two agents controlled by a randomized FA solve the synchronous or asynchronous version of the ANTS problem? Is there an effective FA-protocol for async-ANTS for three agents when no random bits are available?

As a last remark, we point out that all our algorithms can be easily adapted to guarantee that upon finding the treasure, the agents can locate the initial starting cell and bring the treasure back to it with a constant multiplicative overhead in terms of the runtime.

References

1. Emek, Y., Langner, T., Uitto, J., Wattenhofer, R.: Solving the ANTS Problem with Asynchronous Finite State Machines. In: Esparza, J., Fraigniaud, P., Husfeldt, T., Koutsoupias, E. (eds.) ICALP 2014, Part II. LNCS, vol. 8573, pp. 471–482. Springer, Heidelberg (2014)
2. Feinerman, O., Korman, A., Lotker, Z., Sereni, J.S.: Collaborative Search on the Plane Without Communication. In: Proceedings of the 31st ACM Symposium on Principles of Distributed Computing (PODC), pp. 77–86 (2012)
3. Feinerman, O., Korman, A.: Memory Lower Bounds for Randomized Collaborative Search and Implications for Biology. In: Aguilera, M.K. (ed.) DISC 2012. LNCS, vol. 7611, pp. 61–75. Springer, Heidelberg (2012)
4. Fraigniaud, P., Ilcinkas, D., Peer, G., Pelc, A., Peleg, D.: Graph Exploration by a Finite Automaton. Theoretical Computer Science 345(2-3), 331–344 (2005)
5. Lenzen, C., Lynch, N., Newport, C., Radeva, T.: Trade-offs between Selection Complexity and Performance when Searching the Plane without Communication. In: Proceedings of the 33rd Symposium on Principles of Distributed Computing, PODC (to appear, 2014)
6. Albers, S., Henzinger, M.: Exploring Unknown Environments. SIAM Journal on Computing 29, 1164–1188 (2000)
7. Deng, X., Papadimitriou, C.: Exploring an Unknown Graph. Journal of Graph Theory 32, 265–297 (1999)
8. Diks, K., Fraigniaud, P., Kranakis, E., Pelc, A.: Tree Exploration with Little Memory. Journal of Algorithms 51, 38–63 (2004)
9. Panaite, P., Pelc, A.: Exploring Unknown Undirected Graphs. In: Proceedings of the 9th Annual ACM-SIAM Symposium on Discrete Algorithms (SODA), 316–322 (1998)
10. Reingold, O.: Undirected Connectivity in Log-Space. Journal of the ACM (JACM) 55, 17:1–17:24 (2008)
11. Aleliunas, R., Karp, R.M., Lipton, R.J., Lovasz, L., Rackoff, C.: Random Walks, Universal Traversal Sequences, and the Complexity of Maze Problems. In: Proceedings of the 20th Annual Symposium on Foundations of Computer Science (SFCS), pp. 218–223 (1979)
12. Aigner, M., Fromme, M.: A Game of Cops and Robbers. Discrete Applied Mathematics 8, 1–12 (1984)
13. Baeza-Yates, R.A., Culberson, J.C., Rawlins, G.J.E.: Searching in the Plane. Information and Computation 106, 234–252 (1993)
14. López-Ortiz, A., Sweet, G.: Parallel Searching on a Lattice. In: Proceedings of the 13th Canadian Conference on Computational Geometry (CCCG), pp. 125–128 (2001)
15. Angluin, D., Aspnes, J., Diamadi, Z., Fischer, M.J., Peralta, R.: Computation in Networks of Passively Mobile Finite-State Sensors. Distributed Computing, 235–253 (2006)
16. Aspnes, J., Ruppert, E.: An Introduction to Population Protocols. In: Garbinato, B., Miranda, H., Rodrigues, L. (eds.) Middleware for Network Eccentric and Mobile Applications, pp. 97–120. Springer (2009)
17. Emek, Y., Wattenhofer, R.: Stone Age Distributed Computing. In: Proceedings of the 32nd ACM Symposium on Principles of Distributed Computing (PODC) (2013)
18. Hopcroft, J.E., Ullman, J.D.: Introduction to Automata Theory, Languages, and Computation. Addison-Wesley (1979)

What Do We Need to Know to Elect in Networks with Unknown Participants?

Jérémie Chalopin, Emmanuel Godard and Antoine Naudin*

LIF, Université Aix-Marseille and CNRS, France

Abstract. A network with unknown participants is a communication network where the processes have very partial knowledge of the system. Nodes do not know the full set of participating nodes and some nodes do not even know the full set of nodes they can communicate directly with. It is a "contact list" like network where the initial communication is possibly asymmetric and one can communicate with an unknown neighbour only if one has been first contacted by this neighbour. This model is quite natural and of important theoretical interest. It has also proved useful for the study of bootstrapping mobile ad hoc networks. In this paper, we investigate the classical Leader Election problem in general networks with unknown participants.

We give the first necessary and sufficient condition on global knowledge that nodes should be provided in order to solve Election problem. Since Election problem is a useful benchmark in distributed computability investigations, this result could lead to a complete characterisation of what is solvable in networks with unknown participants.

Keywords: Distributed Algorithm, Message Passing, Leader Election, Distributed Computability, Unknown Participants, Structural Knowledge.

1 Introduction

A Natural Model for Distributed Computations. Distributed systems are pervasive and recently more and more interest has been in studying systems that range from dynamic to highly dynamic. Surprisingly there have been few studies of some models that are static but where the local connectivity evolves in a light way during the computation. The following "contact list model" is a fairly natural model related to the communication namespace necessitated by distributed computing.

Consider a set of participants that communicate with phones. Initially, everybody knows a subset of the phone numbers of other people via its personal contact list. It could even be not symmetric. Namely Alice could have the phone number of Bob, whereas Bob would not know the one of Alice. In this situation, Bob cannot call Alice, he does not even know that Alice is participating. Only when Alice has first contacted Bob (and the phone number of Alice is registered

---

* Work partially supported by Macaron project (ANR JCJC 13-JS02-0002-01).

M. Halldórsson (Ed.): SIROCCO 2014, LNCS 8576, pp. 279–294, 2014.

by the phone of Bob), then Bob can possibly call Alice. In this setting, the contact list (that is the set of neighbours) of Bob increases during the computation. This is a fairly natural model that exhibit general and interesting properties : connectivity is directed, adjacency is initially limited but increases over time, this increase is not automatic and depends of the communications that take place, everybody calling at its own pace. That is the system is asynchronous.

This model is natural and realistic, it is a slight variation of the model introduced in [CSS04] to investigate the self-organized bootstrapping of mobile ad hoc networks (MANETs). Conjointly with the fact that this model has not been very much studied, such systems exhibit new interesting properties related to the theory of distributed computability. We describe them more precisely later.

The Formal Model. The underlying communication graph is an arbitrary undirected graph denoted by G. Nodes are endowed with *identities*, they communicate by *messages*. Their neighbours are addressed with *port numbers* but this port numbering is not explicitly available to the nodes. Initially, a node can send messages only to a (*possibly strict*) *subset* of its actual neighbours in G: its *contact list* of neighbours, this defines the initial *directed* graph G^0. This contact list of neighbours will be extended whenever a message is received from an "unknown" in-neighbour. Therefore, at the end of the computation, the possible communication graph corresponds to G, as G is the undirected version of the initial digraph G^0. Note that the network is reliable but asynchronous: messages are always delivered but they can have unpredictable delays. In particular, a neighbour in G can be unpredictably long to appear in the contact list.

Solving the Leader Election Problem. We aim at a general distributed computability study of this model. We therefore introduce partial global knowledge in order to overcome the single sink condition of [CSS04] and we look for necessary and sufficient knowledge to solve a given problem. The leader election problem is a fundamental problem in distributed programming. It has also proved to be a good benchmark for distributed computability characterisation [YK89,YK96b,BCG+96,BV99,GM02,CGM08].

Our Results. In this paper, we give a simple characterisation of the partial knowledge that enables to solve Election problem in networks with unknown participants. From our results, it appears that in the unknown participants model, the partial knowledge that a node can have initially about the structure of the underlying network has a dramatic impact.

As a consequence, we prove that knowing the size of the network enables to solve Election problem on every network whereas knowing only an upper bound on the size is not enough. We also prove that knowing the number of sink components in the initial graph is also a sufficient condition to solve it.

Related Work. The network with unknown participants model presented here is a slight variation of a model that has been formally introduced and studied in [CSS04]. In [CSS04], processes are endowed with a participant detector returning a set of initial out-neighbours. Initial values from such a participant detector determines the initial communication network G^0. Moreover, the communication networks end up being a complete graph by allowing direct communication as soon as the network name of a node is learnt from a received message. In the model of this paper, the communication graph G becomes at most the undirected version of the initial graph G^0, but, computability is equivalent. Indeed, it is always possible to add a "routing layer" to build an end-to-end communication overlay in order to address a process which is not a neighbour. The meaning of "*knowledge*" is also different. In [CSS04], it is meant to correspond to the initial contact lists, that is G^0 with our notations. In this paper "knowledge" denotes the partial information that a node can have about the global structure.

In [CSS04], Cavin, Sasson and Schiper have investigated the Consensus Problem and showed a necessary and sufficient condition for computability of Consensus. They do not assume any partial knowledge about the initial graph and they show that the existence of at most one sink in the strongly connected components of the initial graph is both necessary and sufficient in order to solve Consensus. From a computability point of view, in this model, to the best of our knowledge, only the solvability of the Consensus Problem has been considered so far [CSS04]. Fault-tolerant versions of the Consensus problem were considered : in [GT07], the precise link in the unknown participant model between synchrony and fault-tolerance is given; in [ABFG08], byzantine faults are investigated; in [GSAS12], an eventually strong failure detector is presented.

So only [CSS04] seems to consider reliable networks with unknown participants. But going further in history, such studies for reliable communication networks, but with a partial knowledge, were actually introduced by Angluin [Ang80] in her seminal work for *anonymous networks*. The precise impact of some specific knowledge on distributed computability in anonymous networks has been thoroughly investigated by Yamashita and Kameda [YK96b,YK96a]. Boldi and Vigna have presented general computability results in [BV99].

We show here that, for Election problem, even without failure and with identities, there are still impossibility results. Our principal lemma used in proof for impossibility is a "Angluin's like" lemma called *Isolation Lemma*. It is extended to get a complete characterisation of which partial knowledge are sufficient and necessary to solve Election problem. It seems very likely that it is possible to leverage this lemma to get full computability results.

The problem of describing which arbitrary knowledge enables to solve the Election Problem was introduced in [GM02], where it is solved for a specific model. It has been solved for the standard message passing model in [CGM12]. These papers use quasi-simulation techniques introduced in [MMW97] but contrary to the model investigated in [YK96b,BV99,CGM12], it should be noted that unknown participants networks are communication networks where the difficulty arises from asynchrony and not from synchronous executions. It is therefore a qualitatively different model where computability mostly relates to termination detection and not to symmetry breaking.

The paper is organized as follows. First we present standard graph notation that we will use to describe formally the unknown participant model. We first present an algorithm that enables every node to compute the set of vertices one can reach from this node in the initial digraph. We then present the general

Isolation Lemma and derive our necessary condition on knowledge. Building on our first algorithm, we give an Election algorithm that proves that the necessary condition is actually sufficient. We conclude with some applications of our main theorem.

2   Graphs Properties and Reachable Vertices

Definitions are standard [RM00]. Let G be a directed graph (resp. undirected), where $V(G)$ is its set of vertices, and $E(G)$ is its set of arcs denoted (u, v) (resp. edges denoted $\{u, v\}$). A directed graph is called a digraph. We identify undirected graphs with symmetric digraphs and use graph and digraph interchangeably. Let $deg(v)$, *the degree of the vertex* v. We denote by $pred(v) = \{u|(u, v) \in E(G)\}$ (resp. $next(v) = \{w|(v, w) \in E(G)\}$) the set of predecessors (resp. successors) of v. A directed *path* c (resp. an undirected path $\bar{c}$) linking u and v is a sequence of disjoint vertices $\{s_1, ... s_k\} \subseteq V(G)$ where for all $i < k$, $(s_i, s_{i+1}) \in E(G)$ (resp. $(s_i, s_{i+1}) \in E(G)$ or $(s_{i+1}, s_i) \in E(G)$), $s_1 = u$ and $s_k = v$. The *length of a path* c, denoted by $|c|$, is equal to the numbers of arcs composing it and the directed *distance* d (resp. undirected distance $\bar{d}$) between vertices is the length of the smallest directed (resp. undirected) path in G between u and v. A *strongly connected* (resp. connected) digraph is a digraph where the directed (resp. undirected) distance between any two vertices is always defined. We will consider only connected digraphs. Any digraph can be decomposed in strongly connected subgraphs (called components). A component that has no successor is called a *sink*. A vertex v is *reachable* from u in G if there is a directed path from u to v. We denote by $Reach_G(v)$, the *set of vertices reachable from v* in G.

Remark 2.1. *The following propositions are equivalent:*
  (i) G is strongly connected.
  (ii) $\forall v, v' \in V, Reach_G(v) = Reach_G(v')$
  (iii) $\forall v, v' \in V, v' \in Reach_G(v)$
  (iv) $\forall v \in V, Reach_G(v) = V$

A graph H is a *subgraph* of a graph G if $V(H) \subseteq V(G)$ and $E(H) \subseteq E(G)$. A *labelled graph* $\mathbf{G} = (G, \lambda)$ is a graph G endowed with a labelling $\lambda : V \to \Lambda$ on its vertices or edges where Λ, is the set of labels. Note that if $\mathbf{H} = (H, \lambda_H)$ is a subgraph of $\mathbf{G} = (G, \lambda_G)$, then each node $v \in V(H)$ has the same label in $\mathbf{H}$ and in $\mathbf{G}$.

A *homomorphism* φ from H to G is a function $\varphi : V(H) \to V(G)$ such that for every $(u, v) \in E(H)$, there is $(\varphi(u), \varphi(v)) \in E(G)$. An *isomorphism* φ is a bijective homomorphism such that φ^{-1} is a homomorphism. A homomorphism (resp. isomorphism) φ from $\mathbf{G} = (G, \lambda_G)$ to $\mathbf{H} = (H, \lambda_H)$ is a homomorphism (resp. isomorphism) from G to H such that for each $v \in V(G)$, v and $\varphi(v)$ have the same label, i.e., $\lambda_G(v) = \lambda_H(\varphi(v))$.

Definition 2.2. *A subgraph $\mathbf{H}$ of $\mathbf{G}$ is a subgraph closed by successors of $\mathbf{G}$, denoted by $\mathbf{H} \sqsubseteq_\downarrow \mathbf{G}$, if for every $(u, v) \in E(G)$, if $u \in V(H)$ then $v \in V(H)$ and $(u, v) \in E(H)$.*

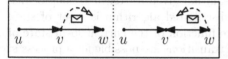 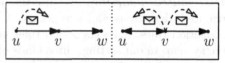

(a) v sends a message. w receives it. After the reception, w can talk to v

(b) u sends a message. v receives it. After the reception, v can talk to both u and its initial neighbour w

Fig. 1. Different executions for the same initial graph

This relation $\sqsubseteq\downarrow$ is extended for two arbitrary graphs $\mathbf{H'}$ and $\mathbf{G}$ where $\mathbf{H'}$ is isomorphic to a graph $\mathbf{H}$ and $\mathbf{H} \sqsubseteq\downarrow \mathbf{G}$. Note that if $\mathbf{H} \sqsubseteq\downarrow \mathbf{G}$ then for every $v \in V(H)$, $Reach_H(v) = Reach_G(v)$.

3   Model

The message passing model. A network is defined by a (possibly symmetric) digraph G, where $V(G)$ is the set of processes, and $E(G)$ is the set of communication channels. Processes communicate by sending and receiving messages via some ports. The communication channels linking ports between processes are asynchronous but reliable and *FIFO*.

Processes identities. Each process v is endowed with a unique label, $id_G(v)$, the identity of the process. We denote it by id_v if the context permits it. As there are several such labellings for a same graph (permutation, renaming), we consider all of them using an injective function $id_G : V \to \mathbb{N}$ giving a unique identity to every process. We denote by (G, id_G) such a graph and let $\mathcal{G}_{id}$ be the set of all connected graphs and their assignations of possible identities. For every family of graphs $\mathcal{F}$, we denote the family $\mathcal{F}_{id} = \{(G, id_G) | G \in \mathcal{F} \wedge (G, id_G) \in \mathcal{G}_{id}\}$.

Port labelling. Each process v can address its different neighbours using a bijective port numbering function $\delta_v : V \to \mathbb{N}$ giving a unique number to every port of v. When v receives a message from a neighbour w, it receives the message via the port $\delta_v(w)$. Since a process does not initially know all its neighbours, processes have access to a local variable denoted by CONTACTS containing the port numbers corresponding to their known neighbours. We explain its usage later. In order to ease notation, we consider a port labelling such that for all neighbours u, v in G, the port $\delta_u(v)$ corresponding to the channel linking u to v is denoted by $id_G(v)$, the identity of v. Therefore, CONTACTS contains the identities of known neighbours.

Graph labelling. The state of each process is represented by a label $\lambda(v)$ associated to the corresponding vertex $v \in V(G)$. Note that $\lambda(v)$ initially contains the identity of v and its known neighbours list CONTACTS$_v$ cited above. Let $\mathbf{G} = (G, \lambda)$, such a labelled graph. For all arcs $(u, v) \in E(G)$, let $B(u, v)$ be the queue containing the messages in transit from u to v. Initially, $B(u, v)$ is empty for all arcs (u, v).

Distributed algorithm. We use the definition given by Tel in [Tel00] for distributed algorithms and executions. A distributed algorithm is a set of state transition rules. Such transition rules are function of the current local state of the system. In our setting, three kinds of transitions are possible for a process v: it can modify its state, it can receive a message from a neighbour or it can send a message to all its known neighbours. See Figure 1.

Let $(\lambda_v, in, m) \vdash (\lambda'_v, send, m')$, denote a recursive relation on state transition of a process v (the current state is function of the previous), with λ_v, the state of v before transition and λ'_v, its state after, m and m' are messages, in is the incoming port number (or $\bot$) and $send$ is either $\bot$ or all. When v modify its state, $in = send = \bot$ and $m = m' = \bot$; the state of v becomes λ'_v after this transition. When v receives a message m sent by u, $in = id_u$, $send = \bot$, $m \neq \bot$ and $m' = \bot$; the message m is in $B(u, v)$ and v receives m via the port id_u. If id_u is not known, v updates CONTACTS$_v$, the list of the neighbours it knows. The state of v becomes λ'_v after this transition. When v sends a message m', it sends the same message to all its known neighbours using a primitive SendAll. In this case, $in = \bot$, $send = all$, $m = \bot$ and $m' \neq \bot$; the message m' is added to all queues $B(v, w)$ such that $id_w \in$ CONTACTS$_v$, and the state of v becomes λ'_v after this transition.

A distributed algorithm $\mathcal{A}$ in the message passing model is a set of algorithms $(\mathcal{A}_v)_{v \in V(G)}$ distributed over the nodes of the network. A transition of the algorithm is a transition of a process v according to its local algorithm $\mathcal{A}_v$.

Distributed algorithm execution representation. An execution ρ of a distributed algorithm $\mathcal{A}$ is a sequence of changes on vertices state. An execution is represented by a sequence of couples $[(\lambda^0, B^0), (\lambda^1, B^1), ..., (\lambda^n, B^n)]_\rho$ where, at step i, λ^i is the state of the system and B^i, the set of messages in transit. Let λ^i_v, the state of vertex v at step i. The initial state λ^0_v is the state of process v before the execution. A transition from step i to step $i+1$ is performed by one and only one process which executes a transition cited above in its local algorithm. This transition leads to the next state of v: λ^{i+1}_v and B^{i+1} and all other processes keep the same state as in λ^i. Note that any asynchronous execution (including the synchronous execution) can be represented this way.

Problem Specification. A specification of a problem has to describe the expected relations on the initial and final labelling of the graph where the problem has to be solved. As processes have to take a decision or compute some values in order to address a problem, we use a dedicated label. Let $out_G(v)$ be the label indicating the *decision computed* by a process v in G, denoted by $out(v)$ if the context permits. Note that a *final labelling* of a graph, denoted by $\mathbf{G}^{out} = (G, \lambda^{out})$ contains, for every process v, its final decision $out(v)$ ($\bot$ if the process has not decided any value). We define a *specification* as a *relabelling relation* S between initial labelled graphs and final labelled graphs.

Execution and Algorithm Properties. An execution *stabilises* if there is a step i_0 where no process can progress in its local algorithm and no message is in transit. In an execution, a process v *decides* if it eventually writes a value in *out* and if it does it only once during the execution. An execution terminates if it stabilises and if every process decides. An algorithm terminates on (G, λ^{in}) with if every execution of the algorithm on (G, λ^{in}) terminates. In an execution ρ that terminates, each process v has an output value $out_v \neq \bot$; in this case, we say that out_v is the final label of v in ρ.

Let S be the specification of a problem. An execution ρ of $\mathcal{A}$ in a graph $(G, \lambda^{in}) \in \mathcal{G}^{in}$ satisfies the *Correction Property* if ρ terminates and (G, λ^{out}), the final labelling computed by ρ satisfies $(G, \lambda^{in})S(G, \lambda^{out})$. An algorithm $\mathcal{A}$ is *valid for a specification* S in a graph $(G, \lambda^{in}) \in \mathcal{G}^{in}$ if every execution ρ satisfies the Correction Property. We will say that $\mathcal{A}$ solves S on (G, λ^{in}) in such a case.

Knowledge and Family. As we will see, some problems need additional global information or knowledge to be solved. This information about the underlying network (e.g. a bound on the size of the system) is inserted in the initial label. Consider a function κ that encodes an arbitrary knowledge. An algorithm $\mathcal{A}$ solves S with knowledge κ, if for all G, $\mathcal{A}$ solves S on $(G, \kappa(G))$. Equivalently we have that, for any $\alpha \in \kappa^{-1}(\mathcal{G}_{id})$, there exists an algorithm $\mathcal{A}_\alpha$ that solves S on the family $\mathcal{F} = \kappa^{-1}(\alpha)$.

Solving a problem with partial knowledge is simply, for any possible value α of knowledge, solving the problem within the family of networks whose knowledge value is α. Considering arbitrary families of labelled graphs enables to represent any initial knowledge: e.g. if the processes initially know the size n of the network, then in the corresponding family $\mathcal{F}^{(n)}$, for each $\mathbf{G} \in \mathcal{F}^{(n)}$ and each $v \in V(G)$, $n = |V(\mathbf{G})|$ is a component of the initial label of v.

Universal Algorithm. We say that an algorithm solving a specification on all graphs of $\mathcal{G}_{id}$ is a universal algorithm. An algorithm is $\mathcal{F}-universal$ if the algorithm solves S for all graphs of the family $\mathcal{F}$. Abusively, We will say that an algorithm $\mathcal{A}$ is $\mathcal{F} - universal$ if $\mathcal{A}$ is $\mathcal{F}_{id} - universal$.

An algorithm is not universal when it is not correct for all graphs but it can be $\mathcal{F} - universal$. So, for every problem without a *universal* algorithm, we look for the necessary and sufficient condition on the knowledge, *i.e.* on the family $\mathcal{F}$, such that the problem can be solved with a $\mathcal{F} - universal$ algorithm.

4   Cartography of Reachable Vertices

REACH or "Cartography of Reachable Vertices Problem" is the problem consisting, for every vertex v in the digraph G^0, to compute a graph isomorphic to subgraph induced by the network initially accessible from v. We denote by $\mathbf{G}_{|Reach(v)}$ such a subgraph of G^0. This problem is investigated because its solution will be used as the basis of our main Election Algorithm. Interestingly, it also admits a universal algorithm (Algorithm 1).

Description of the algorithm. We first introduce the variables used by the algorithm. Each process v initially knows id_v, its identity and $Succ_v = \{id_{v'} \mid v' \in next(v)\}$, the set of the ids of its neighbours in the network it initially knows. These variables are not modified during the execution of the algorithm. Our algorithm is a flooding algorithm where each node v eventually collects the value of $(id_u, Succ_u)$ for every process u.

Algorithm 1. REACH Algorithm.

Output: *out*, Graph induced by M_v

1 **I:**(Initial Procedure) **begin**
2 $\quad\lfloor$ Send $<id, M>$ to all identities of vertices into CONTACTS;

3 **R:**(Receiving a message $<id_u, M_u>$ from u:) **begin**
4 $\quad$ **if** $id_u \notin$ CONTACTS *or* $M_u \setminus M \neq \emptyset$ **then**
5 $\quad\quad$ $M \leftarrow M \cup M_u$;
6 $\quad\quad$ CONTACTS $\leftarrow$ CONTACTS $\cup \{id_u\}$ if $id_u \notin$ CONTACTS;
7 $\quad\quad\lfloor$ Send $<id, M>$ to all identities of vertices into CONTACTS;
8 $\quad$ **if** $View(M) = Covered(M) \wedge out = \perp$ **then**
9 $\quad\quad\lfloor$ $out \leftarrow \mathcal{C}(M)_{|Reach(id_v)}$;

Each process v has also a variable CONTACTS$_v$ containing the list of the ids of its neighbours it knows, either because it initially knows them, or because it receives a message from them. Initially CONTACTS$_v = Succ_v$ and CONTACTS$_v$ is updated each time v receives a message from a neighbour u such that $id_u \notin$ CONTACTS$_v$. Finally, each process v has a mailbox M_v containing pairs of the form $(id, Succ)$. Intuitively, the mailbox contains all the information v has about the network. Initially, $M_v = \{(id_v, Succ_v)\}$. When a vertex v sends a message to its neighbours, it always sends a message of the form $< id_v, M_v >$, i.e., it sends all the information it has on the network.

Our algorithm is a flooding algorithm described by two rules. Initially, each process applies the rule **I** to send its initial mailbox (containing only $(id_v, Succ_v)$) to all its neighbours. The rule **R** is executed whenever a process receives a message $< id_u, M_u >$ from a neighbour u. If the received mailbox provides new entries, then the process learns new information about the network and it updates its mailbox. Moreover, if id_u is not in CONTACTS, then id_u is added to CONTACTS. Then, if the process has learned new information (i.e., if its mailbox or CONTACTS has changed), it sends a copy of its new mailbox to all its neighbours.

Computing a map from a mailbox. In order to explain the rule allowing a process v to writes a value in out_v, we need to first explain how to use the content of a mailbox to construct a digraph similar to the network communication graph. To do so, we define three functions: $View$, $Covered$ and $\mathcal{C}$ as follows.

- $Covered(M) = \{id_v | (id_v, Succ_v) \in M\}$
- $View(M) = \{id_v | \exists (id_u, Succ_u) \in M \wedge id_v \in Succ_u\} \cup Covered(M)$
- $\mathcal{C}(M) = (V_{\mathcal{C}}, E_{\mathcal{C}})$ is a digraph such that $V_{\mathcal{C}} = View(M)$ and
  $E_{\mathcal{C}} = \{(id, id') \mid (id, Succ) \in M \text{ and } id' \in Succ\}$

With those functions, we can prove that if M_v contains the list of successors of every node in the network, the reconstructed graph is isomorphic to the initial communication graph. By construction, the following lemma is proved.

Lemma 4.1. $\mathcal{C}\left(\{(id_v, Succ_v) \mid v \in V(G)\}\right) \simeq \mathbf{G}$

During the execution of the algorithm, as long as $Covered(M_v) \neq View(M_v)$, v can detect that it has not yet received the initial information from all the processes. When $Covered(M_v) = View(M_v)$, v can reconstruct a graph from M_v and it is possible that $\mathcal{C}(M_v)$ is isomorphic to $\mathbf{G}$ but it is not necessary. However, we will show that in this case, $\mathbf{G}_{|Reach(v)} \sqsubseteq_\downarrow \mathcal{C}(M_v)$, and consequently, v can compute $\mathbf{G}_{|Reach(v)}$ by performing a depth-first traversal of $\mathcal{C}(M_v)$ from id_v and v can decide this value. Note that a mailbox can satisfy the constraint $View(M) = Covered(M)$ several times; this is due to the asynchrony of communications. We will elaborate on this interesting property later.

Properties of the algorithm. In order to prove the termination property and the correction of the algorithm, we start by some lemmas on the properties about the content of the mailbox. First, since processes have unique identities, we get the following lemma.

Lemma 4.2 (bounded content). *For every step i and process v, $\mathcal{C}(M_v^i)$ is a subgraph of $\mathcal{C}(\{(id_v, Succ_v) \mid v \in V(G)\})$.*

The next lemma shows that there exists an increasing order on the mailbox content during an execution.

Lemma 4.3. *For every execution ρ of an algorithm, for every process v that executes a transition at step i, $M_v^i \subseteq M_v^{i+1}$, $\text{CONTACTS}_v^i \subseteq \text{CONTACTS}_v^{i+1}$ and v sends a message if and only if $M_v^i \subset M_v^{i+1}$ or $\text{CONTACTS}_v^i \subset \text{CONTACTS}_v^i$.*

Proof. Processes update their local state only when a message is received. Let $m = <id_u, M_u>$, the message received by v from u at step i. If $id_u \notin \text{CONTACTS}_v^i$, then id_u is added to CONTACTS_v^{i+1} (line 6). So, $\text{CONTACTS}_v^i \subset \text{CONTACTS}_v^{i+1}$ and v sends messages to its neighbours. If $M_u \setminus M_v^i \neq \emptyset$, then an update of M_v^i is operated by v, and $M_v^i \subset M_v^{i+1}$. After this update (procedure R at line 5), v will send messages to its neighbours. Otherwise, we get $M_u \subseteq M_v^i$ and $id_u \in \text{CONTACTS}_v^i$. Thus, v performs no action during this step. $\square$

By Lemma 4.2, the mailbox's content can only take a finite number of values, and by Lemma 4.3, it is increasing during any execution. Since, by Lemma 4.3, messages are only sent when the content of a mailbox is modified, there is a step i where the algorithm stabilises. We show in the next lemma that eventually each process gathers all available information.

Lemma 4.4 (Reception). *For all v, v', there is a step i where $(id_{v'}, Succ_{v'}) \in M_v^i$.*

Proof. We prove this lemma by an induction hypothesis on $\overline{d}(v, v')$. First, assume that $\overline{d}(v, v') = 1$. Consider a step $h \leq i$ such that $M_v^h = M_v^i$ and $M_v^{h-1} \neq M_v^h$. At step h, v sends its mailbox M_v^h to its neighbourhood. We distinguish two cases: either v knows v' at step h or not.

Case 1: $\mathbf{id_{v'}} \in \mathbf{contacts^h(id_v)}$. Since $id_{v'} \in \text{CONTACTS}^h(id_v)$, v sends a message $< id_v, M_v^h >$ to v' at step h. Since the channel are reliables, there exists a step $j > h$ where v' receives $< id_v, M_v^h >$ and thus, $M_v^i = M_v^h \subseteq M_{v'}^j$.

Case 2: $\mathbf{id_{v'}} \notin \mathbf{contacts^h(id_v)}$. Since $id_{v'} \notin \text{CONTACTS}^i(id_v)$ and $\overline{d}(v, v')$, $id_v \in \text{CONTACTS}^0(id_{v'})$. Since v' applies eventually the rule **I** of the algorithm, and since the channels are reliable, there is a step $j > 0$ where v receives a message from v'. Since $id_{v'} \notin \text{CONTACTS}_v^h$, it implies that $j > h$. Thus, $M_v^i = M_v^h \subseteq M_v^j$. At step j, when v receives the message from v', the algorithm ensures that v' is added to CONTACTS_v and that a message $< id_v, M_v^j >$ is sent to all the known neighbours of v including v'. By the previous case, there exists a step j' such that $M_v^j \subseteq M_{v'}^{j'}$ and since $M_v^i \subseteq M_v^j$, we are done.

Suppose now that $\bar{d}(v, v') > 1$. Let w be a neighbour of v such that $\bar{d}(w, v') = \bar{d}(v, v') - 1$. From the case where $\bar{d}(v, v') = 1$, we know that there exists a step j such that $M_v^i \subseteq M_w^j$. By induction hypothesis, there exists a step j' such that $M_w^j \subseteq M_{v'}^{j'}$. So $M_v^i \subseteq M_{v'}^{j'}$. □

From Lemmas 4.3 and 4.4, there exists a step i such that for all v, v', $(id_{v'}, Succ_{v'}) \in M_v^i$. Thus, the condition on line 8 is eventually satisfied for every process, proving that *every execution of the algorithm terminates*. It remains to prove that when v decides a value out_v isomorphic to $\mathbf{G}_{|Reach(v)}$.

Lemma 4.5 (Correction). *For every process v, if $View(M_v) = Covered(M_v)$ then for every $w \in Reach_G(v)$, $id_w \in Covered(M_v)$. Consequently, $\mathcal{C}(M_v) \sqsubseteq\downarrow \mathbf{G}$.*

Proof. By contradiction, let $w \in Reach_G(v)$ be the closest process to v in $Reach_G(v)$ such that $id_w \notin Covered(M_v)$. Let $w' \in Reach_G(v)$ be a predecessor of w belonging to a shortest path from v to w. By the choice of w, $id_{w'} \in Covered(M_v)$. Thus, $id_w \in Succ_{w'}$, we get $id_w \in Covered(M_v) = View(M_v)$, a contradiction. □

Now, from Lemmas 4.4 and 4.5 which prove that every process v decides $\mathbf{G}_{|Reach(v)}$, we can give a first theorem on the computability of the REACH Problem:

Theorem 1. *There is an universal algorithm for* REACH.

5 Isolation Lemma

In this section, we present an isolation lemma to prove impossibility results caused by isolated executions in a subset of the network. As it has proved for anonymous network to be the basis for all impossibility proofs [Ang80,Cha06], the isolation lemma is presented like a lifting lemma. Initially, a process v knows only its outgoing neighbourhood. Any other neighbour u of v cannot receive a message from v before v received a message from u. If $H \sqsubseteq\downarrow G$ and if all messages sent from processes in $V(G) \setminus V(H)$ to processes in $V(H)$ are arbitrary delayed (the communication is asynchronous), the processes of H are isolated from the rest of the network and execute the algorithm as if they were only in H, without discovering their neighbours outside H before deciding. If such an execution terminates, isolated processes decide a final value for an execution in H and not in G.

We introduce a new notation in order to represent an extended labelled graph with messages in transit. Let $\underline{\mathbf{H}} = (H, \lambda_H, B_H)$ where (H, λ_H) is a labelled graph and B_H, its queues of messages. The relation $\sqsubseteq\downarrow$ is extended between such graphs as follow, $(H, \lambda_H, B_H) \sqsubseteq\downarrow (G, \lambda_G, B_G)$ if $(H, \lambda_H) \sqsubseteq\downarrow (G, \lambda_G)$ and for every $(u, v) \in E(H)$, $B_H(u, v) = B_G(u, v)$. First, we remark that by isomorphism, a labelling of a graph $\mathbf{G}$ induces a labelling for every subgraph $\mathbf{H}$ of $\mathbf{G}$. Such initial labellings are independent of the algorithm used and satisfies the relation $\sqsubseteq\downarrow$ between $\mathbf{H}$ and $\mathbf{G}$.

Remark 5.1 (Initialisation). *For every digraphs G and H such that $H \sqsubseteq\downarrow G$, for every initial labelling λ_G^0 of G, there is a labelling λ_H^0 of H, such that $(H, \lambda_H^0, \emptyset) \sqsubseteq\downarrow (G, \lambda_G^0, \emptyset)$.*

Next, we prove that any step of an execution on a graph $\mathbf{H}$ can be executed on every graph $\mathbf{G}$ satisfying $\mathbf{H} \sqsubseteq\downarrow \mathbf{G}$.

Lemma 5.2 (One step of execution). *For all $\underline{\mathbf{H}} = (H, \lambda_H, B_H)$ and $\underline{\mathbf{G}} = (G, \lambda_G, B_G)$, if $\underline{\mathbf{H}} \sqsubseteq\downarrow \underline{\mathbf{G}}$ then every transition $(\lambda_H, in, m) \vdash (\lambda_H', send, m')$ executed on $\underline{\mathbf{H}}$ can be executed on $\underline{\mathbf{G}}$. The graphs $\underline{\mathbf{H}}' = (H', \lambda_H', B_H')$ and $\underline{\mathbf{G}}' = (G', \lambda_G', B_G')$ obtained after the transition satisfy $\underline{\mathbf{H}}' \sqsubseteq\downarrow \underline{\mathbf{G}}'$.*

Proof. We prove this lemma by constructing a similar execution in $\underline{\mathbf{H}}$ and $\underline{\mathbf{G}}$ which preserves the relation $\sqsubseteq\downarrow$. A step of the execution corresponds to a local transition of a process v. If a process v sends a message m' to its neighbours in H, since for all $v \in V(H)$, $\text{CONTACTS}_H(v) = \text{CONTACTS}_G(v)$, v can send the same message to the same nodes in G, and consequently, for all $(v, v') \in E(H)$, $B_G(v, v') = B_H(v, v')$. Since v ends up in the same state in H and in G, $\underline{\mathbf{H}}' \sqsubseteq\downarrow \underline{\mathbf{G}}'$. Suppose now that a message $m \in B_H(v', v)$ is received from a process $v' \in V(H)$ via a port in in H. Since $(v', v) \in E(H)$; $B_G(v', v) = B_H(v', v)$ and thus v can also receive the message m from v' in G. Since m is removed from both queues, $B_G(v', v) = B_H(v', v)$. Note that v ends up in the same state in H and in G. In particular, if v' was not known by v, then v can now communicate with v' in H and in G. Consequently, $(H', \lambda_H', B_H') \sqsubseteq\downarrow (G', \lambda_G', B_G')$. $\qquad\square$

For any graphs $\mathbf{G}$ and $\mathbf{H}$ such that $\mathbf{H} \sqsubseteq\downarrow \mathbf{G}$, for any algorithm $\mathcal{A}$ and any execution ρ_H on $\mathbf{H}$, we can apply the previous lemma iteratively to construct an execution ρ_G on $\mathbf{G}$ such that only the vertices in $V(H)$ are active in ρ_G and they behave exactly like in ρ_H. In such a case, we say that ρ_H is $\sqsubseteq\downarrow$ −lifted on $\mathbf{G}$. When considering executions that terminate, we get the following lemma.

Lemma 5.3 (Isolation Lemma). *For all $\mathbf{G}, \mathbf{H}$ such that $\mathbf{H} \sqsubseteq\downarrow \mathbf{G}$, for every execution ρ_H of any algorithm $\mathcal{A}$ on H which terminates, there is an execution ρ_G of $\mathcal{A}$ on G such that $\forall v \in V(H)$, $out_{\rho_H}(v) = out_{\rho_G}(v)$.*

6 Election Algorithm

In this section, we study the classical election problem, denoted by ELEC: one and only one process has to decide LEADER and all others should decide FOLLOWER.

We prove that contrary to the classical model, even if nodes have identities, it is not possible to solve ELEC with a *universal algorithm*. Therefore, we give a necessary and sufficient condition that determines which additional knowledge enables to solve ELEC. The impossibility proof uses a standard simulation technique, based on the Isolation Lemma. To show that it is a sufficient condition, we show how processes can avoid $\sqsubseteq\!\downarrow$-lifted executions by delaying their decision when they are provided some additional knowledge satisfying this condition.

A Necessary and Sufficient Condition. If we have isolated executions, we might get more than one processes elected. So the condition on knowledge below enables to somehow forbid disjoint isolated executions.

Definition 6.1 ($\mathbb{C}_{\text{ELEC}}$). *A family of graphs $\mathcal{F}$ satisfies $\mathbb{C}_{\text{ELEC}}$ if for every graphs* $\mathbf{G}, \mathbf{H_1}, \mathbf{H_2} \in \mathcal{F}$ *such that* $\mathbf{H_1}, \mathbf{H_2} \sqsubseteq\!\downarrow \mathbf{G}$, *we have* $V(H_1) \cap V(H_2) \neq \emptyset$.

We first prove that $\mathbb{C}_{\text{ELEC}}$ is necessary,

Lemma 6.2. *If there is an $\mathcal{F}$-universal algorithm solving* ELEC *then $\mathcal{F}$ satisfies* $\mathbb{C}_{\text{ELEC}}$.

Proof. Let $\mathcal{F}$ a family of labelled graphs that does not satisfy $\mathbb{C}_{\text{ELEC}}$ and $\mathcal{A}$, a $\mathcal{F}$-universal algorithm solving ELEC. There are three graphs $\mathbf{H_1}$, $\mathbf{H_2}$ and $\mathbf{G}$ in $\mathcal{F}$ such that $\mathbf{H_1}, \mathbf{H_2} \sqsubseteq\!\downarrow \mathbf{G}$ and $V(H_1) \cap V(H_2) = \emptyset$. We build an execution ρ_G on G as follows: the Isolation Lemma can be applied for H_1 and G, thus ρ_G can begin by a $\sqsubseteq\!\downarrow$ –lifted execution on H_1. Since $\mathcal{A}$ is $\mathcal{F}$-universal and $H_1 \in \mathcal{F}$, there is one elected process v_1. As H_1 and H_2 are vertex-disjoint and Isolation Lemma can also be applied for H_2 and G, we can extend ρ_G by taking a second $\sqsubseteq\!\downarrow$ –lifted execution on H_2. Since $\mathcal{A}$ is $\mathcal{F}$-universal and $H_2 \in \mathcal{F}$, there is one elected process v_2 with $v_2 \neq v_1$ because $V(H_1) \cap V(H_2) = \emptyset$. At this step, the labelling of $\mathbf{G}$ is not valid for ELEC because there are two elected vertices and their decisions are final for the execution. □

To prove that $\mathbb{C}_{\text{ELEC}}$ is a sufficient condition, we propose a $\mathcal{F}$-universal algorithm for any family $\mathcal{F}$ satisfying $\mathbb{C}_{\text{ELEC}}$. The algorithm presented below is an extension of Algorithm 1.

Description of the algorithm. In the algorithm, a process v does not only broadcast $Succ_v$, but it also broadcasts the tuple $(id_v, M_v, status_v)$ where $status_v \in \{\text{LEADER}, \text{FOLLOWER}, \bot\}$ is the content of out_v. In such a way, a process u can detect whether v has been elected or not and it can learn what v knows about the other processes. To do so, each process v has now a "super" mailbox, denoted $\mathcal{V}_v$, containing tuples of the form $(id, M, status)$. When a process v sends a message to its neighbours, it always sends a message of the form $< id_v, \mathcal{V}_v >$ (i.e., it sends its super mailbox instead of its mailbox). This structure gives, to processes, a view of the local states of the known processes. In order to avoid some $\sqsubseteq\!\downarrow$ –lifted executions, we have to delay the decisions of the processes for a sufficiently long period using an additional knowledge. Before a process writes in out, it checks that the reconstructed graph is in $\mathcal{F}$; to do so, we assume that each process knows the *characteristic function* $\chi_{\mathcal{F}}$ of the family $\mathcal{F}$ (the additional knowledge). When called on $\mathcal{C}(M)$, this function returns true if $\mathcal{C}(M) \in \mathcal{F}$ and false otherwise. To prevent two processes to be elected at the same time, we add an additional condition related to the supermailbox of all known processes at line 9. This condition ensures that two processes that want

---

Algorithm 2. A $\mathcal{F}$-universal Election Algorithm

Input: $\chi_{\mathcal{F}}$, Characteristic function of $\mathcal{F}$

1 **I:**(Initial Procedure) **begin**
2 | SendAll $<id, \mathcal{V}>$;

3 **R:**(Receiving a message $<id_u, \mathcal{V}_u>$ from u) **begin**
4 | **if** $id_u \notin$ CONTACTS or $\mathcal{V}_u \setminus \mathcal{V} \neq \emptyset$ **then**
5 | $M \leftarrow M \cup \bigcup_{\{M_w | \exists (id_w, M_w, status_w) \in \mathcal{V}_u\}} M_w$;
6 | $\mathcal{V} \leftarrow \mathcal{V} \cup \mathcal{V}_u \cup \{(id, M, out)\}$;
7 | CONTACTS $\leftarrow$ CONTACTS $\cup \{id_u\}$ if $id_u \notin$ CONTACTS ;

8 | **if** $out = \bot \wedge \chi_{\mathcal{F}}(\mathcal{C}(M)) \wedge View(M) = Covered(M) \wedge \forall id_w \in$
      $View(M), \exists (id_w, M, status_w) \in \mathcal{V}$ **then**
9 | **if** $id = \min\{id' \mid id' \in View(M)\} \wedge \nexists (id_w, M_w, \text{LEADER}) \in \mathcal{V}$ **then**
10 | $out \leftarrow$ LEADER;
11 | **else**
12 | $out \leftarrow$ FOLLOWER;
13 | $\mathcal{V} \leftarrow \mathcal{V} \cup \{(id, M, out)\}$;

14 | **if** $\mathcal{V}$ or CONTACTS *have changed* **then**
15 | SendAll $<id, \mathcal{V}>$;

---

to decide LEADER have to actually know the state of each other. Changes between the previous and the next algorithm are the content of the messages (line 2), the manipulation of the new structure (lines 5 and 6) and the condition of termination (lines 8 and 9) as seen above.

Properties of the algorithm. Consider an execution ρ_G of the algorithm on G. Note that as in the previous algorithm M_v can only take a finite number of values and can only increase during the execution of the algorithm. Consequently, as before, there exists a step i where all processes have the same mailbox $M = \{(id_v, Succ_v) \mid v \in V(G)\}$.

The following lemma is immediate and ensures that for each process v, $\mathcal{V}_v$ can only take a finite number of values.

Lemma 6.3. *For every process v, for every step i, for every $(id_w, M, status) \in \mathcal{V}_v^i$, there is a step $i' \leq i$ such that $M_w^{i'} = M$ and $out_w^{i'} = status$.*

Similarly as for mailboxes in the previous algorithm, we can show that for each step i, $\mathcal{V}_v^i \subseteq \mathcal{V}_v^{i+1}$ and v sends some messages at step i if and only if $\mathcal{V}_v^i \subsetneq \mathcal{V}_v^{i+1}$. Since for every v, the content of $\mathcal{V}_v$ is increasing and can only take a finite number of values, the execution stabilises. An eventually, each process knows the state of all other processes. The proof is similar to the proof of Lemma 4.4.

Lemma 6.4. *For every processes v, v', for every step i, there is a step $i' > i$ such that $(id_v, M_v^i, out_v^i) \in \mathcal{V}_{v'}^{i'}$.*

Consequently, there exists a step where all processes have the same supermailbox. Note that in such a step, all processes have the same mailbox $M = \{(id_v, Succ_v) \mid v \in V(G)\}$, and for all $v, w \in V(G)$, there exists $(id_v, M, status) \in$

$\mathcal{V}_w$. Note that in this case, the condition on line 8 is satisfied and thus, eventually, each process decides a value. Therefore, the execution terminates. It remains to show that in each execution, there is always one and only one elected process.

Lemma 6.5. *For every execution, there is at least one elected process.*

Proof. Let id_{min} be the minimum identity present in the network and let v be the unique process such that $id_v = id_{min}$. Since every process eventually decides, there is a step i where the state of v satisfies the condition on line 8. We consider two cases. Either there exists some $(id_w, M_v, \text{LEADER}) \in \mathcal{V}_v^i$ and by Lemma 6.3, w has been elected and the lemma is proved. Or, there is no $(id_w, M_v, \text{LEADER}) \in \mathcal{V}_v^i$ and then v writes LEADER in $out(v)$. In both cases, there is always at least one elected process. $\qquad\square$

Lemma 6.6. *For every execution ρ, there is at most one elected process.*

Proof. Suppose that there are two processes u and v elected at step i and j in an execution ρ. Wlog, we assume that $i \leq j$. Let $H_u = \mathcal{C}(M_u^i)$ and $H_v = \mathcal{C}(M_v^j)$. Since u and v are elected respectively at step i and j, $View(M_u^i) = Covered(M_u^i)$, $H_u \in \mathcal{F}$, $View(M_v^j) = Covered(M_v^j)$ and $H_v \in \mathcal{F}$. Consequently, $H_u \sqsubseteq\downarrow G$ and $H_v \sqsubseteq\downarrow G$. Two cases are possible when u is elected at step i: either $id_v \in View(M_u^i)$, or not.

Case 1: $id_v \in View(M_u^i)$. In this case, line 9 ensures that $id_u < id_v$. Moreover, lines 8 and 9 and Lemma 6.3 ensure that there is a step $h < i$ such that $M_v^h = M_u^i$ and $out_v^h = \bot$. Since $id_u \in View(M_u^i) = View(M_v^h)$, and since M_v can only increase during the execution (recall that $h < i < j$), $id_u \in View(M_v^j)$. Consequently, the condition on line 9 is not satisfied by the state of v at step j, and v is not elected in this case.

Case 2: $id_v \notin View(M_u^i)$. Since $H_u \sqsubseteq\downarrow G$ and $H_v \sqsubseteq\downarrow G$ and since $\mathcal{F}$ satisfies $\mathbb{C}_{\text{ELEC}}$, there exists w such that $id_w \in V(H_u) \cap V(H_v)$. By Lemma 6.3 and since $id_w \in V(H_u)$, we know that there is a step $i' \leq i$ where $M_w^{i'} = M_u^i$. Similarly, we know there exists a step $j' \leq j$ where $M_w^{j'} = M_v^j$. Note that $id_v \notin View(M_w^{i'})$ and that $id_v \in View(M_w^{j'})$. By Lemma 4.3, it implies that $i' < j'$ and that $M_u^i = M_w^{i'} \subsetneq M_w^{j'} = M_v^j$. Therefore $id_u \in View(M_v^j)$ and since v is elected at step j, there is some $(id_u, M_v^j, status) \in \mathcal{V}_v^j$. By Lemma 6.3, it implies that there exists a step $i'' < j$ such that $M_u^{i''} = M_v^j$ and $out_u^{i''} = status$. Since $id_v \in View(M_u^{i''})$ and $id_v \notin View(M_u^i)$, Lemma 4.3 implies that $i < i''$. Since $out_u^{i''} = out_u^{i+1} = \text{LEADER}$, $(id_u, M_v^j, \text{LEADER}) \in \mathcal{V}_v^j$. Thus, the condition on line 9 is not satisfied by the state of v at step j, and v is not elected in this case. $\qquad\square$

Consequently, any execution of Algorithm 2 terminates and leads to one elected process if $\mathcal{F}$ satisfies $\mathbb{C}_{\text{ELEC}}$. Together with Lemma 6.2, we get

Theorem 2. *There is an $\mathcal{F}$-universal algorithm for ELEC if and only if $\mathcal{F}$ satisfies $\mathbb{C}_{\text{ELEC}}$.*

Applications. As first example, given $n \in \mathbb{N}$, consider the family of graphs with n vertices, denoted by $\mathcal{G}^{(n)}$. As every strict subgraph H of a graph $G \in \mathcal{G}^{(n)}$ has strictly less than n vertices, $H \notin \mathcal{G}^{(n)}$ and $\mathcal{G}^{(n)}$ trivially satisfies $\mathbb{C}_{\text{ELEC}}$. It is also possible to directly design an Election algorithm by simply waiting until *Covered* is of size n. Another, maybe less obvious, example where ELEC is possible is the family of graphs with n sink components, denoted by $\mathcal{P}^{(n)}$. This family satisfies $\mathbb{C}_{\text{ELEC}}$ because every two subgraphs H, H' of a graph $G \in \mathcal{P}^{(n)}$ have to share the n sink components if they also belong to $\mathcal{P}^{(n)}$. Thus, H and H' can not be disjoint.

Note that in [CSS04], the authors consider only families that are closed by $\sqsubseteq\downarrow$. Such families satisfy $\mathbb{C}_{\text{ELEC}}$ if and only if their graphs have only one sink. In this case, we obtain for ELEC a "one sink" condition similar to the one given in [CSS04].

Families defined by having a bound on the size are also closed by $\sqsubseteq\downarrow$. The corresponding families contains graphs with two sinks and it is therefore impossible to elect knowing a bound. An integer k is a *tight bound* for the size of G if $|V(G)| \leq k < 2|V(G)|$. Given $k \in \mathbb{N}$, the family $\mathcal{B}^{(k)}$ of graphs with a tight bound k admits an Election algorithm because $\mathcal{B}^{(k)}$ satisfies $\mathbb{C}_{\text{ELEC}}$. Indeed, consider $\mathbf{G}, \mathbf{H}, \mathbf{H}' \in \mathcal{B}^{(k)}$, k being a tight bound for $\mathbf{H}, \mathbf{H}'$, and $\mathbf{G}$ implies that $|V(H)| + |V(H')| > |V(G)|$. So when $\mathbf{H} \sqsubseteq\downarrow \mathbf{G}, \mathbf{H}' \sqsubseteq\downarrow \mathbf{G}$, we get $\mathbf{H} \cap \mathbf{H}' \neq \emptyset$. This majority argument also applies to the family of graphs where a tight bound is known for the number of sinks, so ELEC is also solvable in this case.

One major consequence of Theorem 2 is that ELEC cannot be solved on $\mathcal{G}_{id}$, and, furthermore, there is no maximum family, *i.e.* maximum knowledge, for which ELEC is solvable. Given any graph $\mathbf{G}$ with subgraphs $\mathbf{H_1}, \mathbf{H_2}$ such that $\mathbf{H_1}, \mathbf{H_2} \sqsubseteq\downarrow \mathbf{G}$, and $V(H_1) \cap V(H_2) = \emptyset$, the families $\{\mathbf{G}, \mathbf{H_1}\}$ and $\{\mathbf{G}, \mathbf{H_2}\}$ are incomparable and ELEC is solvable on both, whereas this problem has no solution on their union $\{\mathbf{G}, \mathbf{H_1}, \mathbf{H_2}\}$.

7   Conclusion

We investigated the computability of Election problem in the unknown participants model introduced in [CSS04]. Our result gives a simple condition on the partial knowledge that has to be provided to processes in order to solve this problem. This condition extends and improves the previous results known for the model of reliable unknown participants.

Before obtaining a general computability result, it is already possible to see that some other problems can be investigated with the same tools. For example, the k-Consensus can be solved with a similar algorithm. We do not give a proof but the condition on knowledge would be to forbid more than k disjoint $\sqsubseteq\downarrow$-subgraphs having the same knowledge value as the whole graph.

An interesting open problem is to consider unknown participants in anonymous networks. The conditions given in this paper would remain true. But, from [BV01], it is expected that additional conditions will be necessary to overcome specific impossibilities related to anonymous networks.

References

ABFG08. Alchieri, E.A.P., Bessani, A.N., da Silva Fraga, J., Greve, F.: Byzantine consensus with unknown participants. In: Baker, T.P., Bui, A., Tixeuil, S. (eds.) OPODIS 2008. LNCS, vol. 5401, pp. 22–40. Springer, Heidelberg (2008)

Ang80.   Angluin, D.: Local and global properties in networks of processors. In: STOC, pp. 82–93 (1980)

BCG⁺96. Boldi, P., Codenotti, B., Gemmell, P., Shammah, S., Simon, J., Vigna, S.: Symmetry breaking in anonymous networks: Characterizations. In: Proc. 4th Israeli Symp. on Theory of Computing and Systems, pp. 16–26 (1996)

BV99.   Boldi, P., Vigna, S.: Computing anonymously with arbitrary knowledge. In: PODC, pp. 181–188 (1999)

BV01.   Boldi, P., Vigna, S.: An effective characterization of computability in anonymous networks. In: Welch, J.L. (ed.) DISC 2001. LNCS, vol. 2180, p. 33. Springer, Heidelberg (2001)

CGM08.   Chalopin, J., Godard, E., Métivier, Y.: Local terminations and distributed computability in anonymous networks. In: Taubenfeld, G. (ed.) DISC 2008. LNCS, vol. 5218, pp. 47–62. Springer, Heidelberg (2008)

CGM12.   Chalopin, J., Godard, E., Métivier, Y.: Election in partially anonymous networks with arbitrary knowledge in message passing systems. Distrib. Comput (2012)

Cha06.   Chalopin, J.: Algorithmique distribue, calculs locaux et homorphismes de graphes. PhD thesis, Universit Bordeaux 1 (2006)

CSS04.   Cavin, D., Sasson, Y., Schiper, A.: Consensus with unknown participants or fundamental self-organization. In: Nikolaidis, I., Barbeau, M., An, H.-C. (eds.) ADHOC-NOW 2004. LNCS, vol. 3158, pp. 135–148. Springer, Heidelberg (2004)

GM02.   Godard, E., Métivier, Y.: A characterization of families of graphs in which election is possible (extended abstract). In: Nielsen, M., Engberg, U. (eds.) FOSSACS 2002. LNCS, vol. 2303, p. 159. Springer, Heidelberg (2002)

GSAS12. Greve, F., Sens, P., Arantes, L., Simon, V.: Eventually strong failure detector with unknown membership. The Computer Journal 55(12), 1507–1524 (2012)

GT07.   Greve, F., Tixeuil, S.: Knowledge connectivity vs. synchrony requirements for fault-tolerant agreement in unknown networks. In: DSN 2007, pp. 82–91 (2007)

MMW97.   Métivier, Y., Muscholl, A., Wacrenier, P.A.: About the local detection of termination of local computations in graphs. In: SIROCCO (1997)

RM00.   Rosen, K.H., Michaels, J.G.: Handbook of Discrete and Combinatorial Mathematics (2000)

Tel00.   Tel, G.: Introduction to Distributed Algorithms. Cambridge U.P (2000)

YK89.   Yamashita, M., Kameda, T.: Electing a leader when processor identity numbers are not distinct. In: Bermond, J.-C., Raynal, M. (eds.) WDAG 1989. LNCS, vol. 392, pp. 303–314. Springer, Heidelberg (1989)

YK96a.   Yamashita, M., Kameda, T.: Computing on anonymous networks. I. characterizing the solvable cases. IEEE TPDS 7, 69–89 (1996)

YK96b.   Yamashita, M., Kameda, T.: Computing on anonymous networks. II. decision and membership problems. IEEE TPDS 7, 90–96 (1996)

Rendezvous of Distance-Aware Mobile Agents in Unknown Graphs*

Shantanu Das[1], Dariusz Dereniowski[2],
Adrian Kosowski[3,4], and Przemysław Uznański[1]

[1] LIF, Aix-Marseille University and CNRS, Marseille, France
[2] Dept. of Algorithms and System Modeling, Gdańsk University of Technology, Gdańsk, Poland
[3] GANG Project, Inria Paris, France
[4] LIAFA, Paris Diderot University and CNRS, France

Abstract. We study the problem of rendezvous of two mobile agents starting at distinct locations in an unknown graph. The agents have distinct labels and walk in synchronous steps. However the graph is unlabelled and the agents have no means of marking the nodes of the graph and cannot communicate with or see each other until they meet at a node. When the graph is very large we want the time to rendezvous to be independent of the graph size and to depend only on the initial distance between the agents and some local parameters such as the degree of the vertices, and the size of the agent's label. It is well known that even for simple graphs of degree Δ, the rendezvous time can be exponential in Δ in the worst case. In this paper, we introduce a new version of the rendezvous problem where the agents are equipped with a device that measures its distance to the other agent after every step. We show that these *distance-aware* agents are able to rendezvous in any unknown graph, in time polynomial in all the local parameters such the degree of the nodes, the initial distance D and the size of the smaller of the two agent labels $l = \min(l_1, l_2)$. Our algorithm has a time complexity of $O(\Delta(D + \log l))$ and we show an almost matching lower bound of $\Omega(\Delta(D + \log l / \log \Delta))$ on the time complexity of any rendezvous algorithm in our scenario. Further, this lower bound extends existing lower bounds for the general rendezvous problem without distance awareness.

Keywords: Mobile Agent, Rendezvous, Synchronous, Anonymous Networks, Distance Oracle, Lower Bounds.

* This work was done when the second and the third author were visiting the LIF laboratory in Marseille. Research partially supported by the Polish National Science Center grant DEC-2011/02/A/ST6/00201, by the ANR project DISPLEXITY (ANR-11-BS02-014) and by the ANR project MACARON (ANR-13-JS02-0002-01). Dariusz Dereniowski was partially supported by a scholarship for outstanding young researchers funded by the Polish Ministry of Science and Higher Education.

M. Halldórsson (Ed.): SIROCCO 2014, LNCS 8576, pp. 295–310, 2014.

1   Introduction

1.1   Overview

Suppose two friends travel to a distant land and arrive at a city where all road signs are written in a language unknown to either of them. If the friends get separated and can no longer communicate with each-other, how could the two friends get together again without any help from a third person. This problem of gathering two autonomous mobile agents, called the *rendezvous* problem has been studied in many different contexts, for example for two ships lost in the sea, two astronauts that land in separate parts of a planet and so on. Initial studies on the problem were restricted to finding probabilistic strategies for movement of the two agents that minimize the expected time to rendezvous (See [1] for a survey of such results). In recent years, the deterministic version of the problem has received a lot of attention especially by the distributed computing community. The rendezvous problem for two agents moving along the edges of a graph is a typical problem of symmetry breaking and it is a primitive for distributed coordination among autonomous mobile robots. The solution to the problem depends on the structure of the graph, the capabilities of the agents and the initial knowledge available to the agents. In this paper, we consider the problem for deterministic agents with local vision moving in an initially unknown graph; the problem is solved when the two agents are simultaneously located in the same vertex. We are interested in the worst case time complexity for rendezvous. In general, rendezvous cannot always be solved deterministically when the underlying graph is highly symmetric and the agents follow identical strategies. A typical example is that of a ring network with unlabelled nodes where the two agents are placed on opposite vertices on any diameter. In this case, the distance between the two agents may never decrease if the agents use identical strategies, moving left or right at the same time.

Known solutions to the rendezvous problem are based on one of the following two approaches. The first type of solutions relies on finding a point of asymmetry in the graph and meeting at a unique point of asymmetry (e.g. such a point of asymmetry always exists in graphs where the nodes are labelled uniquely). The second type of solutions assumes that the agents are provided with distinct labels and thus, they can execute distinct strategies and ensure rendezvous. The former type of results require the agents to traverse every edge of the graph in the worst case and the time to rendezvous depends on the size of the graph. On the other hand, the latter type of solutions allow the agents to rendezvous in graphs of arbitrary size or even infinite graphs when the agents are located a finite distance apart. It has been shown that rendezvous of agents with distinct labels can be achieved in arbitrary finite graphs in time polynomial in the size of the graph and the size of the smaller of the labels assigned to the agents both in the synchronous case [17,20] and the asynchronous case [10]. For infinite graphs, the only known results are for very specific graphs such as lines [19] or grids [3]. For a Δ-dimensional infinite grid, the optimal time to rendezvous is $\Theta(D^\Delta)$ which is already exponential in the maximum degree of the graph. In fact, in unknown graphs of degree Δ, where the agents start a distance of D apart, an agent may have to visit all vertices at a distance of D from its initial location. Since there could be D^Δ such vertices, the time cost of rendezvous would be exponential in Δ, even if the agents have complete knowledge of the graph as well as the initial distance between them. Thus the question is what additional capabilities would enable the agents to rendezvous in polynomial time.

In this paper we are interested in designing the simplest mechanism that can help the agents to rendezvous in a large (possibly infinite) graph in time polynomial in the other parameters of the problem, e.g. the initial distance, the maximum degree of the graph and the labels assigned to the agents. We achieve this by equipping the agents with a device that can measure the distance to the other agent in the graph after each step of the algorithm[1]. In fact, our algorithm does not require knowledge of the exact distance between the agents, but instead it is sufficient if the agent can detect whether the distance to the other agent increased or decreased after each move. We assume time to be discretized into rounds; in each round an agent can either traverse one edge of the graph or stay at its current location, and at the end of each round the agent can determine whether the distance to the other agent increased, decreased or remained unchanged during this round. Note that the agents have no means of detecting the direction leading to the other agent. We call agents equipped with the above device *distance-aware* agents. We show that distance-aware agents can rendezvous in arbitrary graphs in time polynomial in the initial distance, in the degree of the graph, and in the size of the smaller of the two agent labels.

1.2   Our Contributions

We show that two distance-aware agents, starting from an initial distance D apart from each other in a connected graph with maximum degree Δ, can rendezvous in time $O(\Delta \cdot D + \Delta \cdot \log l)$ rounds, where $l = \min(l_1, l_2)$ and l_1 and l_2 are the labels of the agents. The proof is constructive and provides a deterministic algorithm for the agent that takes as input the label of the agent. The algorithm does not require any prior knowledge of the graph and works for any connected graph. We also show that our algorithm is almost optimal by providing a lower bound of $\Omega(\Delta(D + \log l / \log \Delta))$ for rendezvous of distance-aware agents. Thus, our algorithm is asymptotically optimal when the maximum degree Δ is not extremely large.

The lower bound presented in this paper holds even for agents that can compute the exact distance to each other at every step, while the algorithm requires only the knowledge of changes in distance. Moreover, this lower bound extends existing lower bounds for the general problem of rendezvous of labelled agents in unknown graphs. In terms of the size of agent labels, the previous lower bound (without distance awareness) was $\Omega(\log l \cdot D)$ which already holds for the ring;

---

[1] Such a device can be implemented in practice e.g. by emitting specific signals at periodic intervals and measuring the intensity of the signal received from the other device held by the other agent.

no generalizations of this lower bound to graphs of arbitrary degree have been presented before. Our results show that the lower bound for rendezvous must be at least $\Omega(\log l(D + \Delta/\log \Delta))$.

1.3   Related Work

This paper considers the deterministic version of rendezvous (for randomized solutions see e.g. [1]). The problem of rendezvous of autonomous mobile agents has been studied for agents moving in a discrete space i.e. a graph [18] or those moving on a continuous space (e.g. two dimensional plane [11]). In the graph setting, rendezvous of *identical* agents is possible only if the graph is asymmetric or the agents are placed in asymmetric positions on the graph. There exists a characterization of such instances (graphs and initial positions of agents) where rendezvous is solvable [21]. If the agents are asynchronous, they can take advantage of the asymmetry in their initial positions by marking their initial position by a pebble [2,16]. On the other hand, if the agents have distinct labels then rendezvous is possible in any graph and any starting positions, without the need to mark nodes. The first deterministic synchronous rendezvous algorithm for agents with distinct labels was presented by Dessmark et al. [9] and the time complexity of the algorithm was $O(n^5\sqrt{\tau \log l}\log n + n^{10}\log^2 n \log l)$ for a graph of n nodes, where τ is the delay in the starting time of the agents. Subsequent studies [17,20] improved this result and removed the parameter τ from the time complexity allowing for rendezvous in time polynomial in both n and $\log l$. For the asynchronous case, De Marco et al. [12] provided an algorithm for rendezvous with a cost of $O(D \cdot \log l)$ rounds, when the graph is known. For unknown arbitrary graphs, Czyżowicz et al. [6] gave the first algorithm for asynchronous rendezvous but the cost of this algorithm is at least exponential in the distance D and the degree Δ of the graph. Recently, Dieudonné et al. [10] provided an improvement over this result achieving asynchronous rendezvous in time polynomial in n and $\log l$. Rendezvous of agents starting from a finite distance D in an infinite graph has been studied for the special cases when the graph is a line [12,19] or a grid [3], assuming that the agents have a sense of orientation and they know their own location in the labelled grid.

There have been several studies on the minimum capabilities needed by the agents to solve rendezvous. For example, the minimum memory required by an agent to solve rendezvous is known to be $\Theta(\log n)$ for arbitrary graphs. Czyżowicz et al. [7] have provided a memory optimal algorithm for rendezvous, and there are studies on the tradeoff between time and space requirements for rendezvous [8]. In some papers, additional capacities are assumed for the agents to overcome other limitations, e.g. global vision is assumed to overcome memory limitations [15] or the capability to mark nodes using tokens [16] or whiteboards [5] is often used to break symmetry. The model used in this paper can be seen as a special case of the oracle model for computation [13] where the agent is allowed to query an oracle that has global knowledge of the environment. However in our case, since the only queries are distance queries, the oracle can be implemented without complete knowledge of the graph topology.

2   Model and Notations

We model the environment as an undirected connected (possibly infinite) graph $G(V, E)$. The nodes of V are unlabelled such that vertices of the same degree look identical to any agent (i.e. the nodes are anonymous). At each node of the graph, the edges incident to it are locally labelled, so that an agent arriving at a node can distinguish among them[2]. We assume that edges incident to a node v are labelled by distinct integers (called port numbers) from the set $\{1, 2, \ldots, d(v)\}$, where $d(v)$ is the degree of node v. The degree of each node is finite and bounded by the parameter Δ (which is unknown to the agent). For any two distinct vertices $u, v \in V$, the distance between them, denoted by $\text{dist}(u, v)$, is the number of edges in any shortest path from u to v in G.

There are exactly two agents a_1 and a_2, and each agent a_i has a distinct label $\ell(A_i) \in \{0, 1, \ldots, L - 1\}$ for some integer $L \geq 1$. An agent knows its own label but not that of the other agent. The agents have no prior knowledge of the graph. Each agent starts from a distinct node of the graph and moves along the edges of the graph in synchronous steps following a deterministic algorithm. In other words, time is discretized into regular intervals called rounds; in each round, an agent at a node v can either move to an adjacent node of G or remain stationary at v. If the agent moves to an adjacent node w, the agent becomes aware of the port number of the edge through which it entered w. The agent has no means of marking a node that it visits and the agents cannot communicate with each other. An agent can see the other agent only when both agents are on the same node (in particular the agents do not see each other if they cross on the same edge from opposite directions). The two agents start the algorithm in the same round (called round 0) and rendezvous is achieved in the earliest round T when the two agents are at the same node. We denote by D the distance between the starting locations of the agents.

Contrary to previous studies on rendezvous, we assume that the agent is equipped with a device that measures the distance to the other agent. An agent at a node v in round t can make a query to this device (modelled as a function call $\mathtt{distance}()$) which returns the value $\text{dist}(v, u)$, where node u is the location of the other agent in this round. In each round the agent can make one call to $\mathtt{distance}()$ and depending on the value returned, the value of the agent's label, the current state of the agent, and the degree of the current node, the agent chooses a number between 0 and $d(v)$ and leaves the current node v through this port. We assume that the port number 0 corresponds to a self-loop at node v and if the agent chooses 0 it remains at the same node v. In this paper we do not restrict the memory of an agent in any way. Thus, the agent can memorize its complete history of moves up to the current round and store this as its internal state.

3   Lower Bound for Distance-Aware Rendezvous

In this section we provide lower bounds on the rendezvous time for distance-aware agents. Observe that a trivial lower bound is $\Omega(D)$ since at least one of

---

[2] Such an assumption is necessary to allow the agent to navigate in the graph.

the two agents must traverse $D/2$ edges to achieve rendezvous. We could easily obtain a better lower bound for graphs of degree Δ. The result below is folklore and we include it only for completeness.

Lemma 1. *Rendezvous of two agents that are initially at a distance of D apart in an unknown graph of maximum degree Δ requires $\Omega(D \cdot \Delta)$ rounds in the worst case.*

Proof. Consider the caterpillar graph shown in Figure 1 obtained by taking a line of $D+1$ nodes and replacing each node with a star of size $\Delta-1$. Suppose that the two agents start at the endpoints of the path of length D in this graph, as shown. If an agent traverses an edge leading to one of the leaves, it has no other option than to return and try another port. For any deterministic algorithm, an adversary can assign the port numbers in such a way that the agent needs to traverse all the Δ incident edges before it gets any closer to the other agent. Thus after $2\Delta - 1$ rounds the agents can reduce the distance between them by 2. Using a repetition of this argument, the result follows.

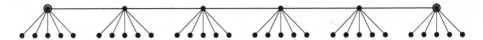

Fig. 1. The two agents are located at the specially marked nodes at the two ends of a Caterpillar graph

The above lower bound is independent of the agent's labels. We now present a bound which includes the parameter L which bounds the size of the set of possible agent labels. It is known that rendezvous of two agents with distinct labels requires $\Omega(\log L)$ rounds [9]. However this lower bound is for the simplest graph consisting of two nodes and a single edge connecting them. We provide below a better lower bound for arbitrary graphs of maximum degree Δ.

Our method is constructive, i.e., given a deterministic rendezvous algorithm $\mathcal{A}$, we describe a graph and a procedure for the adversary to label the ports in a graph and to choose starting positions that will keep the agents executing $\mathcal{A}$ from meeting for the desired time period.

Definition 1. *A k-clique-p-butterfly is a $(k+3)$-regular graph on $(k \times p)$ vertices (denoted by $v_{0,0}, \ldots, v_{k-1,p-1}$), constructed as follows:*

- *for each $0 \leq j < p$, we connect all pairs of vertices from $v_{0,j}, v_{1,j}, \ldots, v_{k-1,j}$ (so they form a k-vertex clique, named j-th clique),*
- *for any given $0 \leq j, j' < p$, $0 \leq i, i' < k$, we connect $v_{i,j}$ and $v_{i',j'}$ if $j' = (j+1) \bmod p$ and $i' \in \{2i \bmod k, (2i+1) \bmod k\}$.*

For two vertices, v from j-th clique and v' from j'-th clique, we say that their horizontal distance is $\min((j - j') \bmod p, (j' - j) \bmod p)$. By the properties of

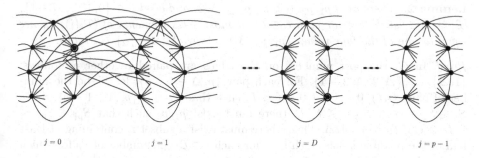

$j = 0$ $j = 1$ $j = D$ $j = p - 1$

Fig. 2. A k-clique-p-butterfly graph where the two agents are located at the marked nodes at a distance $D > \log k$

the butterfly-type interconnections between the cliques, we observe that, if for two vertices the horizontal distance is at least $\log k$, then their actual distance in the graph is equal to the horizontal distance. See Figure 2 for an example of a k-clique-p-butterfly graph where the agents are at a distance $D > \log k$.

Theorem 1. *For any odd $k \geq 3$ and $p \geq 2 \cdot \lfloor \log k \rfloor$, given a k-clique-p-butterfly (denoted further as G), for any deterministic algorithm $\mathcal{A}$, and integers $L > 0$ and $D \geq \log k = \log(\Delta - 3)$, there exist labels $0 \leq \ell(1), \ell(2) < L$ and a port numbering of G such that if two agents with these labels start at two vertices at distance D in this graph, they will remain at distance D for at least $\Theta(\Delta \log L / \log \Delta)$ steps.*

Proof. The main idea of the proof is as follows. We will choose four special port numbers that will be assigned in every vertex to the "bridge" edges connecting it to an adjacent clique (two ports to advance to the next clique and two ports to go backward). Note that moving inside the clique doesn't change the distance to the other agent. By carefully picking the agent labels, we will ensure that both agents will choose the forward ports at the same rounds and the backward ports also at the same rounds, thus they will maintain a constant distance between themselves, for a sufficient amount of time and during this time period, any queries to the distance oracle would be useless.

Before we proceed to prove the theorem, we need to fix some notation and provide some basic lemmas. We will consider the k-clique-p-butterfly graphs with a special type of port numbering where the ports on the two ends of each edge always form one of the pairs of values: $\{1, \Delta\}, \{2, \Delta - 1\}, \dots, \{\Delta/2, \Delta/2 + 1\}$. In other words, whenever an agent chooses to leave a node by port j, it arrives at the adjacent node by port $\Delta + 1 - j$. In a Δ-regular graph with such a port numbering, an agent can never distinguish between any two vertices and it can learn nothing new about the graph by traversing it. Thus, the algorithm $\mathcal{A}$ must choose a predefined sequence of ports to follow during the first $t > 0$ rounds when the distance between the agents remains D. For every $0 \leq \ell < L$ we denote the sequence of ports chosen by algorithm $\mathcal{A}$ (with input label ℓ) by $(\mathcal{P}_\ell(i))_{i=0}^{i<t}$.

Lemma 2. *There exist* p_1, p_2, $0 < p_1, p_2 \leq \Delta/2$, *and a set* $\mathcal{L} \subset \{0, 1, \ldots, L-1\}$, $|\mathcal{L}| \geq L/2$, *such that for each* $\ell \in \mathcal{L}$, *at most* $8t/\Delta$ *elements of* $(\mathcal{P}_\ell(i))_{i=0}^{i \leq t}$ *are equal to one of the four values* $\{p_1, p_2, \Delta + 1 - p_1, \Delta + 1 - p_2\}$ *(in total).*

Proof. In G, the port numbers on the edges are paired together as follows: $\{1, \Delta\}, \ldots, \{\Delta/2, \Delta/2 + 1\}$. For each pair $\{p, \Delta + 1 - p\}$, let S_p be the number of pairs (i, ℓ), $0 \leq i < t, 0 \leq \ell < L$, such that $\mathcal{P}_\ell(i) \in \{p, \Delta + 1 - p\}$. Since $S_1 + S_2 + \cdots + S_{\Delta/2} \leq t \cdot L$, there must exist p_1, p_2 such that $S_{p_1} + S_{p_2} \leq t \cdot L \cdot 2/(\Delta/2) = t \cdot L \cdot 4/\Delta$. Thus, there must exist a subset $\mathcal{L}$ containing at least half of the possible labels, such that for each $\ell \in \mathcal{L}$, the number of i's for which $\mathcal{P}_\ell(i) \in \{p_1, p_2, \Delta + 1 - p_1, \Delta + 1 - p_2\}$ is not greater than $2 \cdot \frac{t \cdot L \cdot 4/\Delta}{L} = 8t/\Delta$.

Given the ports p_1 and p_2 as in the lemma above, the adversary can assign the pairs of port numbers $(p_1, \Delta + 1 - p_1)$ and $(p_2, \Delta + 1 - p_2)$ to those edges of G that connect two adjacent cliques, such that ports p_1 and p_2 will take an agent forward to the next clique, ports $\Delta + 1 - p_1$ and $\Delta + 1 - p_2$ would take an agent backward to the previous clique and any other port will keep an agent within the same clique. We partition the set of port numbers into the subsets:

$$A = \{p_1, p_2\},$$
$$B = \{\Delta + 1 - p_1, \Delta + 1 - p_2\},$$
$$C = \{0, 1, \ldots \Delta\} \setminus (A \cup B).$$

Informally, the ports in A take an agent one step ahead, the ports in B take it one step back, and the C ports keep it in the same clique.

Lemma 3. *There exist* $0 \leq \ell(1) < \ell(2) < L$ *such that* $\forall i \leq \lfloor \frac{\log L}{2 \log \Delta} \rfloor \cdot \frac{\Delta}{8}$,

$$\mathcal{P}_{\ell(1)}(i) \in A \text{ iff } \mathcal{P}_{\ell(2)}(i) \in A, \text{ and} \tag{1}$$

$$\mathcal{P}_{\ell(1)}(i) \in B \text{ iff } \mathcal{P}_{\ell(2)}(i) \in B. \tag{2}$$

Proof. Let $t > 0$ be the first round such that for $i = t$, either (1) or (2) does not hold for any possible pair of labels $\ell(1)$ and $\ell(2)$ from the set $\{0, 1, \ldots, L-1\}$. We will bound the value of t. Note that the sequence of ports $(\mathcal{P}_\ell(i))_{i=0}^{i \leq t}$ can be written as a sequence of A, B and C's. We choose the set of labels $\mathcal{L}$ from Lemma 2 and for any $\ell \in \mathcal{L}$, let us count the number of possible sequences $(\mathcal{P}_\ell(i))_{i=0}^{i \leq t}$ which correspond to distinct words from the alphabet $\{A, B, C\}$; let us denote this number as X. Recall that any such sequence can have at most $8t/\Delta$ A-ports and B-ports in total (cf. Lemma 2). Assuming for the sake of notation that $T = 8t/\Delta$ is an integer and using Stirling's approximation, we get:

$$X = \sum_{i=0}^{\lfloor T \rfloor} \binom{t}{i} \cdot 2^i \leq 1 + T \cdot \binom{t}{T} \cdot 2^T \leq 1 + T \cdot \frac{(2t)^T}{(8t/\Delta)!}$$

$$\leq 1 + T \cdot \left(\frac{2t}{8t/\Delta \cdot \frac{1}{e}} \right)^T = 1 + T \left(\frac{e}{4} \Delta \right)^T.$$

Since for any two labels in $\mathcal{L}$ the algorithm chooses distinct sequences, we have $L/2 \le |\mathcal{L}| \le X$, which gives us $T \ge \lfloor \frac{\log L}{2 \log \Delta} \rfloor$ and thus $t \ge \lfloor \frac{\log L}{2 \log \Delta} \rfloor \cdot \frac{\Delta}{8}$.

We now return to the proof of the Theorem 1. From Lemma 3, we know that there exist two agents with labels $\ell(1), \ell(2)$ such that these agents, when executing algorithm $\mathcal{A}$, will use the ports going forward or backward simultaneously during the first $t = \lfloor \frac{\log L}{2 \log \Delta} \rfloor \cdot \frac{\Delta}{8}$ rounds. Since the distance in the graph is independent of relative positions inside cliques (for $D \ge \log k$), the distance between agents is maintained to be D for at least $t = \Theta(\Delta \log L / \log \Delta)$ steps.

The results of this section provide a lower bound of $\Omega(\Delta(D + \log L / \log \Delta))$ rounds for rendezvous of distance-aware agents.

4   Upper Bound and Algorithm for Rendezvous

In this section we provide an algorithm that guarantees rendezvous of two distance-aware agents in $O(\Delta(D + \log l))$ rounds. The algorithm is divided into four procedures. For each procedure, first we provide some intuitions on its behavior, then we give its formal description in the form of pseudo-code. After the descriptions of all procedures, the proof of correctness and analysis of rendezvous time of the algorithm are given.

We start with a procedure TestPorts (see Algorithm 1), which attempts to decrease the distance between the two agents. The input consists of a positive integer δ and a bit $b \in \{0, 1\}$. Recall that the command distance() performs the oracle query and provides the current distance from the other agent. We also introduce a command move(x), where x is an integer, which behaves as follows. If $x \in \{1, \dots, d(v)\}$, where v is the node currently occupied by the executing agent, then move(x) forces the executing agent to move from v by taking port x, and the value returned by the move(x) command is the entry port at the arrival node. Otherwise, that is when $x \notin \{1, \dots, d(v)\}$, the executing agent stays idle in the given round, and move(x) returns 0. (We pass 0 as an argument to deliberately make an agent idle.)

The interpretation of the input variable b is that whenever $b = 0$, then the agent does not perform any movements during the execution of TestPorts. Note that if $b = 1$, then move($i \cdot b$) forces an agent to move only if there exists an edge with port i at v.

Suppose that, in the same round, both agents perform calls to procedure TestPorts with input parameters δ_1, b_1 and δ_2, b_2, respectively. We will always ensure that $\delta = \delta_1 = \delta_2$. Informally, $b_1 = b_2 = 1$ implies that both agents iteratively take ports $1, \dots, \delta$ (skipping the ones not present at the current node), ending the process if the distance between them decreases. Clearly, if $b_1 = b_2 = 0$, then both agents just stay idle during the 2δ rounds and the procedure returns failure. If $b_1 \ne b_2$, then one agent stays idle while the other 'tests ports'. If we ensure that δ exceeds the degree of the node occupied by the agent that performs the movements, then the procedure will return success whenever $b_1 \ne b_2$.

Algorithm 1. Procedure TestPorts(δ, b)

Input: Two integers $\delta \geq 1$ and $b \in \{0, 1\}$.
Output: success if the distance between agents decreases in some round; failure otherwise.

    for $i \leftarrow 1$ **to** δ **do**
      $x \leftarrow$ distance()
      $p \leftarrow$ move($i \cdot b$)
      $y \leftarrow$ distance()
      if $y < x$ **then**
         return success {When the distance to the other agent decreased.}
      end if
      move($p \cdot b$) {Going back along the same edge.}
    end for
    return failure {When the distance to the other agent never decreased.}

We now describe a procedure BoundDegrees with input variable $b \in \{0, 1\}$ (see Algorithm 2). Informally speaking, for each 2^l such that $2^l < d(v)$, where v is the node occupied at the beginning of the execution of the procedure, the executing agent stays idle for 2^{l+1} rounds (this is achieved by the call to TestPorts($2^l, 0$). This part is independent of b. Then, TestPorts($2^{\lceil \log_2 d(v) \rceil}, b$) is called. If $b = 0$, then the agent stays idle for another $2^{\lceil \log_2 d(v) \rceil + 1}$ rounds. If $b = 1$, then the agent sequentially explores all ports at v (Note that the value of $2^{\lceil \log_2 d(v) \rceil}$ exceeds the degree of v). The above process is interrupted whenever the agents observe that the distance between them decreased, and the procedure returns success in that case.

Algorithm 2. Procedure BoundDegrees(b)

Input: An integer $b \in \{0, 1\}$.
Output: success or failure.

    Let v be the currently occupied node.
    for $l \leftarrow 0$ **to** $\lceil \log_2 d(v) \rceil - 1$ **do**
      $s \leftarrow$ TestPorts($2^l, 0$)
      if $s =$ success **then**
         return success
      end if
    end for
    $s \leftarrow$ TestPorts($2^{\lceil \log_2 d(v) \rceil}, b$)
    return s

The observation given below follows directly from the formulation of procedure BoundDegrees.

Observation 1. *Suppose that agent A_i occupies node v_i, $i \in \{1, 2\}$, and executes procedure BoundDegrees at the beginning of round r. If procedure TestPorts does not return success in the first j iterations of BoundDegrees, then:*

(i) A_i occupies v_i at the end of the j-th iteration of BoundDegrees,
(ii) both agents end the execution of the j-th iteration in round $r + 2^j - 1$.

Denote $I_0 = (0, 1]$ and $I_j = (2^{j-1}, 2^j]$ for $j \geq 1$. We say that two nodes u and v are *similar* if there exists $j \geq 0$ such that $d(u) \in I_j$ and $d(v) \in I_j$.

Lemma 4. *Let r be some integer. Suppose that agent A_i is present at v_i, $i \in \{1, 2\}$, and calls in round r procedure BoundDegrees with input value b_i. Then:*

(i) *for $b_1 = b_2 = 1$, if both calls to BoundDegrees return* failure, *then the nodes v_1 and v_2 are similar, and*
(ii) *if $b_1 \neq b_2$ and v_1 and v_2 are similar, then both calls to BoundDegrees return* success.

Proof. We first prove (ii). Suppose without loss of generality that $b_1 = 1$ and $b_2 = 0$. Since the nodes v_1 and v_2 are similar, $x = \lceil \log_2 d(v_1) \rceil = \lceil \log_2 d(v_2) \rceil$ and therefore the executions of procedure BoundDegrees have the same number of iterations of the 'for' loop. If, in one of those iterations, the execution of procedure BoundDegrees ends, then (ii) holds and hence suppose that this is not the case. Thus, the calls to TestPorts($2^x, b_1$) and TestPorts($2^x, b_2$) are made by the agents. Moreover, by Observation 1, both calls are made in the same round and when agent A_i is at v_i, $i \in \{1, 2\}$. During these calls, A_2 stays idle during 2^{x+1} rounds (because $b_2 = 0$) while A_1 explores all ports at v_1 during the same 2^{x+1} rounds (because $b_1 = 1$ and $2^{\lceil \log_2 d(v_1) \rceil} \geq d(v_1)$). This guarantees that the latter calls to TestPorts return success, which completes the proof of (ii).

We now prove (i). Let $b_1 = b_2 = 1$. We argue that if the nodes v_1 and v_2 are not similar, then the call to BoundDegrees results in returning success. Suppose without loss of generality that $d(v_i) \in I_{j_i}$, $i \in \{1, 2\}$, where $j_1 < j_2$. The number of iterations of the 'for' loop of procedure BoundDegrees executed by A_i is $\lceil \log_2 d(v_i) \rceil = j_i$, $i \in \{1, 2\}$. Thus, by Observation 1(i), A_i occupies v_i at the end of j_1-th iteration of BoundDegrees for each $i \in \{1, 2\}$. Moreover, by Observation 1(ii), after finishing the execution of the 'for' loop, A_1 calls TestPorts($2^{j_1}, b_1$) while A_2 calls TestPorts($2^{j_1}, 0$) in the $(j_1 + 1)$-st iteration of the 'for' loop of BoundDegrees. Also, both of the above-mentioned calls to TestPorts are made in the same round r'. By similar arguments as when proving (ii), we obtain that condition (i) holds.

Note that if procedure BoundDegrees returns success, then the agents get closer, and we can repeat the same process for at their current locations. However, for some nodes procedure BoundDegrees may return failure and then procedure CompareLabels described below (see Algorithm 3) helps to break the symmetry. Procedure CompareLabels uses the notion of extended labels. The *extended label* $\xi(\ell)$ of a label ℓ is an integer whose j-th bit is defined as follows:

$$\xi(\ell)_j = \begin{cases} \ell_{\lceil j/2 \rceil}, & \text{for } j \in \{1, 3, 5, \ldots, 2\lceil \log_2 \ell \rceil - 1\}, \\ 1, & \text{for } j = 2\lceil \log_2 \ell \rceil, \\ 0, & \text{otherwise.} \end{cases}$$

Algorithm 3. Procedure CompareLabels

Input: None.
Output: $\xi(\ell)_i$ such that i is distinguishing for the extended labels of the agents.
  Let $\xi(\ell)$ be the extended label of the executing agent.
  for $i \leftarrow 1$ **to** $\lceil \log_2 \xi(\ell) \rceil$ **do**
    $s \leftarrow$ BoundDegrees$(\xi(\ell)_i)$
    if $s =$ success **then**
      return $\xi(\ell)_i$
    end if
  end for

The index $j = 2\lceil \log_2 \ell \rceil$ is called the *terminating bit* of $\xi(\ell)$ (this is the last bit set to 1). Informally, the odd positions of the extended label are the the bits of ℓ while the even positions are all zeros, except for the terminating bit. We say that an index i is *distinguishing* for two extended labels $\xi(\ell)$ and $\xi(\ell')$ if $\xi(\ell)_i \neq \xi(\ell')_i$.

Informally, procedure CompareLabels iterates over the bits of the extended label of the executing agent in order to find an index i that is distinguishing for the extended labels of the two agents. The construction of extended labels guarantees that there exists a distinguishing index i not greater than the smaller length of the two extended labels and hence BoundDegrees returns success at the latest in the i-th iteration of the 'for' loop of procedure CompareLabels (the formal proof is given later; we remark here that if we used the label instead of the extended label, then the number of iterations of the 'for loop' would have to be equal to the length of the greater label to ensure rendezvous).

We postpone the analysis of procedure CompareLabels (given in Lemma 5) as it depends on the context at which it is called by the main procedure.

We finally describe the main procedure Rendezvous (see Algorithm 4). We

Algorithm 4. Procedure Rendezvous

$s \leftarrow$ success
while rendezvous not achieved **and** $s =$ success **do**
  $s \leftarrow$ BoundDegrees(1)
end while
$b \leftarrow$ CompareLabels
while rendezvous not achieved **do**
  BoundDegrees(b)
end while

start with its intuitive description. The first 'while' loop iteratively calls procedure BoundDegrees(1) as long as its execution gets the agents closer to each other. If a call to BoundDegrees(1) does not achieve that, then (as we formally prove later) the agents observed the same distance between each other while both explored all ports at their respective locations. This is significant as both agents learn that they occupy nodes whose degrees are in the same interval I_j for some $j \geq 0$. In other words, the agents learn an asymptotically tight upper bound on both degrees. Then, procedure CompareLabels is called and uses the above fact as well as the labels of the agents to break the symmetry that occurs at the current agents' nodes. Note that CompareLabels returns either 0 or 1 and in this case different values are returned for both agents (see Lemma 5 below). Thus, the agent whose execution of CompareLabels returned 0 stays idle from now on. The other agents continues making calls to BoundDegrees(1) and since each execution results in exploring all ports at the currently occupied node, each execution gets the agent one step closer to the one that is idle. We also remark that the respective calls to procedure BoundDegrees in the second 'while' loop of CompareLabels are not necessarily 'synced', that is, the j-th of those calls can be made in different rounds by the agents. This, however, is not important as one of the agents stays idle and the other one performs appropriate movements.

Lemma 5 analyzes the only call to procedure CompareLabels made by procedure Rendezvous. Then, Theorem 2 provides the upper bound on the rendezvous time for distance-aware agents in arbitrary networks.

Lemma 5. *Whenever procedure* CompareLabels *is called by both agents during the execution of procedure* Rendezvous, *both agents finish the execution of* CompareLabels *in the same round and the values returned by* CompareLabels *are different for the two agents.*

Proof. Since both agents call CompareLabels, none of the preceding calls to procedure BoundDegrees returns success and the agents do not rendezvous prior to the call to CompareLabels. By Observation 1 and a simple inductive argument, the calls to procedure CompareLabels made by both agents end in the same round r.

It remains to prove that the calls to CompareLabels return different values. By Lemma 4(i), at the beginning of round r the agents A_1 and A_2 are, respectively, at two similar nodes v_1 and v_2. Let $j \geq 0$ be the minimum distinguishing index for agents' labels, i.e., $\xi(\ell(A_1))_j \neq \xi(\ell(A_2))_j$ and $\xi(\ell(A_1))_{j'} \neq \xi(\ell(A_2))_{j'}$ for each $1 \leq j' < j$. Such an index j exists because the labels of the agents are different and hence the extended labels have a distinguishing index. Moreover,

$$j \leq \min \left\{ \lceil \log_2 \xi(\ell(A_1)) \rceil, \lceil \log_2 \xi(\ell(A_2)) \rceil \right\}.$$

Indeed, if the two labels are of the same length, then the extended labels are of the same length. If, on the other hand, the labels have different lengths, then the terminating bits are at different positions, which in particular implies that the terminating bit of the smaller label is at position that is distinguishing for the two extended labels.

By assumption, Observation 1 and an inductive argument, A_i is at v_i at the beginning of the j'-th call to procedure BoundDegrees during the execution of procedure CompareLabels, where $j' \leq j$. The last execution of procedure BoundDegrees preceding the call to CompareLabels returns failure. Hence, if $\xi(\ell(A_1))_{j'} = \xi(\ell(A_2))_{j'} = 1$, then the j'-th call to BoundDegrees returns failure. Clearly, if $\xi(\ell(A_1))_{j'} = \xi(\ell(A_2))_{j'} = 0$, then the agents stay idle during the j'-th call to procedure BoundDegrees which also implies that it returns failure.

Thus, the above proves that the j-th call to procedure BoundDegrees takes place during the execution of procedure CompareLabels and, by Lemma 4(ii), it returns success for both agents. Thus, procedure CompareLabels returns the respective bits of the extended label at position that is distinguishing, which completes the proof.

Theorem 2. *Suppose that agent A_i with label $\ell(A_i)$ initially occupies node v_i, $i \in \{1,2\}$. Procedure Rendezvous guarantees that A_1 and A_2 rendezvous within $O(\Delta \cdot (D + \min_i\{\log \ell(A_i)\}))$ rounds where $D = \text{dist}(v_1, v_2)$.*

Proof. We first prove that the execution of procedure Rendezvous guarantees rendezvous. If the agents rendezvous during the execution of the first 'for' loop, then the claim follows and hence suppose that this is not the case. Denote by b_i the value of the variable b returned by the call to procedure CompareLabels by agent A_i, $i \in \{1,2\}$. By Lemma 5, $b_1 \neq b_2$. Let without loss of generality, $b_1 = 0$ and $b_2 = 1$. This implies that the agent A_1 stays idle indefinitely. The agent A_2, during execution of procedure BoundDegrees(b_2) called in the second 'for' loop of procedure Rendezvous, explores all ports of the currently occupied node. Thus, the distance between agents decreases in some round which implies that each such call to procedure BoundDegrees(b_2) returns success. Thus, the agents rendezvous eventually.

Now we analyze the rendezvous time. Each call to BoundDegrees takes $O(\Delta)$ rounds. Moreover, each such call, except for at most one, made directly by procedure Rendezvous ensures that the distance between the agents decreases. This follows immediately for calls to BoundDegrees preceding the call to procedure CompareLabels since those calls, possibly except for the last one, return success. As for the remaining calls to BoundDegrees, the above claim is due to the fact that the input values are different for both agents due to Lemma 5. Thus, the total number of rounds due to all calls to BoundDegrees made directly by procedure Rendezvous is $O(\Delta \cdot \text{dist}(v_1, v_2))$. The number of iterations of the 'for' loop of procedure CompareLabels is $O(\min\{\log \ell(A_1), \log \ell(A_2)\})$, each resulting in $O(\Delta)$ rounds (the call to BoundDegrees).

5   Conclusions

This paper presented a new model for mobile agent computation by providing the agents with the capability of measuring distances to each other (or detecting changes in distances) at each step. We show that this simple mechanism allows us to reduce the time to rendezvous from exponential to polynomial in the degree of the graph. Assuming that such a distance measuring device is available to the agents, one could ask what other problems can be solved more easily using this additional capability. For example, the agents could use this mechanism for communication at distance by moving back and forth, when there are no other means of communication. This opens up a new area of research which is worth investigating.

References

1. Alpern, S., Gal, S.: The Theory of Search Games and Rendezvous. Kluwer (2003)
2. Baston, V., Gal, S.: Rendezvous search when marks are left at the starting points. Naval Research Logistics 48(8), 722–731 (2001)
3. Bampas, E., Czyzowicz, J., Gąsieniec, L., Ilcinkas, D., Labourel, A.: Almost optimal asynchronous rendezvous in infinite multidimensional grids. In: Lynch, N.A., Shvartsman, A.A. (eds.) DISC 2010. LNCS, vol. 6343, pp. 297–311. Springer, Heidelberg (2010)
4. Chalopin, J., Das, S., Santoro, N.: Rendezvous of Mobile Agents in Unknown Graphs with Faulty Links. In: Pelc, A. (ed.) DISC 2007. LNCS, vol. 4731, pp. 108–122. Springer, Heidelberg (2007)
5. Chalopin, J., Das, S., Widmayer, P.: Rendezvous of Mobile Agents in Directed Graphs. In: Lynch, N.A., Shvartsman, A.A. (eds.) DISC 2010. LNCS, vol. 6343, pp. 282–296. Springer, Heidelberg (2010)
6. Czyżowicz, J., Labourel, A., Pelc, A.: How to meet asynchronously (almost) everywhere. In: Proc. 21st Annual ACM-SIAM Symposium on Discrete Algorithms (SODA), pp. 22–30 (2010)
7. Czyżowicz, J., Kosowski, A., Pelc, A.: How to meet when you forget: Log-space rendezvous in arbitrary graphs. Distributed Computing 25, 165–178 (2012)
8. Czyżowicz, J., Kosowski, A., Pelc, A.: Time vs. space trade-offs for rendezvous in trees. Distributed Computing 27(2), 95–109 (2014)
9. Dessmark, A., Fraigniaud, P., Kowalski, D., Pelc, A.: Deterministic rendezvous in graphs. Algorithmica 46, 69–96 (2006)
10. Dieudonné, Y., Pelc, A.: Vincent Villain. How to meet asynchronously at polynomial cost. In: Proc 32nd Annual ACM Symposium on Principles of Distributed Computing (PODC), pp. 92–99 (2013)
11. Flocchini, P., Prencipe, G., Santoro, N., Widmayer, P.: Gathering of asynchronous robots with limited visibility. Theoretical Computer Science 337(1-3), 147–168 (2005)
12. De Marco, G., Gargano, L., Kranakis, E., Krizanc, D., Pelc, A., Vaccaro, U.: Asynchronous deterministic rendezvous in graphs. Theoretical Computer Science 355(3), 315–326 (2006)
13. Fraigniaud, P., Ilcinkas, D., Pelc, A.: Oracle size: A new measure of difficulty for communication problems. In: Proc. 25th Ann ACM Symposium on Principles of Distributed Computing (PODC), pp. 179–187 (2006)
14. Fraigniaud, P., Pelc, A.: Deterministic Rendezvous in Trees with Little Memory. In: Taubenfeld, G. (ed.) DISC 2008. LNCS, vol. 5218, pp. 242–256. Springer, Heidelberg (2008)
15. Klasing, R., Markou, E., Pelc, A.: Gathering asynchronous oblivious mobile robots in a ring. Theoretical Computer Science 390(1), 27–39 (2008)
16. Kranakis, E., Krizanc, D., Markou, E.: The mobile agent rendezvous problem in the ring. Morgan and Claypool Publishers (2010)
17. Kowalski, D.R., Malinowski, A.: How to meet in anonymous network. Theoretical Computer Science 399(1-2), 141–156 (2008)

18. Pelc, A.: Deterministic rendezvous in networks: A comprehensive survey. Networks 59, 331–347 (2012)
19. Stachowiak, G.: Asynchronous deterministic rendezvous on the line. In: Nielsen, M., Kučera, A., Miltersen, P.B., Palamidessi, C., Tůma, P., Valencia, F. (eds.) SOFSEM 2009. LNCS, vol. 5404, pp. 497–508. Springer, Heidelberg (2009)
20. Ta-Shma, A., Zwick, U.: Deterministic rendezvous, treasure hunts and strongly universal exploration sequences. In: Proc 18th Annual ACM-SIAM Symposium on Discrete Algorithms (SODA), pp. 599–608 (2007)
21. Yamashita, M., Kameda, T.: Computing on anonymous networks: Part I–Characterizing the solvable cases. IEEE Transactions on Parallel and Distributed Systems 7(1), 69–89 (1996)
22. Yu, X., Yung, M.: Agent rendezvous: A dynamic symmetry-breaking problem. In: Meyer auf der Heide, F., Monien, B. (eds.) ICALP 1996. LNCS, vol. 1099, pp. 610–621. Springer, Heidelberg (1996)

Rendezvous of Heterogeneous Mobile Agents in Edge-Weighted Networks*

Dariusz Dereniowski[1], Ralf Klasing[2], Adrian Kosowski[3,4], and Łukasz Kuszner[1]

[1] Department of Algorithms and System Modeling,
Gdańsk University of Technology, Poland
[2] LaBRI, CNRS and University of Bordeaux, France
[3] GANG Project, Inria Paris, France
[4] LIAFA, CNRS and Paris Diderot University, France

Abstract. We introduce a variant of the deterministic rendezvous problem for a pair of heterogeneous agents operating in an undirected graph, which differ in the time they require to traverse particular edges of the graph. Each agent knows the complete topology of the graph and the initial positions of both agents. The agent also knows its own traversal times for all of the edges of the graph, but is unaware of the corresponding traversal times for the other agent. The goal of the agents is to meet on an edge or a node of the graph. In this scenario, we study the time required by the agents to meet, compared to the meeting time T_{OPT} in the offline scenario in which the agents have complete knowledge about each others speed characteristics. When no additional assumptions are made, we show that rendezvous in our model can be achieved after time $O(nT_{OPT})$ in a n-node graph, and that such time is essentially in some cases the best possible. However, we prove that the rendezvous time can be reduced to $\Theta(T_{OPT})$ when the agents are allowed to exchange $\Theta(n)$ bits of information at the start of the rendezvous process. We then show that under some natural assumption about the traversal times of edges, the hardness of the heterogeneous rendezvous problem can be substantially decreased, both in terms of time required for rendezvous without communication, and the communication complexity of achieving rendezvous in time $\Theta(T_{OPT})$.

1 Introduction

Solving computational tasks using teams of agents deployed in a network gives rise to many problems of coordinating actions of multiple agents. Frequently, the communication capabilities of agents are extremely limited, and the exchange

* Research partially supported by the Polish National Science Center grant DEC-2011/02/A/ST6/00201 and by the ANR project DISPLEXITY (ANR-11-BS02-014). This study has been carried out in the frame of the "Investments for the future" Programme IdEx Bordeaux CPU (ANR-10-IDEX-03-02). Dariusz Dereniowski was partially supported by a scholarship for outstanding young researchers funded by the Polish Ministry of Science and Higher Education.

M. Halldórsson (Ed.): SIROCCO 2014, LNCS 8576, pp. 311–326, 2014.
© Springer International Publishing Switzerland 2014

of large amounts of information between agents is only possible while they are located at the same network node. In the rendezvous problem, two identical mobile agents, initially located in two nodes of a network, move along links from node to node, with the goal of occupying the same node at the same time. Such a question has been studied in various models, contexts and applications [1].

In this paper we focus our attention on heterogeneous agents in networks, where the time required by an agent to traverse an edge of the network depends on the properties of the traversing agent. In the most general case we consider, the traversal time associated with every edge and every agent operating in the graph may be different. Scenarios in which traversal times depend on the agent are easy to imagine in different contexts. In a geometric setting, one can consider a road connection network, with agents corresponding to different types of vehicles moving in an environment. One agent may represent a typical road vehicle which performs very well on paved roads, but is unable to traverse other types of terrain. By contrast, the other agent may be a specialized mobile unit, such as a vehicle on caterpillars or an amphibious vehicle, which is able to traverse different types of terrain with equal ease, but without being capable of developing a high speed. In a computer network setting, agents may correspond to software agents with different structure, and the transmission times of agents along links may depend on several parameters of the link being traversed (transmission speed, transmission latency, ability to handle data compression, etc.).

In general, it may be the case that one agent traverses some links faster than the other agent, but that it traverses other links more slowly. We will also analyze more restricted cases, where we are given some a priori knowledge about the structure of the problem. Specially, we will be interested in the case of *ordered agents*, i.e., where we assume that one agent is always faster than the other one, and the case of *ordered edges*, where we assume that if in a fixed pair of links, one agent takes more time to traverse the first link, the same will also be true for the other agent.

We study the rendezvous problem under the assumption that each agent knows the complete topology of the graph and its traversal times for all edges, but knows nothing about the traversal times or the initial location of the other agent. In all of the considered cases, we will ask about the best possible time required to reach rendezvous, compared to that in the "offline scenario", in which each of the agents also has complete knowledge of the parameters of the other agent. We will also study how this time can be reduced by allowing the agents to communicate (exchange a certain number' of bits at a distance) at the start of the rendezvous process.

1.1   The Model and the Problem

Let us consider a simple graph $G = (V, E)$ and its weight functions $w_A : E \mapsto \mathbb{N}_+$ and $w_B : E \mapsto \mathbb{N}_+$, where $\mathbb{N}_+$ is the set of positive integers. Let $s_A, s_B \in V$, $s_A \neq s_B$, be two distinguished nodes of G – the agents' A and B starting nodes. We assume that initially an agent $K \in \{A, B\}$ knows the graph G, s_A, s_B and w_K. Thus, A knows w_A but it does not know w_B, and B knows w_B but it does not now w_A. We assume that the nodes of G have unique identifiers and that G is given to each agent together with the identifiers. The latter in particular implies that the agents have unique identifiers – they can 'inherit' the identifiers of the nodes s_A and s_B. Also, the agents do not see each other unless they meet.

The weight functions indicate the time required for A and B to move along edges. That is, given an edge $e = \{u, v\}$, an agent $K \in \{A, B\}$ needs $w_K(e)$ units of time to move along e (in any direction). We assume that both agents start their computation at time 0 by exchanging messages. The time required to send and to receive a message is negligible.

Once an agent $K \in \{A, B\}$ is located at a node v, it can do one of the following *actions*:

– the agent can wait $t \in \mathbb{N}_+$ units of time at v; after time t the agent will decide on performing another action,
– the agent can start a movement from v to one of its neighbors u; in such case the agent moves with the uniform speed from v to u along the edge $\{u, v\}$ and after $w_K(\{v, u\})$ units of time K arrives at u and then performs its next action.

While an agent is performing its local computations preceding an action, it has access to all messages sent by the other agent at time 0. We assume that the time of agent's computations preceding an action is negligible.

We say that A and B *rendezvous at time t* (or simply *meet*) if they share the same location at time t,

– they both are located at the same node at time t, or
– $K \in \{A, B\}$ started a movement from u_K to v_K at time $t_K < t$, $u_A = v_B$, $v_A = u_B$, $e = \{u_A, v_A\}$, $t_K + w_K(e) < t$ and $\frac{t - t_A}{w_A(e)} = 1 - \frac{t - t_B}{w_B(e)}$ (informally speaking, the agents 'pass' each other on e as they start from opposite endpoints of e), or
– $K \in \{A, B\}$ started a movement from u to v at time $t_K < t$, $e = \{u, v\}$, $t_K + w_K(e) < t$ and $\frac{t - t_A}{w_A(e)} = \frac{t - t_B}{w_B(e)}$ (informally speaking, both agents start at the same endpoint but the one of them 'catches up' the other: $t_A < t_B$ and $w_A(e) > w_B(e)$, or $t_A > t_B$ and $w_A(e) < w_B(e)$).

Observe that the last case is not possible in an optimum offline solution, as the agents could rendezvous earlier in the vertex u.
We are interested in the following problem:

Given two integers b and t, does there exist an algorithm whose execution by A and B guarantees that the agents send to each other at time 0 messages consisting of at most b bits in total, and A and B meet after time at most t?

Given an algorithm for the agents, we refer to the total number of bits sent between the agents as the *communication complexity* of the algorithm. The *rendezvous time* of an algorithm is the minimum time length t such that the agents meet at time t as a result of the execution of the algorithm.

1.2   Related Work

The rendezvous problem has been thoroughly studied in the literature in different contexts. In a general setting, the rendezvous problem was first mentioned in [26]. Authors investigating rendezvous (cf. [1] for an extensive survey) considered either the geometric scenario (rendezvous in an interval of the real line, see, e.g., [5, 6, 18], or in the plane, see, e.g., [2, 3]) or the graph scenario (see, e.g., [14, 17, 23]). A natural extension of the rendezvous problem is that of gathering [16, 20, 24, 28], when more than two agents have to meet in one location.

Rendezvous in anonymous graphs. In the anonymous graph model, the agents rely on local knowledge of the graph topology, only. Nodes have no unique identifiers, and maintain only a local labeling of outgoing edges (ports) leading to their neighbors. When studying the feasibility and efficiency of deterministic rendezvous in anonymous graphs, a key problem which needs to be resolved is that of breaking symmetry. Without resorting to marking nodes, this can be achieved by taking advantage of the different labels of agents [14, 23, 25]. Labeled agents allowed to mark nodes using whiteboards were considered in [29]. Rendezvous of labeled agents using variants of Universal Exploration Sequences was also investigated in [23, 27] in the synchronous model, who showed that such meeting can be achieved in time polynomial in the number of nodes of the graph and in the length of the smaller of the labels of the agents. For the case of unlabeled agents, rendezvous is not always feasible when the agents move in synchronous rounds and are allowed only to meet on nodes. However, for any feasible starting configuration, rendezvous of anonymous agents can be achieved in polynomial time, and even more strongly, using only logarithmic memory space of the agent [12]. In the asynchronous scenario, it has recently been shown that agents can always meet within a polynomial number of moves if they have unique labels [15]. For the case of anonymous agents, the class of instances for which asynchronous rendezvous is feasible is quite similar to that in the synchronous case, though under the assumption that agents are also allowed to meet on edges (which appears to be indispensable in the asynchronous scenario), certain configurations with a mirror-type symmetry also turn out to be gatherable [19].

Location-aware rendezvous. The anonymous scenario may be sharply contrasted with the case in which the agent has full knowledge of the map of the environment, and knows its position within it. Such assumption, partly fueled by the availability and the expansion of the Global Positioning System (GPS), is sometimes called the *location awareness* of agents or nodes of the network. Thus, the only unknown variable is the initial location of the other agent. In [4, 9] the authors study the rendezvous problem of location-aware agents in the asynchronous case. The authors of [9] introduced the concept of covering sequences that permitted location aware agents to meet along the route of polynomial length in the initial distance d between the agents for the case of multi-dimensional grids. Their result was further advanced in [4], where the proposed algorithm provides a route, leading to rendezvous, of length being only a polylogarithmic factor away from the optimal rendezvous trajectory. The synchronous case of location-aware rendezvous was studied in [8], who provided algorithms working in linear time with respect to the initial distance d for trees and grids, also showing that for general networks location-aware rendezvous carried a polylogarithmic time overhead with respect to n, regardless of the initial distance d.

Problems for heterogeneous agents. Scenarios with agents having different capabilities have been also studied. In [13] the authors considered multiple colliding robots with different velocities traveling along a ring with a goal to determine their initial positions and velocities. Mobile agents with different speeds were also studied in the context of patrolling a boundary, see e.g. [11, 22]. In [10] agents capable of traveling in two different modes that differ with maximal speeds were considered in the context of searching a line segment. We also mention that speed, although very natural, is not the only attribute that can be used to differentiate the agents. For example, authors in [7] studied robots with different ranges or, in other words, with different battery sizes limiting the distance that a robot can travel.

1.3   Additional Notation

Let $T_K(u, v, w)$, $K \in \{A, B\}$, denote the minimum time required by agent K to move from u to v in G with a weight function w. If $w = w_K$, then we write $T_K(u, v)$ in place of $T_K(u, v, w_K)$, $K \in \{A, B\}$. In other words $T_K(u, v)$ is the length of the shortest path from u to v in G with weight function w_K, where the length of a path composed of edges $e_1, \ldots, e_l$ is $\sum_{j=1}^{l} w_K(e_j)$. We use the symbol $T_{\text{OPT}}(s_A, s_B)$ to denote the minimum time for rendezvous in the off-line setting where agents that are initially placed on s_A and s_B know all parameters. We will skip starting positions if it will not lead to confusion writing simply T_{OPT}. Denote also $M_K := \max\{w_K(e) \mid e \in E\}$, $K \in \{A, B\}$, and let $M := \max\{M_A, M_B\}$. All logarithms have base 2, i.e., we write for brevity log in place of $\log_2$.

The following lemma, informally speaking, implies that we do not have to consider scenarios in which rendezvous occurs on edges, and by doing so we restrict ourselves to solutions among which there exists one that is within a constant factor from an optimal one. Let $T_{\text{RV}}(s_A, s_B, v)$ denote the minimum time for rendezvous at v, that is, $T_{\text{RV}}(s_A, s_B, v) = \max\{T_A(s_A, v), T_B(s_B, v)\}$. Let any node u that minimizes the $T_{\text{RV}}(s_A, s_B, u)$ be called a *rendezvous node*.

Lemma 1. *For each graph $G = (V, E)$ and for each $s_A, s_B \in V$, if $u \in V$ is the rendezvous node, then $T_{\text{RV}}(s_A, s_B, u) \leq 2T_{\text{OPT}}(s_A, s_B)$.*

Proof. If the two agents can achieve rendezvous on a node in time $T_{\text{OPT}}(s_A, s_B)$, then the lemma follows and hence we assume in the following that rendezvous occurs on an edge. For $K \in \{A, B\}$, let v_K be the last node visited by K prior to rendezvous that the two agents achieve in time $T_{\text{OPT}}(s_A, s_B)$. Observe that $v_A \neq v_B$ and $e = \{v_A, v_B\} \in E$.

In an optimum solution at least one of the agents traversed at least half of e, so

$$2T_{\text{OPT}}(s_A, s_B) \geq \min\{T_A(s_A, v_B), T_B(s_B, v_A)\}. \tag{1}$$

Moreover, $T_A(s_A, v_B) > T_B(s_B, v_B)$ and $T_B(s_B, v_A) > T_A(s_A, v_A)$, so

$$\min\{T_A(s_A, v_B), T_B(s_B, v_A)\} = \min\{\max\{T_A(s_A, v_B), T_B(s_B, v_B)\},$$
$$\max\{T_B(s_B, v_A), T_A(s_A, v_A)\}\} \tag{2}$$
$$= \min\{T_{\text{RV}}(s_A, s_B, v_B), T_{\text{RV}}(s_A, s_B, v_A)\}.$$

If u is a rendezvous node, then

$$\min\{T_{\text{RV}}(s_A, s_B, v_B), T_{\text{RV}}(s_A, s_B, v_A)\} \geq T_{\text{RV}}(s_A, s_B, u).$$

This, together with (1) and (2), prove the lemma.   □

1.4   Possible Restrictions on Weight Functions

Arbitrary weight functions might cause very bad performance of rendezvous (see Theorems 2 and 7). Thus, beside the arbitrary case, we will be interested in restricted cases, namely:

1. w_A and w_B are *arbitrary functions*,
2. $\forall_{e_1, e_2 \in E} \ w_A(e_1) < w_A(e_2) \iff w_B(e_1) < w_B(e_2)$,
3. $\forall_{e \in E} \ w_A(e) \leq w_B(e)$ or $\forall_{e \in E} \ w_B(e) \leq w_A(e)$.

Case 1 reflects the situation where both agents and edges are not related in terms of time needed to move along them. Whenever two functions have the property case 2, we will refer to the problem instance as the case of *ordered edges*. Informally, in such scenario both agents obtain the same ordering of edges (up to resolving ties) with respect to their weights. The last case reflects the situation where one of the agents is always at least as fast as the other one. Instances with this property are referred to as the cases of *ordered agents*.

1.5   Our Results

In this work we analyze the following two extreme scenarios. In the first scenario (the middle column in Table 1) we consider the communication complexity of algorithms that guarantee that rendezvous occurs in time $\Theta(T_{\text{OPT}})$ regardless of the starting positions. In the second scenario (the third column) we provide bounds on the rendezvous time in case when the agents send no messages to each other.

2   Communication Complexity for $\Theta(T_{\text{OPT}})$ Time

In this section we determine upper and lower bounds for communication complexity of algorithms that achieve rendezvous in asymptotically optimal time. Section 2 is subdivided into three parts reflecting the three cases of weight functions we consider.

Table 1. Summary of results (n is the number of nodes of the input graph)

	communication complexity for rendezvous in time $\Theta(T_{OPT})$	rendezvous time in case of no communication
Case 1: arbitrary	$O(n \cdot (\log \log(M \cdot n)))$ (Thm. 1) $\Omega(n)$ (Thm. 2)	$\Theta(n \cdot T_{OPT})$ (Thms 6, 7)
Case 2: ordered edges	$O(\log \log M + \log^2 n)$ (Thm. 3) $\Omega(\log n)$ (Thm. 4)	$O(n \cdot T_{OPT})$ (Thm. 6) $\Omega(\sqrt{n} \cdot T_{OPT})$ (Thm. 8)
Case 3: ordered agents	none (Thm. 5)	$\Theta(T_{OPT})$ (Thm. 5)

2.1   The Case of Arbitrary Functions

We start by giving an upper bound on communication complexity of asymptotically optimal rendezvous. Our method is constructive, i.e., we provide an algorithm for the agents (see proof of Theorem 1). Then, (cf. Theorem 2) we give the corresponding lower bound.

Theorem 1. *There exists an algorithm that guarantees rendezvous in $\Theta(T_{OPT})$ time and has communication complexity $O(n(\log \log(M \cdot n)))$ for arbitrary functions.*

Proof. Let $I_0 = [0, 1]$, and for $j > 0$ let $I_j = (2^{j-1}, 2^j]$. Denote $V = \{v_1, \ldots, v_n\}$, where the vertices are ordered according to their identifiers. We first formulate an algorithm and then we prove that it has the required properties. We assume that A is the executing agent and B is the other agent (the algorithm for B is analogous).

1. For each $j = 1, \ldots, n$ (in this order) send to B the integer $r(A, j)$ such that $T_A(s_A, v_j) \in I_{r(A,j)}$.
2. After receiving the corresponding messages from B, construct $T' \colon V \mapsto \mathbb{N}_+$ such that

$$T'(v_j) := \max\{2^{r(A,j)}, 2^{r(B,j)}\}, \quad j \in \{1, \ldots, n\}.$$

3. Find a node v_ρ with minimum value of $T'(v_\rho)$. If more than one such node v_ρ exists, then take v_ρ to be the one with minimum identifier.
4. Go to v_ρ along a shortest path and stop.

Note that both agents compute the same function T'. This in particular implies that the same vertex v_ρ, to which each agent goes, is selected by both agents. Hence, the agents rendezvous at v_ρ. The transmission of $r(K, j)$ requires $O(\log \log(M \cdot n))$ bits because $r(K, j) = O(\log(M \cdot n))$ for each $K \in \{A, B\}$ and $j \in \{1, \ldots, n\}$. Thus, the communication complexity of the algorithm is $O(n \log \log(M \cdot n))$.

We now give an upper bound on the rendezvous time at v_ρ. By definition, for each $j \in \{1, \ldots, n\}$ and for each $K \in \{A, B\}$ we have

$$2^{r(K,j)-1} \leq T_K(s_K, v_j) \leq 2^{r(K,j)}.$$

Thus, having in mind that $T_{\mathrm{RV}}(s_A, s_B, v) = \max\{T_A(s_A, v), T_B(s_B, v)\}$, we obtain:

$$\frac{1}{2}T'(v_j) \leq T_{\mathrm{RV}}(s_A, s_B, v_j) \leq T'(v_j), \quad j \in \{1, \ldots, n\}. \tag{3}$$

Now, let u be a rendezvous node. By (3), the choice of index ρ, again by (3) and by Lemma 1 we obtain

$$T_{\mathrm{RV}}(s_A, s_B, v_\rho) \leq T'(v_\rho) \leq T'(u) \leq 2T_{\mathrm{RV}}(s_A, s_B, u) \leq 4T_{\mathrm{OPT}}(s_A, s_B),$$

which completes the proof.   □

Theorem 2. *Each algorithm that guarantees rendezvous in time $\Theta(T_{\mathrm{OPT}})$ has communication complexity $\Omega(n)$ for some n-node graphs.*

Proof. Let $\mathcal{G}$ be a class of graph such that each $G \in \mathcal{G}$ is a complete bipartite graph $K_{2,n}$ with $V = \{s_A, s_B, v_1, v_2, \ldots v_n\}$ and $E = E_A \cup E_B$, where $E_K = \{\{s_K, v_j\} \mid j \in \{1, 2, \ldots n\}\}$, $K \in \{A, B\}$, and, for each $K \in \{A, B\}$, $w_K(e) = X$ for each $e \in E \setminus E_K$ and $w_K(e) \in \{1, X\}$ for each $e \in E_K$, where X is a sufficiently big integer, e.g., $X = n$.

Note that for each $G \in \mathcal{G}$, $T_{\mathrm{OPT}} \in \{1, X\}$. Moreover, $T_{\mathrm{OPT}} = 1$ if and only if there exists an index $j \in \{1, \ldots, n\}$ such that $w_A(\{s_A, v_j\}) = w_B(\{s_B, v_j\}) = 1$. A problem to find such an index j is equivalent to a known problem of set intersection [21] and requires $\Omega(n)$ bits to be transmitted between A and B.   □

2.2   The Case of Ordered Edges

Theorem 3. *There exists an algorithm that guarantees rendezvous in $\Theta(T_{\mathrm{OPT}})$ time and has communication complexity $O(\log \log M + \log^2 n)$ in case of monotone edges.*

Proof. Let $I_0 = [0, 1]$, and for $j > 0$ let $I_j = (2^{j-1}, 2^j]$. For $K \in \{A, B\}$ and a function $w_K : E \mapsto \mathbb{N}_+$ let $m(w_K)$ be the maximum integer such that the removal of all edges from G with weights greater than $m(w_K)$ disconnects G in such a way that s_A and s_B belong to different connected components. For $K \in \{A, B\}$ and $j \geq 0$, define $r_j^K = |\{e \in E \mid w_K(e) \in I_j\}|$.

We now give a statement of an algorithm with communication complexity $O(\log \log M + \log^2 n)$. Then, we prove that its execution by each agent guarantees rendezvous in time $\Theta(T_{\mathrm{OPT}})$.

1. Let A be the executing agent and let B be the other agent (the statement for B is analogous). Send to B the index c_A such that $m(w_A) \in I_{c_A}$ (this requires sending $\log c_A \leq \log \log m(w_A) = O(\log \log M)$ bits). Set $c := \min\{c_A, c_B\}$ (c_B is in the corresponding message received from B).
2. Send to B the value of r_c^A and, for each $j \in \{1, \ldots, \lceil \log n \rceil\}$ send to B the values of r_{c+j}^A and r_{c-j}^A (this requires sending $O(\log^2 n)$ bits in total).
3. Send to B the value of $r^A := r_0^A + r_1^A + \cdots + r_{c-\lceil \log n \rceil - 1}^A$ (this requires sending $O(\log n)$ bits).

4. After receiving the corresponding messages from B construct a weight function $\widetilde{w}_B \colon E \mapsto \mathbb{N}_+$ as follows. First, sort the edges so that $w_A(e_j) \leq w_A(e_{j+1})$ for each $j \in \{1, \ldots, |E| - 1\}$. Denote

$$E_0^B = \{e_1, \ldots, e_{r^B}\} \quad \text{and} \quad E_\infty^B = E \setminus \{e \mid \widetilde{w}_B(e) \in I_0 \cup \cdots \cup I_{c + \lceil \log n \rceil}\}.$$

Then, $\widetilde{w}_B(e) := 0$ for each $e \in E_0^B$; $\widetilde{w}_B(e) = +\infty$ for each $e \in E_\infty^B$; and for each edge $e \in E \setminus (E_0^B \cup E_\infty^B)$ set $\widetilde{w}_B(e) := 2^{j'-1}$ if $e \in I_{j'}$ (this can be deduced from messages received from B).

5. Calculate the function $\widetilde{w}_A$ (i.e., the function that B constructs based on the information sent to B).

6. Find a node $v_\rho \in V$ such that $\max\{T_A(s_A, v_\rho, \widetilde{w}_A), T_B(s_B, v_\rho, \widetilde{w}_B)\}$ is minimum. If more than one such node exists, then take v_ρ to be the one with minimum identifier.

7. Go to v_ρ along a shortest path and stop.

Note that the communication complexity of the above algorithm is $O(\log \log M + \log^2 n)$. Also, both agents calculate $\widetilde{w}_A$ and $\widetilde{w}_B$ and hence the node v_ρ is the same for both agents, which implies that the algorithm guarantees rendezvous.

Therefore, it remains to prove that

$$\max\{T_A(s_A, v_\rho), T_B(s_B, v_\rho)\} = O(T_{\text{OPT}}(s_A, s_B)).$$

Due to Lemma 1, it is enough to show that

$$\max\{T_A(s_A, v_\rho), T_B(s_B, v_\rho)\} = O(T_{\text{RV}}(s_A, s_B, u)), \tag{4}$$

where u is a rendezvous node. For $K \in \{A, B\}$ and $x \in \{u, v_\rho\}$, let P_K^x be the set of edges of a shortest path from s_K to x in G with weight function w_K and let $\widetilde{P}_K^x$ be the set of edges of a shortest path from s_K to x in G with weight function $\widetilde{w}_K$. Note that (4) follows from

$$\max\left\{w_A(P_A^{v_\rho}), w_B(P_B^{v_\rho})\right\} = O\left(\max\{w_A(P_A^u), w_B(P_B^u)\}\right). \tag{5}$$

Due to the lack of space we skip the proof of the latter equation[1]. $\qquad\square$

Theorem 4. *Each algorithm that guarantees rendezvous in time $\Theta(T_{\text{OPT}})$ has communication complexity $\Omega(\log n)$ for some n-node graphs in case of ordered edges.*

Proof. Let k be a positive integer. We first define a family of graphs $\mathcal{G} = \{G_1, \ldots, G_k\}$. Each graph in $\mathcal{G}$ has the same structure, namely, the vertices s_A and s_B are connected with k node-disjoint paths but the graphs in $\mathcal{G}$ have different weight functions associated with them. Each of those paths consists of exactly $k + 2$ edges (see Figure 1).

Thus, each graph in $\mathcal{G}$ has $k^2 + k + 2$ nodes and $m = k^2 + 2k$ edges. The edges are denoted by $e_1, \ldots, e_m$. The location of each edge in $\mathcal{G}$ is shown in Figure 1, where

---

[1] The complete proof can be found at http://hal.inria.fr/hal-01003010

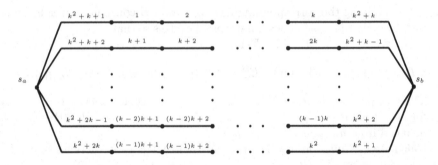

Fig. 1. The structure of the graphs in the proof of Theorems 4 and 8; the numbers give ordering of edges with respect to agents' weight functions

for improving the presentation we write i in place of e_i for each $i \in \{1, \ldots, m\}$. We will set the labels of the edges so that $w_K(e_1) < \cdots < w_K(e_m)$ for each $K \in \{A, B\}$. Now, for each graph in $\mathcal{G}$, we put

$$w_A(e_i) := X + i \quad \text{for each } i \in \{1, \ldots, m\},$$

where $X = k^4$. For $i \in \{1, \ldots, m\}$ and $j \in \{1, \ldots, k\}$, we set the weight function w_B for G_j as follows:

$$w_B(e_i) := \begin{cases} i, & \text{for } i \leq jk, \\ X + i, & \text{for } jk < i \leq k^2 + k - j + 1, \\ kX + i, & \text{for } k^2 + k - j + 1 < i. \end{cases}$$

Note that $w_A(e_1) < \cdots < w_B(e_m)$ and $w_B(e_1) < \cdots < w_B(e_m)$ in each graph $G_j \in \mathcal{G}$ which ensures that all problem instances are cases of ordered edges.

Let $G_j \in \mathcal{G}$. Denote by $H_1, \ldots, H_k$ the k edge-disjoint paths connecting s_A and s_B, where $H_{j'}$ is the path containing the edge $e_{k^2+k+j'}$ incident to s_A, $j' \in \{1, \ldots, k\}$. We argue that if A and B rendezvous on a path $H_{j'}$ in time at most $kX/2$, then $j' = j$. First note that A is not able to reach any vertex adjacent to s_B in time $kX/2$. Also, $j' < j$ is not possible for otherwise B would traverse one of the edges $e_{k^2+k-j+2}, \ldots, e_{k^2+k}$, each of weight at least kX — a contradiction. Now, suppose for a contradiction that $j' > j$. Then, one of the agents traverses at least half of the path $H_{j'}$, i.e., it traverses at least $k/2 + 1$ of its edges. If this agent is A, then clearly rendezvous occurs not earlier than $(k/2+1)X$ — a contradiction. If this agent is B, then it does not traverse any of the edges $e_1, \ldots, e_{jk}$, since those belong to paths $H_1, \ldots, H_j$ and we have $j' > j$. Hence, by the definition of w_B, we also have rendezvous after more than $kX/2$ time units, which gives the required contradiction. We have proved that, in H_j, rendezvous is obtained before time $kX/2$ only if it occurs on the path H_j.

Observe that for n large enough it holds $T_{\text{OPT}}(s_A, s_B) < 2X$ for each $G_j \in \mathcal{G}$. To see that, let A traverse the edge e_{k^2+k+j}, and let B traverse the remaining edges of H_j, i.e., $e_{(j-1)k+1}, e_{(j-1)k+2}, \ldots, e_{(j-1)k+k}$ and $e_{k^2+k-j+1}$. We have $w_A(e_{k^2+k+j}) = X + \Theta(k^2)$ and

$$w_B(e_{k^2+k-j+1}) + \sum_{l=1}^{k} w_B(e_{(j-1)k+l}) = X + k^2 + k - j + 1 + \sum_{l=1}^{k}((j-1)k+l)$$

$$= X + O(k^3).$$

Since $X = k^4$, we obtain that $T_{OPT}(s_A, s_B) < 2X$ for n large enough.

Suppose for a contradiction that there exists an algorithm $\mathcal{A}$ that guarantees rendezvous in time $\Theta(T_{OPT})$ and has communication complexity $o(\log n)$. Let C be such a constant that the rendezvous time guaranteed by $\mathcal{A}$ is bounded by CT_{OPT}. We will show that for $n > C^4$, the algorithm $\mathcal{A}$ that sends at most $C_2 \log n$ bits, where $C_2 = \frac{3}{8}$, cannot guarantee rendezvous in time CT_{OPT}, which will give the desired contradiction.

Note that in each algorithm, and thus in particular in $\mathcal{A}$, if the agent A receives the same message from B in two different graphs in $\mathcal{G}$, then A must traverse the same sequence of edges for both graphs.

The number of all possible messages that A might receive using $C_2 \log n$ bits is at most $2^{C_2 \log n}$. Observe that $2^{C_2 \log n} < \frac{k}{C}$. Indeed,

$$C_2 = \frac{3}{8} = \frac{\log n^{3/4}}{2 \log n} < \frac{\log(n/C)}{2 \log n} < \frac{\log(k/C)}{\log n}$$

implies the required inequality. Thus, there must exist $\mathcal{G}'$, a subset of $\mathcal{G}$, with at least $C + 1$ elements such that A traverses the same sequence of edges for all graphs in $\mathcal{G}'$.

To traverse an edge adjacent to s_A from H_i, the agent A requires $X + i > X$ units of time. In order to traverse any of the remaining edges, A requires to traverse back the mentioned edge first using $X + i > X$ units of time again. Hence, during $CT_{OPT} < 2CX$ time, A is able to traverse at most C edges adjacent to s_A. It implies that there must exist an index j such that $G_j \in \mathcal{G}'$ and agent A does not traverse an edge from H_j adjacent to s_A. It means that A has not met B in time $2CX > CT_{OPT}$, a contradiction.   □

2.3   The Case of Ordered Agents

Now we will present a general solution which achieves $\Theta(T_{OPT}(s_A, s_B))$ time without communication for the case of ordered agents. This property allows us to obtain our asymptotically tight bounds both for communication complexity of optimal-time rendezvous and for optimal rendezvous time with no communication. We point out that, unlike in previous cases, the algorithms for the agents are different, i.e., A and B perform different (asymmetric) actions. We assume that the algorithm for the agent A (respectively, B) is executed by the agent whose starting node has smaller (bigger, respectively) identifier. Note that, since both agents know the graph and both starting nodes, they can correctly decide on executing an algorithm.

Simple Algorithm Tasks of agent A:

1. wait $T_A(s_A, s_B)$ units of time,
2. go to s_B along an arbitrarily chosen shortest path (according to the weight function w_A) from s_A to s_B, and return to s_A along the same path and stop.

Tasks of agent B:

1. wait $T_B(s_A, s_B)$ units of time,
2. go to s_A along an arbitrarily chosen shortest path (according to the weight function w_B) from s_B to s_A and stop.

Lemma 2. *For the case of ordered agents, Simple Algorithm guarantees rendezvous in time* $6\min\{T_A(s_A, s_B), T_B(s_A, s_B)\}$.

Proof. For sure, A and B will eventually rendezvous, as both of them reach s_A and stay there. Let y be the time point at which the agents rendezvous. Let us consider agent A. It might meet B while:

1. waiting $T_A(s_A, s_B)$ units of time at s_A. In this case $y = 2T_B(s_A, s_B) \leq T_A(s_A, s_B)$.
2. moving towards s_B or on the way back to s_A. Clearly $y \leq 3T_A(s_A, s_B)$. Also, agent B at time point y is either at s_B or is moving from s_B to s_A. Thus, $y \leq 2T_B(s_A, s_B)$.
3. arriving at s_A, i.e., rendezvous occurs at s_A at the moment when A returns to s_A. Clearly, $y \leq 3T_A(s_A, s_B)$. As the agents have not met at s_A before A started moving we have $T_A(s_A, s_B) \leq 2T_B(s_A, s_B)$. So, $y \leq 6T_B(s_A, s_B)$.
4. waiting for B at s_A after the path traversals. Clearly, $y = 2T_B(s_A, s_B)$. In this case, as the agents have not met at s_B, we have $T_B(s_A, s_B) \leq 2T_A(s_A, s_B)$. So, $y \leq 4T_A(s_A, s_B)$.

□

We remark that the constant 6 from Lemma 2 might be reduced to $2\sqrt{2} + 3$ if we would allow both agents A and B to wait a little longer in the initial state: $\sqrt{2}T_A(s_A, s_B)$ and $\sqrt{2}T_B(s_A, s_B)$ respectively.

Lemma 3. *In the case of ordered agents we have*

$$\min\{T_A(s_A, s_B), T_B(s_A, s_B)\} \leq 2T_{\mathrm{OPT}}(s_A, s_B)$$

Proof. Suppose that both agents rendezvous at x after $T_{\mathrm{OPT}}(s_A, s_B)$ units of time. If rendezvous does not occur at a node, then with a slight abuse of notation we write $T_K(u, x)$ to denote the time an agent K needs to go from a node u to x. Suppose without loss of generality that A is a 'faster' agent, i.e., $w_A(e) \leq w_B(e)$ for each edge e. This in particular implies that $T_A(s_A, s_B) \geq T_B(s_A, s_B)$ and hence it remains to provide the upper bound on $T_A(s_A, s_B)$. Moreover, by first using the triangle inequality we have $T_A(s_A, s_B) \leq T_A(s_A, x) + T_A(x, s_B) \leq T_A(s_A, x) + T_B(x, s_B) \leq 2T_{\mathrm{OPT}}(s_A, s_B)$. □

Now, due to Lemmas 2 and 3, we are ready to conclude:

Theorem 5. *In the case of ordered agents (case 3) there exists an algorithm that guarantees rendezvous in time* $\Theta(T_{\mathrm{OPT}}(s_A, s_B))$ *without performing any communication.* □

3   Rendezvous with No Communication

3.1   The Case of Arbitrary Functions

Theorem 6 below gives the upper bound on rendezvous time without communication. Then, Theorem 7 provides our lower bound for this case.

Theorem 6. *There exists an algorithm that without performing any communication guarantees rendezvous in time $O(n \cdot T_{OPT}(s_A, s_B))$, where n is the number of nodes of the network.*

Proof. We start by giving an algorithm. Its first step in an initialization and the remaining steps form a loop. Denote $V = \{v_1, \ldots, v_n\}$.

1. Let initially $x := 1$. Let K be the executing agent.
2. For each $j \in \{1, \ldots, n\}$ do:
   2.1. If $T_K(s_K, v_j) \leq x$, then set $x' := T_K(s_K, v_j)$ and go to v_j along a shortest path. Otherwise, set $x' := 0$.
   2.2. Wait $x - x'$ time units at the current node.
   2.3. Return to s_K along a shortest path. (This step is vacuous if $x' = 0$.)
   2.4. Wait $x - x'$ time units.
3. Set $x := 2x$ and return to Step 2.

Let us introduce some notation regarding the above algorithm. We divide the time into *phases*, where the p-th phase, $p \geq 0$, consists of all time units in which both agents were performing actions determined in Step 2 for $x = 2^p$. Then, each phase is further subdivided into *stages*, where the s-th stage, $s \in \{1, \ldots, n\}$, of the p-th phase consists of all time units in which both agents were performing actions determined in Step 2 for $x = 2^p$ and $j = s$. Note that these definitions are correct since both agents simultaneously start at time 0.

First observe, by a simple induction on the total number of stages, that at the beginning of each stage each agent $K \in \{A, B\}$ is present at s_K. We now prove that both agents are guaranteed to rendezvous at a rendezvous node v in the p-th phase, where $2^p \geq \max\{T_A(s_A, v), T_B(s_B, v)\}$. Consider the s-th stage of p-th phase such that $v_s = v$. Since $2^p \geq \max\{T_A(s_A, v), T_B(s_B, v)\}$, both agents reach v in at most 2^p moves. Due to the waiting time of $2^p - T_K(s_K, v)$ of agent $K \in \{A, B\}$ after reaching v, we obtain that both agents are present at v at the end of the 2^p-th time unit of the s-stage in the p-th phase. This completes the proof of the correctness of our algorithm.

It remains to bound the time in which the agents rendezvous. The duration of the p-th phase is $O(n2^p)$. The total number of phases is at most $P = \lceil \log \max\{T_A(s_A, v), T_B(s_B, v)\} \rceil$. Thus, the agents rendezvous in time

$$O(n \sum_{p=1}^{P} 2^p) = O(n2^P) = O\left(n \cdot \max\{T_A(s_A, v), T_B(s_B, v)\}\right).$$

Lemma 1 implies that the agents rendezvous in time $O(n \cdot T_{OPT}(s_A, s_B))$ as required. $\square$

Theorem 7. *Any algorithm that without performing any communication guarantees rendezvous uses time $\Omega(n \cdot T_{\text{OPT}}(s_A, s_B))$, where n is the number of nodes of the network.*

Proof. Let us consider the complete bipartite graph G given in the proof of Theorem 2 with $V(G) = \{s_A, s_B, v_1, v_2, \dots v_n\}$ and $E = E_A \cup E_B$, where

$$E_K = \{\{s_K, v_j\} \mid j \in \{1, 2, \dots n\}\}, \quad K \in \{A, B\}.$$

Let $w_A(e) = X$ for each $e \in E_B$ and $w_A(\{s_A, v_i\}) = 1$ for each $e \in E_A$, where X is some sufficiently big integer, say $X = n$. We will now give a partial definition of w_B, starting with $w_B(e) = X$ for each $e \in E_A$. This weight functions will be constructed in such a way that rendezvous at time 1 is possible. Informally, we will set only one edge in E_B to have weight 1 for the agent B while the remaining edges will have weight X. This is done by analyzing possible moves of the agent A.

Now, we consider an arbitrary sequence of moves of agent A during the first n time units. Clearly, after this time, agent A is not able to reach s_B. There also exists an edge $\{s_A, v_j\} \in E_A$ that agent A performed no move along it, i.e., A did not visit v_j. We set $w_B(\{s_A, v_j\}) := 1$ and $w_B(\{s_A, v_i\}) := X$ for all $i \neq j$.

It is easy to observe that $T_{\text{OPT}}(s_A, s_B)$ is equal to 1 and this time can be achieved only by a meeting at v_j. However, A and B did not rendezvous during the first n time units. Thus, there exists no algorithm that guarantees rendezvous in time $o(n \cdot T_{\text{OPT}}(s_A, s_B))$. $\qquad\square$

3.2 Lower Bound for the Case of Ordered Edges without Communication

Theorem 8. *In the case of ordered edges, any algorithm that guarantees rendezvous without performing any communication uses time $\Omega(\sqrt{n} \cdot T_{\text{OPT}}(s_A, s_B))$, where n is the number of nodes of the network.*

Proof. We will use the same family of graphs $\mathcal{G}$ as constructed in the proof of Theorem 4; see also Figure 1. Recall that if A and B rendezvous on a path $H_{j'}$ in time at most $kX/2$, then $j' = j$ and $T_{\text{OPT}} = O(X)$ for each $G_j \in \mathcal{G}$.

Note that, the agent A has the same input for each graph in $\mathcal{G}$ since w_A is the same for all graphs in $\mathcal{G}$. Thus, for any algorithm $\mathcal{A}$, the agent A traverses the same sequence of edges for each graph in $\mathcal{G}$. Moreover, rendezvous time bounded by $kX/2$ (see the proof of Theorem 4) implies that there exist edges adjacent to s_A that A does not traverse. In other words, there exists $j \in \{1, \dots, k\}$ such that A traverses no edge of H_j. Therefore, we obtain that A and B cannot rendezvous in G_j in time less than $kX/2$. Since $k = \Theta(\sqrt{n})$ and rendezvous can be achieved in time $O(X)$ for each graph in $\mathcal{G}$ the proof has been completed. $\qquad\square$

4 Final Remarks

It seems that the most interesting and challenging among the analyzed cases is the one of ordered edges without communication. There is still a substantial gap between the lower and the upper bounds we have provided and we leave it an interesting open question whether there exists an algorithm with a better approximation ratio than that of $O(nT_{OPT})$. It is also interesting if the upper bound M on the weights of the edges affects the communication complexity for arbitrary functions and the cases of ordered edges.

Another interesting research direction is to analyze scenarios in which we allow agents to communicate at any time. To point out an advantage that the agents may gain in such case, note that the agents can rendezvous very quickly in graphs that we used for a lower bound in the proof of the Theorem 4. Indeed, transmitting just one bit in the moment correlated with the index of the preferred (optimal for rendezvous) path would help the agents to learn which path they should follow.

References

1. Alpern, S., Gal, S.: The theory of search games and rendezvous. International Series in Operations Research and Managment Science. Kluwer Academic Publishers, Boston (2003)
2. Anderson, E., Fekete, S.: Asymmetric rendezvous on the plane. In: Proceedings of 14th Annual ACM Symposium on Computational Geometry (SoCG), pp. 365–373 (1998)
3. Anderson, E., Fekete, S.: Two-dimensional rendezvous search. Operations Research 49(1), 107–118 (2001)
4. Bampas, E., Czyzowicz, J., Gąsieniec, L., Ilcinkas, D., Labourel, A.: Almost optimal asynchronous rendezvous in infinite multidimensional grids. In: Lynch, N.A., Shvartsman, A.A. (eds.) DISC 2010. LNCS, vol. 6343, pp. 297–311. Springer, Heidelberg (2010)
5. Baston, V., Gal, S.: Rendezvous on the line when the players' initial distance is given by an unknown probability distribution. SIAM Journal on Control and Optimization 36(6), 1880–1889 (1998)
6. Baston, V., Gal, S.: Rendezvous search when marks are left at the starting points. Naval Research Logistics 48(8), 722–731 (2001)
7. Chalopin, J., Das, S., Mihalák, M., Penna, P., Widmayer, P.: Data delivery by energy-constrained mobile agents. In: Flocchini, P., Gao, J., Kranakis, E., Meyer auf der Heide, F. (eds.) ALGOSENSORS 2013. LNCS, vol. 8243, pp. 111–122. Springer, Heidelberg (2013)
8. Collins, A., Czyzowicz, J., Gąsieniec, L., Kosowski, A., Martin, R.: Synchronous rendezvous for location-aware agents. In: Peleg, D. (ed.) DISC 2011. LNCS, vol. 6950, pp. 447–459. Springer, Heidelberg (2011)
9. Collins, A., Czyzowicz, J., Gąsieniec, L., Labourel, A.: Tell me where I am so I can meet you sooner. In: Abramsky, S., Gavoille, C., Kirchner, C., Meyer auf der Heide, F., Spirakis, P.G. (eds.) ICALP 2010, Part II. LNCS, vol. 6199, pp. 502–514. Springer, Heidelberg (2010)
10. Czyzowicz, J., Gasieniec, L., Georgiou, K., Kranakis, E., MacQuarrie, F.: The beachcombers' problem: Walking and searching with mobile robots. CoRR, abs/1304.7693 (2013)
11. Czyzowicz, J., Gąsieniec, L., Kosowski, A., Kranakis, E.: Boundary patrolling by mobile agents with distinct maximal speeds. In: Demetrescu, C., Halldórsson, M.M. (eds.) ESA 2011. LNCS, vol. 6942, pp. 701–712. Springer, Heidelberg (2011)

12. Czyzowicz, J., Kosowski, A., Pelc, A.: How to meet when you forget: log-space rendezvous in arbitrary graphs. Distributed Computing 25(2), 165–178 (2012)
13. Czyzowicz, J., Kranakis, E., Pacheco, E.: Localization for a system of colliding robots. In: Fomin, F.V., Freivalds, R., Kwiatkowska, M., Peleg, D. (eds.) ICALP 2013, Part II. LNCS, vol. 7966, pp. 508–519. Springer, Heidelberg (2013)
14. Dessmark, A., Fraigniaud, P., Kowalski, D.R., Pelc, A.: Deterministic rendezvous in graphs. Algorithmica 46(1), 69–96 (2006)
15. Dieudonné, Y., Pelc, A., Villain, V.: How to meet asynchronously at polynomial cost. CoRR, abs/1301.7119 (2013)
16. Flocchini, P., Prencipe, G., Santoro, N., Widmayer, P.: Gathering of asynchronous robots with limited visibility. Theor. Comput. Sci. 337(1-3), 147–168 (2005)
17. Fraigniaud, P., Pelc, A.: Deterministic rendezvous in trees with little memory. In: Taubenfeld, G. (ed.) DISC 2008. LNCS, vol. 5218, pp. 242–256. Springer, Heidelberg (2008)
18. Gal, S.: Rendezvous search on the line. Operations Research 47(6), 974–976 (1999)
19. Guilbault, S., Pelc, A.: Asynchronous rendezvous of anonymous agents in arbitrary graphs. In: Fernàndez Anta, A., Lipari, G., Roy, M. (eds.) OPODIS 2011. LNCS, vol. 7109, pp. 421–434. Springer, Heidelberg (2011)
20. Israeli, A., Jalfon, M.: Token management schemes and random walks yield self-stabilizing mutual exclusion. In: Dwork, C. (ed.) PODC, pp. 119–131. ACM (1990)
21. Kalyanasundaram, B., Schintger, G.: The probabilistic communication complexity of set intersection. SIAM J. Discret. Math. 5(4), 545–557 (1992)
22. Kawamura, A., Kobayashi, Y.: Fence patrolling by mobile agents with distinct speeds. In: Chao, K.-M., Hsu, T.-S., Lee, D.-T. (eds.) ISAAC 2012. LNCS, vol. 7676, pp. 598–608. Springer, Heidelberg (2012)
23. Kowalski, D.R., Malinowski, A.: How to meet in anonymous network. Theoretical Computer Science 399(1-2), 141–156 (2008)
24. Lim, W., Alpern, S.: Minimax rendezvous on the line. SIAM Journal on Control and Optimization 34(5), 1650–1665 (1996)
25. Marco, G.D., Gargano, L., Kranakis, E., Krizanc, D., Pelc, A., Vaccaro, U.: Asynchronous deterministic rendezvous in graphs. Theoretical Computer Science 355(3), 315–326 (2006)
26. Schelling, T.: The strategy of conflict. Oxford University Press, Oxford (1960)
27. Ta-Shma, A., Zwick, U.: Deterministic rendezvous, treasure hunts and strongly universal exploration sequences. In: Proceedings of 18th Annual ACM-SIAM Symposium on Discrete Algorithms (SODA), pp. 599–608 (2007)
28. Thomas, L.: Finding your kids when they are lost. Journal of the Operational Research Society 43(6), 637–639 (1992)
29. Yu, X., Yung, M.: Agent rendezvous: A dynamic symmetry-breaking problem. In: Meyer auf der Heide, F., Monien, B. (eds.) ICALP 1996. LNCS, vol. 1099, pp. 610–621. Springer, Heidelberg (1996)

Move-Optimal Partial Gathering
of Mobile Agents in Asynchronous Trees

Masahiro Shibata, Fukuhito Ooshita, Hirotsugu Kakugawa,
and Toshimitsu Masuzawa

Graduate School of Information Science and Technology, Osaka University
{m-sibata,f-oosita,kakugawa,masuzawa}@ist.osaka-u.ac.jp

Abstract. In this paper, we consider the partial gathering problem of
mobile agents in asynchronous tree networks. The partial gathering problem is a new generalization of the total gathering problem, which requires
that all the agents meet at the same node. The partial gathering problem requires, for given input g, that each agent should move to a node
and terminate so that at least g agents should meet at the same node.
The requirement for the partial gathering problem is weaker than that
for the (well-investigated) total gathering problem, and thus, we clarify
the difference on the move complexity between them. We assume that
n is the number of nodes and k is the number of agents. We propose
two algorithms to solve the partial gathering problem. First, we consider
the strong multiplicity detection and non-token model. In this model, we
show that agents require $\Omega(kn)$ total moves to solve the partial gathering
problem and we propose an algorithm to achieve the partial gathering in
$O(kn)$ total moves. Second, we consider the weak multiplicity detection
and removable-token model. In this model, we propose an algorithm to
achieve the partial gathering in $O(gn)$ total moves. It is known that the
partial gathering problem requires $\Omega(gn)$ total moves. Hence, the second
algorithm is asymptotically optimal in terms of total moves.

Keywords: distributed system, mobile agent, gathering problem, partial gathering problem.

1 Introduction

1.1 Background and Our Contribution

A *distributed system* is a system that consists of a set of computers (*nodes*) and
communication links. In recent years, distributed systems have become large
and design of distributed systems has become complicated. As a way to design
efficient distributed systems, (mobile) agents have attracted a lot of attention
[1–5]. Agents can traverse the system and process tasks on each node, hence they
can simplify design of distributed systems.

The total gathering problem is a fundamental problem for agents' cooperation
[1, 6–9]. The total gathering problem requires all agents to meet at a single node
in finite time. The total gathering problem is useful because, by meeting at a

M. Halldórsson (Ed.): SIROCCO 2014, LNCS 8576, pp. 327–342, 2014.

single node, all agents can share information or synchronize behaviors among them.

In this paper, we consider a variant of the gathering problem, called the *partial gathering problem* [10]. The partial gathering problem does not always require all agents to gather at a single node, but requires agents to gather partially at several nodes. More precisely, we consider the problem which requires, for given input g, that each agent should move to a node and terminate so that at least g agents should meet at the same node. We define this problem as the g-*partial gathering problem*. Clearly, if $k/2 < g \le k$ holds, the g-partial gathering problem is equal to the total gathering problem. If $2 \le g \le k/2$ holds, the requirement for the g-partial gathering problem is weaker than that for the total gathering problem, and thus it seems possible to solve the g-partial gathering problem with smaller total moves. In addition, the g-partial gathering problem is still useful especially in large-scale networks. This is because, in the total gathering problem, it requires high costs for all agents to meet at a single node. On the other hand, the cost for at least g agnets to meet at several nodes is not as high as the cost for all agents to meet at a single node. Moerover, agents can share information and process tasks cooperatively among at least g agents.

The g-partial gathering problem in unidirectional ring networks is studied in [10]. For distinct agents (i.e., agents have distinct IDs), the paper proposes a deterministic algorithm to solve the g-partial gathering problem in $O(gn)$ total moves without knowledge of k. For anonymous agents (i.e., agents have no IDs), the paper proposes a randomized algorithm to solve the g-partial gathering problem in $O(n \log k + gn)$ expected total moves. Since the total gathering problem requires $\Omega(kn)$ total moves, these results show that the g-partial gathering problem can be solved in smaller total moves compared to the total gathering problem. Moreover, since the g-partial gathering problem requires $\Omega(gn)$ total moves if $g \ge 2$, the paper showed that the deterministic algorithm is asymptotically optimal in terms of total moves.

In this paper, we consider the g-partial gathering problem for asynchronous tree networks for the case $2 \le g \le k/2$. The contribution of this paper is summarized in Table 1. We consider two multiplicity detection models and two token models. First, we consider the case of the strong multiplicity detection and non-token model, where in the strong multiplicity detection model each agent can count the number of agents at the current node. In this case, we show that agents require $\Omega(kn)$ total moves to solve the g-partial gathering problem. In addition, we propose a deterministic algorithm to solve the g-partial gathering problem in $O(kn)$ total moves, that is, our algorithm is asymptotically optimal in terms of the total moves. Second, we consider the case of the weak multiplicity detection and removable-token model, where in the weak multiplicity detection model each agent can detect whether another agent exists at the current node or not but cannot count the exact number of agents. In this case, we propose a deterministic algorithm to solve the g-partial gathering problem in $O(gn)$ total moves. This result shows that the total moves can be reduced by using tokens. Since agents require $\Omega(gn)$ total moves to solve the g-partial gathering problem,

Table 1. Proposed algorithms in asynchronous trees

Model	Algorithm 1	Algorithm 2
Multiplicity detection	Strong	Weak
Removable-token	Not available	Available
The total moves	$\Theta(kn)$	$\Theta(gn)$

this algorithm is also asymptotically optimal in terms of the total moves. Note that, due to limitation of space, we omit discriptions of pseudocode and proofs of lemmas and theorems.

1.2 Related Works

Many fundamental problems for cooperation of mobile agents have been studied in literature. For example, the searching problem [2, 5, 11], the gossip problem [3], the election problem [12], the map construction problem [4], and the total gathering problem [1, 6–9] have been studied.

In particular, the total gathering problem has received a lot of attention and has been extensively studied in many topologies, which include lines [13, 14], trees [1, 3, 7–9, 15], tori [1, 16], arbitrary graphs [13, 17, 18] and rings [1, 3, 6, 13, 19]. The total gathering problem for trees has been extensively studied because tree networks are utilized in a lot of applications. To solve the total gathering problem, it is necessary to select exactly one gathering node, i.e., a node where all agents meet. There are many ways to select the gathering node. For example, in [1, 16, 19–22], agents leave marks (tokens) on their initial nodes and select the gathering node based on every distance of neighboring tokens. In [2, 11], agents have distinct IDs and select the gathering node based on the IDs. In [6], agents can use random numbers and select the gathering node based on IDs generated randomly. In [1, 3, 12], agents execute the leader agent election and the elected leader decides the gathering node. In [7–9, 15, 17], agents explore graphs and decide which node they meet at.

2 Preliminaries

2.1 Network and Agent Model

A *tree network* T is a tuple $T = (V, L)$, where V is a set of nodes and L is a set of communication links. We denote by $n \ (= |V|)$ the number of nodes. Let d_v be the degree of v. We assume that each link l incident to v is uniquely labeled at v from the set $\{0, 1, \ldots, d_v - 1\}$. We call this label *port number*. Since each communication link connects to two nodes, it has two port numbers. However, port numbering is *local*, that is, there is no coherence between two port numbers. The *path* $P(v_0, v_k) = (v_0, v_1, \ldots, v_k)$ with length k is a sequence of nodes from v_0 to v_k such that $\{v_i, v_{i+1}\} \in L \ (0 \leq i < k)$ and $v_i \neq v_j$ if

$i \neq j$. Note that, for any u, $v \in V$, $P(u,v)$ is unique in a tree. The *distance* from u to v, denoted by $dist(u,v)$, is the length of the path from u to v. The *eccentricity* $r(u)$ of node u is the maximum distance from u to an arbitrary node, i.e., $r(u) = \max_{v \in V} dist(u,v)$. The *radius* R of the network is the minimum eccentricity in the network. A node with eccentricity R is called a *center*. We use the following theorem about a center later [23].

Theorem 1. *There exist one or two center nodes in a tree. If there exist two center nodes, they are neighbors.*

Let $A = \{a_1, a_2, \ldots, a_k\}$ be a set of agents. We assume that each agent does not know the number n of nodes and the number k of agents. We consider the *strong multiplicity detection model* and the *weak multiplicity detection model* in tree networks. In the strong multiplicity detection model, each agent can count the number of agents at the current node. In the weak multiplicity detection model, each agent can recognize whether another agent stays at the same node or not, but cannot count the number of agents on its current node. However, in both models, each agent cannot detect the states of agents at the current node. Moreover, we consider the *non-token model* and the *removable-token model*. In the non-token model, agents cannot mark the nodes or the edges in any way. In the removable-token model, each agent initially has a token and can leave it on a node, and agents can remove such tokens. We assume that agents are anonymous (i.e., agents have no IDs) and execute a deterministic algorithm. We model an agent as a finite automaton $(S, \delta, s_{initial}, s_{final})$. The first element S is the set of all states of agents, which includes initial state $s_{initial}$ and final state s_{final}. When an agent changes its state to s_{final}, the agent terminates the algorithm. The second element δ is the state transition function. In the strong multiplicity detection and non-token model, δ is described as $\delta : S \times M_T \times N \to S \times M_T$. In the definition, set $M_T = \{\bot, 0, 1, \ldots, \Delta - 1\}$ represents the agent's movement, where Δ is the maximum degree of the tree. In the left side of δ, the value of M_T represents the port number of the current node that the agent observes in visiting the current node (The value is $\bot$ in the first activation). In the right side of δ, the value of M_T represents the port number through which the agent leaves the current node to visit the next node. If the value is $\bot$, the agent does not move and stays at the current node. In addition, N represents the number of other agents at the current node. In the weak multiplicity detection and removable-token model, δ is described as $\delta : S \times M_T \times EX_A \times EX_T \to S \times EX_T \times M_T$. In the definition, $EX_A = \{0,1\}$ represents whether another agent stays at the current node or not. The value 0 represents that no other agents stay at the current node, and the value 1 represents that another agent stays at the current node. In addition, in the left side of δ, $EX_T = \{0,1\}$ represents whether a token exists at the current node or not. The value 0 of EX_T represents that there does not exist a token at the current node, and the value 1 of EX_T represents that there exists a token at the current node. In the right side of δ, $EX_T = \{0,1\}$ represents whether an agent remove a token at the current node or not. If the value of EX_T in the left side is 1 and the value of EX_T in the right side is 0, it means that an agent removes a token at the current node. Otherwise, it means that an agent does not remove a token at the current node. Note that, in both models, we assume that each agent is not imposed any restriction on the memory.

2.2 System Configuration

In the non-token model, (global) *configuration* c is defined as a product of states of agents and locations of agents. In the removable-token model, configuration c is defined as a product of states of agents, states of nodes (tokens), and locations of agents. Moreover, in the initial configuration c_0, we assume that node v_j has a token if there exists an agent at v_j, and v_j does not have a token if there exists no agent at v_j. In both models, we assume that no pair of agents stay at the same node in the initial configuration c_0.

Let A_i be an arbitrary non-empty set of agents. When configuration c_i changes to c_{i+1} by a step of every agent in A_i, we denote the transition by $c_i \xrightarrow{A_i} c_{i+1}$. In the both models, each $a_j \in A_i$ reaches some node (if a_j exists in some link), executes local computation, leaves the node or stays at the node as one common atomic step. Concretely, in the strong multiplicity detection and non-token model, each $a_j \in A_i$ reaches some node (if a_j exists in some link), counts the number of agents at the current node, executes local computation, decides the port number, and moves to the node through the port number or stays at the current node. In the weak multiplicity detection and the removable-token model, each $a_j \in A_i$ reaches some node (if a_j exists in some link), detects whether there exists another agent at the current node or not, detects whether there exists a token at the current node or not, executes local computation, decides whether the a_j removes the token or not (if any), decides the port number, and moves to the node through the port number or stays at the current node. When a_j completes this series of events, we say that a_j takes one step. If multiple agents at the same node are included in A_i, the agents take steps in an arbitrary order. When $A_i = A$ holds for any i, all agents take steps. This model is called the *synchronous model*. Otherwise, the model is called the *asynchronous model*. Note that, in asynchronous model, the period that an agent takes a step is finite but there is no assumption of the upper bound on the length of the period. Moreover, agents move through a link in a FIFO manner, that is, when an agent a_i leaves v_j after a_h leaves v_j through the same communication link as a_h, then a_i reaches v_i's neighboring node $v_{i'}$ after a_h reaches $v_{i'}$. In addition, if a_h reaches v_j before a_i reaches v_j through the same link as a_h, a_h takes a step before a_i takes a step.

If sequence of configurations $E = c_0, c_1, \ldots$ satisfies $c_i \xrightarrow{A_i} c_{i+1}$ $(i \geq 0)$, E is called an *execution* starting from c_0. Execution E is infinite, or ends in final configuration c_{final} where every agent's state is s_{final}.

2.3 Partial Gathering Problem

The requirement of the partial gathering problem is that, for a given input g, each agent should move to a node and terminate so that at least g agents should meet at the node. Formally, we define the g-partial gathering problem as follows.

Definition 1. *Execution E solves the g-partial gathering problem when the following conditions hold:*

- *Execution E is finite.*
- *In the final configuration, for any node v_j such that there exist some agents on v_j, there exist at least g agents on v_j.*

From [10], agents require $\Omega(gn)$ total moves to solve the g-partial gathering problem in unidirectional ring networks. This lower bound also holds for line networks, hence we have the following theorem.

Theorem 2. *The total moves required to solve the g-partial gathering problem for tree networks are $\Omega(gn)$ if $g \geq 2$.*

3   Strong Multiplicity Detection and Non-Token Model

In this section, we consider a deterministic algorithm to solve the g-partial gathering problem for the strong multiplicity detection and non-token model. First, we have the following theorem.

Theorem 3. *In the strong multiplicity detection and non-token model, agents require $\Omega(kn)$ total moves to solve the g-partial gathering problem even if agents know k.*

Next, we propose a deterministic algorithm to solve the g-partial gathering problem in $O(kn)$ total moves for the strong multiplicity detection and non-token model for the case $g \leq k/2$. Remind that, in the strong multiplicity detection model, each agent can count the number of agents at the current node. After starting the algorithm, each agent performs *a basic walk* [7]. In the basic walk, each agent a_h leaves the initial node through the port 0. Later, when a_h visits a node v_j through the port p, a_h leaves v_j through the port $(p + 1) \mod d_{v_j}$. In the basic walk, each agent traverses the tree in the DFS-traversal. Hence, when each agent visits nodes $2(n - 1)$ times, it visits the all nodes in the tree and returns to the initial node. Note that, we assume that agents do not know the number n of nodes. However, if an agent records the topology of the tree every time it visits nodes, it can know the time when it returns to the initial node.

The idea of the algorithm is as follows: First, each agent performs the basic walk until it obtains the whole topology of the tree. Next, each agent computes a center node of the tree and moves there to meet other agents. If the tree has exactly one center node, then each agent moves to the center node and terminates the algorithm. If the tree has two center nodes, then each agent moves to one of the center nodes so that at least g agents meet at each center node. Concretely, agent a_h first moves to the closer center node v_j. If there exist at most g agents at v_j, including a_h, then a_h terminates the algorithm at v_j. Otherwise, a_h moves to another center node $v_{j'}$ and terminates the algorithm.

We have the following theorem.

Theorem 4. *In the strong multiplicity detection and non-token modelC agents solve the g-partial gathering problem in $O(kn)$ total moves.*

4   Weak Multiplicity Detection and Removable-Token Model

In this section, we propose a deterministic algorithm to solve the g-partial gathering problem for the weak multiplicity detection and removable-token model. We show that our algorithm solves the g-partial gathering problem in $O(gn)$ total moves. Remind that, in the removable-token model, each agent has a token. In the initial configuration, each agent leaves a token at the initial node. We define *a token node* (resp., *a non-token node*) as a node that has a token (resp., does not have a token). In addition, when an agent visits a token node, the agent can remove the token.

The idea of the algorithm is similar to [10], but in [10], the network is a unidirectional ring. In this section, we make agents perform the basic walk and regard a tree network as a unidirectional ring network. Concretely, if agent a_h starts the basic walk at node v_0 and continues it until a_h visits nodes $2(n-1)$ times, then each communication link is passed twice and a_h returns to v_0. Thus, when a_h visits nodes $v_1, v_2, \ldots, v_{2(n-1)}$ in this order, then we consider that a_h moves in the unidirectional ring network with $2(n-1)$ nodes. Later, we call this ring *the virtual ring*. In the virtual ring, we define the direction from v_i to v_{i+1} as a *forward* direction, and the direction from v_{i+1} to v_i as a *backward* direction. Moreover, when a_h visits a node v_j through a port p from a node v_{j-1} in the virtual ring, agents also use p as the port number at (v_{j-1}, v_j). For example, let us consider a tree in Fig. 1(a). Agent a_h performs the basic walk and visits nodes a, b, c, b, d, b in this order. Then, the virtual ring of Fig. 1(a) is represented in Fig. 1(b). Each number in Fig. 1(b) represents the port number through which a_h visits each node in the virtual ring. Next, we define a token node in a virtual ring as follows. First, the initial token node in the tree network is also the token node in the virtual ring. In addition, when agent a_h visits a token node v_j in the tree, we define that a_h visits a token node in the virtual ring if it visits v_j through the port $(d_{v_j} - 1)$. In Fig. 1(a), if nodes a and b are token nodes, then in Fig. 1(b), nodes a and b'' are token nodes. By this definition, a token node in the tree network is mapped to one token node in the virtual ring. Thus, by performing the basic walk, we can assume that each agent moves in the same virtual ring. Moreover, in the virtual ring, each agent also moves in a FIFO manner, that is, when an agent a_h leaves some node v_j before another agent a_i leaves v_j, a_h takes a step before a_i does it.

The algorithm consists of two parts. In the first part, agents execute the leader agent election partway and elect some leader agents. In the second part, leader agents instruct the other agents which node they meet at, and the other agents move to the node by the instruction. In the following section, we explain the algorithm by using a virtual ring.

4.1   The First Part: Leader Election

In the leader agent election, the states of agents are divided into the following three types:

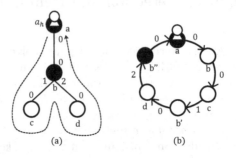

Fig. 1. An example of the basic walk

- *active*: The agent is performing the leader agent election as a candidate of leaders.
- *inactive*: The agent has dropped out from the candidate of leaders.
- *leader*: The agent has been elected as a leader.

The aim of the first part is similar to [10], that is, to elect some leaders and satisfy the following three properties: 1) At least one agent is elected as a leader, 2) at most $\lfloor k/g \rfloor$ agents are elected as leaders, and 3) in the virtual ring, there exist at least $g - 1$ inactive agents between two leader agents.

At first, we explain the idea of the leader election in [10] to adopt it in this paper. In [10], the network is a unidirectional ring, each agent is distinct, and each node has a whiteboard. First, we explain the idea under the assumption that the ring is bidirectional. Later, we apply the idea to the unidirectional ring. The algorithm consists of several phases. In each phase, each active agent compares its own ID with IDs of its forward and backward neighboring active agents. More concretely, each active agent writes its ID on the whiteboard of its current node, and then moves forward and backward to observe IDs of the forward and backward active agents. If its own ID is the smallest among the three agents, the agent remains active (as a candidate of leaders) in the next phase. Otherwise, the agent drops out from the candidate of leaders and becomes inactive. Note that, in each phase, neighboring active agents never remain as candidates of leaders. Hence, at least half active agents become inactive and the number of inactive agents between two active agents at least doubles in each phase. After executing j phases, there exists at least $2^j - 1$ inactive agents between two active agents. Thus, after executing $\lceil \log g \rceil$ phases, the following properties are satisfied: 1) At least one agent remains as a candidate of leaders, 2) at most $\lfloor k/g \rfloor$ agents remain as candidates of leaders, and 3) the number of inactive agents between two active agents is at least $g - 1$. Therefore, all remaining active agents become leaders.

Next, we implement the above algorithm in asynchronous unidirectional rings. In [10], agents use a traditional approach [24] to implement the above algorithm in a unidirectional ring. Let us consider the behavior of active agent a_h. In unidirectional rings, a_h cannot move backward and cannot observe the ID of its backward active agent. Instead, a_h moves forward until it observes IDs of two active agents. Then, a_h observes IDs of three successive active agents. We assume a_h observes id_1, id_2, id_3 in this order. Note that id_1 is the ID of a_h. Here this situation is similar to that the active agent with ID id_2 observes id_1 as its backward active agent and id_3 as its forward active agent in bidirectional rings. For this reason, a_h behaves as if it would be an active agent with ID id_2 in bidirectional rings. That is, if id_2 is the smallest among the three IDs, a_h remains active as a candidate of leaders. Otherwise, a_h drops out from the candidate of leaders and becomes inactive. This is the idea of the leader election in [10].

In the following, we explain the way to apply the above leader election [10] to this paper. In [10], each agent is distinct and each node has whiteboard. However, in this paper, we assume that each agent is anonymous and nodes have tokens. First, we explain the treatment about IDs. For explanation, let *active nodes* be nodes where active agents start execution of each phase. In this section, agents use *virtual IDs* in the virtual ring. Concretely, when agent a_h moves from an active node v_j to v_j's forward active node $v_{j'}$, a_h observes port sequence $p_1, p_2, \ldots p_l$, where p_m is the port number through which a_h visits the node by the m-th movement after leaving v_j. In this case, a_h uses this port sequence $p_1, p_2, \ldots p_l$ as its virtual ID. For example, in Fig. 1(b), when a_h moves from a to b'', a_h observes the port numbers $0, 0, 1, 0, 2$ in this order. Hence, a_h uses 00102 as a virtual ID from a to b''. Similarly, a_h uses 0 as a virtual ID from b'' to a. Note that, multiple agents may have the same virtual IDs, and we explain the behavior in this case later. Next, we explain the treatment about whiteboards. In [10], each node has a whiteboard, while in this paper, each node is allowed to have an only token. Fortunately, we can easily overcome this problem by using virtual IDs. Concretely, each active agent a_h moves until a_h visits three active nodes. Then, a_h observes its own virtual ID, the virtual ID of a_h's forward active agent a_i, and the virtual ID of a_i's forward active agent a_j. Thus, a_h can obtain three virtual IDs id_1, id_2, id_3 without using whiteboards. Therefore, agents can use the above approach [24], that is, a_h behaves as if it would be an active agent with ID id_2 in bidirectional rings. In the rest of this paragraph, we explain how agents detect active nodes. In the beginning of the algorithm, each agent starts the algorithm at a token node and all token nodes are active nodes. After each agent a_h visits three active nodes, a_h decides whether a_h remains active or drops out from the candidate of leaders at the active (token) node. If a_h remains active, then a_h starts the next phase and leaves the active node. Thus, in some phase, when some active agent a_h visits a token node v_j with no agent, a_h knows that a_h visits an active node and the other nodes are not active nodes in the phase.

After observing three virtual IDs id_1, id_2, id_3, each active agent a_h compares virtual IDs and decides whether a_h remains active (as a candidate of leaders) in the next phase or not. Different from [10], multiple agents may have the same IDs. To treat this case, if $id_2 < \min(id_1, id_3)$ or $id_2 = id_3 < id_1$ holds, then a_h remains active as a candidate of leaders. Otherwise, a_h becomes inactive and drops out from the candidate of leaders. For example, let us consider the initial configuration like Fig. 2(a). In the figure, black nodes are token nodes

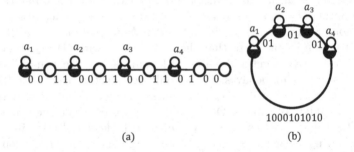

Fig. 2. An example that agents observe the same port sequence

and the numbers near communication links are port numbers. The virtual ring of Fig. 2(a) is represented in Fig. 2(b). For simplicity, we omit non-token nodes in Fig. 2(b). The numbers in Fig. 2(b) are virtual IDs. Each agent a_h continues to move until a_h visits three active nodes. By the movement, a_1 observes three virtual IDs (01,01,01), a_2 observes three virtual IDs $(01, 01, 1000101010)$, a_3 observes three virtual IDs $(01, 1000101010, 01)$, and a_4 observes three virtual IDs $(1000101010, 01, 01)$ respectively. Thus, a_4 remains as a candidate of leaders, and a_1, a_2, and a_3 drop out from the candidates of leaders. Note that, like Fig. 2, if an agent observes the same virtual IDs three times, it drops out from the candidate of leaders. This implies, if all active agents have the same virtual IDs, all agents become inactive. However, we can show that, when there exist at least three active agents, it does not happen that all active agents observe the same virtual IDs. Moreover, if there are only one or two active agents in some phase, then the agents notice the fact during the phase. In this case, the agents immediately become leaders. By executing $\lceil \log g \rceil$ phases, agents complete the leader agent election.

Pseudocode. The pseudocode to elect leaders is given in Algorithm 1. All agents start the algorithm with active states. The pseudocode describes the behavior of active agent a_h, and v_j represents the node where agent a_h currently stays. If agent a_h becomes an inactive state or a leader state, a_h immediately moves to the next part and executes the algorithm for an inactive state or a leader state in section 4.2. Agent a_h uses variables id_1, id_2, and id_3 to store three virtual IDs. Variable *phase* stores the phase number of a_h. In Algorithm 1, each active agent a_h moves until a_h observes three virtual IDs and decides whether a_h remains active as a candidate of leaders or not on the basis of virtual IDs. Note that, since each agent moves in a FIFO manner, it does not happen that some active agent passes another active agent in the virtual ring, and each active agent correctly observes three neighboring virtual IDs in the phase. In Algorithm 1, a_h uses procedure *NextActive()*, by which a_h moves to the next active node and returns the port sequence as a virtual ID. The pseudocode of *NextActive()* is described in Algorithm 2. Agent a_h uses variable *port* to store a virtual ID while moving, and a_h uses variable *move* to store the number of nodes it visits. Note that, if there exist only one or two active agents in some phase,

---

Algorithm 1. The behavior of active agent a_h (v_j is the current node of a_h.)

Variables in Agent a_h
int $phase = 0$;
int id_1, id_2, id_3;
Main Routine of Agent a_h
1: $phase = phase + 1$
2: $id_1 = NextActive()$
3: $id_2 = NextActive()$
4: $id_3 = NextActive()$
5: **if** there exist at most two active agents in the tree **then**
6:      change its state to a leader state
7: **end if**
8: **if** $(id_2 < min(id_1, id_3)) \vee (id_2 = id_3 < id_1)$ **then**
9:      **if** $(phase = \lceil \log g \rceil)$ **then**
10:        change its state to a leader state
11:     **else**
12:        return to line 1
13:     **end if**
14: **else**
15:     change its state to an inactive state
16: **end if**

---

then the agent moves around the virtual ring before getting three virtual IDs. In this case, the active agent knows that there exist at most two active agents in the phase and they become leaders. To do this, agents record the topology every time they visit nodes, but we omit the description of this behavior in Algorithm 1 and Algorithm 2.

First, we show the following lemma to show that at least one agent remains active or becomes a leader in each phase.

Lemma 1. *When there exist at least three active agents, at least one agent has a virtual ID different from another agent.*

Next, we have the following lemmas about Algorithm 1.

Lemma 2. *Algorithm 1 eventually terminates, and satisfies the following three properties.*

- *There exists at least one leader agent.*
- *There exist at most $\lfloor k/g \rfloor$ leader agents.*
- *In the virtual ring, there exist at least $g - 1$ inactive agents between two leader agents.*

Lemma 3. *Algorithm 1 requires $O(n \log g)$ total moves.*

4.2   The Second Part: Leaders' Instruction and Agents' Movement

In this section, we explain the second part, i.e., an algorithm to achieve the g-partial gathering by using leaders elected by the algorithm in Section 4.1. Let

---

Algorithm 2. int *NextActive*() (v_j is the current node of a_h.)

Main Routine of Agent a_h
array *port*[ ];
int *move*;
Main Routine of Agent a_h
 1: $move = 0$
 2: leave v_j through the port 0
    // arrive at the forward node
 3: let p be the port number through which a_h visits v_j
 4: $port[move] = p$
 5: $move = move + 1$
 6: **while** (there does not exist a token) $\vee$
    $(p \neq d_{v_j} - 1) \vee$ (there exists another agent) **do**
 7:    leave v_j through the port $(p + 1) \mod d_{v_j}$
       // arrive at the forward node
 8:    let p be the port number through which a_h visits v_j
 9:    $port[move] = p$
10:    $move = move + 1$
11: **end while**
12: return *port*[ ]

---

leader nodes (resp., inactive nodes) be the nodes where agents become leaders (resp., inactive agents). Note that all leader nodes and inactive nodes are token nodes. In this part, states of agents are divided into the following three types:

- *leader*: The agent instructs inactive agents where they should move.
- *inactive*: The agent waits for the leader's instruction.
- *moving*: The agent moves to its gathering node.

We explain the idea of the algorithm in the virtual ring. The basic movement is also similar to [10], that is, to divide agents into groups with at least g agents. In [10], each node has a whiteboard, while in this paper, each node is allowed to have an only token. In this section, agents achieve the g-partial gathering by using removable tokens. Concretely, each leader agent a_h moves to the next leader node, and while moving a_h repeats the following behavior: a_h removes tokens of inactive nodes $g - 1$ times consecutively and then a_h does not remove a token of the next inactive node. After that, agents move to token nodes and meet at least g agents there.

First, we explain the behavior of leader agents. Whenever leader agent a_h visits an inactive node v_j, it counts the number of inactive nodes that a_h has visited. If the number plus one is not a multiple of g, a_h removes a token at v_j. Otherwise, a_h does not remove the token and continues to move. Agent a_h continues this behavior until a_h visits the next leader node $v_{j'}$. After that, a_h removes a token at $v_{j'}$. After completing this behavior, there exist at least $g - 1$ inactive agents between two token nodes. Hence, agents solve the g-partial gathering problem by going to the nearest token node (This is done by changing their states to moving states). For example, let us consider the configuration like

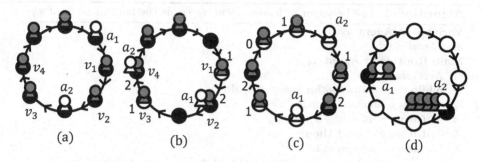

Fig. 3. Partial gathering for removable-token model for the case $g = 3$ (a_1 and a_2 are leaders, and black nodes are token nodes)

Fig. 3(a) ($g = 3$). We assume that a_1 and a_2 are leader agents and the other agents are inactive agents. In Fig. 3(b), a_1 visits the node v_2 and a_2 visits the node v_4 respectively. The numbers near nodes represent the number of inactive nodes that a_1 and a_2 observed respectively. Agents a_1 and a_2 remove tokens at v_1 and v_3, and do not remove tokens at v_2 and v_4 respectively. After that, a_1 and a_2 continue this behavior until they visit the next leader nodes. At the leader nodes, they remove the tokens (Fig. 3(c)).

When a token at v_j is removed, an inactive agent at v_j changes its state to a moving state and starts to move. Concretely, each moving agent moves to the nearest token node v_j. Note that, since each agent moves in a FIFO manner, it does not happen that a moving agent passes a leader agent and terminates at some token node before the leader agent removes the token. After all agents complete their own movements, the configuration changes from Fig. 3(c) to Fig. 3(d) and agents can solve the g-partial gathering problem. Note that, since each agent moves in the same virtual ring in a FIFO mannner, it does not happen that an acitve agent executing the leader agent elaction passes a leader agent and that a leader agent passes an active agent.

Pseudocode. In the following, we show the pseudocode of the algorithm. The pseudocode of leader agents is described in Algorithm 3. Variable $tCount$ is used to count the number of inactive nodes a_h visits. When a_h visits a token node v_j with another agent, v_j is an inactive node because an inactive agent becomes inactive at a token node and agents move in a FIFO manner. Whenever each leader agent a_h visits an inactive node, a_h increments the value of $tCount$. At inactive node v_j, a_h removes a token at v_j if $tCount \neq g - 1$ and continues to move otherwise. This means that, if a token is not removed at inactive node v_j, at least g agents meet at v_j. When a_h removes a token at v_j, an inactive agent at v_j changes its state to a moving state. When a_h visits a token node $v_{j'}$ with no agents, $v_{j'}$ is the next leader node. This is because agents at token nodes are in leader or inactive states, and each inactive agent does not leave the token node until the token is removed. When leader agent a_h moves to the next leader node $v_{j'}$, a_h removes a token at $v_{j'}$ and changes its state to a moving state. In Algorithm 3, a_h uses the procedure *NextToken*(), by which a_h moves to the

---

Algorithm 3. The behavior of leader agent a_h (v_j is the current node of a_h)

Variable in Agent a_h
int $tCount = 0$;
Main Routine of Agent a_h
1: *NextToken()*
2: **while** there exists another agent at v_j **do**
3:     //this is an inactive node
4:     $tCount = (tCount + 1) \mod g$
5:     **if** $tCount \neq g - 1$ **then**
6:         remove a token at v_j
7:         //an inactive agent at v_j changes its state to a moving state
8:     **end if**
9:     *NextToken()*
10: **end while**
11: remove a token at v_j
12: change its state to a moving state

Algorithm 4. void *NextToken()* (v_j is the current node of a_h.)

Main Routine of Agent a_h
1: leave v_j through the port 0
2: let p be the port number through which a_h visits v_j
3: **while** (there dose not exist a token) $\vee$ ($p \neq d_{v_j} - 1$) **do**
4:     leave v_j through the port $(p + 1) \mod d_{v_j}$
5:     let p be the port number through which a_h visits v_j
6: **end while**

---

next token node. The pseudocode of *NextToken()* is described in Algorithm 4. In Algorithm 4, a_h performs the basic walk until a_h visits a token node v_j through the port $(d_{v_j} - 1)$.

We omit the pseudocode of inactive agents due to limitation of space. Inactive agent a_h waits at v_j until either a token at v_j is removed or a_h observes another agent. If the token is removed, a_h changes its state to a moving state. If a_h observes another agent, the agent is a moving agent and terminates the algorithm at v_j. This means v_j is selected as a token node where at least g agents meet in the end of the algorithm. Hence, a_h terminates the algorithm at v_j.

We also omit the pseudocode of moving agents due to limitation of the space. In the virtual ring, each moving agent a_h moves to the nearest token node by using *NextToken()*.

We have the following lemma about algorithms in Section 4.2.

Lemma 4. *After the leader agent election, agents solve the g-partial gathering problem in $O(gn)$ total moves.*

From Lemma 3 and Lemma 4, we have the following theorem.

Theorem 5. *In the weak multiplicity detection and the removable-token model-Cour algorithm solves the g-partial gathering problem in $O(gn)$ total moves.*

5   Conclusion

In this paper, we proposed two move-optimal algorithms to solve the g-partial gathering problem in asynchronous tree networks. First, in the strong multiplicity detection and non-token model, we showed that agents require $\Omega(kn)$ total moves to solve the g-partial gathering problem and proposed a deterministic algorithm to solve the g-partial gathering problem in $O(kn)$ total moves. Second, in the weak multiplicity detection and removable-token model, we proposed a deterministic algorithm to solve the g-partial gathering problem in $O(gn)$ total moves. As a future work, we want to consider the weak multiplicity detection and non-token model. We conjecture that the g-partial gathering problem is not solvable in this model. If the conjecture is correct, we can show that agents require strong multiplicity detection or removable token to solve the g-partial gathering problem. In particular, by using tokens, agents can solve the g-partial gathering problem with smaller total moves compared to the total gathering problem.

References

1. Kranakis, E., Krozanc, D., Markou, E.: The mobile agent rendezvous problem in the ring, vol. 1. Morgan & Claypool Publishers (2010)
2. Dobrev, S., Flocchini, P., Prencipe, G., Santoro, N.: Mobile search for a black hole in an anonymous ring. Algorithmica 48(1), 67–90 (2007)
3. Suzuki, T., Izumi, T., Ooshita, F., Kakugawa, H., Masuzawa, T.: Move-optimal gossiping among mobile agents. Theoretical Computer Science 393(1), 90–101 (2008)
4. Chalopin, J., Das, S., Kosowski, A.: Constructing a map of an anonymous graph: Applications of universal sequences. In: Lu, C., Masuzawa, T., Mosbah, M. (eds.) OPODIS 2010. LNCS, vol. 6490, pp. 119–134. Springer, Heidelberg (2010)
5. Gasieniec, L., Pelc, A., Radzik, T., Zhang, X.: Tree exploration with logarithmic memory. In: Proc. of SODA, pp. 585–594 (2007)
6. Kawai, S., Ooshita, F., Kakugawa, H., Masuzawa, T.: Randomized rendezvous of mobile agents in anonymous unidirectional ring networks. In: Even, G., Halldórsson, M.M. (eds.) SIROCCO 2012. LNCS, vol. 7355, pp. 303–314. Springer, Heidelberg (2012)
7. Elouasbi, S., Pelc, A.: Time of anonymous rendezvous in trees: Determinism vs. randomization. In: Even, G., Halldórsson, M.M. (eds.) SIROCCO 2012. LNCS, vol. 7355, pp. 291–302. Springer, Heidelberg (2012)
8. Baba, D., Izumi, T., Ooshita, H., Kakugawa, H., Masuzawa, T.: Linear time and space gathering of anonymous mobile agents in asynchronous trees. Theoretical Computer Science, 118–126 (2013)
9. Czyzowicz, J., Kosowski, A., Pelc, A.: Time vs. space trade-offs for rendezvous in trees. In: Proc. of SPAA, pp. 1–10 (2012)
10. Shibata, M., Kawai, S., Ooshita, F., Kakugawa, H., Masuzawa, T.: Algorithms for partial gathering of mobile agents in asynchronous rings. In: Baldoni, R., Flocchini, P., Binoy, R. (eds.) OPODIS 2012. LNCS, vol. 7702, pp. 254–268. Springer, Heidelberg (2012)
11. Dobrev, S., Flocchini, P., Prencipe, G., Santoro, N.: Multiple agents rendezvous in a ring in spite of a black hole. In: Papatriantafilou, M., Hunel, P. (eds.) OPODIS 2003. LNCS, vol. 3144, pp. 34–46. Springer, Heidelberg (2004)

12. Barriere, L., Flocchini, P., Fraigniaud, P., Santoro, N.: Rendezvous and election of mobile agents: impact of sense of direction. Theory of Computing Systems 40(2), 143–162 (2007)

13. De Marco, G., Gargano, L., Kranakis, E.., Krizanc, D., Pelc, A., Vaccaro, U.: Asynchronous deterministic rendezvous in graphs. In: Jedrzejowicz, J., Szepietowski, A. (eds.) MFCS 2005. LNCS, vol. 3618, pp. 271–282. Springer, Heidelberg (2005)

14. Guilbault, S., Pelc, A.: Gathering asynchronous oblivious agents with restricted vision in an infinite line. In: Higashino, T., Katayama, Y., Masuzawa, T., Potop-Butucaru, M., Yamashita, M. (eds.) SSS 2013. LNCS, vol. 8255, pp. 296–310. Springer, Heidelberg (2013)

15. Collins, A., Czyzowicz, J., Gąsieniec, L., Kosowski, A., Martin, R.: Synchronous rendezvous for location-aware agents. In: Peleg, D. (ed.) DISC 2011. LNCS, vol. 6950, pp. 447–459. Springer, Heidelberg (2011)

16. Kranakis, E., Krizanc, D., Markou, E.: Mobile agent rendezvous in a synchronous torus. In: Correa, J.R., Hevia, A., Kiwi, M. (eds.) LATIN 2006. LNCS, vol. 3887, pp. 653–664. Springer, Heidelberg (2006)

17. Guilbault, S., Pelc, A.: Asynchronous rendezvous of anonymous agents in arbitrary graphs. In: Fernàndez Anta, A., Lipari, G., Roy, M. (eds.) OPODIS 2011. LNCS, vol. 7109, pp. 421–434. Springer, Heidelberg (2011)

18. Dieudonne, Y., Pelc, A., Peleg, D.: Gathering despite mischief. In: Proc. of SODA, pp. 527–540 (2012)

19. Flocchini, P., Kranakis, E., Krizanc, D., Luccio, F.L., Santoro, N., Sawchuk, C.: Mobile agents rendezvous when tokens fail. In: Kralovic, R., Sýkora, O. (eds.) SIROCCO 2004. LNCS, vol. 3104, pp. 161–172. Springer, Heidelberg (2004)

20. Gąsieniec, L., Kranakis, E., Krizanc, D., Zhang, X.: Optimal memory rendezvous of anonymous mobile agents in a unidirectional ring. In: Wiedermann, J., Tel, G., Pokorný, J., Bieliková, M., Štuller, J. (eds.) SOFSEM 2006. LNCS, vol. 3831, pp. 282–292. Springer, Heidelberg (2006)

21. Kranakis, E., Santoro, N., Sawchuk, C., Krizanc, D.: Mobile agent rendezvous in a ring. In: Proc. of ICDCS, pp. 592–599 (2003)

22. Flocchini, P., Kranakis, E., Krizanc, D., Santoro, N., Sawchuk, C.: Multiple mobile agent rendezvous in a ring. In: Farach-Colton, M. (ed.) LATIN 2004. LNCS, vol. 2976, pp. 599–608. Springer, Heidelberg (2004)

23. Korach, E., Rotem, D., Santoro, N.: Distributed algorithms for finding centers and medians in networks. TOPLAS 6(3), 380–401 (1984)

24. Peterson, G.L.: An $O(n \log n)$ unidirectional algorithm for the circular extrema problem. TOPLAS 4(4), 758–762 (1982)

A Recursive Approach
to Multi-robot Exploration of Trees

Christian Ortolf and Christian Schindelhauer

University of Freiburg, Department of Computer Science, Computer Networks, Germany
{ortolf,schindel}@informatik.uni-freiburg.de

Abstract. The multi-robot exploration problem is to explore an unknown graph of size n and depth d with k robots starting from the same node. For known graphs a traversal of all nodes takes at most $\mathcal{O}(d + n/k)$ steps. The ratio between the time until cooperating robots explore an unknown graph and the optimal traversal of a known graph is called the competitive exploration time ratio.

It is known that for any algorithm this ratio is at least $\Omega\left((\log k)/\log\log k\right)$. For $k \le n$ robots the best algorithm known so far achieves a competitive time ratio of $\mathcal{O}\left(k/\log k\right)$.

Here, we improve this bound for trees with bounded depth or a minimum number of robots. Starting from a simple $\mathcal{O}(d)$-competitive algorithm, called Yo-yo, we recursively improve it by the Yo-star algorithm, which for any $0 < \alpha < 1$ transforms a $g(d, k)$-competitive algorithm into a $\mathcal{O}((g(d^\alpha, k)\log k + d^{1-\alpha})(\log k + \log n))$-competitive algorithm. So, we achieve a competitive bound of $\mathcal{O}\left(2^{\mathcal{O}(\sqrt{(\log d)(\log\log k)})}(\log k)(\log k + \log n)\right)$. This improves the best known bounds for trees of depth d, whenever the number of robots is at least $k = 2^{\omega(\sqrt{(\log d)(\log\log d)})}$ and $n = 2^{\mathcal{O}(2^{\sqrt{\log d}})}$.

Keywords: competitive analysis, robot, collective graph exploration.

1 Introduction

Maintenance robots are nowadays common for households and every day life. One of the most basic tasks such robotic lawn mowers, vacuum cleaners, and underwater cleaning robots is to explore a new environment. Besides the technical problem of localization, orientation and communication it is not clear how to make use of the full potential of the parallel exploration. This is the problem setting of multi-robot exploration.

The Model. The multi-robot exploration takes place on a labeled, connected, undirected graph $G = (V, E)$ with $|V| = n$ and diameter d. In each round an algorithm has to decide for each of the k robots which edge it traverses to a neighboring node. The algorithm knows the positions of all robots and is not computationally restricted. The goal is to visit all nodes of the graph as fast as possible at least once.

An offline algorithm has full knowledge of G, while an online algorithm can only use the induced subgraph defined by the already visited nodes of V and their neighboring nodes. The efficiency of an online algorithm is measured by comparing its run-time

M. Halldórsson (Ed.): SIROCCO 2014, LNCS 8576, pp. 343–354, 2014.

against the asymptotically optimal offline algorithm. The maximum of the ratio between the run-time of the online and the offline exploration time is called the competitive ratio of an algorithm.

Our Results. In this paper we present two algorithms. The first, called Yo-yo, achieves competitive ratio of $4d$, while the second, called Yo*, recursively improves this bound up to a ratio of $\mathcal{O}\left(2^{\mathcal{O}(\sqrt{(\log d)(\log \log k)})}(\log k)(\log k + \log n)\right)$. This is an improvement of the best known results for a minimum number of $k = d^c$ robots for $c > 0$ up to a ratio of $k^\epsilon \log n$ for any $\epsilon > 0$.

Related Work. Exploration is a more than a century old problem (for a survey we recommend [18]) closely related to the Traveling Salesman and the Hamiltonian cycle problem. But most work on exploration only handles various cases concerning a single robot.

For the single robot case asymptotically optimal exploration up to a factor of two is possible with depth-first-search DFS. Using a map the exploration of a line or tree can be improved by preventing double traversal of edges. Desmark et al. show in [8] various competitive constants that can be gained depending on if an anchored, unanchored or no map at all is available.

If graphs are not labeled, DFS cannot be directly used. M.A. Bender presents solutions for this scenario in [3,4] using a pebble or a second robot for bookkeeping.

In 2006, Fraigniaud et al. consider the multi-robot exploration problem for trees in [12]. They present an algorithm that with a run-time of $\mathcal{O}(k/\log k)$ is far apart from their lower bound of $\Omega(2 + 1/k)$ and quite close the trivial upper bound of $\mathcal{O}(k)$ achieved by executing a depth first search using a single robot. While the lower bound is improved by Dynia et al. in [10] to $\Omega(\frac{\log k}{\log \log k})$ the upper bound still is the state of the art and the exponential gap remains open between these two bounds.

Several restrictions for the exploration can improve the bounds. If algorithms are restricted to greedy exploration an even stronger bound of $\Omega(k/\log k)$ is shown by Higashikawa et al. [15].

For restricted graphs several better algorithms exist. Dynia et al. showed in [9] a faster exploration for trees restricted by a *density* parameter p, enforcing a minimum depth for any subtree depending on its size. For example trees embeddable in p-dimensional grids could be explored with competitiveness of $\mathcal{O}(d^{1-1/p})$.

For 2-dimensional grids with only convex obstacles we improved the competitive bound to $\mathcal{O}(\log^2 n)$ in [17]. Also note that despite this strong restrictions to a graph the same lower bound of $\Omega(\frac{\log k}{\log \log k})$ as for trees holds.

In Brass et al.'s work an upper bound of $\mathcal{O}(\frac{n}{k} + d^{k-1})$ [5] is shown, they implement an algorithm that moves robots similar to the method of Fraigniaud et al. [12], but also works on graphs using only a local communication model with bookkeeping devices.

Dereniowski et al. discuss in their work very large values of k. They show how many more robots need to be invested to explore in asymptotically optimal time. They show a minimum of $k = dn^{1+\epsilon}$ for an $\epsilon > 0$ robots to be necessary and improve with this the trivial bound of $\mathcal{O}(n^d)$ required to explore any graph in time d with flooding [7].

Exploration of directed graphs is not discussed here. Competitive analysis done by Albers et al. [1], Fleischer et al. [11], Papadimitriou et al. [6] and Förster et al. [13] indicates this to be a harder problem than the undirected case.

Some works model the exploration geometrically, this is useful if robots have a sense of sight enabling to see additional nodes before visiting them [14,16] or having to move around corners to make everything visible in case of unlimited vision [2].

A constant factor offline approximation. A very basic observation is the constant factor offline approximation presented in Algorithm 1, which establishes a constant approximation factor of four.

Algorithm 1. Offline 4-competitive multi-robot exploration of trees for robot j

---

1: Compute a cycle of length $2n$ using DFS covering the tree
2: Divide the cycle into k intervals of size at most $\lceil 2n/k \rceil$
3: Go to the j-th interval
4: Traverse the interval

Lemma 1. *Algorithm 1 needs at most $d + \lceil 2n/k \rceil$ robot moves and has a competitive factor of four.*

Proof. Every exploration algorithm needs at least $\max\{\lceil n/k \rceil, d\}$ steps. The number of robot moves of Algorithm 1 is $d + \lceil 2n/k \rceil \leq 2\max\{d, \lceil 2n/k \rceil\} \leq 4\max\{d, \lceil n/k \rceil\}$. Hence, it is 4-competitive.

2 The Yo-yo Exploration

The basic idea of the Yo-yo exploration algorithm is to successively explore every set of nodes in the tree with the same depth. After each exploration step all robots return to the root and are perfectly rebalanced for the next exploration step. For most trees this algorithm is not very efficient, since most of the time the robots commute between the root and the leafs of the so far known sub-tree. We denote the number of nodes in depth i by n_i.

Algorithm 2. The Yo-yo Algorithm: $4d$-competitive multi-robot exploration of a tree

---

1: All robots start at the root of the tree
2: **for** $i \leftarrow 2, \ldots, d$ **do**
3:    Partition all n_i nodes in depth i into k subsets $V_{i,1}, \ldots, V_{i,k}$ with $|V_{i,j}| \leq \lceil \frac{n_i}{k} \rceil$.
4:    **for all** $j \leftarrow 1, \ldots, k$ **do in parallel**
5:      **for all** $u \in V_{i,j}$ **do**
6:        Move robot j to u
7:        Move robot j to the root
8:      **end for**
9:    **end for**
10: **end for**

The main motivation of this algorithm is that the competitive ratio $4d$ only depends on the depth d, which we can improve later on by a technique which does not work with a competitive factor only depending on k. Note that for at least $k = \Omega(d \log d)$ robots Yo-yo is asymptotically at least as good as the best known algorithm of Fraigniaud et al. [12].

Lemma 2. *The Yo-yo algorithm needs at most $d(d + 1) + 2dn/k$ rounds to explore a graph with n nodes, depth d and k robots, and thus has a competitive exploration ratio of at most $4d$.*

Proof. The success of the exploration algorithm follows by an easy induction over the tree depth. For the number of rounds, note that in lines 6 and 7 each of the k robots moves for $2i$ rounds in order to explore a node in $V_{i,j}$ and return to the root. This is repeated in the loop starting at line 5 for at most $\lceil n_i/k \rceil$ times. Therefore, the overall number of rounds for the outer loop starting at line 2 is the following.

$$\sum_{i=1}^{d} 2i \left\lceil \frac{n_i}{k} \right\rceil \leq \sum_{i=1}^{d} 2i \left(1 + \frac{n_i}{k} \right)$$

$$= d(d + 1) + \sum_{i=1}^{d} 2i \frac{n_i}{k}$$

$$\leq d(d + 1) + 2d \frac{n}{k} ,$$

Where we use $\sum_{i=0}^{d} n_i = n$. Now, every exploration algorithm needs at least $\max\{d, \lceil n/k \rceil\} \geq \frac{1}{2}(d + n/k)$ rounds. So, the competitive factor is at most

$$\frac{d(d + 1) + 2d \frac{n}{k}}{\max\{d, \lceil n/k \rceil\}} \leq \frac{2d^2 + 2d \frac{n}{k}}{\frac{1}{2}(d + n/k)} \leq 4d$$

$\square$

3   The Yo* Algorithm

Starting from the Yo-yo algorithm (Algorithm 3) we use a recursive approach to improve the efficiency of the exploration. To avoid the rebalancing step passing the root in each step we divide the graph into the uppermost segment of depth c and b segments of depth a such that $d = ab + c$ which values are to be chosen later on, see Fig. 1. The first segment will be explored by the base algorithm, e.g. the Yo-yo algorithm. One can easily see that if the competitive ratio grows with the depth of the tree we can bound the ratio with a smaller term now.

All deeper unexplored segments will be handled together with the last explored segment, see Fig. 2. These two segments form a forest of trees. If the number of trees is greater than the number of robots, we can use DFS to efficiently explore them. However, the size of the trees can differ and therefore, we rebalance the robots if half of the trees have been explored by DFS. The rebalancing costs at most d steps and this has to be repeated at most $\log n$ times.

If the number of trees has been reduced to be smaller than the number of robots, we use the base algorithm and rebalance again, if half of the trees have been completely explored. So, we have $\log k$ iterations for all the b segments.

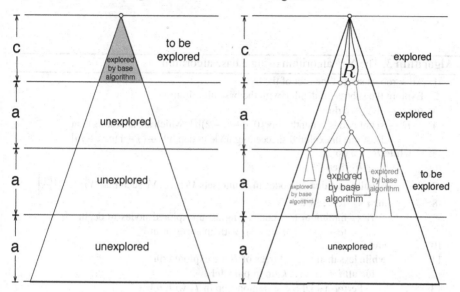

Fig. 1. The first round of the Yo-star algorithm   **Fig. 2.** The principle of the Yo-star algorithm

Taking the Yo-yo algorithm and choosing segments of depth $a = c = d^{\frac{1}{2}}$ a back-on-the-envelope calculation gives us a competitive ratio of $O(d^{\frac{1}{2}})$ for the first segment and a ratio of $O(d^{\frac{1}{2}}(\log n + \log k))$ for all the other segments. So, after one iteration of the Yo* algorithm we improve the depth-dependent factor in the ratio from d to $d^{\frac{1}{2}}$. Now, if we take this new algorithm as base algorithm and choose segments of size $d^{\frac{1}{3}}$ we improve the ratio to $d^{\frac{1}{3}}$. However, there is an overhead in the iteration, where constant factors grow exponentially over the number of iterations and thus must be carefully analyzed.

So, we assume we start from a $g(d, k)(d + \frac{n}{k})$ time bounded algorithm and try to turn it into a more efficient one using the Yo* algorithm. From $g(d, k)$ we only know that it is a monotone increasing function with respect to d and k, e.g. for the Yo-yo algorithm we have $g(d, k) = 4d$.

Lemma 3. *Given a $g(d, k)(d + \frac{n}{k})$-time bounded base algorithm for a graph with unknown number of nodes n, given depth $d = ab + c$, $a, b, c \in \mathbb{N}$, and k exploring robots, then the Yo* algorithm can explore such a tree within the following number of rounds*

$$\left(d + \frac{n}{k}\right)(8g(2a, k)\log k + g(c, k) + 2b(\log k + \log n) + 4\log n) .$$

Proof. We denote by n_0 the number of nodes in depth at most c. By n_j we denote the number of nodes of the tree with depth in the interval $[1 + c + (j - 1)a, c + ja]$. By definition $\sum_{j=0}^{b} n_j = |V|$.

Algorithm 3. The Yo* algorithm using a base algorithm

1: All k robots start at the root of the tree
2: Explore the subtree of depth c with the base algorithm
3: **for** $j \leftarrow 1, \ldots, b$ **do**
4:     $R \leftarrow$ set of nodes in depth $\max\{0, c + (j-2)a\}$, which are ancestors to at
        least one unexplored succeeding node in depth $[c + (j-1)a, c + ja]$
5:     **while** $R \neq \emptyset$ **do**
6:         **if** $k \leq |R|$ **then**
7:             Equally partition all nodes in R into sets $V_1, \ldots, V_k$ such that $|V_i| \leq \left\lceil \frac{|R|}{k} \right\rceil$
8:             **for** $i \leftarrow 1, \ldots, k$ **do**
9:                 $T_i \leftarrow$ minimum tree connecting all unexplored nodes in depth
                    $[c + (j-1)a, c + ja]$ with an ancestor in V_i
10:            **end for**
11:            **while** less than $k/2$ subtrees of R are explored **do**
12:                **for all** $i \leftarrow 1, \ldots, k$ **do in parallel**
13:                    Perform a DFS exploration step in T_i with robot i
14:                **end for**
15:            **end while**
16:        **else**
17:            **for** $i \leftarrow 1, \ldots, |R|$ **do**
18:                Equally assign k_i robots to node v_i of R such that $k_i \in \left\{ \left\lfloor \frac{k}{|R|} \right\rfloor, \left\lceil \frac{k}{|R|} \right\rceil \right\}$
19:                $T_i \leftarrow$ minimum tree connecting all unexplored nodes in depth
                    $[c + (j-1)a, c + ja]$ with ancestor v_i
20:            **end for**
21:            **while** less than $k/2$ subtrees of R are fully explored **do**
22:                **for all** $i \leftarrow 1, \ldots, |R|$ **do in parallel**
23:                    Perform one step of the base exploration algorithm
                        on T_i with k_i robots
24:                **end for**
25:            **end while**
26:        **end if**
27:        $R \leftarrow$ set of nodes in depth $\max\{0, c + (j-2)a\}$, which are ancestors to at
            least one unexplored succeeding node in depth $[c + (j-1)a, c + ja]$
28:    **end while**
29: **end for**

We use the base algorithm to explore the first segment, which needs at most

$$t_1 = g(c, k) \left(d + \frac{n_0}{k} \right) \tag{1}$$

rounds.

In all other rounds we use the base algorithm several times when $k > |R|$. After each iteration of the while-loop from lines 21-25 the number of $|R|$ is reduced by a factor of 2, which implies at most $\log k$ iterations. The variable $\nu = 1, \ldots, \log k$ counts the iterations of this loop. Let $R_{j,\nu}$ be the variable R in the j-th loop and the ν-th iteration. Let $k_{j,\nu}$ be the smallest number of robots in this phase, i.e. $k_{j,\nu} = \lfloor k/|R_{j,\nu}| \rfloor$.

The trees connecting all unexplored nodes with an ancestor node in $R_{j,\nu}$ are named $T_{j,\nu,i}$ for $i \in \{1, \ldots, |R_{j,\nu}|\}$. Now define

$$n_{j,\nu} := \mathrm{median}(|V(T_{j,\nu,i})| , \ i \in \{1, \ldots, |R_{j,\nu}|\})$$

where for even number m the median refers to the $m/2$-th largest element. Note that the median implies that

$$n_{j,\nu} \frac{|R_{j,\nu}|}{2} \leq \sum_{i=1}^{|R_{j,\nu}|} |V(T_{j,\nu,i})| \leq n_{j-1} + n_j$$

So, we can conclude that

$$\sum_{\nu=1}^{\log k} n_{j,\nu} |R_{j,\nu}| \leq 2(n_j + n_{j-1}) \log k .$$

The run-time of one invocation the base algorithm is by definition at most

$$g(2a, k_{j,\nu}) \left(2a + \left\lceil \frac{n_{j,\nu}}{k_{j,\nu}} \right\rceil \right)$$

by design of the loop in line 21. Since $k_{j,\nu} \geq |R_{j,\nu}|$ and $n_{j,\nu} \geq 1$ we can use $\left\lceil \frac{x}{\lfloor y \rfloor} \right\rceil \leq 2\frac{x}{y} + 1$ for $x, y \geq 1$.

$$\left\lceil \frac{n_{j,\nu}}{k_{j,\nu}} \right\rceil = \left\lceil \frac{n_{j,\nu}}{\lfloor k/|R_{j,\nu}| \rfloor} \right\rceil \leq 2\frac{n_{j,\nu}|R_{j,\nu}|}{k} + 1$$

The run-time over all invocations of all these loops is therefore

$$t_2 \leq \sum_{j=1}^{b} \sum_{\nu=1}^{\log k} g(2a, k_{j,\nu})(2a + \lceil n_{j,\nu}/k_{j,\nu} \rceil)$$

$$\leq g(2a,k)\left(2ab\log k + b\log k + 2\sum_{j=1}^{b}\sum_{\nu=1}^{\log k}\frac{n_{j,\nu}|R_{j,\nu}|}{k}\right)$$

$$\leq g(2a,k)\left(2ab\log k + b\log k + 2\sum_{j=1}^{b}\frac{2(n_{j-1}+n_j)\log k}{k}\right)$$

$$\leq g(2a,k)\left(2(d-c)\log k + b\log k + 8\frac{n}{k}\log k\right)$$

$$\leq \left(2d + b + 8\frac{n}{k}\right)g(2a,k)\log k .$$

It remains to count all rebalancing moves of the robots. It takes at most $2d$ steps to reassign a robot to its new tree. For the case $k > |R|$, this iterates at most $b\log k$ times, resulting in

$$t_3 \leq 2bd\log k$$

steps.

Now we analyze the case $k \leq |R|$. After each iteration of the loop of line 11 the number of nodes in $|R|$ is halved. Hence, the number of loops is bounded by $\log n$. The sum of all iterations of the loop 11 is bounded by $4(n_{j-1}+n_j)/k$ rounds, since $k/2$ robots successfully explore the graph in parallel. So, summing over all j we get

$$t_4 \leq \sum_{j=1}^{b}\frac{4(n_{j-1}+n_j)}{k} \leq 8\frac{n}{k}\log n$$

for the DFS-exploration. Again we have to rearrange the robots between the trees which costs at most

$$t_5 \leq 2bd\log n$$

additional steps.

So, for $c \leq 2a$ and $d \geq 1$ the cost is bounded by:

$$\text{run-time} = t_1 + t_2 + t_3 + t_4 + t_5$$
$$\leq g(c,k)\left(d + \frac{n_0}{k}\right) + g(2a,k)\left(2d + b + 8\frac{n}{k}\right)\log k$$
$$\quad + 2bd\log k + 4\frac{n}{k}\log n + 2bd\log n$$
$$\leq d\left(g(c,k) + \left(2 + \frac{b}{d}\right)g(2a,k)\log k\right)$$
$$\quad + \frac{n}{k}\left(g(c,k) + 8g(2a,k)\log k + 4\log n\right) + 2db(\log k + \log n)$$
$$\leq \left(d + \frac{n}{k}\right)\left(g(c,k) + 8g(2a,k)\log k + 2b(\log k + \log n) + 4\log n\right)$$

□

We use polynomials of d for a and b, which results in the following Lemma.

Lemma 4. *Given a $g(d,k)(d + \frac{n}{k})$-time bounded base algorithm for a graph with unknown number of nodes n, given depth d and k exploring robots, then the Yo* algorithm provides a $(d + \frac{n}{k})(9g(2d^\alpha, k) \log k + 8d^{1-\alpha}(\log k + \log n))$-time bounded robot exploration algorithm.*

Proof. We choose $a = \lfloor d^\alpha \rfloor$, $b = \lfloor d/a \rfloor$, and $c = n - ab$. Note that $c \le a$ and $b \le 2d^{1-\alpha}$. From Lemma 3 it follows that the run-time of Yo* is the following:

$$\text{time} \le \left(d + \frac{n}{k}\right)\left(g(c,k) + 8g(2a,k)\log k + 2b(\log k + \log n) + 4\log n\right)$$

$$\le \left(d + \frac{n}{k}\right)\left(g(d^\alpha,k) + 8g(2d^\alpha,k)\log k + 4d^{1-\alpha}(\log k + \log n) + 4\log n\right)$$

$$\le \left(d + \frac{n}{k}\right)\left(9g(2d^\alpha,k)\log k + 8d^{1-\alpha}(\log k + \log n)\right)$$

Starting from the $4d$-competitive Yo-yo algorithm we choose $\alpha = \frac{1}{2}$ and obtain by the last Lemma a $\mathcal{O}(d^{\frac{1}{2}}(\log k + \log n))$-competitive multi-robot exploration algorithm. This algorithm can be also asymptotically improved by the same lemma. For this we can choose $\alpha = \frac{2}{3}$ and get a $\mathcal{O}(d^{\frac{1}{3}}(\log k)(\log k + \log n))$ algorithm. Of course this process can be iterated using the following lemma.

Lemma 5. *For $k \ge 2$, $c \ge 4$, $\beta \in [0,1]$, $\gamma \ge 0$ and a base exploration algorithm with a run-time of $(d + \frac{n}{k}) cd^\beta(\log k)^\gamma(\log k + \log n)$, the Yo-star algorithms can achieve an exploration time bound of $(d + \frac{n}{k}) 20c \cdot d^{\beta/(\beta+1)}(\log k)^{\gamma+1}(\log k + \log n)$.*

Proof. We choose $\alpha = \frac{1}{1+\beta}$ such that $\alpha\beta = 1 - \alpha$. This observation will be used for the run-time of an iteration the Yo* algorithm.

$$\frac{\text{time}}{d + \frac{n}{k}} \le 9g(2d^\alpha, k)\log k + 8d^{1-\alpha}(\log k + \log n)$$

$$\le 9c2^\beta d^{\alpha\beta}(\log k)^{\gamma+1}(\log k + \log n) + 8d^{1-\alpha}(\log k + \log n)$$
$$\le 18cd^{\beta/(\beta+1)}(\log k)^{\gamma+1}(\log k + \log n) + 8d^{\beta/(\beta+1)}(\log k + \log n)$$
$$\le 20cd^{\beta/(\beta+1)}(\log k)^{\gamma+1}(\log k + \log n) ,$$

where we use $2^\beta \le 2$ and $18c + 8 \le 20c$ for $c \ge 4$.   □

Note that the iteration $\beta \mapsto \beta/(\beta+1)$ with starting point $\beta = 1$ results in the series $1, \frac{1}{2}, \frac{1}{3}, \frac{1}{4}, \dots$. Let $\beta_1 := 1$ and $\beta_{i+1} := \beta_i/(\beta_i + 1)$. If $\beta_i = \frac{1}{i}$, then

$$\beta_{i+1} = \frac{\frac{1}{i}}{\frac{1}{i} + 1} = \frac{1}{i + 1} .$$

So, after ℓ iterations of the Yo-star algorithm, starting from the Yo-yo algorithm we have the following ratio:

$$\frac{\text{time}}{\frac{n}{k} + d} \le 4 \cdot 20^\ell d^{\frac{1}{\ell+1}}(\log k)^\ell(\log k + \log n) .$$

So far, we have assumed to know the depth. This is not necessary, since we use exponential doubling to find it. This introduces an additional factor of $\log d$ which we will take into account from now on. This competitive factor is only taken into account once, since after a correct guess the recursive approach invokes the next exploration algorithm with the correct depth value.

Theorem 1. *The Yo-star multi-robot exploration algorithm with ℓ iterations can explore an unknown tree with depth d and size n with a competitive ratio of at most*

$$\mathcal{O}\left(20^\ell d^{\frac{1}{\ell+1}}(\log k)^\ell (\log k + \log n)(\log d)\right) .$$

Proof. Since the depth of tree is unknown we iteratively restart the Yo-star exploration algorithm with an assumed depth of $d' = 1, 2, 4, \ldots$. An exploration is canceled if a node with depth larger than d' has been found, then the exploration starts from scratch. In the final step the time for the exploration is therefore at most (assuming $d' = 2d - 1$ in the worst case)

$$\left(2d - 1 + \frac{n}{k}\right) 4 \cdot 20^\ell (2d - 1)^{\frac{1}{\ell+1}}(\log k)^\ell (\log k + \log n) .$$

Now $2d - 1 + n/k \leq 2(d + n/k)$ and $(2d - 1)^{\frac{1}{\ell+1}} \leq (2d)^{\frac{1}{\ell+1}} \leq 2d^{\frac{1}{\ell+1}}$ results in an additional factor of 4. It takes $\log d$ iterations until $d' \geq d$ and therefore we have a total run-time of at most

$$16\left(d + \frac{n}{k}\right) \cdot 20^\ell d^{\frac{1}{\ell+1}}(\log k)^\ell (\log k + \log n) \log d$$

The competitive factor originates from the observation that the minimal time for offline exploration is $\max\{d, n/k\} \geq \frac{1}{2}(d + n/k)$.   □

This is the main result of this paper. What follows is a discussion of how many iterations are necessary to achieve best possible asymptotical bounds. It turns out that the relationship between the depth and the number of robots is crucial. If $d = \mathcal{O}((\log k)^c)$ then already the Yo-yo algorithm provides a competitive ratio of $\mathcal{O}((\log k)^c)$. If the depth is larger with respect to the number of robots, then Yo* provides better bounds.

Theorem 2. *The Yo-star algorithm can achieve a competitive factor of*

$$2^{(2+o(1))\sqrt{(\log d)(\log \log k)}}(\log k)(\log k + \log n)$$

for a k-multi-robot exploration of graphs of size n and depth d.

Proof. Again we test the depth of the tree by performing the ℓ iterations of the Yo* algorithm, where we double a depth parameter d' every time we finde a node in depth $d' + 1$. Then, we relaunch the exploration. As the iteration depth of Yo* we choose $\ell = \left\lceil \sqrt{\frac{\log d'}{\log \log k}} \right\rceil$, because

$$(\log k)^\ell = 2^{\left\lceil \sqrt{\frac{\log d'}{\log \log k}} \right\rceil \log \log k} \leq 2^{\sqrt{(\log d')(\log \log k)}} \log k$$

Table 1. Competitive exploration time ratios for the Yo-yo and the Yo* algorithm

k	d	n	competitive factor	algorithm
$2^{d^{1/c}}$	$(\log k)^c$		$4(\log k)^c$	Yo-yo
$d^{\Omega(1)}$	$k^{\mathcal{O}(1)}$		$k^{o(1)}\log n$	Yo*
$d^{\Omega(1)}$	$k^{\mathcal{O}(1)}$	$2^{d^{o(1)}}$	$k^{o(1)}$	Yo*
	$2^{\frac{c^2}{4}(\log k)^2/\log\log k}$		$k^{c(1+o(1))}\log n$	Yo*
$2^{\omega(\sqrt{\log d\log\log d})}$		$2^{\mathcal{O}(2^{\sqrt{\log d}})}$	$k^{o(1)}$	Yo*
		$2^{2^{\mathcal{O}(\sqrt{(\log d)(\log\log k)})}}$	$2^{\mathcal{O}(\sqrt{(\log d)\log\log k})}$	Yo*

and

$$d'^{2/(2\ell+1)} \leq 2^{(\log d)\sqrt{\log\log k}/\sqrt{\log d}} = 2^{\sqrt{(\log d)(\log\log k)}}$$

Now

$$20^{\ell}(\log d') \leq 2^{\sqrt{\log d}+\log\log d} = 2^{o(1)\sqrt{(\log d)(\log\log k)}}$$

for large enough k. So, the only remaining relevant factor is $\log k + \log n$ which implies the result. □

This bound is not always smaller than the best known competitive ratio of $\mathcal{O}(k/\log k)$. Yet, for trees with depth $d = 2^{(\frac{1}{4}-\epsilon)\frac{(\log k)^2}{\log\log k}}$, $\epsilon > 0$ and size $n = 2^{\mathcal{O}(2^{\sqrt{d}})}$, which includes the interesting cases of $d = \mathcal{O}(k^c)$ and $n = 2^{d^{o(1)}}$ for any $c > 0$, the Yo-star algorithms is currently the best available multi-robot exploration algorithm.

4   Conclusions

We discussed in this work the collaborative multi-robot exploration of trees. The algorithms know the positions of all robots and are not restricted in computation. Until now a wide gap was open between upper bound of $\mathcal{O}(\frac{k}{\log k})$ and the lower bound of $\Omega(\frac{\log k}{\log\log k})$. While these bounds could not been improved for nearly a decade and the only improvements have been happening on more restricted models, we finally were able to present new upper borders for the collective tree exploration.

The first, rather simple, Yo-yo algorithm has a competitive ratios of $4d$ and improves this bound for $d = o(k/\log k)$ robots. If the number of robots is smaller, then our Yo-star algorithm provides a new bound of

$$2^{(2+o(1))\sqrt{(\log d)(\log\log k)}}(\log k)(\log k + \log n) .$$

This hard to understand bound needs some interpretation and one can be derived the bounds for $k \leq n$ shown in Table 1 for any constant $c > 0$.

We presented in this work the first collaborative exploration for trees reaching sub-polynomial competitive ratio of $o(k^{\epsilon})$ for any constant $\epsilon > 0$ and $k < n$, especially this is the case for the Jellyfish tree with $k = d$ and $n/k = d$, which is used to establish the lower bound of $\mathcal{O}(\log k/\log\log k)$ [10]. This is an important step closing the gap between the upper and lower bound. Future work will be dedicated to the multi-robot exploration of general graphs using the Yo-yo and Yo* approach.

Acknowledgments. We are very grateful for the comments of the anonymous reviewers pointing out numerous mistakes in the first version.

References

1. Albers, S., Henzinger, M.R.: Exploring Unknown Environments. SIAM Journal on Computing 29(4), 1164 (2000)
2. Albers, S., Kursawe, K., Schuierer, S.: Exploring unknown environments with obstacles. In: Proceedings of the Tenth Annual ACM-SIAM Symposium on Discrete Algorithms, SODA 1999, pp. 842–843. Society for Industrial and Applied Mathematics, Philadelphia (1999)
3. Bender, M.A.: The power of team exploration: Two robots can learn unlabeled directed graphs. In: Proceedings of the Thirty Fifth Annual Symposium on Foundations of Computer Science, pp. 75–85 (1994)
4. Bender, M.A., Fernández, A., Ron, D., Sahai, A., Vadhan, S.: The power of a pebble: Exploring and mapping directed graphs. Information and Computation 176(1), 1–21 (2002)
5. Brass, P., Cabrera-Mora, F., Gasparri, A., Xiao, J.: Multirobot tree and graph exploration. IEEE Transactions on Robotics 27(4), 707–717 (2011)
6. Deng, X., Papadimitriou, C.: Exploring an unknown graph. In: Proceedings of the 31st Annual Symposium on Foundations of Computer Science, vol. 1, pp. 355–361 (October 1990)
7. Dereniowski, D., Disser, Y., Kosowski, A., Pająk, D., Uznański, P.: Fast collaborative graph exploration. In: Fomin, F.V., Freivalds, R., Kwiatkowska, M., Peleg, D. (eds.) ICALP 2013, Part II. LNCS, vol. 7966, pp. 520–532. Springer, Heidelberg (2013)
8. Dessmark, A., Pelc, A.: Optimal graph exploration without good maps. Theor. Comput. Sci. 326, 343–362 (2004)
9. Dynia, M., Kutyłowski, J., Meyer auf der Heide, F., Schindelhauer, C.: Smart robot teams exploring sparse trees. In: Královič, R., Urzyczyn, P. (eds.) MFCS 2006. LNCS, vol. 4162, pp. 327–338. Springer, Heidelberg (2006)
10. Dynia, M., Łopuszański, J., Schindelhauer, C.: Why robots need maps. In: Prencipe, G., Zaks, S. (eds.) SIROCCO 2007. LNCS, vol. 4474, pp. 41–50. Springer, Heidelberg (2007)
11. Fleischer, R., Trippen, G.: Exploring an unknown graph efficiently. In: Brodal, G.S., Leonardi, S. (eds.) ESA 2005. LNCS, vol. 3669, pp. 11–22. Springer, Heidelberg (2005)
12. Fraigniaud, P., Gąsieniec, L., Kowalski, D.R., Pelc, A.: Collective tree exploration. Netw. 48, 166–177 (2006)
13. Förster, K.-T., Wattenhofer, R.: Directed graph exploration. In: Baldoni, R., Flocchini, P., Binoy, R. (eds.) OPODIS 2012. LNCS, vol. 7702, pp. 151–165. Springer, Heidelberg (2012)
14. Gabriely, Y., Rimon, E.: Competitive on-line coverage of grid environments by a mobile robot. Comput. Geom. Theory Appl. 24(3), 197–224 (2003)
15. Higashikawa, Y., Katoh, N., Langerman, S., Tanigawa, S.-I.: Online graph exploration algorithms for cycles and trees by multiple searchers. Journal of Combinatorial Optimization, 1–16 (2012)
16. Kolenderska, A., Kosowski, A., Małafiejski, M., Żyliński, P.: An improved strategy for exploring a grid polygon. In: Kutten, S., Žerovnik, J. (eds.) SIROCCO 2009. LNCS, vol. 5869, pp. 222–236. Springer, Heidelberg (2010)
17. Ortolf, C., Schindelhauer, C.: Online multi-robot exploration of grid graphs with rectangular obstacles. In: Proceedings of the Twenty-fourth Annual ACM Symposium on Parallelism in Algorithms and Architectures, SPAA 2012, pp. 27–36. ACM, New York (2012)
18. Rao, N.S.V., Kareti, S., Shi, W., Iyengar, S.S.: Robot navigation in unknown terrains: Introductory survey of non-heuristic algorithms. Technical Report ORNL/TM-12410:1–58, Oak Ridge National Laboratory (July 1993)

Improved Periodic Data Retrieval in Asynchronous Rings with a Faulty Host

Evangelos Bampas[1,*], Nikos Leonardos[2], Euripides Markou[3,**],
Aris Pagourtzis[4,***], and Matoula Petrolia[5]

[1] Univ. Bordeaux, LaBRI, UMR 5800, F-33400 Talence, France
evangelos.bampas@labri.fr
[2] Department of Informatics and Telecommunications,
National and Kapodistrian University of Athens, Greece
nikos.leonardos@gmail.com
[3] Department of Computer Science and Biomedical Informatics,
University of Thessaly, Lamia, Greece
emarkou@ucg.gr
[4] School of Electrical and Computer Engineering,
National Technical University of Athens, Greece
pagour@cs.ntua.gr
[5] LINA, University of Nantes, France
stamatina.petrolia@univ-nantes.fr

Abstract. The exploration problem has been extensively studied in unsafe networks containing malicious hosts of a highly harmful nature, called *black holes*, which completely destroy mobile agents that visit them. In a recent work, Královič and Miklík [SIROCCO 2010, LNCS 6058, pp. 157–167] considered various types of malicious host behavior in the context of the *Periodic Data Retrieval* problem in asynchronous ring networks with exactly one malicious host. In this problem, a team of initially co-located agents must report data from all safe nodes of the network to the homebase, infinitely often. The malicious host can choose whether to kill visiting agents or allow them to pass through (gray hole). In another variation of the model, the malicious host can, in addition, alter its whiteboard contents in order to deceive visiting agents. The goal is to design a protocol for Periodic Data Retrieval using as few agents as possible.

In this paper, we present the first nontrivial lower bounds on the number of agents for Periodic Data Retrieval in asynchronous ring networks. Specifically, we show that at least 4 agents are needed when the

* Partial support by the ANR project DISPLEXITY (ANR-11-BS02-014).
** This research has been co-financed by the European Union (European Social Fund — ESF) and Greek national funds through the Operational Program "Education and Lifelong Learning" of the National Strategic Reference Framework (NSRF) — Research Funding Program: THALIS-UOA (MIS 375891).
*** This research has been co-financed by the European Union (European Social Fund – ESF) and Greek national funds through the Operational Program "Education and Lifelong Learning" of the National Strategic Reference Framework (NSRF) — Research Funding Program: THALIS-NTUA (MIS 379414).

M. Halldórsson (Ed.): SIROCCO 2014, LNCS 8576, pp. 355–370, 2014.

malicious host is a gray hole, and at least 5 agents are needed when the malicious host whiteboard is unreliable. This improves the previous lower bound of 3 in both cases and answers an open question posed in the aforementioned paper.

On the positive side, we propose an optimal protocol for Periodic Data Retrieval in asynchronous rings with a gray hole, which solves the problem with only 4 agents. This improves the previous upper bound of 9 agents and settles the question of the optimal number of agents in the gray-hole case. Finally, we propose a protocol with 7 agents when the whiteboard of the malicious host is unreliable, significantly improving the previously known upper bound of 27 agents. Along the way, we set forth a detailed framework for studying networks with malicious hosts of varying capabilities.

Keywords: periodic data retrieval, malicious host, gray hole, red hole, unreliable whiteboard.

1   Introduction

In distributed mobile computing, one of the main issues is the security of both the agents that explore a network and the hosts. Various methods of protecting mobile agents against malicious nodes as well as of protecting hosts against harmful agents have been proposed (see, e.g., [19] and references therein).

In particular, the exploration problem has been extensively studied in unsafe networks which contain malicious hosts of a highly harmful nature, called *black holes*. A black hole is a node which contains a stationary process destroying all mobile agents visiting that node, without leaving any trace. In the *Black Hole Search* problem (BHS in short) the goal for the agents is to locate the black hole within finite time. More specifically, at least one agent has to survive knowing all edges leading to the black hole. The problem has been introduced by Dobrev, Flocchini, Prencipe, and Santoro in [7,10]. Since any agent visiting a black hole vanishes without leaving any trace, the location of the black hole must be deduced by some communication mechanism employed by the agents. Four such mechanisms have been proposed in the literature: a) the *whiteboard* model [5,9,10,2,16] in which there is a whiteboard at each node of the network where the agents can leave messages, b) the *pure token* model [14,1] where the agents carry tokens which they can leave at nodes, c) the *enhanced token* model [6,11,23] in which the agents can leave tokens at nodes or edges, and d) the time-out mechanism (only for synchronous networks) in which one agent explores a new node and then, after a predetermined fixed time, informs another agent who waits at a safe node [21].

In an asynchronous network, the number of nodes of the network must be known to the agents, otherwise the problem is unsolvable [10]. If the graph topology is unknown, at least $\Delta + 1$ agents are needed, where Δ is the maximum node degree in the graph [9]. Furthermore, the network should be 2-connected. It is also not possible to answer the question of *whether* a black hole exists in the network. If the agents have a map of the network or at least a *sense of direction* [17,18] and can use whiteboards, then two agents with memory suffice to solve the problem. In asynchronous networks with dispersed agents (i.e., not initially located at the same node), the problem has been investigated for the ring topology [8,10] and for arbitrary networks [15,3] in the whiteboard model, while in the enhanced token model it has been studied for rings [12,13] and for some interconnected networks [23]. The problem has been also studied in synchronous networks. For a survey on BHS the reader is referred to [21].

As already mentioned, a black hole is a particular type of malicious host with a very simple behavior: killing every agent instantly without leaving any trace. In reality, a host may have many more ways to harm the agents: it may introduce fake agents, change the contents of the whiteboard, or even confuse agents by directing them to ports different from the requested ones.

In [20,22], Královič and Miklík studied how the various capabilities of a malicious host affect the solvability of exploration problems in asynchronous networks with whiteboards. They first consider networks with a malicious host (called *gray hole*) which can at any time choose whether to behave as a black-hole or as a safe node. Since the malicious behavior may never appear, the agents might not be able, in certain cases, to decide the location of the malicious host. Hence, they introduce and study the so called *Periodic Data Retrieval* problem in which, on each safe node of the network, an infinite sequence of data is generated over time and these data have to be gathered in the homebase. The goal is to design a protocol for a team of initially co-located agents so that data from every safe node are reported to the homebase, infinitely often, minimizing the total number of agents used. One agent can solve the problem in networks without malicious hosts, where the problem reduces to the *Periodic Exploration* problem (e.g., see [4] and references therein) in which the goal is to minimize the number of moves between two consecutive visits of a node. When the malicious host is a black hole, the Periodic Data Retrieval and the Periodic Exploration problem are solved by the same number of agents. As observed in [20], $n - 1$ agents are sufficient for solving the Periodic Data Retrieval problem in any 2-connected network of n nodes with one malicious host when the topology is known to the agents: each of the $n - 1$ agents selects a different node of the network and periodically visits all other nodes. The authors show that two agents are not sufficient to solve the problem in a ring with a gray hole and they present a protocol which solves the problem using 9 agents. They also consider a second type of malicious host which behaves as a gray hole and, in addition, can alter the contents of its whiteboard; they show that 27 agents are sufficient to solve the Periodic Data Retrieval problem in a ring, under this type of malicious host.

Our contribution. In this paper, we study and refine the model of [20]. We present the first nontrivial lower bounds on the number of agents for Periodic Data Retrieval in asynchronous rings. Specifically, we show that at least 4 agents are needed when the malicious host is a gray hole, and at least 5 agents are needed when the malicious host whiteboard is unreliable. This improves the previous lower bound of 3 agents in both cases and answers an open question posed in [20]. On the positive side, we propose an optimal protocol for Periodic Data Retrieval in asynchronous rings with a gray hole, which solves the problem with only 4 agents. This improves the previous upper bound of 9 agents and settles the question of the optimal number of agents in the gray-hole case. Finally, we propose a protocol with 7 agents when the whiteboard of the malicious host is unreliable, significantly improving the previously known upper bound of 27 agents. Along the way, we set forth a detailed framework for studying networks with malicious hosts of varying capabilities.

In order to derive the lower bounds, we make extensive use of certain configurations which the adversary can enforce in a benign execution (i.e., an execution in which the malicious host obeys the protocol), in particular 2-traversals and 3-traversals (informally, configurations in which some agent traverses an edge "with the intention" to eventually advance one or two more edges in the same direction, respectively). We are then able to exploit the fact that we can think of the adversary as not having to commit to a particular location of the malicious host as long as the execution remains benign. For the upper bound in the case of the gray hole, we use the well known *cautious step* technique, which is also employed in [20]. However, in our case the agent marks both nodes involved in the cautious step, thus considerably reducing the number of agents that can enter the same link from the opposite direction. When the malicious host whiteboard is unreliable, we employ a natural extension of the cautious step, the *cautious double step*.

Due to lack of space, all missing proofs, as well as the detailed pseudocode for the proposed algorithms, are deferred to the full version of the paper.

2   Preliminaries

2.1   System Model

The agents operate in a ring network where each node contains one host (we will use the terms "host" and "node" interchangeably). Each host is identified by a unique label, and is connected to each of its two neighbors via labeled communication ports. Each port is associated with two order-preserving queues: one for incoming agents and a second one for outgoing agents. Additionally, each host contains a whiteboard, i.e., a piece of memory that is shared among the agents present in the node at any given time, and a queue of agents who are waiting to acquire access to the whiteboard. Neighboring hosts are connected via bidirected asynchronous FIFO links, forming an undirected graph G.

The agents are modeled as deterministic three-tape Turing machines: the first tape serves as the private memory of the agent, the second tape holds the label of the port to which the agent wishes to be transferred, and the third tape holds a copy of the whiteboard of the current node, if the agent has acquired access to the whiteboard. All agents are initially located on the same node of the network, which we will call "the homebase." Each agent possesses a distinct identifier and knows the complete map of the network. The only way for agents to interact with each other is through the whiteboards: they are not aware of the presence of other agents on the same node or on the same link, and they cannot exchange private messages.

Each host is responsible for removing agents from the front of its incoming queue and *executing* them, i.e., advancing each agent's state according to its transition function until the agent requests to be transferred. We assume that this happens in one atomic step, i.e., as soon as one agent A is removed from the front of an incoming queue, no other agent in that node can execute a transition before A executes its own first transition. The host is also responsible for executing the agent that is at the front of the whiteboard queue. Finally, the host is responsible for removing agents from the front of its outgoing queues and transmitting them over the link to the neighboring node (the whiteboard tape is not transmitted). The host has to perform these tasks while ensuring that no queue is neglected for an infinite amount of time. Each host is capable of executing multiple agents concurrently. The set of states of each agent contains special states corresponding to the following actions:

1. *Request the whiteboard lock* (q_{req}): When an agent enters this state, it is inserted in the whiteboard queue. We assume that this happens atomically, i.e., any other agent who subsequently enters this state will be placed in the whiteboard queue *behind* this agent. Its execution is suspended until it reaches the front of the queue. When this happens, the host continues to execute this agent (possibly concurrently with other agents who are not accessing the whiteboard) without removing him from the whiteboard queue. Simultaneously with the transition from q_{req}, the whiteboard of the node is copied to the third tape of the agent.

2. *Release the whiteboard lock* (q_{rel}): When an agent enters this state, its whiteboard tape is copied back to the whiteboard of the node and the agent is removed from the whiteboard queue.

3. *Leave through a specified port* (q_{port}): When an agent enters this state, it is atomically inserted in the outgoing queue of the port indicated on its second tape. If the agent has not yet released the whiteboard lock, its whiteboard tape is also copied back to the whiteboard of the node and the agent is removed from the whiteboard queue.

Note that an agent actually traverses a link only when the source host decides to remove it from the outgoing queue and transmit it to the target host. Link traversal is not instantaneous. Its duration is determined by the adversary.

The system is asynchronous, meaning that any agent can be stalled for an arbitrary but finite amount of time while executing any computation or traversing any link. We assume that the system contains exactly one malicious host which may deviate from the system specification in several ways:

Definition 1 (Malicious behavior from the malicious host). *The malicious host in the system may choose to:*

1. Kill *any agent which is stored in any of its queues or is being executed. In this case, the agent disappears without leaving any trace, apart from what it may have already written on the whiteboards.*

2. *Operate without fairness, i.e., it can neglect one or more of its queues forever.*
3. *Transmit an agent to a node different from the one that it requested to be transmitted to, or it can transmit an agent without the agent asking for a transmission, or misreport its own node label to agents requesting it.*
4. *Execute (resp. forward) any agent in the incoming or the whiteboard (resp. outgoing) queues, without respecting the queue order.*
5. *Create and execute multiple copies of an agent at any stage.*
6. *Provide to each agent that requests access to the whiteboard an arbitrary whiteboard tape, possibly erroneous or inconsistent with the whiteboard tapes that it has provided to the other agents.*

We classify the various types of malicious host behavior in order of increasing power as follows:

Definition 2. *The malicious host is called:*

- 1-malicious *or* black hole *if it kills every agent that appears in any of its queues at every time $t \geq 0$.*
- 2-malicious *if it kills every agent that appears in any of its queues or is being executed at every time $t \geq t_0$, where $t_0 \geq 0$ is chosen by the adversary. Until time t_0, which may even be equal to $+\infty$, it acts as a safe node.*
- 3-malicious *or* gray hole *if it can choose whether to deviate (or not) from the protocol in the way described in item 1 of Definition 1 at any time $t \geq 0$.*
- 4-malicious *if it can choose whether to deviate (or not) from the protocol in any of the ways described in items 1-5 of Definition 1 at any time $t \geq 0$.*
- 5-malicious *or* red hole *if it can choose whether to deviate (or not) from the protocol in any of the ways described in items 1-6 of Definition 1 at any time $t \geq 0$.*

The agents do not have any information on the location of the malicious host, except from the fact that the homebase is safe.

2.2 Periodic Data Retrieval

We assume that every host in the system generates over time an infinite sequence of data items, all of which have to eventually reach the homebase. The agents operate in the network and their aim is to deliver the data from any safe node to the homebase infinitely often. Once an agent has acquired a chunk of data items from a host, the data may be stored at an intermediate node and possibly read by another agent before reaching the homebase. This problem is known as the *Periodic Data Retrieval* problem [20].

Definition 3. *An* instance *of Periodic Data Retrieval is a tuple $\langle G, \lambda, H, k, \omega, m \rangle$, where G is an undirected graph, λ is a function that assigns labels to nodes and local ports of the nodes, $H \in V(G)$ is the homebase, k is a positive integer representing the number of agents starting on the homebase, $\omega \in V(G) \setminus \{H\}$ is the malicious host, and $m \in \{1, 2, 3, 4, 5\}$ is the maliciousness level of ω as per Definition 2.*

Definition 4. *An execution of an algorithm on an instance is completely determined by a sequence of choices made by the adversary. The adversary can choose which agents are activated at any given time, the speed at which agents are executed and the speed at which they perform each edge traversal, as well as any malicious behavior on the part of the malicious host. An execution $\mathcal{E}'$ is a continuation of an execution $\mathcal{E}$ from time t_0 if $\mathcal{E}'$ is identical to $\mathcal{E}$ up to time t_0. An execution is called* benign *if the malicious host exhibits no malicious behavior.*

During an execution, we will say that an agent is *frozen*, either on an edge or at a node, if the adversary has decided to delay the actions of that agent. If an agent is frozen at some time t, the adversary has to unfreeze it at some finite time $t' > t$.

Definition 5. *Given an execution of an algorithm, a node v is said to be t-reported if there exists a time $t' > t$ such that at time t' the homebase whiteboard contains all the data items that v has generated up to time t.*

Definition 6. *An algorithm $\mathcal{A}$ is (k, m)-correct if for every Periodic Data Retrieval instance $\mathcal{I} = \langle G, \lambda, H, k, \omega, m \rangle$, for every execution $\mathcal{E}$ of $\mathcal{A}$ on $\mathcal{I}$, for every node $v \in V(G) \setminus \{\omega\}$, and for every time t, node v is t-reported.*

Remark 1. A necessary condition for v to be t-reported is that there exist a natural number r, a sequence of (not necessarily distinct) agents $A_0, \ldots, A_r$, a sequence of nodes $v_0, \ldots, v_r$, and an increasing sequence of times $t_0 < \cdots < t_r$, such that v_0 is v, v_r is the homebase, $t \leq t_0$, and, for each i, agent A_i visits node v_i at time t_i and node v_{i+1} at time t'_i, where $t_i < t'_i < t_{i+1}$. If ω is a red hole, then in addition we must have that $\omega \notin \{v_0, \ldots, v_r\}$.

Propositions 1-3 follow directly from the definitions.

Proposition 1. *Let $\mathcal{A}$ be any algorithm. Every execution of $\mathcal{A}$ on some instance $\mathcal{I} = \langle G, \lambda, H, k, \omega, m \rangle$ is also an execution of $\mathcal{A}$ on $\mathcal{I}' = \langle G, \lambda, H, k, \omega, m' \rangle$, where $m' \geq m$.*

Proposition 2. *If an algorithm is (k, m)-correct, then it is (k, m')-correct for all $m' \leq m$.*

Proposition 3. *Let $\mathcal{A}$ be any algorithm. Every benign execution of $\mathcal{A}$ on some instance $\mathcal{I} = \langle G, \lambda, H, k, \omega, m \rangle$, where $m \geq 2$, is also a benign execution of $\mathcal{A}$ on $\mathcal{I}' = \langle G, \lambda, H, k, \omega', 2 \rangle$, for all $\omega' \in V(G) \setminus \{H\}$.*

3   Lower Bounds on the Number of Agents

In this section, we give lower bounds on the number of agents required to achieve Periodic Data Retrieval in rings with gray holes (Section 3.1) and red holes (Section 3.2). We give two more definitions before presenting the results. Let C_n denote an undirected ring with n nodes.

Definition 7 (Waiting). *Let $\mathcal{E}$ be an execution of an algorithm $\mathcal{A}$ on instance $\mathcal{I} = \langle G, \lambda, H, k, \omega, m \rangle$. Let W be a set of nodes that induces a connected subgraph $G(W)$ of G. We say that an agent A is* waiting on W *at time t_0 under $\mathcal{E}$ if the agent is in $G(W)$ at time t_0 and, under any continuation of $\mathcal{E}$ from t_0 in which agent A does not perceive any changes in the whiteboard contents of the nodes in W (with respect to their contents at time t_0) except for those made by itself, agent A never leaves $G(W)$.*

When $W = \{v\}$, we will say that agent A is waiting on the node v. *When $W = \{u, v\}$, we will say that agent A is* waiting on the edge (u, v).

Definition 8 (ℓ-traversal). *Let $\mathcal{E}$ be an execution of $\mathcal{A}$ on $\mathcal{I} = \langle C_n, \lambda, H, k, \omega, m \rangle$ and let $\ell \geq 1$. We say that an agent A performs an ℓ-traversal from node v_0 at time t_0 under $\mathcal{E}$ if all of the following hold:*

1. *Nodes $v_0, v_1, \ldots, v_\ell$ are successive on the ring and none of them is the homebase.*
2. *At time t_0, agent A traverses the edge (v_0, v_1).*
3. *At time t_0, no other agent is located on nodes $v_1, \ldots, v_{\ell-1}$ or their incident edges.*
4. *Under any continuation of $\mathcal{E}$ from t_0 in which agent A is not killed and the only changes in the whiteboards of nodes $v_1, \ldots, v_{\ell-1}$ (with respect to their contents at time t_0) that are observed by agent A until it reaches node v_ℓ are the changes made by itself, agent A reaches node v_ℓ in finite time without visiting node v_0 in the meantime.*

Note that a 1-traversal is simply a traversal of an edge that is not incident to the homebase. A direct corollary of Definition 8 is the following:

Corollary 1. *If there exists an execution $\mathcal{E}$ of $\mathcal{A}$ on $\mathcal{I} = \langle C_n, \lambda, H, k, \omega, m \rangle$ such that properties 1–3 of Definition 8 hold and, in addition, there exists a continuation of $\mathcal{E}$ from t_0 such that agent A reaches node v_ℓ in finite time without visiting node v_0 in the meantime and no other agent traverses any of the edges (v_0, v_1) and $(v_{\ell-1}, v_\ell)$ from t_0 up to the first time when agent A reaches node v_ℓ, then agent A performs an ℓ-traversal from node v_0 at time t_0 under $\mathcal{E}$.*

3.1 Three Agents Are Not Enough for Gray Holes

The inexistence of $(1, 3)$-correct or $(2, 3)$-correct algorithms has already been demonstrated in [20]. In this section, we show that no algorithm can be $(3, 3)$-correct. We achieve this by proving that, if there existed a $(3, 3)$-correct algorithm, then the adversary would be able to force one of the agents to perform a 2-traversal (Lemma 1). However, we also prove that if any agent performs a 2-traversal while executing a $(3, 3)$-correct algorithm, then the adversary can kill all three agents (Lemma 2). This establishes that a $(3, 3)$-correct algorithm cannot exist.

Lemma 1. *Let $\mathcal{A}$ be a $(3, 3)$-correct algorithm and let $\mathcal{I} = \langle C_n, \lambda, H, 3, \omega, 3 \rangle$ with $n \geq 6$. There exists a benign execution of $\mathcal{A}$ on $\mathcal{I}$ under which some agent performs a 2-traversal.*

Lemma 2. *Let $\mathcal{A}$ be a $(3,3)$-correct algorithm and let $\mathcal{I} = \langle C_n, \lambda, H, 3, \omega, 3 \rangle$. Under any benign execution of $\mathcal{A}$ on $\mathcal{I}$, no agent can ever perform a 2-traversal.*

By Lemmas 1 and 2, the existence of a $(3,3)$-correct algorithm yields a contradiction. Therefore, we have proved the following:

Theorem 1. *There does not exist a $(3,3)$-correct algorithm.*

3.2 Four Agents Are Not Enough for Red Holes

In view of Proposition 2, the impossibility result in [20] together with Theorem 1 imply that there do not exist $(1,5)$-correct, $(2,5)$-correct, or $(3,5)$-correct algorithms. In this section, we show that no algorithm can be $(4,5)$-correct. To this end, we first prove in Lemma 3 that, under any $(4,5)$-correct algorithm, the adversary can force some agent to perform a 3-traversal (in fact, this can even be enforced under any $(4,3)$-correct algorithm). Then, we derive a contradiction by showing in Lemma 4 that if an agent performs a 3-traversal, then four agents can die in the red hole and thus the algorithm cannot be $(4,5)$-correct.

Lemma 3. *Let $\mathcal{A}$ be a $(4,3)$-correct algorithm and let $\mathcal{I} = \langle C_n, \lambda, H, 4, \omega, 3 \rangle$ with $n \geq 9$. There exists a benign execution of $\mathcal{A}$ on $\mathcal{I}$ under which some agent performs a 3-traversal.*

Lemma 4. *Let $\mathcal{A}$ be a $(4,5)$-correct algorithm and let $\mathcal{I} = \langle C_n, \lambda, H, 4, \omega, 5 \rangle$. Under any benign execution of $\mathcal{A}$ on $\mathcal{I}$, no agent performs a 3-traversal.*

By Lemmas 3 and 4 and Propositions 1 and 2, the existence of a $(4,5)$-correct algorithm yields a contradiction. Therefore, we have proved the following:

Theorem 2. *There does not exist a $(4,5)$-correct algorithm.*

4   An Optimal Algorithm for Rings with a 4-Malicious Host

In view of Theorem 1, no algorithm can achieve Periodic Data Retrieval on a ring with a 4-malicious host using only three agents (in fact, not even on a ring with a 3-malicious host). In this section, we present algorithm PDR_RINGS_4-MALICIOUS, which solves the problem in the presence of a 4-malicious host in a ring, using an optimal number of four agents.

Remark 2. In order to simplify the presentation, we will not make explicit the part of the algorithm that is responsible for picking up the data from nodes and delivering it to the homebase or to an intermediate node to be picked up by another agent. We assume that each agent, after getting access to the whiteboard of any node, reads all the node data that has been generated from the node or left there from other agents and also leaves a copy of the node data that it is already carrying but is not present in the node. In the following, we will deal explicitly only with the part of the algorithm that ensures that four agents are sufficient to ensure Periodic Data Retrieval in the presence of a 4-malicious host.

Before presenting the algorithm, we outline the interface exposed by the nodes to visiting agents:

- Each node exposes to the agents two functions: *getNodeID()* and *transfer(port)*. The former returns the ID of the current node (recall that this may be misreported by ω). The latter places the agent in the outgoing queue of the port specified in its argument, releasing the whiteboard lock if necessary.
- Additionally, each node exposes to the agents which it is executing the whiteboard object *WB*, which has two members, *WB.list* and *WB.flags*, and two methods, *WB.access()* and *WB.release()*. *WB.access()* requests the whiteboard lock and thus results in the agent being placed in the whiteboard queue. The agent remains inactive until it reaches the front of the queue. At that point, it gains access to *WB.list* and *WB.flags*. *WB.list* contains quadruples of the form $\langle id, op, port, s \rangle$, where id is an agent identifier, op is one of the constants $\{ARR, DEP\}$, port is a port number, and s is a non-negative integer. If op = ARR, the entry means that the agent with the specified id arrived from the specified port after traversing s edges. If op = DEP, the entry means that the agent with the specified id departed from the specified port before traversing its $(s+1)$-st edge. *WB.flags* contains pairs of the form $\langle id, dir \rangle$, where id is an agent identifier and dir $\in \{+, -\}$. The meaning of an entry in *WB.flags* will become apparent when we describe the algorithm.

While moving from node to node, agents perform several low-level operations outlined below:

- When arriving at a node, the agent requests whiteboard access and, when this is granted, it inserts a quadruple $\langle id, ARR, p, s \rangle$ into *WB.list*. The agent releases the whiteboard lock just before it leaves the node, after inserting a quadruple $\langle id, DEP, p', s \rangle$ into *WB.list*. However, if the agent is granted whiteboard access and it detects that some other agent has inserted its ARR-quadruple but not the corresponding DEP-quadruple, it releases the whiteboard lock without writing anything and requests whiteboard access *de novo*, waiting for the other agent to conclude its computation on the node.
- Additionally, before leaving each node, the agent keeps a copy of *WB.list*. When arriving at the destination node, after being granted whiteboard access for the first time, the agent checks the following conditions and halts if any of them is true: *(a)* The current node, as reported by *getNodeID()*, is not the same as its intended destination node. *(b)* A DEP-quadruple by the agent itself at its current step already exists on the node. *(c)* The ARR-quadruple that the agent wishes to insert into *WB.list* is already there. *(d)* One of the agents which reported their departure from the previous node has not reported its arrival at the current node.

Note that, by waiting for agents already present at the node to conclude their interaction with the whiteboard before initiating its own, the algorithm guarantees that an agent which is killed by the malicious host while holding the whiteboard lock will also cause future agents visiting the host to effectively kill themselves, as they will keep requesting the whiteboard lock forever. Moreover, the conditions *(a)-(c)* ensure that if the malicious host forwards an agent to the wrong node, or does not forward an agent at all and pretends to be the destination node, or attempts to re-forward a duplicate copy of an agent, then the offended agent will detect this and kill itself instead of continuing the protocol erroneously and disrupting the entire system. Finally, condition *(d)* ensures that if the malicious host disrupts the FIFO order of its queues, the agents which are pushed forward in the queues will detect this and kill themselves.

We now give a high-level description of the algorithm. An agent is always in one of two modes: *clockwise* $(+)$ or *counterclockwise* $(-)$. In any configuration of the system in which a node u contains an entry of the form $\langle id, + \rangle$ (resp. $\langle id, - \rangle$) in *WB.flags*, we will say that u contains the flag u^+ (resp. u^-), or that the flag u^+ (resp. u^-) is present. An agent in clockwise mode performs consecutive *cautious steps* in the clockwise direction, until it detects a node w with a flag (either w^+ or w^-), at which point it *bounces* and starts performing cautious steps in the counterclockwise direction. Let u be a node and v its clockwise neighbor. A cautious step starting from u in the clockwise direction entails the following sequence of operations:

- An Explore$(+)$ step: The agent inserts a flag $\langle id, + \rangle$ and moves to v.
- A Return$(+)$ step: The agent inserts a flag $\langle id, - \rangle$ and moves back to u.
- A Finish$(+)$ step: The agent removes its $\langle id, + \rangle$ flag, moves to v, removes its $\langle id, - \rangle$ flag, and then starts an Explore$(+)$ step from v.

However, if after the Explore$(+)$ step the agent detects a flag at v, then it performs a Bounce$(+)$ step instead: The agent moves back to u without inserting a flag in the whiteboard of v, removes its $\langle id, + \rangle$ flag, and then either switches to counterclockwise mode and starts an *explore* step in the counterclockwise direction if there is no u^- flag, or remains in clockwise mode and starts an *explore* step in the clockwise direction if there is a u^- flag.

An agent in counterclockwise mode operates in a completely symmetric fashion and performs consecutive cautious steps in the counterclockwise direction, until it bounces and switches to clockwise mode. Note that an agent can start the algorithm in clockwise or counterclockwise mode: this is decided when the agent begins its execution, depending on which flags are present on the homebase. Figure 1 illustrates the high-level workings of the algorithm.

The next lemma follows by a straightforward adaptation of the proof of Theorem 1 in [20].

Lemma 5 ([20]). *Under any execution of* PDR_RINGS_4-MALICIOUS *in which not all agents are killed, Periodic Data Retrieval is achieved.*

In order to show that PDR_RINGS_4-MALICIOUS works with four agents, we reason as follows: First, we show that under any benign execution, at most three agents can be in the queues of the same node at the same time (Lemma 7 below). For this, we take advantage of the flags left by the agents during the cautious

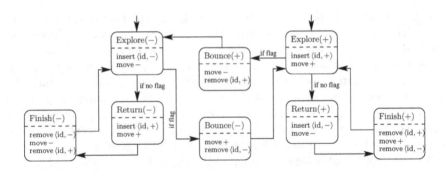

Fig. 1. A diagram of the basic operations of Algorithm PDR_RINGS_4-MALICIOUS. "move +" (resp. "move −") stands for traversing an edge in the clockwise (resp. counterclockwise) direction.

step. Then, we show how to convert any execution in which four agents die into a benign execution in which all four agents are in the queues of the malicious host at the same time. This contradicts the immediately preceding statement, thus the malicious host cannot kill four agents, and thus, by Lemma 5, the (4, 4)-correctness of the algorithm follows (Theorem 3 below). The low-level operations of the algorithm play a crucial role in the proof of Theorem 3.

Definition 9. *We say that an agent is going from u to v if it has written its DEP-quadruple on u and requested to be transferred to v, but has not yet written its corresponding ARR-quadruple on v. This means that the agent could be either in the outgoing queue of u, or in the process of being transferred to v, or in the incoming queue of v, or in the whiteboard queue of v. An agent is traversing an edge (u, v) if it is going from u to v or from v to u. An agent is on a node u if it has written its ARR-quadruple on u but has not yet written its DEP-quadruple.*

Proposition 4 below states an easy to check property of the algorithm.

Proposition 4. *Under any benign execution of* PDR_RINGS_4-MALICIOUS, *at most one agent can be on a given node at a given time.*

Let A be an agent which is making a move from node u to a neighboring node v, i.e., A has inserted a DEP-quadruple at u but has not yet inserted the corresponding ARR-quadruple at v. If this move is part of an Explore(+) step, we assign to agent A the tag E^+. Similarly, we use the tags R^+, F^+, and B^+ for the Return(+), Finish(+), and Bounce(+) steps, respectively, and the tags E^-, R^-, F^-, and B^- for the symmetric counterclockwise-mode steps.

By a careful case analysis, we can show that if two agents are traversing the same edge in any direction, the only possible combinations of tags are: $\{E^+, B^-\}$, $\{E^+, F^+\}$, $\{E^-, B^+\}$, $\{E^-, F^-\}$, $\{E^+, E^-\}$, and $\{B^+, B^-\}$. Using this characterization, we can prove Lemma 6:

Lemma 6. *Under any benign execution of* PDR_RINGS_4-MALICIOUS, *it is not possible for three agents to traverse the same edge at the same time.*

Lemma 6 and Proposition 4 considerably limit the candidate configurations of four agents in the queues of the same node. By a more elaborate case analysis, we can also eliminate the remaining possibilities and arrive at a contradiction in all cases, thus obtaining the following:

Lemma 7. *For any node v other than the homebase, under any benign execution of* PDR_RINGS_4-MALICIOUS, *a total of at most three agents can be in the queues of v at the same time.*

Theorem 3. PDR_RINGS_4-MALICIOUS *is* $(4, 4)$-*correct in rings.*

5   An Efficient Algorithm for Rings with a Red Hole

Note that, irrespective of the number of agents, the PDR_RINGS_4-MALICIOUS algorithm fails if the malicious host is a red hole. Indeed, the red hole can kill every clockwise (resp. counterclockwise) agent that approaches it after it has removed the + (resp. −) flag from the neighboring node and while it is concluding its Finish(+) (resp. Finish(−)) step on the red hole, by presenting to it a whiteboard which shows that previous clockwise (resp. counterclockwise) agents were not killed but continued their intended trajectory.

In order to remedy this situation, we propose algorithm PDR_RINGS_RED, which employs a natural extension of the cautious step idea: the *cautious double step*. Let u, v, w be consecutive nodes in clockwise order. A cautious double step starting from u in the clockwise direction entails the following sequence of operations:

- An Explore1(+) step: The agent inserts a flag $\langle id, + \rangle$ and moves to v.
- An Explore2(+) step: The agent moves to w.
- A Return2(+) step: The agent moves back to v.
- A Return1(+) step: The agent moves back to u.
- A Finish(+) step: The agent removes its $\langle id, + \rangle$ flag, moves to v, and then starts an Explore1(+) step from v.

However, if after the Explore1(+) step the agent detects a + flag at v, then it bounces but first it goes to w anyway. More specifically, in this case the agent performs the following sequence of operations after the Explore1(+) step:

- An Explore*2(+) step: The agent moves to w.
- A Bounce2(+) step: The agent moves back to v.
- A Bounce1(+) step: The agent moves back to u, removes its $\langle id, + \rangle$ flag, and then starts an Explore1(−) step from u.

Under PDR_RINGS_RED, a clockwise agent performs consecutive cautious double steps in the clockwise direction, until it bounces and switches to counterclockwise mode. A counterclockwise agent operates completely symmetrically.

We should mention at this point that the low-level operations performed by this algorithm when an agent moves from node to node are somewhat more contrived than in the previous case. We highlight the differences below:

- The ARR- and DEP-tuples now contain more information, namely the mode of the agent (clockwise or counterclockwise) and the name of the step which it is currently executing (one of Explore1, Explore2, Return2, Return1, Finish, Explore*2, Bounce2, Bounce1).
- The agent keeps copies of *WB.list* from each node before every step, and after every step checks its stored copies against the whiteboard of the current node for inconsistencies. This allows the agent to verify that each whiteboard is consistent with the whiteboards of its neighbors, as well as that it reports a correct execution of the protocol. If any inconsistency is detected, the agent halts (kills itself). This check supersedes the simpler check in PDR_RINGS_4-MALICIOUS, whereby the agent simply checked whether all agents previously departed from the same port had reached the destination.

As in the case of PDR_RINGS_4-MALICIOUS, the agent never walks too far away from its flag. The agent is always at distance at most 2 from a flag that it has left behind. In fact, if the agent does not return to pick up the flag, then this must have happened because it was killed as a result of malicious activity from the red hole. Therefore, any flag which remains forever on a node after a given point in time is at distance at most 2 from the red hole. This, together with the fact that even when an agent decides to bounce, it still goes one step further in its intended direction (step Explore*2), implies that a straightforward adaptation of the proof of Theorem 1 in [20] yields the following:

Lemma 8. *Under any execution of* PDR_RINGS_RED *in which not all agents are killed, Periodic Data Retrieval is achieved.*

A feature of the algorithm is that clockwise agents ignore the flags of counterclockwise agents (i.e., they do not bounce upon detecting such a flag) and vice versa. This leads to an algorithm which is likely suboptimal, but can be analyzed more easily by considering the deaths of clockwise agents separately from those of counterclockwise agents. In fact, one can show that under any execution, we can have at most three deaths of clockwise agents and, symmetrically, at most three deaths of counterclockwise agents.

Lemma 9. *Under any execution of* PDR_RINGS_RED, *at most three clockwise agents die.*

By Lemmas 8 and 9, and by the symmetric of Lemma 9 for counterclockwise agents, we obtain that PDR_RINGS_RED achieves Periodic Data Retrieval with seven agents:

Theorem 4. PDR_RINGS_RED *is* $(7,5)$-*correct in rings.*

Remark 3. Note that the red hole might not interfere with the agents in any way, except by modifying the data items that they store in its whiteboard. In this case, it could happen that altered or corrupted data from certain nodes reach the homebase, thus rendering the algorithm incorrect. However, the cautious double step ensures that any agent which leaves a data item on the red hole will also leave a copy of it on at least one of its neighbors. Therefore, by enforcing agents to pick up data items only if they find them twice on two neighboring nodes, we ensure that an agent will never pick up a corrupted data item from the whiteboard of the red hole.

6 Concluding Remarks

We gave the first nontrivial lower bounds on the number of agents for Periodic Data Retrieval in asynchronous rings with either one gray hole or one red hole, answering an open question posed in [20]. Moreover, we proposed an optimal, with respect to the number of agents, protocol for Periodic Data Retrieval in asynchronous rings with a gray hole, improving the previous upper bound of 9 agents and settling the question of the optimal number of agents in the gray-hole case. Finally, we proposed a protocol working with 7 agents in the presence of a red hole, significantly improving the previously known upper bound of 27 agents.

We made no effort to optimize the amount of data stored on the whiteboards of the hosts. Indeed, since the protocol is executed indefinitely, the amount of data stored in every host under both PDR_RINGS_4-MALICIOUS and PDR_RINGS_RED grows unbounded. However, it should be clear that this amount can be reduced to a reasonable function of the number of nodes and the number of agents, by deprecating and removing information which is known to be no longer useful. We defer the implementation of this mechanism to the full version of the paper.

Algorithm PDR_RINGS_RED is almost certainly suboptimal. In principle, we should be able to further reduce the total number of agents killed by suitably marking all of the nodes involved in a cautious double step, and then having clockwise and counterclockwise agents *not* ignore each other's flags. We conjecture that an algorithm along these lines would work with an optimal number of 5 agents in the presence of a red hole.

One important research direction which remains completely open is the case of a malicious host which can alter the state of an agent, its memory, or even its program. It would be particularly interesting to develop algorithms that cope with this kind of malicious behavior. Another question that remains open is what happens in other network topologies under the various malicious host models.

References

1. Balamohan, B., Dobrev, S., Flocchini, P., Santoro, N.: Asynchronous exploration of an unknown anonymous dangerous graph with $O(1)$ pebbles. In: Even, G., Halldórsson, M.M. (eds.) SIROCCO 2012. LNCS, vol. 7355, pp. 279–290. Springer, Heidelberg (2012)
2. Balamohan, B., Flocchini, P., Miri, A., Santoro, N.: Time optimal algorithms for black hole search in rings. Discrete Math., Alg. and Appl. 3(4), 457–472 (2011)
3. Chalopin, J., Das, S., Santoro, N.: Rendezvous of mobile agents in unknown graphs with faulty links. In: Pelc, A. (ed.) DISC 2007. LNCS, vol. 4731, pp. 108–122. Springer, Heidelberg (2007)

4. Czyzowicz, J., Dobrev, S., Gąsieniec, L., Ilcinkas, D., Jansson, J., Klasing, R., Lignos, I., Martin, R., Sadakane, K., Sung, W.K.: More efficient periodic traversal in anonymous undirected graphs. Theor. Comput. Sci. 444, 60–76 (2012)
5. Dobrev, S., Flocchini, P., Královič, R., Ruzicka, P., Prencipe, G., Santoro, N.: Black hole search in common interconnection networks. Networks 47(2), 61–71 (2006)
6. Dobrev, S., Flocchini, P., Královič, R., Santoro, N.: Exploring an unknown graph to locate a black hole using tokens. In: Navarro, G., Bertossi, L.E., Kohayakawa, Y. (eds.) IFIP TCS. IFIP, vol. 209, pp. 131–150. Springer, Boston (2006)
7. Dobrev, S., Flocchini, P., Prencipe, G., Santoro, N.: Mobile search for a black hole in an anonymous ring. In: Welch, J.L. (ed.) DISC 2001. LNCS, vol. 2180, pp. 166–179. Springer, Heidelberg (2001)
8. Dobrev, S., Flocchini, P., Prencipe, G., Santoro, N.: Multiple agents rendezvous in a ring in spite of a black hole. In: Papatriantafilou, M., Hunel, P. (eds.) OPODIS 2003. LNCS, vol. 3144, pp. 34–46. Springer, Heidelberg (2004)
9. Dobrev, S., Flocchini, P., Prencipe, G., Santoro, N.: Searching for a black hole in arbitrary networks: optimal mobile agents protocols. Distributed Computing 19(1), 1–19 (2006)
10. Dobrev, S., Flocchini, P., Prencipe, G., Santoro, N.: Mobile search for a black hole in an anonymous ring. Algorithmica 48(1), 67–90 (2007)
11. Dobrev, S., Královič, R., Santoro, N., Shi, W.: Black hole search in asynchronous rings using tokens. In: Calamoneri, T., Finocchi, I., Italiano, G.F. (eds.) CIAC 2006. LNCS, vol. 3998, pp. 139–150. Springer, Heidelberg (2006)
12. Dobrev, S., Santoro, N., Shi, W.: Scattered black hole search in an oriented ring using tokens. In: IPDPS, pp. 1–8. IEEE (2007)
13. Dobrev, S., Santoro, N., Shi, W.: Using scattered mobile agents to locate a black hole in an un-oriented ring with tokens. Int. J. Found. Comput. Sci. 19(6), 1355–1372 (2008)
14. Flocchini, P., Ilcinkas, D., Santoro, N.: Ping pong in dangerous graphs: Optimal black hole search with pebbles. Algorithmica 62(3-4), 1006–1033 (2012)
15. Flocchini, P., Kellett, M., Mason, P.C., Santoro, N.: Map construction and exploration by mobile agents scattered in a dangerous network. In: IPDPS, pp. 1–10. IEEE (2009)
16. Flocchini, P., Kellett, M., Mason, P.C., Santoro, N.: Searching for black holes in subways. Theory Comput. Syst. 50(1), 158–184 (2012)
17. Flocchini, P., Mans, B., Santoro, N.: Sense of direction: Definitions, properties, and classes. Networks 32(3), 165–180 (1998)
18. Flocchini, P., Mans, B., Santoro, N.: Sense of direction in distributed computing. Theor. Comput. Sci. 291(1), 29–53 (2003)
19. Flocchini, P., Santoro, N.: Distributed security algorithms for mobile agents. In: Cao, J., Das, S.K. (eds.) Mobile Agents in Networking and Distributed Computing, ch. 3, pp. 41–70. John Wiley & Sons, Inc., Hoboken (2012)
20. Královič, R., Miklík, S.: Periodic data retrieval problem in rings containing a malicious host. In: Patt-Shamir, B., Ekim, T. (eds.) SIROCCO 2010. LNCS, vol. 6058, pp. 157–167. Springer, Heidelberg (2010)
21. Markou, E.: Identifying hostile nodes in networks using mobile agents. Bulletin of the EATCS 108, 93–129 (2012)
22. Miklík, S.: Exploration in faulty networks. Ph.D. thesis, Comenius University, Bratislava (2010)
23. Shi, W.: Black hole search with tokens in interconnected networks. In: Guerraoui, R., Petit, F. (eds.) SSS 2009. LNCS, vol. 5873, pp. 670–682. Springer, Heidelberg (2009)

Author Index